Lecture Notes in Computer Science 8402

Commenced Publication in 1973
Founding and Former Series Editors:
Gerhard Goos, Juris Hartmanis, and Jan van Leeuwen

T. V. Gopal   Manindra Agrawal
Angsheng Li   S. Barry Cooper (Eds.)

Theory and Applications of Models of Computation

11th Annual Conference, TAMC 2014
Chennai, India, April 11-13, 2014
Proceedings

 Springer

Volume Editors

T. V. Gopal
Anna University, College of Engineering
Department of Computer Science and Engineering,
Chennai, India
E-mail: gopal@annauniv.edu

Manindra Agrawal
Indian Institute of Technology
Department of Computer Science and Engineering
Kanpur, India
E-mail: manindra@iitk.ac.in

Angsheng Li
Chinese Academy of Sciences, Institute of Software
State Key Laboratory of Computer Science
Zhongguancun, Haidian District, Beijing, China
E-mail: angsheng@ios.ac.cn

S. Barry Cooper
University of Leeds, School of Mathematics, UK
E-mail: pmt6sbc@maths.leeds.ac.uk

ISSN 0302-9743   e-ISSN 1611-3349
ISBN 978-3-319-06088-0   e-ISBN 978-3-319-06089-7
DOI 10.1007/978-3-319-06089-7
Springer Cham Heidelberg New York Dordrecht London

Library of Congress Control Number: 2014934828

LNCS Sublibrary: SL 1 – Theoretical Computer Science and General Issues

Typesetting: Camera-ready by author, data conversion by Scientific Publishing Services, Chennai, India

Printed on acid-free paper

Springer is part of Springer Science+Business Media (www.springer.com)

Kamal Lodaya	Institute of Mathematical Sciences, India
Kazuhisa Makino	University of Tokyo, Japan
R. Nadarajan	PSG College of Technology, India
Y. Narahari	Indian Institute of Science, India
Naijun Zhan	State Key Laboratory for Computer Science (LCS), China
Navin Goyal	Microsoft Research, India
Pan Peng	Institute of Software, Chinese Academy of Sciences, China
C. Pandurangan	Indian Institute of Technology, Madras, India
Rajagopal Srinivasan	Tata Consultancy Services, India
Rajeeva Karandikar	Chennai Mathematical Institute, India
Richard Banach	University of Manchester, UK
R.K. Shyamasundar	Tata Institute of Fundamental Research (TIFR), India
Somenath Biswas	Indian Institute of Technology, Kanpur, India
Toshihiro Fujito	Toyohashi University of Technology, Japan
Venkat Chakaravarthy	IBM Research, India
Vincent Duffy	Purdue University, USA
Wenhui Zhang	State Key Laboratory of Computer Science, China
Xiaoming Sun	Institute of Computing Technology, Chinese Academy of Sciences, China

Steering Committee

Manindra Agrawal	Indian Institute of Technology, Kanpur, India
Jin-Yi Cai	University of Wisconsin - Madison, USA
S. Barry Cooper	University of Leeds, UK
John Hopcroft	Cornell University, USA
Angsheng Li	Chinese Academy of Sciences
Zhiyong Liu	Institute of Computing Technology, Chinese Academy of Sciences

Organizing Committee

C. Chellappan	Dean, CEG, Anna University – Chair
T.V. Geetha	HOD, Department of Computer Science and Engineering, Anna University
Shekhar Sahasrabuddhe	Regional Vice President (RVP VI), Computer Society of India
S.P. Soman	Regional Vice President (RVP VII), Computer Society of India

Organization

TAMC 2014 was organized at the Vivekananda Auditorium, Anna University, College of Engineering Guindy [CEG] Campus, Chennai, India.

Patron

M. Rajaram — Vice Chancellor, Anna University, Chennai, India

Conference Chair

T.V. Gopal — Anna University, Chennai, India

Conference Co-chair

Manindra Agrawal — Indian Institute of Technology, Kanpur, India

Program Committee

Aaron D. Jaggard	U.S. Naval Research Laboratory, USA
Ajith Abraham	Machine Intelligence Research Labs (MIR Labs), USA
Bakhadyr Khoussainov	University of Auckland, New Zealand
Carlo Alberto Furia	ETH Zurich, Switzerland
Chaitanya K. Baru	University of California, San Diego, USA
Christel Baier	Technische Universität Dresden, Germany
Cristian S. Calude	University of Auckland, New Zealand
Dimitris Fotakis	National Technical University of Athens, Greece
Dipti Deodhare	Centre for Artificial Intelligence and Robotics (CAIR), India
Hongan Wang	State Key Laboratory for Computer Science (LCS), China
Jacques Sakarovitch	Ecole Nationale Superieure des Telecommunications, France and International Federation for Information Processing (IFIP) TC - 1 - Foundations of Computer Science
Jianxin Wang	Central South University (CSU), China
Jose R. Correa	Universidad de Chile, Chile

committees. The success of TAMC 2014 is due to the unstinted efforts of various committees and the hard work of the invited speakers and the authors.

TAMC 2014 was the beginning of the next decade of this series of conference. The 11th International Conference on Theory and Applications of Models of Computation (TAMC 2014) team has attempted to provide a roadmap for at least the next decade of TAMC conferences.

February 2014

Gopal T.V.
Manindra Agrawal
Angsheng Li
S. Barry Cooper

Preface

Theory and Applications of Models of Computation (TAMC) is an international conference series with an interdisciplinary character, bringing together researchers working in computer science, mathematics, and the physical sciences. It is this, together with its predominantly computational and computability theoretic focus, which gives the series its special character.

Models of computation span across the traditional discipline boundaries of computer science, logic and applied mathematics, communication systems, computer networks and the control of dynamical systems to best leverage this expansive view of information and computation in natural and engineered systems.

The TAMC series explores the algorithmic foundations, computational methods, and computing devices to meet today's and tomorrow's challenges of complexity, scalability, and sustainability, with wide-ranging impacts on everything from the design of biological systems to the understanding of economic markets and social networks. The TAMC series is distinguished by an appreciation for mathematical depth, scientific—rather than heuristic – approaches and the integration of theory and implementation.

The TAMC conference series clearly indicates a growing influence in the regional and international scientific endeavors in this field. The quality of the conference has caught the attention of professionals all over the world, who eagerly look forward to the TAMC series of conferences carefully nurtured over the past ten years by the Steering Committee. All the Steering Committee members guided this edition of TAMC at Anna University, Chennai, India.

The 11th International Conference on Theory and Applications of Models of Computation (TAMC 2014) attracted 112 quality submissions by active researchers from 18 countries, from which the Program Committee carefully reviewed and selected the best papers for inclusion in this LNCS volume.

The review process was rigorous. There were at least two reviews for every paper. At least one third of the submissions had three reviews. There were 30 Program Committee members and 35 additional reviewers involved in the review process. The Program Committee members were actively involved in ranking the papers, after which 27 excellent papers were selected for inclusion in this volume.

We are very grateful to the Program Committee, and the many external reviewers they called on, for the hard work and expertise that they brought to the difficult selection process. We thank all those authors who submitted their work for our consideration. We thank the members of the Editorial Board of *Lecture Notes in Computer Science* and the editors at Springer for their encouragement and cooperation throughout the preparation of this conference.

Dr. M. Rajaram, Honorable Vice-Chancellor, Anna University, Chennai, India, was the patron of TAMC 2014. He ensured that all the necessary facilities for the successful organization of TAMC 2014 were available for the various

Dipti Prasad Mukherjee Regional Vice President (RVP II), Computer
           Society of India

V. Rhymend Uthariaraj   Professor and Director, Ramanujan Computing
           Center, Anna University

M. Chandrasekhar   Professor and Head, Department of
           Mathematics, Anna University

R.S. Bhuvaneswaran   Deputy Director - Web Admin, Ramanujan
           Computing Center, Anna University

A. Azad   Director i/c, Centre for International Affairs,
           Anna University

Table of Contents

A Roadmap for TAMC

T V Gopal[1], Manindra Agrawal[2], Angsheng Li[3], and S. Barry Cooper[4]

[1] Department of Computer Science and Engineering, College of Engineering
Anna University, Chennai - 600 025, India
gopal@annauniv.edu, gayamadhgop@hotmail.com
[2] Department of Computer Science and Engineering, Dean, Faculty Affairs
Indian Institute of Technology Kanpur, India – 208016
manindra@iitk.ac.in
[3] State Key Laboratory of Computer Science, Institute of Software,
Chinese Academy of Sciences, 3rd Floor, Building 5, Software Park, No.4,
South 4th Street, Zhongguancun, Haidian District, Beijing, China
angsheng@ios.ac.cn
[4] School of Mathematics, University of Leeds, Leeds LS2 9JT, United Kingdom
pmt6sbc@maths.leeds.ac.uk

Computability is certainly one of the most interesting and fundamental concepts in Mathematics and Computer Science. **Computation** is any type of calculation or use of computing technology in information processing. Computation is a process following a well-defined model understood and expressed as, for example, an algorithm, or a protocol.

Models of Computation began as an outgrowth of mathematical logic and information theory in the 1960s. They evolved into addressing the classical problems with the aesthetics of computational complexity and asking the fundamental questions concerning non-determinism, randomness, approximation, interaction, and locality. Models of Computation have a foundational role in addressing challenges arising in computer systems and networks, such as error-free communication, cryptography, routing, and search. They are now a rising force in the exact, life, and social sciences.

Models of Computation maintain a core of fundamental ideas and problems such as the famous P vs. NP problem or speeding up algorithms for traditional problems in graph theory, algebra, and geometry. The Models of Computation have increasingly been branching out with fantastic application in biology, economics, physics, and many other fields.

Computation and computational problems are understood in their most general, *interactive* sense, and are precisely seen as interactions between a *machine* (computer, agent, robot) with its environment (user, nature, or the angel / devil itself).

A computation can be seen as a purely physical phenomenon occurring inside a closed physical system called a computer. Examples of such physical systems include digital computers, mechanical computers, quantum computers, DNA computers, molecular computers, analog computers or wetware computers. This point of view is the one adopted by the branch of theoretical physics called the physics of computation. An even more radical point of view is the postulate of digital physics that the evolution of the universe itself is a computation.

T V Gopal et al. (Eds.): TAMC 2014, LNCS 8402, pp. 1–6, 2014.
© Springer International Publishing Switzerland 2014

"The problems of language here are really serious. We wish to speak in some way about the structure of the atoms... But we cannot speak about atoms in ordinary language."

– **Werner Heisenberg**, Physics and Philosophy, 1963

In 1623, Galileo Galilei published "The Assayer" in which he observed "Nature's great book is written in mathematical language".

"All our reasoning is nothing but the joining and substituting of characters, whether these characters be words or symbols or pictures, ... if we could find characters or signs appropriate for expressing all our thoughts as definitely and as exactly as arithmetic expresses numbers or geometric analysis expresses lines, we could in all subjects in so far as they are amenable to reasoning accomplish what is done in Arithmetic and Geometry."

- **Gottfried Wilhelm von Leibniz**, *On Reasoning*, **1677**

Mathematics helps us determine the meaning of what is being communicated with minimum ambiguity and distortion. Clearly mathematics does not have the same fluency as a natural language and, even more obviously, it is rarely spoken aloud. This suggests that mathematics is really a more restrictive limited form of language. Mathematics is an abstract system of ordered and structured thought, existing for its own sake.

Mathematics is the study of systems of elementary objects, conceived independently of our world, and whose only nature is to be exact, unambiguous (two objects are equal or different, related or not; an operation gives an exact result, and so on). Mathematics is split into diverse branches, frameworks of any mathematical work, implicit or explicit that may be formalized as (axiomatic) theories. Each theory is the study of a supposedly fixed system of mathematical objects, whose kind was initially specified (selected from the whole of possible mathematical systems) by a mathematical description called the *foundation* of this theory. There are possible hierarchies between theories, where some can play a foundational role for others. For instance, the foundations of several theories may have a common part forming a simpler theory, whose developments are applicable to all.

The study of the foundations of mathematics as a whole was developed as a branch of mathematics called *mathematical logic*, made of definitions and theorems about systems of objects, answering many philosophical questions and providing frameworks for all mathematics.

"Pure mathematics consists entirely of assertions to the effect that, if such and such a proposition is true of anything, then such and such another proposition is true of that thing. It is essential not to discuss whether the first proposition is really true, and not to mention what the anything is, of which it is supposed to be true. [...] Thus mathematics may be defined as the subject in which we never know what we are talking about, nor whether what we are saying is true. People who have been puzzled by the beginnings of mathematics will, I hope, find comfort in this definition, and will probably agree that it is accurate."

-      **Bertrand Russell, Mysticism and Logic, *1917***

"The skeptic will say: "It may well be true that this system of equations is reasonable from a logical standpoint. But this does not prove that it corresponds to nature." You are right, dear skeptic. Experience alone can decide on truth."

-    **Albert Einstein**

Mathematical logic is dominated by two theories: Set Theory and Model Theory.

Set theory studies the universe of "all mathematical objects", from the simplest to the most complex such as infinite systems. But in details it has a **limitless** diversity of possible variants (not always equivalent to each other).

Model theory is the general theory of theories (describing their formalism as systems of symbols), and of *systems* (worlds) of objects they may describe, called their *models* (their possible interpretations). It is completed by proof theory (describing the rules of proofs). It is essentially unique, giving a clear meaning to the concepts of theory and theorem in each theory.

Etymologically, the word "geometry" means "measure of the earth". There were classically two geometries of interest: the studies of "the plane" and "the space" as they naturally appear. In modern mathematics, geometries are a wide and fuzzy range of mathematical theories describing more general systems also intuitively thought of as "spaces", whose basic objects are "points", and other objects are built over them. Geometry has a formal language (vocabulary) that is a list of symbols (names) of structures. It has axioms which are the claims assumed to be true and are expressed in this language.

The process of understanding often unveils structure; and this, in turn, entails deeper understanding. The structure is formally articulated in mathematical terms. The mathematical structure typically plays a clarifying role providing new insight and leading to new results. Ultimately theories are built; and then specialized, generalized, or unified.

Theory of Computation begins with "What is a model of computation?"

The theory of computation that we have inherited from the 1960's focuses on algorithmic computation as embodied in the Turing Machine to the exclusion of other types of computation that Turing had considered.

In the theory of computation, a diversity of mathematical models of computers have been developed. Typical mathematical models of computers are the following:

- State models including Turing machine, push-down automaton, finite state automaton, and PRAM
- Functional models including lambda calculus
- Logical models including logic programming
- Concurrent models including actor model and process calculi

The focus of the field changed from the (relatively understood) notion of "computation" to the (much more elusive) notion of "efficient computation". The fundamental notion of *NP-completeness* was formulated and its near universal impact was gradually understood. Long term goals, such as the *P vs NP question*, were set.

The *theory of algorithms* was developed, with the fundamental focus on asymptotic and worst-case analysis. Numerous techniques, maturing in mathematical sophistication, were invented to solve major computational problems.

A variety of computational models, designed to explain and sometimes anticipate existing computer systems and practical problems were developed and studied. Among them are *parallel and distributed models, asynchronous and fault-tolerant computation, on-line algorithms and competitive analysis.*

Randomness was introduced as a key tool and resource. This revolutionized the theory of algorithms. In many cases, probabilistic algorithms and protocols can achieve goals which are impossible deterministically. In other cases they enable much more efficient solutions than deterministic ones. Following this, a series of *derandomization techniques* developed to convert in general cases probabilistic algorithms to deterministic ones.

The emergence of the (complexity based) notion of *one-way function*, together with essential use of randomness, has lead to the development of *modern cryptography. Probabilistic proof systems*, with their many variants --- *zero knowledge, Arthur-Merlin, multi-prover, and probabilistically checkable*, have enriched to a tremendous extent our understanding of basic complexity classes which have nothing to do with randomness and interaction, such as *space bounded computation* or *approximate solutions to optimization problems.*

The intimate connection between computational difficulty and pseudo-randomness has brought us closer to understanding the power of randomness in various computational contexts, as well as in purely information theoretic contexts.

Mathematization of reality (and equivalent forms of expressing an experience) has been carried out to the unhappy point where the world begins to disappear behind a ghostly veil of abstraction.

Advances in computation and information technology have already transformed our lives, giving rise to innovations from smart phones, to search engines, to the sequencing of the human genome. However, the greatest transformations lie ahead.

Studying the structures that communicate, store, and process information from this viewpoint will propel computing in new and exciting directions in the years to come. The structures may be expressed:

- in hardware and called machines; or
- in software and called programs; or
- in abstract notation and called mathematics; or
- in nature and society and called biological or social networks and markets.

The power of computing stems from according priority to resource tradeoffs and complexity classifications over the structure of machines and their relationships to languages. It reflects the growing importance of computational models that are more realistic than the abstract ones studied in the 1950s, '60s and early '70s.

Dealing with discrete objects, questions from theoretical computer science inspired much interest in the combinatorics community, and for many of its leaders became a primary scientific goal. This collaboration has been extremely beneficial to both the discrete mathematics and theoretical computer science communities, with wealthy exchange of ideas, problems and techniques. It is extremely important that this conversation between mathematics and theoretical computer science is two-way. More and more mathematicians are considering "computational" aspects of their areas, following theorems like *an object exists* with the problem *how efficiently can this object be constructed?*

Trying to answer them typically reveals more structural questions, and a combination of mathematics and algorithmic theoretical computer science techniques resulted in active research areas like *computational number theory, computational algebra and computational group theory.*

However, the currently available resources of even the newest computer systems are far from being sufficient for solving some of the well defined problems. A whole new type of algorithmic problems from natural sciences in which the required output is not "well defined in advance" are challenging theoretical computer science. Exciting models and solutions in the areas *computational learning, usage of* economic theories to solve problems such as multitudes of autonomous robots, or of independent programs on the Web are now budding. The algorithms are "trying to make sense" of the data, "explain it", "predict future values of it" and so on.

The increasing prominence of the Internet, the Web, and large data-networks in general is profoundly impacting the "Quality of Life" of every individual. It has brought about one of the most challenging shifts in Computer Science.

True technologists just can't stop thinking about tomorrow. The future always looks bright; the question is who and what will help get us there? **TAMC facilitates the creation of technologists both as individuals and as team members through vision and ingenuity.** TAMC enables a strong architectural plan with input from all stakeholders creates a vastly different, participative and delivery working environment. The continued commitment to creating excellence and an atmosphere that embraces change are foundational characteristics of the Computability and Computation in future.

Typically, people are overwhelmed by the wonders of the presented computing technologies. Often times they are oblivious of the Theories and Models that make the technology work for them. TAMC series addresses a wide range of challenges that are natural, man-made and imaginary. Big ideas are lenses for envisioning the future. People have been experiencing "dehumanizing" technology - software or hardware that seems to diminish our ability to communicate with others or to function effectively in the world. TAMC points at technology that creates new boundaries between people rather than erasing old ones. Humanizing entails Safety and Security, Human Relationships and Personal Growth.

The mathematics of relations among objects with which we deal is provides a useful model for our investigation of computing systems. Three desirable properties of such a model are generality, a predictive ability, and appropriateness.

TAMC Series has facilitated better comprehension of the context of Computability and Computation from generalizing the problems and abstracting away unnecessary technological details to ensure forming and learning the structures and the connections to pertinent knowledge. TAMC has changed the way information and knowledge is combined to solve complex problems. TAMC accords a view of proofs set in a meaningful and purposeful context yielding only the validity of the claim they vouch for.

The "formative questions" that are ahead indicate the importance of TAMC. The success of TAMC over the past one decade gives adequate assurance for successful solutions evolving across many disciplines.

References

[1] Korukonda, A.R.: Taking stock of Turing test: a review, analysis, and appraisal of issues surrounding thinking machines. International Journal Human-Computer Studies 58, 240–257 (2003)

[2] Barry Cooper, S., Abramsky, S.: The foundations of computation, physics and mentality: the Turing legacy, Preface. Philosophical Transactions of the Royal Society, A 370, 3273–3276 (2012)

[3] Abramson, D.: Descartes' influence on Turing. Studies in History and Philosophy of Science 42, 544–551 (2011)

[4] Schneider, D.F.: Software Construction: Building a Process Model. Stratus Engineering, Texas (1997)

[5] Elliott Bell, D., LaPadula, L.J.: Secure Computer Systems: Mathematical Foundations, MITRE Technical Report 2547, I (1973)

[6] Eberbach, E., Goldin, D., Wegner, P.: Turing's Ideas and Models of Computation

[7] Savage, J.E.: Models of Computation - Exploring the Power of Computing, Creative Commons License (2008)

[8] Fiore, M.P.: Mathematical Models of Computational and Combinatorial Structures. In: Sassone, V. (ed.) FOSSACS 2005. LNCS, vol. 3441, pp. 25–46. Springer, Heidelberg (2005)

[9] Goldreich, O., Wigderson, A.: Theory of Computation: A Scientific Perspective (1996)

[10] Valdes-Perez, R.E.: A Scientific Basis for Computational Science, CMU-CS-93-162 (1993)

[11] Smale, S.: Mathematical Problems for the Next Century. Mathematical Intelligencer 20(2), 7–15 (1998)

A Tight Lower Bound Instance
for k-means++ in Constant Dimension

Anup Bhattacharya[1], Ragesh Jaiswal[1], and Nir Ailon[2,*]

[1] IIT Delhi, India
{csz128275,rjaiswal}@cse.iitd.ac.in
[2] Technion, Haifa, Israel
nailon@cs.technion.ac.il

Abstract. The k-means++ seeding algorithm is one of the most popular algorithms that is used for finding the initial k centers when using the k-means heuristic. The algorithm is a simple sampling procedure and can be described as follows:

Pick the first center randomly from the given points. For $i > 1$, pick a point to be the ith center with probability proportional to the square of the Euclidean distance of this point to the closest previously $(i - 1)$ chosen centers.

The k-means++ seeding algorithm is not only simple and fast but also gives an $O(\log k)$ approximation in expectation as shown by Arthur and Vassilvitskii [7]. There are datasets [7,3] on which this seeding algorithm gives an approximation factor of $\Omega(\log k)$ in expectation. However, it is not clear from these results if the algorithm achieves good approximation factor with reasonably high probability (say $1/poly(k)$). Brunsch and Röglin [9] gave a dataset where the k-means++ seeding algorithm achieves an $O(\log k)$ approximation ratio with probability that is exponentially small in k. However, this and all other known *lower-bound examples* [7,3] are high dimensional. So, an open problem was to understand the behavior of the algorithm on low dimensional datasets. In this work, we give a simple two dimensional dataset on which the seeding algorithm achieves an $O(\log k)$ approximation ratio with probability exponentially small in k. This solves open problems posed by Mahajan *et al.* [13] and by Brunsch and Röglin [9].

Keywords: Clustering, k-means, k-means++.

1 Introduction

The k-means clustering problem is one of the most important problems in Data Mining and Machine Learning that has been widely studied. The problem is defined as follows:

* Nir Ailon acknowledges the support of a Marie Curie International Reintegration Grant PIRG07-GA-2010-268403, as well as the support of The Israel Science Foundation (ISF) no. 1271/13.

T V Gopal et al. (Eds.): TAMC 2014, LNCS 8402, pp. 7–22, 2014.

(k-means problem): Given a set of n points $X = \{x_1, ..., x_n\}$ in a d-dimensional space, find a set of k points $C = \{c_1, ..., c_k\}$ (these are called *centers*) such that the cost function $\Phi_C(X) = \sum_{x \in X} \min_{c \in C} D(x, c)$ is minimized. Here $D(x, c)$ denotes the square of the Euclidean distance between points x and c. In the *discrete* version of this problem, the centers are constrained to be a subset of the given points X.

The problem is known to be NP-hard even for small values of the parameters such as when $k = 2$ [10] and when $d = 2$ [14,13]. There are various approximation algorithms for the problem. However, in practice, a heuristic known as the k-means algorithm (also known as Lloyd's algorithm) is used because of its excellent performance on real datasets even though it does not give any performance guarantees. This algorithm is simple and can be described as follows:

(k-means Algorithm): (i) Arbitrarily, pick k points C as centers. (ii) Cluster the given points based on the nearest distance to centers in C. (iii) For all clusters, find the mean of all points within a cluster and replace the corresponding member of C with this mean. Repeat steps (ii) and (iii) until convergence.

Even though the above algorithm performs very well on real datasets, it guarantees only convergence to local minima. This means that this *local search* algorithm may either converge to a local optimum solution or may take a large amount of time to converge [5,6]. Poor choice of the initial k centers (step (i)) is one of the main reasons for its bad performance with respect to approximation factor. A number of *seeding* heuristics have been suggested for choosing the initial centers. One such seeding algorithm that has become popular is the k-means++ seeding algorithm. The algorithm is extremely simple and runs very fast in practice. Moreover, this simple randomized algorithm also gives an approximation factor of $O(\log k)$ in expectation [7]. In practice, this seeding technique is used for finding the initial k centers to be used with the k-means algorithm and this ensures a theoretical approximation guarantee. The simplicity of the algorithm can be seen by its simple description below:

(k-means++ seeding): Pick the first center randomly from the given points. After picking $(i - 1)$ centers, pick the ith center to be a point p with probability proportional to the square of the Euclidean distance of p to the closest previously $(i - 1)$ chosen centers.

A lot of recent work has been done in understanding the power of this simple sampling based approach for clustering. We discuss these in the following paragraph.

1.1   Related Work

Arthur and Vassilvitskii [7] showed that the sampling algorithm gives an approximation guarantee of $O(\log k)$ in expectation. They also give an example dataset on which this approximation guarantee is best possible. Ailon *et al.* [4] and Aggarwal *et al.* [3] showed that sampling more than k centers in the manner described above gives a constant *pseudo-approximation*.[1] Ackermann and Blömer [1] showed that the results of Arthur and Vassilvitskii [7] may be extended to a large class of other distance measures. Jaiswal *et al.* [12] showed that the seeding algorithm may be appropriately modified to give a $(1 + \epsilon)$-approximation algorithm for the k-means problem. Jaiswal and Garg [11] and Agarwal *et al.* [2] showed that if the dataset satisfies certain separation conditions, then the seeding algorithm gives constant approximation with probability $\Omega(1/k)$. Bahmani *et al.* [8] showed that the seeding algorithm performs well even when fewer than k sampling iterations are executed provided that more than one center is chosen in a sampling iteration. We now discuss our main results.

1.2 Main Results

The lower-bound examples of Arthur and Vassilvitskii [7] and Aggarwal *et al.* [3] have the following two properties: (a) the examples are high dimensional and (b) the examples lower-bound the *expected* approximation factor. Brunsch and Röglin [9] showed that the k-means++ seeding gives an approximation ratio of at most $(2/3 - \epsilon) \cdot \log k$ only with probability that is exponentially small in k. They constructed a high dimensional example where this is not true and showed that an $O(\log k)$ approximation is achieved with probability exponentially small in k. An important open problem mentioned in their work is to understand the behavior of the seeding algorithm on low-dimensional datasets. This problem is also mentioned as an open problem by Mahajan *et al.* [13] who showed that the *planar* (dimension=2) k-means problem is NP-hard. In this work, we construct a two dimensional dataset on which the k-means++ seeding algorithm achieves an approximation ratio $O(\log k)$ with probability exponentially small in k. More formally, here is the main theorem that we prove in this work.

Theorem 1 (Main Theorem). *Let* $r(k) = \delta \cdot \log k$ *for a fixed real* $\delta \in (0, \frac{1}{120})$. *There exists a family of instances for which k-means++ achieves an* $r(k)$-*approximation with probability at most* $2^{-k} + e^{\left(-(k-1)^{1-120\delta-o(1)}\right)}$.

Note that the theorem refutes the conjecture by Brunsch and Röglin [9]. They conjectured that the k-means++ seeding algorithm gives an $O(\log d)$-approximation for any d-dimensional instance.

1.3 Our Techniques

All the known lower-bound examples [7,3,9] have the following general properties:

---

[1] Here pseudo-approximation means that the algorithm is allowed to output more than k centers but the approximation factor is computed by comparing with the optimal solution with k centers.

(a) All optimal clusters have equal number of points.
(b) The optimal clusters are high dimensional simplices.

In order to construct a counterexample for the two dimensional case, we consider datasets that have different number of points in different optimal clusters. Our counterexample is shown in Figure 2. The optimal clusters (indicated in the figure using shaded areas) are along the vertical lines drawn along the x-axis. In the next section, we will show that these are indeed the optimal clusters. Note that the cluster sizes decrease exponentially going from left to right. We say that an optimal cluster is *covered* by the algorithm if the algorithm picks a center from that optimal cluster. We will use the following two high level observations to show the main theorem:

- **Observation 1**: The algorithm needs to cover more than a certain minimum fraction of clusters to achieve a required approximation.
- **Observation 2**: After any number of iterations, the probability of sampling the next center from an uncovered cluster is not too large compared to the probability of sampling from a covered cluster.

We bound the probability of covering more than a certain minimum fraction of clusters by analyzing a simple Markov chain. This Markov chain is almost the same as the chain used by Brunsch and Röglin [9]. We also borrow the analysis of the Markov chain from [9]. So, in some sense, the main contribution of this paper is to come up with a two dimensional instance the analysis of which may be reduced to the Markov chain analysis in [9].

In the next section, we give the details of our construction and proof.

2   The Bad Instance

We provide a family of 2-dimensional instances on which performance of k-means++ is bad in the sense of Theorem 1. This family is depicted in Figure 2. We first recursively define certain quantities that will be useful in describing the construction. Here m is any positive integer, r is any positive real number, and Δ is a positive real number dependent on k (we will define this dependency later during analysis).

$$r_1 = r \quad \text{and} \quad \forall i, 2 \leq i < k, r_i = 2 \cdot r_{i-1}$$
$$m_1 = m \quad \text{and} \quad \forall i, 2 \leq i < k, m_i = (1/4) \cdot m_{i-1}$$

Note that the input points may overlap in our construction. We will consider k *groups* of points $G_0, ..., G_{k-1}$. These groups are shown as shaded areas in Figure 2. They are located at only k distinct x-coordinates. These k distinct x-coordinates are given by $(x_0, x_1, ..., x_{k-1})$, where $x_0 = 0, x_1 = \Delta \cdot r_1, x_2 = \Delta \cdot (r_1 + r_2), ..., x_{k-1} = \Delta \cdot (r_1 + ... + r_{k-1})$. The ith group, G_i, consists of points that have the x-coordinate x_i. We will later show that $G_0, ..., G_{k-1}$ is actually the optimal k-means clustering for our instance. Group G_0 has $12k2^k m$ points located at $(x_0, 0)$. For all $i \geq 1$, group G_i has $4km_i$ points located at $(x_i, 0)$, and for all $0 \leq j < k$, G_i has $\frac{m_i}{4^j}$ points located at each of $(x_i, 2^j r_i)$ and $(x_i, -2^j r_i)$.

Let the total number of points on i^{th} group be denoted by M_i. Therefore, we can write summing points across all locations on that cluster to get the following:

$$\forall i \geq 1, M_i = 4km_i + 2m_i + 2(m_i/4) + \ldots + 2(m_i/4^{k-1})$$
$$= 4km_i + 2m_i(1 + 1/4 + \ldots + 1/4^{k-1}) \tag{1}$$

Note that $M_{i+1} = M_i/4$.

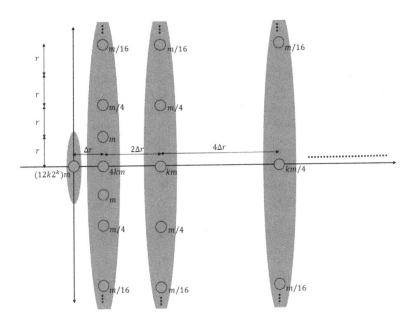

Fig. 1. 2-D example instance showing the 0th, 1st, 2nd, and 3rd optimal clusters only. Note that this figure is not to scale.

2.1   Optimal Solution for Our Instance

We consider the following partitioning of the given points: Let H_0 denote the subset of points on the x-axis and for $|i| \geq 1$, let H_i denote the subset of all points that are located at y-coordinate $sgn(i) \cdot 2^{|i|-1} \cdot r$. For any point $p \in H_i$, we say that point p is in *level* i. Given the above definitions of group and level, the location of a point may be defined by a tuple (i, j), where i denotes the index of the group to which this point belongs and j denotes the level of this point.

Given a set C of centers and a subset of points Y, the potential of Y with respect to C is given by $\Phi_C(Y) = \sum_{y \in Y} \min_{c \in C} D(y, c)$. Furthermore, the potential of a location $l = (i, r)$ with respect to C is defined as $\Phi_C(l) = \sum_{p \text{ located at } l} D(p, C)$. Here, $D(p, C) = \min_{c \in C} D(p, c)$. Given a set of locations

$L = \{l_1, ..., l_s\}$ and a subset of points Y, $\Phi_L(Y)$ denotes the potential of points Y with respect to a set of centers located at locations in L.

We start by showing certain basic properties of our instance.

Lemma 1. *Let $|j| \geq i > 0$. The total number of points at level j in group i is $\frac{m}{4^{|j|-1}}$.*

Proof. The points for group i are located at distance $\pm r_i, \pm 2r_i, \pm 4r_i, \ldots$. Since $r_i = 2^{i-1} \cdot r$, this means that the points in G_i are located at $\pm 2^{i-1}r, \pm 2^i r, \pm 2^{i+1}r, \ldots$. So, the number of points is given by $\frac{m_i}{4^{|j|-i}} = \frac{m}{4^{|j|-1}}$.

Lemma 2. *For all $i > 0$ and $|j| > 0$,*

$$\Phi_{\{(i,0)\}}(H_j) - \Phi_{\{(i,j)\}}(H_j) \leq kmr^2$$

Proof. From Lemma 1, we know that the total number of points at level j of any group is either 0 or $m/4^{|j|-1}$. The net change in the squared Euclidean distance of any point in H_j with respect to locations $(i,0)$ and (i,j) is $(2^{|j|-1}r)^2$. So, the total change in potential is at most $k \cdot \frac{m}{4^{|j|-1}} \cdot (2^{|j|-1}r)^2 = kmr^2$.

Lemma 3. *For all $i > 0$ and $|j| > 0$,*

$$\Phi_{\{(i,0)\}}(H_{j+1 \cdot sgn(j)} \cup H_{j+2 \cdot sgn(j)} \cup ...) \leq$$
$$\Phi_{\{(i,j)\}}(H_{j+1 \cdot sgn(j)} \cup H_{j+2 \cdot sgn(j)} \cup ...) + 2kmr^2.$$

Proof. WLOG assume that $j > 0$. From Lemma 1, we can get an upper bound in the following manner:

$$\Phi_{\{(i,0)\}}(H_{j+1} \cup H_{j+2} \cup \cdots) - \Phi_{\{(i,j)\}}(H_{j+1} \cup H_{j+2} \cup \cdots)$$

$$\leq \sum_{t=1}^{k} \sum_{l=j+1}^{\infty} \frac{m}{4^{l-1}} \cdot r^2 \left((2^{l-1})^2 - (2^{l-1} - 2^{j-1})^2\right)$$

$$= kmr^2 \cdot \sum_{l=j+1}^{\infty} \frac{1}{4^{l-1}} \cdot \left(2^{l+j-1} - 2^{2j-2}\right)$$

$$\leq kmr^2 \cdot \sum_{l=j+1}^{\infty} 2^{j-l+1}$$

$$\leq 2kmr^2$$

Let C denote a set of optimal centers for the k-means problem. Let L denote the set of locations of these centers. We will show that $L = \{(0,0), (1,0), ..., (k-1,0)\}$. We start by showing some simple properties of the set L. We will need the following additional definitions: We say that a group G_i is *covered* with respect to C if C has at least one center from group G_i. Group G_i is said to be *uncovered* otherwise.

Lemma 4. $(0,0) \in L$.

Proof. Let $L' = \{(0,0),(1,0),...,(k-1,0)\}$. Then we have:

$$\Phi_{L'}(X) = \sum_{i=1}^{k-1} 2\left(m_i r_i^2 + \frac{m_i}{4}(2r_i)^2 + ... + \frac{m_i}{4^{k-1}}(2^{k-1}r_i)^2\right)$$

$$= \sum_{i=1}^{k-1} 2k \cdot m_i r_i^2$$

$$= 2k(k-1)mr^2.$$

Let L'' be any set of locations that do not include $(0,0)$, then $\Phi_{L''}(X) \geq 12k2^k mr^2$ (since the nearest location to $(0,0)$ is $(1,0)$). So, L necessarily includes the location $(0,0)$.

Lemma 5. *For any i, if group G_i is covered with respect to C, then $(i,0) \in L$.*

Proof. For the sake of contradiction, assume that $(i,0) \notin L$. Let $(i,j) \in L$ be the location that is farthest from the x-axis among the locations of the form $(i,.) \in L$. Consider the set of locations $L' = (L \setminus \{(i,j)\}) \cup \{(i,0)\}$. We will now show that $\Phi_{L'}(X) < \Phi_L(X)$. WLOG let us assume that j is positive. The change in center location does not decrease the potential of $H_j, H_{j+1}, ...,$ does not increase the potential of $H_{j-1}, H_{j-2}, ...,$ and does not increase the potential of points on the x-axis. From Lemmas 2 and 3, we have that the increase in potential is at most $3kmr^2$. On the other hand, since the contribution of the points located at $(i,0)$ to the total potential changes from $4kmr^2$ to 0, the total decrease in potential is at least $4kmr^2$. So, we have that the total potential decreases and hence $\Phi_{L'}(X) < \Phi_L(X)$. This contradicts the fact that L denotes the location of the optimal centers.

Lemma 6. *All groups are covered with respect to C.*

Proof. For the sake of contradiction, assume that there is a group G_i that is uncovered. This means that there is another group G_j such that there are at least two locations from G_j that is present in L. Note that from the previous lemma $(j,0) \in L$. Let $(j,l) \in L$ for some $l > 0$. We now consider the set of locations $L' = (L \setminus \{(j,l)\}) \cup \{(i,0)\}$. We will now show that $\Phi_{L'}(X) < \Phi_L(X)$. Since $(j,0) \in L$, the change in center location does not decrease the potential of $H_l, H_{l+1}, ...,$ does not increase the potential of $H_{l-1}, H_{l-2}, ...$ and does not increase the potential of points on the x-axis. From Lemmas 2 and 3, we have that the increase in potential is at most $3kmr^2$. On the other hand, since the contribution of the points located at $(i,0)$ to the total potential changes from $4kmr^2$ to 0, the total decrease in potential is at least $4kmr^2$. So, we have that the total potential decreases and hence $\Phi_{L'}(X) < \Phi_L(X)$. This contradicts the fact that L denotes the location of the optimal centers.

The following is a simple corollary of Lemmas 5 and 6.

Corollary 1. *Let C denote the optimal set of centers for our k-means problem instance and let L denote the location of these optimal centers. Then $L = \{(0,0),(1,0),...,(k-1,0)\}$.*

2.2   Potential of the Optimal Solution

Let us denote the potential of the optimum solution by Φ^*. Since optimum chooses its centers only from locations on the x-axis, we can compute Φ^* as follows:

$$\Phi^* = \sum_{i=1}^{k-1} 2 \cdot (m_i r_i^2 + \frac{m_i}{4}(2r_i)^2 + \cdots + \frac{m_i}{4^{k-1}}(2^{k-1}r_i)^2)$$

$$= \sum_{i=1}^{k-1} 2km_i r_i^2$$

$$= 2k(k-1)mr^2 \qquad (2)$$

3   Analysis of k-means++ for Our Instance

We will first show that with very high probability, the first center chosen by the k-means++ seeding algorithm is located at the location $(0,0)$. This is simply due to the large number of points located at the location $(0,0)$ and the fact that the first center is chosen uniformly at random from all the given points.

Lemma 7. *Let p be the location of the first center chosen by the k-means++ seeding algorithm. Then $\mathbf{Pr}[p \neq (0,0)] \leq 2^{-k}$.*

Proof. For any $i \geq 1$ let $\omega(i) = 1 + 1/4 + ... + 1/4^{i-1} = (4/3) \cdot (1 - 1/4^i)$. Since the first center is chosen uniformly at random, we have:

$$\mathbf{Pr}[p = (0,0)] = \frac{M_0}{M_0 + M_1 + ... + M_{k-1}}$$

$$= \frac{M_0}{M_0 + \sum_{i=1}^{k-1} m_i \cdot (4k + 2\omega(k))}$$
$$\text{(since from (1), } M_i = m_i(4k + 2\omega(k)))$$

$$= \frac{M_0}{M_0 + \sum_{i=1}^{k-1} \frac{m}{4^{i-1}} \cdot (4k + 2\omega(k))}$$

$$= \frac{M_0}{M_0 + m \cdot \omega(k-1) \cdot (4k + 2\omega(k))}$$

$$= \frac{(12k) \cdot 2^k}{(12k) \cdot 2^k + \omega(k-1) \cdot (4k + 2\omega(k))}$$

$$\geq \frac{(12k) \cdot 2^k}{(12k) \cdot 2^k + (4/3) \cdot (4k + (8/3))}$$

$$\geq \frac{(12k) \cdot 2^k}{(12k) \cdot 2^k + 12k} \quad \text{(since } k \geq 1)$$

$$\geq 1 - 2^{-k}$$

Let us define the following event:

Definition 1. ξ *denotes the event that the location of the first chosen center is* $(0,0)$.

Lemma 7 shows that ξ happens with a very high probability. We will do the remaining analysis conditioned on the event ξ. We will later use the above lemma to remove the conditioning. The advantage of using this event is that once the first center has the location $(0,0)$, computing an upper-bound on the potential of any location becomes easy. This is because we can compute potential with respect to the center at location $(0,0)$. Computing such upper bounds will be crucial in our analysis.

Our analysis closely follows that of [9]. Let us analyze the situation after $(1+t)$ iterations of the k-means++ seeding algorithm (given that the event ξ happens). Let C_t denote the set of chosen centers. Let $s \le t$ denote the number of optimal clusters among $G_1, ..., G_{k-1}$ that are covered by C_t. Let X_c denote the points in these covered clusters and X_u denote the points in the uncovered clusters. Conditioned on ξ, the probability that the next center will be chosen from X_u is $\frac{\Phi(X_u)}{\Phi(X_u)+\Phi(X_c)}$. So, the probability of covering a previously uncovered cluster in iteration $(t+2)$ depends on the ratio $\frac{\Phi(X_u)}{\Phi(X_c)}$. The smaller this ratio, the smaller is the chance of covering a new cluster. We will show that this ratio is small for most iterations of the algorithm. This means that even when the algorithm terminates, there are a number of uncovered clusters. This implies that the algorithm gives a solution that is worse compared to the optimal solution. In order to upper-bound the ratio $\frac{\Phi(X_u)}{\Phi(X_c)}$, we will upper bound the value of $\Phi(X_u)$ and lower-bound the value of $\Phi(X_c)$. We state these bounds formally in the next two lemmas.

Lemma 8. $\Phi(X_c) \ge (2s-1) \cdot \frac{kmr^2}{4}$.

Proof. For any covered cluster G_i for $i > 0$, we know that G_i has points at levels $0, i-1, -i+1, i, -i, i+1, ...$. For any such location (i,j) (except location $(i,0)$) such that C_t does not have a center at this location, the contribution of the points at this location to $\Phi(X_c)$ is at least $\frac{m_i}{4^{|j|-1}} \cdot (2^{|j|-1} - \max(2^{|j|-2},1))^2 \cdot r_i^2 \ge mr^2/4$. Furthermore, the contribution of points at location $(i,0)$ in case C_t does not contain a center from this location, is at least mr^2. Therefore,

$$\Phi(X_c) \ge ((2k+1) \cdot s - t) \cdot \frac{mr^2}{4}$$

$$\ge (2s-1) \cdot \frac{kmr^2}{4} \quad (\text{since } t \le k-1)$$

Lemma 9. $\Phi(X_u) \le (40k) \cdot (k-s-1)mr^2\Delta^2$.

Proof. Since the number of covered clusters among $G_1, ..., G_{k-1}$ is s, the number of uncovered clusters is given by $(k-s-1)$. Let G_i be any such uncovered cluster. Since ξ happens, there is a center at location $(0,0)$. Therefore, the contribution of G_i to $\Phi(X_u)$ can be upper bounded by the quantity $\Phi_{\{(0,0)\}}(G_i)$. This can be computed in the following manner:

$$\Phi_{\{(0,0)\}}(G_i)$$
$$= \Phi_{\{(i,0)\}}(G_i) + M_i \cdot \Delta^2 \cdot (r_1 + r_2 + \ldots + r_i)^2$$
$$= \Phi_{\{(i,0)\}}(G_i) + M_i \cdot \Delta^2 \cdot (2^i - 1)^2 \cdot r^2$$
$$= \Phi_{\{(i,0)\}}(G_i) + (4k + 2\omega(k))\frac{m}{4^{i-1}} \cdot \Delta^2 \cdot (2^i - 1)^2 \cdot r^2$$
$$\leq \Phi_{\{(i,0)\}}(G_i) + (4k + (8/3)) \cdot (4mr^2\Delta^2)$$
$$= 2 \cdot \sum_{j=1}^{k} \frac{m_i}{4^{j-1}} \cdot (2^{j-1}r_i)^2 + (4k + (8/3)) \cdot (4mr^2\Delta^2)$$
$$= 2kmr^2 + (4k + (8/3)) \cdot (4mr^2\Delta^2)$$
$$\leq (40k)mr^2\Delta^2$$

Hence, the total contribution from the uncovered clusters $\Phi(X_u)$ is upper bounded by $(40k)(k - s - 1)mr^2\Delta^2$.

We will also need a lower bound on $\Phi(X_u)$. This is given in the next lemma.

Lemma 10. $\Phi(X_u) \geq 4k(k - s - 1)mr^2\Delta^2$.

Proof. Let G_i be an uncovered cluster for some $i \geq 1$. For any location (i, j), the contribution of the points at this location to $\Phi(X_u)$ is at least $r_i^2\Delta^2$ times the number of points at that location. So we have:

$$\Phi(X_u) \geq \sum_{\{i | G_i \text{ uncovered}\}} M_i \cdot r_i^2\Delta^2$$
$$= \sum_{\{i | G_i \text{ uncovered}\}} m_i(4k + 2\omega(k)) \cdot r_i^2\Delta^2$$
$$\geq \sum_{\{i | G_i \text{ uncovered}\}} 4k \cdot mr^2\Delta^2$$
$$\geq 4k(k - s - 1) \cdot mr^2\Delta^2$$

Since most of our bounds have the term $k - 1$, we define $\bar{k} = k - 1$ and do the remaining analysis in terms of $\bar{k}$. Note that all the bounds on $\Phi(X_u)$ and $\Phi(X_c)$ are dependent only on s and not on t. This allows us to define the following quantity that will be used in the remaining analysis. This is an upper bound on the ratio $\frac{\Phi_u(X)}{\Phi_c(X)}$ obtained from Lemmas 8 and 9.

$$z_s \stackrel{def}{=} \frac{(\bar{k} - s)(80\Delta^2)}{s - 1/2} = \frac{(k - s - 1)(80\Delta^2)}{s - 1/2} \tag{3}$$

We now get a bound on the number of clusters among $G_1, \ldots, G_{\bar{k}}$ that are needed to be covered to achieve an approximation factor of α for a fixed α. For any such fixed approximation factor α, we define the following quantities that will be used in the analysis.

$$u \stackrel{def}{=} \frac{\alpha}{2\Delta^2} \quad \text{and} \quad s^* \stackrel{def}{=} \lceil \bar{k} \cdot (1 - u) \rceil \tag{4}$$

Lemma 11. *Any α-approximate clustering covers G_0 and at least s^* clusters among $G_1, ..., G_{\bar{k}}$.*

Proof. The optimal potential is given by $\Phi^* = 2k\bar{k}mr^2$ (by (2)). Consider any α-approximate clustering. Suppose this clustering covers s clusters among $G_1, ..., G_{\bar{k}}$. Let the covered and uncovered clusters be denoted by X_c and X_u respectively. Then we have:

$$\alpha = \frac{\Phi(X)}{\Phi^*} \geq \frac{\Phi(X_u)}{\Phi^*} \geq \frac{4k(\bar{k}-s)mr^2\Delta^2}{2k\bar{k}mr^2} \geq \frac{2(\bar{k}-s)\Delta^2}{\bar{k}}$$

The second inequality above is using Lemma 10. This means that the number of covered clusters among $G_1, ..., G_{k-1}$ should satisfy

$$s \geq \left\lceil \bar{k} \cdot \left(1 - \frac{\alpha}{2\Delta^2}\right) \right\rceil = s^*.$$

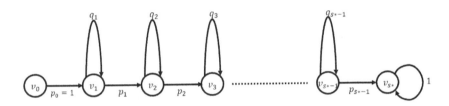

Fig. 2. Markov chain used for analyzing the algorithm

We analyze the behavior of the k-means++ seeding algorithm with respect to the number of covered optimal clusters using a Markov chain (see Figure 2). This Markov chain is almost the same as the Markov chain used to analyze the bad instance by Brunsch and Röglin [9]. In fact, the remaining analysis will mostly mimic that analysis in [9]. The next lemma formally relates the probability that the algorithm achieves an α approximation to the Markov chain reaching its end state. We analyze this Markov chain in the next subsection.

Lemma 12. *Let $p_0 = 1$ and for $s = 1, 2, ..., s^*$, let $p_s = \frac{1}{1+\frac{1}{z_s}}$ We consider the linear Markov chain with states $v_0, v_1, ..., v_{s^*}$ with starting state v_0 (see Figure 2). Edges (v_s, v_{s+1}) have transition probabilities p_s and the self-loops (v_s, v_s) have transition probabilities $q_s = (1 - p_s)$. Then the probability that the k-means++ seeding algorithm gives an α-approximate solution is upper bounded by the probability that the state v_{s^*} is reached by the Markov chain within $\bar{k}$ steps.*

Proof. The proof is trivial from the observation that the probability that a previously uncovered cluster will be covered in iteration $i > 2$ is given by
$$\frac{\Phi(X_u)}{\Phi(X_c)+\Phi(X_u)} \leq \frac{1}{1+\frac{1}{z_s}} = p_s.$$

3.1   Definitions and Inequalities

A number of quantities will be used for the analysis of the Markov chain. The reader is advised to refer to this subsection when reading the next subsection dealing with the analysis of the Markov chain. The following quantities written as a function of $\bar{k}$ will be used in the analysis:

$$\alpha(\bar{k}) = \delta \cdot \log \bar{k} \tag{5}$$

$$\epsilon(\bar{k}) = \frac{1}{120} \cdot \frac{\log \alpha(\bar{k})}{\alpha(\bar{k})} \tag{6}$$

$$\Delta(\bar{k}) = \left\lceil \sqrt{\alpha(\bar{k})} \cdot \exp\left(80 \cdot \alpha(\bar{k}) \cdot \frac{1 + \epsilon(\bar{k})}{4}\right) \right\rceil \tag{7}$$

$$u(\bar{k}) = \frac{\alpha(\bar{k})}{2\Delta^2(\bar{k})} \tag{8}$$

$$s^*(\bar{k}) = \lceil \bar{k} \cdot (1 - u(\bar{k})) \rceil \tag{9}$$

$$z_s(\bar{k}) = \frac{(\bar{k} - s) \cdot (80\Delta^2)}{s - 1/2} \tag{10}$$

$$p_s(\bar{k}) = \frac{1}{1 + \frac{1}{z_s(\bar{k})}} \tag{11}$$

We will also use the following inequalities. Here, whenever we say that $f(\bar{k}) \leq g(\bar{k})$ for two functions f and g, we actually mean to say that $f(\bar{k}) \leq g(\bar{k})$ for all sufficiently large $\bar{k}$.

$$\frac{1}{k} \leq u(\bar{k}) < \frac{1}{2} \tag{12}$$

$$(1 + 40\alpha(\bar{k}))^{\Delta(\bar{k})} \geq \frac{1}{u^2(\bar{k})} \tag{13}$$

$$\frac{1}{\bar{k}} \leq \frac{\epsilon(\bar{k})}{9} \tag{14}$$

$$\frac{1}{80\Delta^2(\bar{k})} \leq \frac{\epsilon(\bar{k})}{3} \cdot u(\bar{k}) \tag{15}$$

$$u(\bar{k}) + \frac{\epsilon(\bar{k})}{3} \cdot \left(1 + \frac{\epsilon(\bar{k})}{3}\right) \cdot u^2(\bar{k}) \leq \left(\frac{\epsilon(\bar{k})}{3}\right)^2 \tag{16}$$

Except for inequality (13), all the inequalities are the same as in [9]. We refer the reader to [9] for the correctness of these inequalities. As for (13), note that $(1 + 40\alpha(\bar{k}))^{\Delta(\bar{k})} \geq 2^{\Delta(\bar{k})} = 2^{\Omega(\sqrt{\alpha(\bar{k})} \cdot e^{\alpha(\bar{k})/4})}$ and $1/u^2(\bar{k}) = O(e^{2\alpha(\bar{k})})$. So, for sufficiently large values of $\bar{k}$, the inequality is true.

For the remaining analysis, we will assume that the value of $\bar{k}$ is fixed such that inequalities (12), (13), (14), (15), and (16) are true. Given this, we will avoid using the functional notation and simply use the name of the quantities. For example, we will use u instead of $u(\bar{k})$ and ϵ instead of $\epsilon(\bar{k})$ etc.

3.2   Analysis of Markov Chain

We now analyze the Markov chain and upper bound the probability of this Markov chain reaching the state v_{s^*} within $\bar{k}$ steps. To be able to do so, we define random variables $X_0, X_1, ..., X_{s^*-1}$, where the X_s denotes the number of steps to move from state v_s to state v_{s+1}. We consider the random variable $X = \sum_{s=0}^{s^*-1} X_s$. We would like to show that the expected value of X is much larger than $\bar{k}$ and then use the Hoeffding inequality to bound the probability. To do this using the well known Hoeffding bound, we need to have a bound on the value of each of the random variables. So, we define related random variables $Y_0, Y_1, ..., Y_{s^*-1}$, where $Y_s = \min(X_s, \Delta)$. We will analyze the random variable $Y = \sum_{s=0}^{s^*-1} Y_s \leq X$. We will use the following lemma from [9].

Lemma 13 (Claim 5 from [9]). *The expected value of X_s is $1/p_s$ and the expected value of Y_s is $(1 - q_s^\Delta)/p_s$.*

The next lemma relates the expected values of X_s and Y_s.

Lemma 14 (Similar to Lemma 6 in [9]). $\frac{E[Y_s]}{E[X_s]} \geq 1 - u^2$.

Proof. First we get a lower bound on z_s in the following manner:

$$z_s = \frac{(\bar{k} - s)(80\Delta^2)}{s - 1/2}$$

$$\geq \frac{u^* \cdot (80\Delta^2)}{1 - u - \frac{1}{2k}} \quad (\text{since } s \leq s^* - 1 \leq \bar{k}(1 - u))$$

$$= \frac{40\alpha}{1 - u - \frac{1}{2k}} \quad (\text{using (8)})$$

$$\geq 40\alpha \quad (\text{using (12)})$$

Also, from the previous lemma we have:

$$\frac{E[Y_s]}{E[X_s]} = 1 - q_s^\Delta = 1 - (1 - p_s)^\Delta = 1 - \left(\frac{1}{1 + z_s}\right)^\Delta$$

$$\geq 1 - \left(\frac{1}{1 + 40\alpha}\right)^\Delta \geq 1 - u^2.$$

The last inequality used (13).

Next, we we get a lower bound on $E[X]$.

Lemma 15 (Similar to Lemma 7 in [9]). $\frac{\mathbf{E}[X]}{k} \geq 1 + \frac{\epsilon}{3} \cdot \left(1 + \frac{\epsilon}{3}\right) \cdot u.$

Proof. We can lower-bound $\mathbf{E}[X]$ in the following manner:

$$\mathbf{E}[X] = \sum_{s=0}^{s^*-1} \frac{1}{p_s}$$

$$= 1 + \sum_{s=1}^{s^*-1} \left(1 + \frac{s - 1/2}{(\bar{k} - s)(80\Delta^2)}\right)$$

$$= s^* + \sum_{i=\bar{k}-s^*+1}^{k-1} \frac{\bar{k} - 1/2 - i}{i \cdot (80\Delta^2)}$$

$$\geq s^* - \frac{s^* - 1}{80\Delta^2} + \frac{\bar{k} - 1}{80\Delta^2} \cdot \sum_{i=\bar{k}-s^*+1}^{\bar{k}-1} \frac{1}{i}$$

$$\geq s^* \cdot \left(1 - \frac{1}{80\Delta^2}\right) + \frac{\bar{k} - 1}{80\Delta^2} \cdot \log\left(\frac{\bar{k}}{\bar{k} - s^* + 1}\right)$$

Since $s^* \geq \bar{k}(1 - u)$, we can write,

$$\mathbf{E}[X]$$

$$\geq \bar{k}(1 - u)\left(1 - \frac{1}{80\Delta^2}\right) + \frac{\bar{k} - 1}{80\Delta^2} \cdot \log\left(\frac{\bar{k}}{\bar{k} - \bar{k}(1 - u) + 1}\right)$$

$$\geq \bar{k}\left(1 - u - \frac{1}{80\Delta^2} + \frac{\bar{k} - 1}{\bar{k}} \cdot \frac{1}{80\Delta^2} \cdot \log\left(\frac{1}{u + \frac{1}{\bar{k}}}\right)\right)$$

Using this, we have:

$$\frac{\mathbf{E}[X]}{\bar{k}}$$

$$\geq \left(1 - u - \frac{\epsilon u}{3} + \frac{\bar{k} - 1}{\bar{k}} \cdot \frac{1}{80\Delta^2} \cdot \log\left(\frac{1}{u + \frac{1}{\bar{k}}}\right)\right) \quad \text{(using (15))}$$

$$\geq \left(1 - u\left(1 + \frac{\epsilon}{3}\right) + \frac{\bar{k} - 1}{\bar{k}} \cdot \frac{1}{80\Delta^2} \cdot \log\left(\frac{1}{2u}\right)\right) \quad \text{(using (12))}$$

$$\geq \left(1 - u\left(1 + \frac{\epsilon}{3}\right) + \left(1 - \frac{\epsilon}{9}\right) \cdot \frac{1}{80\Delta^2} \cdot \log\left(\frac{1}{2u}\right)\right) \quad \text{(using (14))}$$

$$= \left(1 - u\left(1 + \frac{\epsilon}{3}\right) + \left(1 - \frac{\epsilon}{9}\right) \cdot \frac{1}{80\Delta^2} \cdot \log\left(\frac{\Delta^2}{\alpha}\right)\right) \quad \text{(using (8))}$$

$$\geq \left(1 - u\left(1 + \frac{\epsilon}{3}\right) + \left(1 - \frac{\epsilon}{9}\right)\frac{1}{\Delta^2} \cdot \alpha \cdot \frac{1 + \epsilon}{2}\right) \quad \text{(using (7))}$$

$$= \left(1 - u\left(1 + \frac{\epsilon}{3}\right) + \left(1 - \frac{\epsilon}{9}\right)(1 + \epsilon)u\right) \quad \text{(using (8))}$$

$$= 1 + \frac{\epsilon}{3}\left(1 + \frac{\epsilon}{3}\right)u$$

Using the previous two lemmas, we can now obtain a lower bound on $\mathbf{E}[Y]$.

Lemma 16 (Same as Corollary 8 in [9]). $\frac{\mathbf{E}[Y]}{k} \geq 1 + \frac{\epsilon}{3} \cdot u.$

Proof. Using the last two lemmas, we have

$$
\begin{aligned}
\frac{\mathbf{E}[Y]}{\bar{k}} &\geq (1 - u^2) \cdot \frac{\mathbf{E}[X]}{\bar{k}} \\
&\geq (1 - u^2) \cdot \left(1 + \frac{\epsilon}{3}\left(1 + \frac{\epsilon}{3}\right)u\right) \\
&= 1 + u \cdot \left(\frac{\epsilon}{3}\left(1 + \frac{\epsilon}{3}\right) - \left(u + \frac{\epsilon}{3}\left(1 + \frac{\epsilon}{3}\right)u^2\right)\right) \\
&\geq 1 + u \cdot \left(\frac{\epsilon}{3}\left(1 + \frac{\epsilon}{3}\right) - \left(\frac{\epsilon}{3}\right)^2\right) \quad \text{(using (16))} \\
&= 1 + \frac{\epsilon}{3} \cdot u
\end{aligned}
$$

We can finally bound the probability that the Markov chain reaches the state v_{s^*}.

Lemma 17 (Similar to Lemma 9 in [9]). *The probability that the state v_{s^*} is reached within $\bar{k}$ steps is bounded by $\exp(-\bar{k}^{1-120\delta-o(1)})$.*

Proof. The bound on the probability is obtained through the following calculations:

$$
\begin{aligned}
\mathbf{Pr}[X \leq \bar{k}] &\leq \mathbf{Pr}[Y \leq \bar{k}] \quad \text{(since } Y \leq X\text{)} \\
&\leq \mathbf{Pr}\left[\mathbf{E}[Y] - Y \geq \frac{\epsilon}{3} \cdot u \cdot \bar{k}\right] \quad \text{(by Lemma 16)} \\
&\leq \exp\left(-\frac{2 \cdot (\frac{\epsilon}{3} \cdot u \cdot \bar{k})^2}{s^* \Delta^2}\right) \quad \text{(by Hoeffding bound)} \\
&\leq \exp\left(-\frac{2\epsilon^2 u^2 \bar{k}^2}{9k\Delta^2}\right) \\
&= \exp\left(-\bar{k} \cdot \frac{2\epsilon^2 u^2}{9\Delta^2}\right)
\end{aligned}
$$

We will now get a bound on $\frac{2\epsilon^2 u^2}{9\Delta^2}$.

$$
\begin{aligned}
\frac{2\epsilon^2 u^2}{9\Delta^2} &= \frac{\epsilon^2 \alpha^2}{18\Delta^6} \\
&= \frac{\epsilon^2 \alpha^2}{18 \cdot \alpha^3 \cdot \exp\left(80 \cdot 6 \cdot \alpha \cdot \frac{1+\epsilon}{4}\right)} \quad \text{(using (7))} \\
&= \frac{\epsilon^2 \cdot \alpha^{-2} \cdot e^{-120\alpha}}{18} \\
&= \bar{k}^{-o(1)} \cdot \bar{k}^{-o(1)} \cdot \bar{k}^{-120\delta}
\end{aligned}
$$

Now we can put everything together and prove our main theorem.

Proof ((Proof of main theorem)). Given that the event ξ occurs, the probability that the k-means++ seeding algorithm gives an approximation factor of at most $(\delta \cdot \log(k-1))$ is upper bounded by the probability that the Markov chain reaches the state v_{s^*} in at most $(k-1)$ steps. This is bounded by $\exp(-(k-1)^{1-o(1)-120\delta})$ from Lemma 17. Also, from Lemma 7, we know that $\mathbf{Pr}[\neg\xi] \leq 2^{-k}$. Combining these, we get that the probability that the algorithm gives an approximation factor of $(\delta \cdot \log k)$ is at most $2^{-k} + \exp(-(k-1)^{1-o(1)-120\delta})$

Acknowledgements. Ragesh Jaiswal would like to thank Prachi Jain, Saumya Yadav, Nitin Garg, and Abhishek Gupta for helpful discussions.

References

1. Ackermann, M.R., Blömer, J.: Bregman clustering for separable instances. In: Kaplan, H. (ed.) SWAT 2010. LNCS, vol. 6139, pp. 212–223. Springer, Heidelberg (2010)
2. Agarwal, M., Jaiswal, R., Pal, A.: k-means++ under approximation stability. In: Chan, T.-H.H., Lau, L.C., Trevisan, L. (eds.) TAMC 2013. LNCS, vol. 7876, pp. 84–95. Springer, Heidelberg (2013)
3. Aggarwal, A., Deshpande, A., Kannan, R.: Adaptive sampling for k-means clustering. In: Dinur, I., Jansen, K., Naor, J., Rolim, J. (eds.) Approx and Random 2009. LNCS, vol. 5687, pp. 15–28. Springer, Heidelberg (2009)
4. Ailon, N., Jaiswal, R., Monteleoni, C.: Streaming k-means approximation. In: NIPS, pp. 10–18 (2009)
5. Arthur, D., Vassilvitskii, S.: How slow is the k-means method? In: Proceedings of the Twenty-second Annual Symposium on Computational Geometry, SCG 2006, pp. 144–153. ACM, New York (2006)
6. Arthur, D., Vassilvitskii, S.: Worst-case and smoothed analysis of the ICP algorithm, with an application to the k-means method. In: Proceedings of the 47th Annual IEEE Symposium on Foundations of Computer Science, FOCS 2006, pp. 153–164. IEEE Computer Society, Washington, DC (2006)
7. Arthur, D., Vassilvitskii, S.: k-means++: the advantages of careful seeding. In: Proceedings of the Eighteenth Annual ACM-SIAM Symposium on Discrete Algorithms, SODA 2007, pp. 1027–1035. Society for Industrial and Applied Mathematics, Philadelphia (2007)
8. Bahmani, B., Moseley, B., Vattani, A., Kumar, R., Vassilvitskii, S.: Scalable k-means++. Proc. VLDB Endow. 5(7), 622–633 (2012)
9. Brunsch, T., Röglin, H.: A bad instance for k-means++. Theoretical Computer Science (2012)
10. Dasgupta, S.: The hardness of k-means clustering. Technical report, University of California San Diego
11. Jaiswal, R., Garg, N.: Analysis of k-means++ for separable data. In: Gupta, A., Jansen, K., Rolim, J., Servedio, R. (eds.) APPROX/RANDOM 2012. LNCS, vol. 7408, pp. 591–602. Springer, Heidelberg (2012)
12. Jaiswal, R., Kumar, A., Sen, S.: A simple D^2-sampling based PTAS for k-means and other clustering problems. Algorithmica (2013)
13. Mahajan, M., Nimbhorkar, P., Varadarajan, K.: The planar k-means problem is NP-hard. Theoretical Computer Science 442, 13–21 (2012); Special Issue on the Workshop on Algorithms and Computation (WALCOM 2009)
14. Vattani, A.: The planar k-means problem is NP-hard. Manuscript (2009)

An Improved Upper-Bound
for Rivest et al.'s Half-Lie Problem

Bala Ravikumar[1] and Duncan Innes[2]

[1] Department of Computer and Engineering Science
Sonoma State University, Rohnert Park, CA 94928, USA
ravi@cs.sonoma.edu
[2] Department of Computer Science
University of Rhode Island, Kingston, RI 02881, USA
innes@cs.uri.edu

Abstract. Ulam proposed the problem of determining an optimum strategy for finding an integer $x \in \{1, 2, ..., n\}$ using binary queries (i.e., queries with yes/no answer) in which the responses to up to k queries (for a fixed k) can be incorrect. This problem has been extensively studied for the past fifty years. The paper by Rivest et al. [9] that made a major advance in Ulam's problem introduced a restricted type of error in responses known as half-lies. Rivest et al. presented a lower-bound on the minimax complexity of the half-lie version of Ulam's search problem. Here we present a new algorithm that improves the previous upper-bound for the half-lie problem (in the case of $k = 1$) for all sufficiently large values of n. Specifically, we show that the number of queries of the form 'Is $x > s$?' sufficient (in the worst-case) to find an unknown integer $x \in \{1, 2, ..., n\}$, when the responder's 'yes' answers are always true, but at most one of the 'no' answers may be false, is at most $\lceil \log_2((n + 4.5) \ln(n + 4.5) - 4.5 \ln(4.5)) \rceil$. We also present an improvement to Rivest et al.'s lower-bound for the special case of $n = 10^6$.

Keywords: algorithm analysis, upper-bound, decision tree, lower-bound, weight balancing.

1 Introduction

Ulam [14] in his autobiography, *Adventures of a Mathematician* stated the following problem:

- *Hawkins and I have speculated on the following related problem: a variation on the game of Twenty Questions. Someone thinks of a number between one and one million (which is just less than 2^{20}). Another person is allowed to ask up to twenty questions, to each of which the first person is supposed to answer only yes or no. Obviously the number can be guessed by asking first: Is the number in the first half million? then again reduce the reservoir of numbers in the next question by one half, and so on. Finally the number is obtained in less than $\log_2(1,000,000)$. Now suppose one were allowed to*

T V Gopal et al. (Eds.): TAMC 2014, LNCS 8402, pp. 23–38, 2014.
© Springer International Publishing Switzerland 2014

lie once or twice, then how many questions would one need to get the right answer? One clearly needs more than n questions for guessing one of 2^n objects because one does not know when the lie was told. This problem is not solved in general.

In this paper, we study a variation of Ulam's problem stated above in the case of one lie. The variation is as follows: we restrict the questioner to questions of the form 'Is $x > s$?' and the responder's one lie (if she chooses to exercise it) is restricted to only 'no' answers. This variation has been studied before, and the main contribution of this work is to improve asymptotially (i.e., for all large enough values of n) the best known upper-bound on the number of questions needed to find the unknown in the worst-case.

The first major work that addressed the Ulam problem was due to Rivest et al. [9] who gave upper and lower-bounds on the worst-case number of queries needed to find the unknown $x \in \{1, ..., n\}$ as a function of n and k, the number of permitted lies. In that paper, they introduced a variation of Ulam's problem that they called a *half-lie* problem. The half-lie problem as studied in [9] deals with the case in which the questioner is restricted to asking questions of the form "Is $x > c$?" for a constant s. The choice of c on a query can be based on all the previous responses received, a model known as *adaptive*. (In later works [12], such questions were referred to as *cut questions*.) Rivest et al. [9] presented an upper-bound on the number of queries needed for the full-lie problem, and this has remained the upper-bound for the half-lie problem as well. The main idea behind the algorithm presented by Rivest et al. was a clever generalization of the well-known binary search strategy that gives an optimal solution in the case of no lies.

Although Rivest et al. presented a lower-bound for the half-lie problem, they did not present any upper-bound and hence the only known upper-bound for the half-lie problem is the one comes directly from the full-lie problem (since any algorithm for the full-lie also works in the half-lie case). There has been a lot of study of the half-lie problem since 1980 (e.g. [12], [3], [13], [2] etc.). However, all these papers consider the half-lie version in which the questioner is allowed to ask arbitrary yes/no question. Notably, [2] deals with the half-lie problem with $k = 1$, and presents both upper- and lower-bound results. However, since it uses arbitrary yes/no questions (bit questions) rather than cut questions, his upper-bound results do not translate to an upper-bound for the cut question model. However, his lower-bound provides a lower-bound for our restricted 'cut question' model, as we note later (in Section 7).

There are reasons why the cut question model is more interesting than the bit question model. One of them is a result of Rivest et al. [9] that showed that the cut question model of the half-lie problem is closely related to the problem of finding the maximum of a collection of keys using only comparison of keys to constants, but not each other [5] - a problem of interest in its own right (see Section 6 for details). Such connections do not exist in the case of bit question model of the half-lie problem. In any case, in this paper, we consider only *the half-lie problem with cut questions* and present an upper-bound for this problem that is an improvement over the corresponding full-lie problem for all large enough values of n. For the specific case of $n = 10^6$, our new upper-bound is 24 while the previous upper-bound was 25. We also make an improvement in the lower-bound for this problem (from 22 to 23) in the specific case of $n = 10^6$.

The rest of the paper is organized as follows. In Section 2, we present an overview of prior results in this area that are relevant to the current work. In Section 3, we present an overview of the new algorithm that leads to the improved upper-bound. In Section 4, we present an analysis of the algorithm during the opening phase, i.e., until a weight of 512 or less is reached. In Section 5, we present the end-game analysis that shows that at most 9 additional queries are needed to complete the end-game. In Section 6, a lower-bound for the half-lie problem with one lie is prsented in the special case of $n = 10^6$ that improves the bound of [9] by one. In Section 7, a connection between the half-lie problem and a fundamental problem of two-player binary search is used to improve a solution to the latter problem using our new algorithm. In Section 8, some open problems and conclusions are presented.

2 Background

As stated in the Introduction, it was [9] that introduced the half-lie problem. This work also established an upper-bound and a lower-bound for the full-lie problem, while presenting only a lower-bound for the half-lie problem. Since we are mainly interested in k (the number of lies) $= 1$ here, we will state the results of Rivest et al. [9] for this case.

Theorem 1. *([9]) Let $Q(n)$ be the number of cut questions needed in the worst-case to identify an unknown $x \in \{1, 2, ..., n\}$ when at most one of the answers received can be erroneous. Then:*

$$min \; \{Q' \mid 2^{Q'} \geq n(Q' + 1)\} \leq Q(n) \leq min \; \{Q' \mid 2^{Q'-1} \geq nQ'\}.$$

Since the upper-bound on the full-lie case applies to half-lie case as well, the above upper-bound also holds for the half-lie case.

Rivest et al. also showed a lower-bound for the half-lie problem.

Theorem 2. *([9]) Let $Q_H(n)$ be the number of queries needed in the worst-case to find an unknown $x \in \{1, 2, ..., n\}$ using comparison queries of the form 'Is $x > s$?' when none of the 'yes' answers are incorrect, but at most one of the 'no' answers may be incorrect. Then, $Q_F(n) \geq \log_2 n + \log_2(\log_2 n) - 3$.*

The upper-bound presented in 2 has been improved by several researchers. For example, Spencer [10] showed

$$Q(n) \leq min \; \{Q' \mid 5 * 2^{Q'} \geq \frac{8n}{5}(Q' + 1)\}.$$

and the second author of this paper [4] improved it to

$$Q(n) \leq min \; \{Q' \mid 7 * 2^{Q'} \geq \frac{8n}{5}(Q' + 3)\}.$$

Our main result is the upper bound $Q_H(n) \leq \lceil \log_2((n + 4.5) \ln(n + 4.5) - 4.5 \ln(4.5)) \rceil$ which is an improvement over the previous upper-bound stated in Theorem 1 above for all sufficiently large values of n. We show this by using a weight-balancing strategy followed by a detailed end-game analysis. What differentiates this work is the use of a continuous weight function instead of combinatorial weight function used in other works.

3   The Algorithm

In this section, we present an overview of the main algorithm presented in this paper. The analysis of the algorithm is quite complicated, and so in this section, an overview of the algorithm is presented without all the formal details. We begin an informal description of the algorithm, and conclude with a more formal description of it. But even this description that appears at the end of the section is not complete since it refers to some key ideas that are presented in Sections 4 and 5. In particular, we will defer to future sections a formal description of a potential function (the weight function) that plays a key role in showing that the algorithm achieves the claimed upper-bound in the worst-case.

It is perhaps the easiest to think of the searching problem as a guessing game. One person, commonly called Carole, picks an integer $x \in \{1, 2, ..., n\}$. Paul, who wants to find out x, asks questions of the form "Is $x > s$?" for some selected s. Carole is allowed to say "no" once when x is greater than s, but be truthful at all other times. Paul, therefore, knows that "yes" answers are truthful, but is suspicious of all "no" answers. Let $q = \lceil \log_2((n + 4.5) \ln(n + 4.5) - 4.5 \ln(4.5)) \rceil$. We will say that the problem has been solved or that Paul wins if there is exactly one x that is consistent with the responses of Carole and the number of questions Paul has asked is at most q. As is the case with most of the analysis of liar games, Carole need not pick x ahead of time. Instead, she would respond to the questions in a way to prolong the game as long as possible.

At any stage of the game, Paul's knowledge is represented by two adjacent intervals of real numbers in the form $C = < I_1, I_2 >$ where $I_1 = (p, q]$ and $I_2 = (q, r]$ for some real numbers p, q and r. (Note that the half-open interval $(n, n]$ represents the empty set.) The configuration C has the following meaning: the unknown x is in I_1 if Carole has never lied so far, and is I_2 if Carole has lied exactly once. The following simple induction will show that Paul's knowledge can always be represented as a configuration, provided Paul never asks a wasteful question: a question for which Carole can give a truthful answer that will not add any new information to Paul's knowledge.

Lemma 1. *Assuming Paul never wastes a question, Paul's knowledge of the unknown x can always be represented as a configuration as defined above.*

Proof. The proof is by induction on the number of questions t asked thus far. When $t = 0$, Paul's knowledge is represented by the configuration $< (0, n], \varnothing >$. For the induction step, assume that the current configuration is $C = < I_1, I_2 >$ where $I_1 = (p, q]$ and $I_2 = (q, r]$. It is easy to see that a question of the form 'Is $x > s$?' is wasteful if $s > r$, or $c \leq p$. Thus, we can assume that in Paul's next question 'Is $x > s$?', s satisfies the bound $p < s \leq r$. There are two cases to consider.

Case 1. $p \leq x \leq q$. The 'yes' and 'no' answers result in configurations $< (c, q]$, $(q, r] >$ and $< (p, c], (c, q] >$, respectively.

Case 2. $q \leq x \leq r$. The 'yes' and 'no' answers result in configurations $< (c, c]$, $(c, r] >$ and $< (p, q], (q, c] >$, respectively.

This concludes the proof.   □

We define a weight function w that maps each configuration C to a real number. The exact weight function is not necessary to understand the overall strategy so we defer it to next section.

Paul's strategy consists of two phases. In phase 1 (denoted by *opening game*), he picks the split point s at each step in such a way that the weight of a "yes" response will equal the weight of a "no" response and asks the question 'Is $x > s$?'. If the response is 'yes', then Paul can eliminate all numbers less than or equal to s. If the response is 'no', he can only eliminate those numbers greater than x that have already been lied about. This set is the second component of the configuration as described above. To give a concrete example, suppose the current configuration is $< (50, 100], (100, 150] >$. If Paul's next question is 'Is $x > 75$?' If he receives a 'yes' answer, the resulting configuration will be $< (75, 100], (100, 150] >$. If the answer is 'no', the resulting configuration will be $< (50, 75], (75, 100] >$. During the first phase, Paul will balance the weights associated with the 'yes' and 'no' answer and select the query point s which would, in general, be a real number.

Phase 2 (or the *end game*) begins when the weight of those numbers that can be potential candidates for the unknown x is less than or equal to 512. In phase 2, the questioner finds the sets of integers contained in each of the continuous intervals he has been keeping track of. Then by consulting a table constructed by working backwards from the winning positions, Paul determines the split point based on the number of integers in each set.

It will be shown that in phase 1 the questioner can cut the weight in half on each question and can thus be assured that after $j - 9$ questions, the weight will be less than 512 if the initial weight w was between 2^{j-1} and 2^j for some $j \geq 10$. The end-game analysis will show that any configuration with weight at most 512 can be won with at most 9 questions. This allows us to conclude that the worst-case number of questions sufficient to win is at most $\lceil \log_2 w_0 \rceil$ where w_0 is the initial weight.

In summary, Paul's algorithm (which leads to the new upper-bound for the half lie problem and forms the main result of this paper) can be described as follows:

1. The game starts with the state $I_1 = < 1, n >$ and $I_2 = \varnothing$, and the associated weight $w_0 = f(n)$ (where f is as defined in the next section). Carole chooses an integer x in $[1, n]$, and Paul's goal is to find x by asking Carole questions of the form 'Is $x > s$?' for adaptively chosen values of s. Carole will respond

with a 'yes' or 'no' answer subject to the condition that all her answers are true, and at most one of the answers can be false.
2. (opening game) While the weight associated with the current state is less than 512, Paul selects a real number s and asks Carole the question 'Is $x > s$?'. As shown in Section 4, such a real number s always exists with the following property: independent of whether Carole's response is 'yes' or 'no', the weight associated with the next state is exactly one-half of the weight associated with the current state.
3. (end game) When the while loop of Step 2 above ends, the game enters the end game. Paul plays the end game as described in Section 5 (using a look-up table). It will be shown in Section 5 that Paul can determine x with at most 9 queries.

Analysis of this algorithm is presented in detail in the next two sections.

4   Opening-game Analysis

Let $c = 4.5$ and let $f(x) = \int_c^{x+c}(\ln\ t + 1)dt = (x+c)\ln\ (x+c) - c\ \ln\ c$.

If $I_1 = (a, b]$ is the set of numbers which are consistent with the answers given so far, and $I_2 = (b, d]$ is the set of numbers about which one lie has been told, then we let $m = b - a$ and $k = d - b$.

The weight of such a state $< I_1, I_2 >$ will be defined as

$$W(m, k) = f(m) + k. \tag{2}$$

We will also use w_i for the weight of the state reached after i questions asked. In particular, the weight associated with the initial configuration is $w_0 = W(n, 0)$ $= f(n)$. Thus $W(m, k) = \int_r^{m+r}(\ln\ (x - r + c) + 1)dx + \int_{m+r}^{k+m+r} dx = (m + c)\ln\ (m + c) - c\ln\ c + k$

The next query point s will be chosen such that the weight resulting from a 'yes' answer will equal the weight resulting from a 'no' answer.

1. If $k > f(m)$ then s should satisfy:

$$W(m, s - m) = W(0, m + k - 2). \tag{3}$$

Specifically, choose s such that,

$$f(m) + s - m = m + k - s.$$

2. If $k \le f(m)$, s will be chosen so that

$$W(s, m - s) = W(m - s, k).$$

Specifically, choose s such that

$$(s + c)\ln\ (s + c) + m - s = (m - s + c)\ln(m - s + c) + k. \tag{4}$$

The next result (which is the main theorem of this paper) shows that the weight of a configuration (that is at least 512) reduces by a factor of two after a query.

Theorem 3. *Given any weight state after i questions, $512 < w_i \leq 2^j$, $j = 10$, 11, 12, ..., the new weight after one more question will be bounded by $w_{i+1} \leq 2^{j-1}$.*

Proof. **Case 1.**
When $k > f(m)$ then from (3) and (2) we choose s such that

$$f(m) + s - m = m + k - s.$$

Since

$$f(m) + s - m + m + k - s = f(m) + k,$$

it follows that

$$m + k - s = \frac{f(m) + k}{2}.$$

Thus,

$$w_{i+1} = W(m, s - m) = W(0, m + k - s) = m + k - s$$

and $w_i = f(m) + k$, it follows that

$$w_{i+1} \leq \frac{w_i}{2} \leq 2^{j-1}.$$

Case 2.
When $k \leq f(m)$, we need to show that, with $512 \leq w_i \leq 2^j$

$$w_{i+1} = W(s, m - s) = W(m - s, k) \leq 2^{j-1}.$$

or equivalently,

$$w_i - 2w_{i+1} \geq w_i - 2^j.$$

Let the gain in weight from one question to the next , as a function of m and k, be defined as

$$g(m, k) = w_i - 2w_{i+1}.$$

Then from (1) and (2),
$$g(m, k) = (m + c)\ln(m + c) - (s + c)\ln(s + c) - (m - s + c)\ln(m - s + c) - m + s + c\ln c.$$
It is enough to show that $g(m, k) \geq 0$ since it would imply that $w_{i+1} \leq w_i/2 \leq 2^{j-1}$. For those values where it is difficult to show that $g(m, k) \geq 0$, it will be sufficient to show that

$$g(m, k) \geq w_i - 2^j.$$

First consider the partial derivative of g with respect to k. Since s depends on k, we must first find $\frac{\partial s}{\partial k}$. Differentiating implicitly with respect to k we get

$$\frac{\partial s}{\partial k}\ln(s + c) = -\frac{\partial s}{\partial k}\ln(m - s + c) - \frac{\partial s}{\partial k} + 1.$$

Thus,

$$\frac{\partial s}{\partial k} = \frac{1}{\ln(s + c) + \ln(m - s + c) + 1}$$

Notice that, since $s, m, m - s \geq 0$,

$$\frac{\partial s}{\partial k} > 0. \tag{5}$$

Now differentiating g with respect to k, we get

$$\frac{\partial g}{\partial k} = \frac{\partial s}{\partial k}(\ln(m - s + c) - \ln(s + c) + 1).$$

Thus the partial derivative with respect to k will be 0 when

$$\ln(m - s + c) - \ln(s + c) + 1 = 0.$$

Thus the partial derivative with respect to k will be 0 when

$$m - s + c = (s + c)/e.$$

Solving for s,

$$s = \frac{me + ce - c}{1 + e}.$$

Since s must also satisfy (4), we can solve for k in (4), and substitute for s. Thus we get

$$k = \frac{e(m + 2c)}{1 + e} \ln\left(\frac{e(m + 2c)}{1 + e}\right) - \frac{(m + 2c)}{1 + e}.$$

With some simplification, we conclude that $\frac{\partial s}{\partial k} = 0$ when

$$k = \frac{e - 1}{1 + e}\left[(m + 2c)\ln(m + 2c) - (m + 2c)\ln(1 + e)\right] + m + c.$$

From (5) we see that for a fixed value of m as k increases, so does s. As s increases, g will increase until $\ln(m - s + c) - \ln(s + c) + 1 = 0$ and g will decrease thereafter. Since $g(m, f(m)) = 0$ for all m when (m, k) is such that $\ln(m - s + c) - \ln(s + c) + 1 \leq 0$ and $k \leq f(m)$ we know that g must increase as k decreases. Therefore $g(m, k) \geq 0$ for these pairs (m, k).

Figure 1 shows values for the domain of g and the regions where the partial derivative with respect to k is positive or negative. The domain D is given by

$$D = \{(m, k) \mid f(m) > k \ and \ f(m) + k \geq 512 \ and \ k \geq 0\}.$$

The curve described by

$$\{(m, k) \mid k = \frac{e - 1}{1 + e}\left[(m + 2c)\ln(m + 2c) - (m + 2c)\ln(1 + e)\right] + m + c.$$

indicates the points at which $\frac{\partial g}{\partial k} = 0$ and g attains a maximum for a fixed value of m. It has just been shown that for all the points in the domain above the line, g will be non-negative. It remains to be shown that those below the line will also result in a non-negative g.

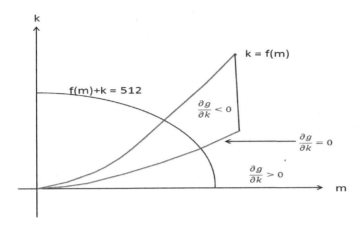

Fig. 1. Domain of $g(m, k)$

Consider the function $\hat{g}(m) = g(m, 0)$. Substituting $k = 0$ and

$$(s + c) \ln(s + c) + m - s = (m - sc + c) \ln(m - s + c),$$

we can write $\hat{g}$ as

$$\hat{g}(m) = (m + c) \ln(m + c) - 2(s + e) \ln(s + c) - 2m + 2s + c \ln c.$$

Differentiating we get,

$$\frac{d\hat{g}}{dm} = \ln(m + c) - 1 - 2 \frac{ds}{dm} \ln(s + c). \tag{6}$$

With $k = 0$, we will choose s such that

$$(s + c) \ln(s + c) + m - s = (m - sc + c) \ln(m - s + c). \tag{7}$$

We can now consider s as a function of m (rather than m and k) and differentiate (7) to obtain:

$$\frac{ds}{dm} = \frac{\ln(m - s + c) - 1}{\ln(m - s + c) + \ln(s + c)}.$$

It is clear from (7) that

$$\ln(s + c) < \ln(m - s + c).$$

This together with the fact that $m > m - s$ allows the conclusion that

$$\frac{ds}{dm} = \frac{\ln(m - s + c) - 1}{\ln(m - sc + c) + \ln(s + c)} < \frac{\ln(m + c) - 1}{2 \ln(s + c)}$$

when $k = 0$. Using (6) we get the following inequality:

$$\frac{d\hat{g}}{dm} > \ln(m+c) - 1 - 2\ln(s+c)\frac{\ln(m+c)-1}{2\ln(s+c)} = 0.$$

Thus $\hat{g}(m) = g(m,0)$ will increase as m increases.

Since $g(94,0) > 0$, we can conclude that $g(m,0) > 0$ for $m \geq 94$. Now since $\frac{\partial g}{\partial k} > 0$ below the line $k = \frac{e-1}{1+e}\big[(m+2c)\ln(m+2c) - (m+2c)\ln(1+e)\big] + m + c$, we know that $g(m,k) \geq 0$ when $m \geq 94$ and (m,k) is below the line. We have now shown that $g(m,k) \geq 0$ for all elements of the domain except the shaded region in Figure 2, i.e.,

$$\{(m,k) \mid k < \tfrac{e-1}{1+e}\big[(m+2c)\ln(m+2c) - (m+2c)\ln(1+e)\big] + m + c,$$
$$f(m) + k \geq 512, \ m < 94\}.$$

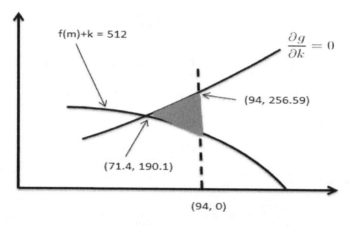

f(m)+k = 512

$\frac{\partial g}{\partial k} = 0$

(94, 256.59)

(71.4, 190.1)

(94, 0)

Fig. 2. Points (m,k) where it is hard to show $w_i \leq 2w_{i+1}$

To deal with this region, we will rely on the fact that $g(m,k)$ only need be greater than $W(m,k) - 2^j$ rather than the harder to prove condition $g(m,k) \geq 0$ (although this seems to be true). It seems obvious from the graph that the largest weight occurring in this interval will be less than $W(94, 256.59) = 701.94$. To show that this is the largest weight, we need to show that all these points will have $k < 256.59$ since we know that $m < 94$ for all points in the region. The slope of the curve where

$$k = \frac{e-1}{1+e}\big[(m+2c)\ln(m+2c) - (m+2c)\ln(1+e)\big] + m + c$$

will be

$$\frac{dk}{dm} = \frac{e-1}{1+e}\big[\ln(m+2c) - \ln(1+e)\big] + \frac{2e}{1+e}.$$

Because $m + 2c > 1 + e$ for all $m \geq 0$,

$$\frac{dk}{dm} = 0.$$

So as m decreases, k will also decrease. All the points in the region have $m < 94$ and $k < 256.59$ and therefore a weight less than 702.

Thus, $W(m, k) < 702 < 2^{10}$ and $j \geq 10$ for these troublesome points. We need only show then that, for all the points in this region, $g(m, k) \geq 702 - 2^{10} = -322$. Notice that $m \geq 70$ for all these points. With $m = 70$ and $k = 0$, $s \approx 30.73$ and $g(70, 0) \approx -2.23$. As before, since $\frac{d\hat{g}}{dm} > 0$, we can conclude that $g(m, 0) > -2.23 > -322$ for all these points. Also since $\frac{\partial g}{\partial k} > 0$ for all the points in this region, $g(m, k) > g(m, 0) > -322$ for all these points. For the other points, we were able to show we could split the weight to half or less after each question, but when the weight is less than or equal to 702, we don't need to split it half exactly to get a weight less than or equal to 512. Thus when $512 \leq w_i \leq 2^j$, we have shown that $w_{i+1} \leq 2^{j-1}$.   □

We have thus shown that until the end-game is reached (when the weight is 512 or less) we are able to balance the weights so that irrespective of how Carole responds, Paul can reduce the weights by a factor of 2. By repeating this weight balancing strategy, we conclude the following:

Theorem 4. *For any initial weight $w_0 \geq 512$ of a configuration, where $2^{j-1} < w_0 \leq 2^j$ for some $j \geq 10$, the weight after i questions using the algorithm presented in Section 3 will be:*

$$w_i \leq 2^{j-i}, \quad j - i \geq 9$$

and thus Paul can reach a weight of 512 or less in no more than $j - 9$ questions.

5   End-game Analysis

In this section, we present the end-game analysis. This involves showing that any configuration with weight at most 512 can be won by Paul with at most 9 questions. Since the total number of such configurations is finite, a brute-force way of showing this claim is to consider every one of these cases. But even for a single starting configuration, establishing the upper-bound involves inspecting a decision tree of depth k where k can be as large as 9, and thus this approach is not feasible. While our approach still requires building a look-up table, the table size we build is small (see Appendix) - and we show that all the cases of the end-game are covered by this table.

Theorem 5. *Paul can win starting from any configuration C with weight at most 512 in nine or fewer questions.*

Proof. Up until now, we have chosen s to be a real number and the intervals represented by the sets M and K have been composed of real numbers. We need to show that the number of integers, m' and k' contained, respectively, in each of these sets is such that the discrete pattern (m', k') guarantees a win for Paul in nine additional questions. The table in Appendix shows for each value of m', the maximum value for k', called 'max', that guarantees a win.

To establish this, we need to make two observations. First, since the sets I_1 and I_2 corresponding to configuration $C = < I_1, I_2 >$ are disjoint,

$$m' + k' \leq \lceil m + k \rceil. \tag{8}$$

The second observation deals with the size of k compared to m. We know that the weight at this point must be no more than 512 and therefore $k \leq 512 - f(m)$. We want to show that any increase in m will decrease the upper bound on k by an even greater amount. That is,

$$m_1 + (512 - f(m_1)) > m_2 + (512 - f(m_2)),$$

if $m_1 < m_2$.

Let $\hat{k}$ be the largest possible k for each value of m, so that

$$\hat{k} = 512 - f(m).$$

Substituting for $f(m)$ and differentiating we get:

$$\frac{d\hat{k}}{dm} = -\ln(m + c) - 1 < -1$$

for all $m \geq 0$.

Therefore, $\hat{k}$ decreases faster than m increases. Thus with $\hat{k}_i = 512 - f(m_i)$

$$m_1 + \hat{k}_1 > m_2 + \hat{k}_2 \ if \ m_1 < m_2.$$

Suppose $L \leq m \leq L + 1$, $L \in \{0, 1, ..., 105\}$. Then

$$m + k \leq m + \hat{k} \leq L + 512 - f(L). \tag{9}$$

From inequalities (7) and (8)

$$m' + k' \leq \lceil L + 512 - f(L) \rceil.$$

This means that either

1. $m' = L = \lfloor m \rfloor$ and $k' \leq \lceil 512 - f(L) \rceil$, or
2. $m' = L + 1 = \lceil m \rceil$ and $k' \leq \lfloor 512 - f(L) \rfloor$

The table below shows $m' = L+1$ and $k' = \lfloor 512 - f(L) \rfloor$ for all $L \in \{1, 2, ..., 105\}$. We can verify that the second possibility above assures a win by comparing k' with max for each L and seeing that it is small enough. Since $\lceil 512 - f(L) \rceil \leq \lfloor 512 - f(L-1) \rfloor$ (L and $L-1$ differ by one, therefore $512 - f(L)$ and $512 - f(L-1)$ differ by more than one), we can verify the first possibility by looking at the previous row. Therefore, we only need to check that all values of k' are no more than the corresponding value for max.

This concludes the proof.   □

L	512-f(L)	m'	k'	max	L	512-f(L)	m'	k'	max	L	512-f(L)	m'	k'	max
0	512	1	511	511	35	373.55	36	373	381	70	197.61	71	197	229
1	509.39	2	509	509	36	368.87	37	368	379	71	192.30	72	192	223
2	506.60	3	506	507	37	364.15	38	364	375	72	186.97	73	186	219
3	503.66	4	503	503	38	359.41	39	359	373	73	181.62	74	181	215
4	500.58	5	500	501	39	354.65	40	354	367	74	176.27	75	176	213
5	497.38	6	497	499	40	349.87	41	349	365	75	170.90	76	170	207
6	494.08	7	494	495	41	345.06	42	345	363	76	165.51	77	165	205
7	490.68	8	490	493	42	340.23	43	340	357	77	160.12	78	160	203
8	487.20	9	487	491	43	335.38	44	335	351	78	154.71	79	154	197
9	483.63	10	483	485	44	330.51	45	330	347	79	149.29	80	149	191
10	479.99	11	479	479	45	325.62	46	325	341	80	143.86	81	143	189
11	476.29	12	476	477	46	320.71	47	320	335	81	138.42	82	138	183
12	472.51	13	472	475	47	315.78	48	315	333	82	132.97	83	132	181
13	468.68	14	468	471	48	310.83	49	310	331	83	127.50	84	127	175
14	464.79	15	464	469	49	305.86	50	305	325	84	122.02	85	122	173
15	460.85	16	460	463	50	300.87	51	300	319	85	116.53	86	116	171
16	456.85	17	456	461	51	295.86	52	295	317	86	111.03	87	111	159
17	452.81	18	452	459	52	290.83	53	290	311	87	105.52	88	105	155
18	448.71	19	448	453	53	285.79	54	285	309	88	100.00	89	100	149
19	444.58	20	444	447	54	280.73	55	280	303	89	94.47	90	94	139
20	440.40	21	440	445	55	275.65	56	275	301	90	88.93	91	88	133
21	436.18	22	436	443	56	270.56	57	270	299	91	83.37	92	83	127
22	431.92	23	431	439	57	265.45	58	265	287	92	77.81	93	77	125
23	427.63	24	427	437	58	260.32	59	260	283	93	72.23	94	72	123
24	423.30	25	423	431	59	255.18	60	255	277	94	66.65	95	66	119
25	418.93	26	418	429	60	250.02	61	250	267	95	61.05	96	61	117
26	414.53	27	414	427	61	244.84	62	244	261	96	55.45	97	55	111
27	410.09	28	410	421	62	239.65	63	239	255	97	49.83	98	49	109
28	405.63	29	405	415	63	234.45	64	234	253	98	44.21	99	44	107
29	401.13	30	401	411	64	229.23	65	229	251	99	38.57	100	38	101
30	396.61	31	396	405	65	224.00	66	223	247	100	32.93	101	32	91
31	392.05	32	392	397	66	218.75	67	218	245	101	27.27	102	27	79
32	387.47	33	387	395	67	213.48	68	213	239	102	21.61	103	21	75
33	382.86	34	382	389	68	208.21	69	208	237	103	15.94	104	15	69
34	378.22	35	378	383	69	202.92	70	202	235	104	10.26	105	10	63
										105	4.56	106	4	61

Fig. 3. Possible values of m' and k' for positions of weight ≤ 512

6 Lower-bound for the Case of $n = 10^6$

Rivest et al. [9] show a lower-bound of $\log_2 n + \log_2 \log_2 n - 3$ for $Q_H(n)$. When $n = 10^6$, this gives a lower-bound of 22. In this section, we will show a slightly stronger result, namely $Q_H(10^6) \geq 23$.

Theorem 6. $Q_H(10^6) \geq 23$.

Proof. The lower-bound for $Q_H(n)$ was presented by Rivest et al. [9] in their Theorem 4. We will briefly present the main ideas of their proof as we suitably adapt it to the case of $k = 1$. (Their proof applies to an arbitrary k that is independent of n.) Consider an arbitrary optimal (in the worst-case) strategy S

of Paul. One can represent such a strategy as a binary search tree T_S in which each internal node has a query "Is $x > c_t$?", the left (right) child of the node is the one that results from a 'yes' ('no') answer. A leaf node is one in which Paul has found the unknown x and the label associated with the node is x. We use $label(n) = x$ to denote this. With each leaf node l, we also associate two more quantities: $yes(l)$ is the set of queries on the path from root to the leaf l that correspond to a 'yes' (left) branch. Also $lies(l)$ is the set of queries on the path from root to l that received an erroneous answer. It is clear that $lies(l)$ is either a singleton or empty. Let Q be the height of the tree. We will also assume that all the paths in the tree (from the root to a leaf) have the same path length Q. This can be accomplished by repeating a fixed query (such as "Is $x > 0$?") $Q - k$ times at a leaf node at depth k and thus replace this node by a tree of height $Q - k$. In this proof, clearly, $n = 10^6$.

Next we define a t-regular path from root to a leaf as a path that has at least t yes branches. We say that an $x \in \{1, 2, ..., n\}$ is t-regular if *all* the paths from the root to a leaf that contains the key x are t-regular.

The proof of the theorem is as follows: suppose the height of the tree T_S is 22 (or less). We will show the following claims.

Claim 1. There are at most $\binom{22}{0} + \binom{22}{1} + ... + \binom{22}{7} = 280599$ paths in T_S that are not 8-regular. Thus there are at least 719401 numbers in $\{1, 2, ..., n\}$ that are 8-regular.

The proof of claim 1 is based on simple counting. Each path that is not 8 regular results in at most one key becoming non-regular.

Claim 2. If a key $x \in \{1, 2, ..., n\}$ is t-regular, then there are t leaves l such that $label(l) = x$ and $|lies(l)| = 1$.

Claim 2 is shown as follows: Consider a leaf node l such that $label(l) = x$ and $|lies(x)| = 0$. (There is always at least one such leaf node that Paul will reach if Carole chooses x as the unknown and never exercises the lying option.) On the path from the root to l, there are at least t internal nodes with 'yes' branch. In each of those internal nodes, if we branch off with a 'no' answer, and then follow the path with no more lies (assuming the unknown as x), we will arrive at a new leaf node with label x and the number of lies $= 1$. Since this can be done t times, and each leaf node we arrive at by following this procedure will result in a new leaf node, the claim follows.

Now we arrive at a contradiction as follows: Since there are at least 719401 numbers that are 8-regular, by Claim 2, there are at least 719401*8 leaf nodes associated with these keys, such that $|lies(l)| = 1$ at each of these nodes. In addition, there are at least 10^6 more leaf nodes each associated with number of lies $= 0$. Thus T_S has at least 719401*8 + 1000000 = 6755208 leaf nodes. However, we started with the assumption that the height of T_S is 22, and hence it has at most $2^{22} = 4194304$ leaf nodes. This contradiction implies that the height of T_S must be at least 23. $\qquad \square$

Let $Q'_H(n)$ denote the number of queries needed in the worst-case for the half-lie problem when the number of lies is 1, and when arbitrary yes/no questions are allowed. It is clear that $Q_H(n) \geq Q'_H(n)$.

It should be noted that [2] presents a tight lower-bound for $Q'_H(n)$. In fact, [2] establishes an upper- and lower-bound for $Q'_H(n)$ that is within 1 of optimum, and hence, Theorem 6 can also be deduced from their paper. However, their general lower-bound argument is considerably more complicated and our proof is much simpler.

Unfortunately, there does not seem to be an easy way to narrow the gap of 1 between the upper- and lower-bounds for this case.

7   Connection to Two-Player Search Game

Consider a fundamental generalization of the standard binary search game. In the standard version, one player chooses a number between 1 and n (for some fixed n known to both players) and the other player tries to find x with the fewest number of questions (in the worst-case) by asking questions of the form "Is $x > s$?" Now there are three players Carole, Loreca and Paul. Carole and Loreca each chooses a number x_C and x_L in $\{1, 2, ..., n\}$. Paul can ask either of them about her number in the form "Is your number greater than c?" He can question them in any order and he chooses his questions based on all the past responses. His goal is to find the *bigger* of the two numbers x_L and x_C. What is his best strategy assuming Loreca and Carole will make his task as hard as possible? Note, however, that neither Carole nor Loreca lie. At the outset, it is not clear how to improve (in the worst-case) the obvious strategy of finding both x_L and x_C separately using binary search - a strategy that gives an upper-bound of $2 \lceil \log_2 n \rceil$.

A remarkable result shown by Rivest et al. [9] was that the number of queries needed by Paul in the worst-case to solve this three-player search game is exactly $Q_H(n)$. See Theorem 3 of their paper. This means the new upper-bound presented in this paper provides an improved upper-bound for this problem. Specifically for the case of 10^6, our improvement in the upper- and the lower-bounds gives an estimate within 1 of the optimal bound for this problem.

8   Conclusions

The main result shown in this work is a new upper-bound $Q_H(n) \leq \lceil \log_2((n + 4.5) \ln(n + 4.5) - 4.5 \ln(4.5)) \rceil$. The main ideas presented in this paper can be extended to the case where number of lies is more than 1, although the analysis becomes even more complicated. To the best of our knowledge, the upper-bound presented here is the best for the half-lie problem for the case of cut questions model with one lie. However, we believe that the upper-bound presented here is not optimal. The current lower-bound for this problem (due to [9]) does not seem to be optimal either - as we were able to improve it at least in the special case of $n = 10^6$. Improving either (or both) of these is the main open problem left.

References

1. Aslam, J., Dhagat, A.: Searching in the presence of linear bounded errors. In: Proc. of 23rd ACM Symp. on Theory of Computing, pp. 487–493 (1991)
2. Cicalese, F., Mundici, D.: Optimal Coding with One Asymmetric Error: Below the Sphere Packing Bound. In: Du, D.-Z., Eades, P., Sharma, A.K., Lin, X., Estivill-Castro, V. (eds.) COCOON 2000. LNCS, vol. 1858, pp. 159–169. Springer, Heidelberg (2000)
3. Ellis, R., Yan, C.: Ulam's pathological liar game with one half-lie. International Journal of Mathematics and Mathematical Sciences 2004(29), 1523–1532 (2004)
4. Innes, D.: Searching with a lie using only comparison questions. In: International Conference on Computing and Information, pp. 87–90 (1992)
5. Gao, F., Guibas, L.J., Kirkpatrick, D.G., Laaser, W.T., Saxe, J.: Finding extrema with unary predicates. Algorithmica 9, 591–600 (1993)
6. Pelc, A.: Solution to Ulam's problem on searching with a lie. Journal of Combinatorial Theory Series A 40(9), 1081–1089 (1991)
7. Pelc, A.: Searching games with errors 50 years of coping with liars. Theoretical Computer Science 270, 71–109 (2002)
8. Ravikumar, B., Lakshmanan, K.B.: Coping with known patterns of lies in a search game. Theoretical Computer Science 33, 85–94 (1984)
9. Rivest, R., Meyer, A.R., Kleitman, D.J., Spencer, J., Winklmann, K.: Coping with errors in binary search procedures. Journal of Computer and System Sciences 20(3), 396–404 (1980)
10. Spencer, J.: Guess a number - with lying. Mathematics Magazine 57(2) (1984)
11. Spencer, J.: Ulam's searching game with a fixed number of lies. Theoretical Computer Science 95, 307–321 (1992)
12. Spencer, J., Winkler, P.: Three thresholds for a liar. Combinatorics, Probability and Computing 1(1), 81–93 (1992)
13. Spencer, J., Yan, C.H.: The half-lie problem. Journal of Combinatorial Theory, Series A 103, 69–89 (2003)
14. Ulam, S.: Adventures of a Mathematician. Scribner, New York (1976)

Reversibility of Elementary Cellular Automata under Fully Asynchronous Update*

Biswanath Sethi[1], Nazim Fatès[2], and Sukanta Das[3]

[1] Department of Computer Science Engineering and Applications
Indira Gandhi Institute of Technology, Sarang
Dhenkanal, Odisha, India-759146
sethi.biswanath@gmail.com
[2] Inria Nancy Grand-Est
LORIA UMR 7503
Université de Lorraine, CNRS
F-54 600, Villers-lès-Nancy, France
nazim.fates@loria.fr
[3] Department of Information Technology
Bengal Engineering and Science University, Shibpur
Howrah, West Bengal, India-711103
sukanta@it.becs.ac.in

Abstract. We investigate the dynamics of Elementary Cellular Automata (ECA) under fully asynchronous update with periodic boundary conditions. We tackle the reversibility issue, that is, we want to determine whether, starting from any initial condition, it is possible to go back to this initial condition with random updates. We present analytical tools that allow us to partition the ECA space into three classes: strongly irreversible, irreversible and recurrent.

Keywords: asynchronous cellular automata, reversibility, recurrence, Markov chain modelling, classification.

1 Introduction

Cellular automata (CA) are spatially-extended dynamical systems which evolve in discrete time and space. They have been extensively studied as models of physical systems and as models of massively parallel computing devices.

Cellular automata are classically defined with a synchronous update, that is, all the cells simultaneously apply the local transition rule to produce the new state of the automaton. This definition has however been questioned in various works and different models of asynchronous cellular automata have been proposed. There are numerous reasons for studying asynchronism, such as: designing robust distributed algorithms (e.g. for self-stabilisation), studying the

---

* This work is supported by DST Fast Track Project Fund (No.SR/FTP/ETA-0071/2011).

T V Gopal et al. (Eds.): TAMC 2014, LNCS 8402, pp. 39–49, 2014.

robustness of discrete models of natural phenomena, obtaining a better under-standing of the dynamics of cellular automata, etc. Interested readers may refer to a recent survey paper for an overview of this field [1].

Our aim is to study how the notion of reversibility in the context of simple asynchronous CA with a *stochastic* updating. We focus on Elementary Cellular Automata (ECA), that is, binary, one-dimensional CA, where the next state of a cell after an update is determined by the current states of the left and right neighbors and the state of the cell itself.

Reversibility of synchronous deterministic cellular automata has been studied for decades [2–6]. However, the study of reversibility of asynchronous cellular au-tomata has been only recently explored. Two different aspects have been studied: on the one hand, the question was asked as to how to update an asynchronous CA so that the system returns to its initial condition. It was shown that it is possible to find an answer for a given subset of one-dimensional asynchronous CA [7–9]. The construction of the arguments was possible under the hypothesis that one may choose the sequence of updates to apply to the cellular automaton. This introduction of update patterns relies on the hypothesis that an external operator is allowed to choose the cells to update in order to return to a given initial condition.

On the other hand, given a CA rule and a type of updating, it was asked to which extent it is possible to construct another rule whose transition graph would be an "inverse" of the transition graph of the original rule. Formally, this means that, given a rule f, we want to know if there is a rule f' such that if for f a state y is reachable from x, then, for f', x is reachable from y [10].

We now tackle a different case: we consider that the ECA are updated in a (stochastic) fully asynchronous mode, that is, at each discrete time step, a single cell is chosen randomly and uniformly for update. In this context, as we will see below, studying reversibility amounts to answering the following question: can we decide whether an asynchronous cellular automaton is *recurrent*, that is, if the system will almost surely return to the initial condition?

Using the definitions from the theory of Markov chains, we propose a full characterisation of the ECA rules into three classes: the strongly irreversible, irreversible, and recurrent rules. Intuitively, these class respectively correspond to the following behaviours: no possibility to return to the initial condition, a possibility to return to the initial condition a finite number of times and, an infinite number of returns to the initial condition.

2   Definitions

The cellular automata we consider use periodic boundary conditions: cells are arranged as a *ring* and we denote by $\mathcal{L} = \mathbb{Z}/n\mathbb{Z}$ the set of cells. The global *state* of the system at a given time will be represented by an element of $\{0, 1\}^{\mathcal{L}}$; for example, for a ring of $n = 6$ cells, we will simply write $x = 011001$ a particular state and denote by x_i the state of a particular cell $i \in \mathcal{L}$. We denote by **0** and **1** the two homogeneous states with cell state 0 and 1, respectively. Similarly, **01** denote a state of even size in which cell states 0 and 1 alternate, **001** a state whose size is a multiple of three, where two 0s are followed by a 1, etc.

An ECA is defined by a local transition function $f : \{0,1\}^3 \mapsto \{0,1\}$; it is common to define such a function with a look-up table (see Table 1). There are $2^8 = 256$ ECA rules, each one referred to with the number that corresponds to the decimal equivalent of the binary number formed by the sequence of its transitions results [11]. Three such rules (87, 99 and 110) are shown in Table 1.

Definition 1. *The association of the neighbourhood x, y, z to the value $f(x, y, z)$, which represents the result of the updating function, is called* Rule Min Term *(RMT). Each RMT is associated to a number $R(x, y, z) = 4x + 2y + z$. An RMT $R(x, y, z)$ is active $f(x, y, z) \neq y$ and otherwise passive.*

For example, for rule 110, RMT 1 is active and RMT 6 is passive (see Table 1).

Table 1. Look-up table for rule 87, 99 and 110

x,y,z	111	110	101	100	011	010	001	000	*Rule*
RMT	(7)	(6)	(5)	(4)	(3)	(2)	(1)	(0)	
f(x,y,z)	0	1	0	1	0	1	1	1	87
f(x,y,z)	0	1	1	0	0	0	1	1	99
f(x,y,z)	0	1	1	0	1	1	1	0	110

We now consider *fully asynchronous updating*, that is, the case where only a single cell is updated randomly and uniformly at each time step. While a synchronous CA is a deterministic system, in an asynchronous CA (ACA), the next state not only depends on the local rule but also on the cells which are updated.

We denote by u_t the cell that updated at time t ; the sequence $U = (u_t)_{t \in \mathbb{N}}$ is called an *update pattern*. For an initial condition x and an update pattern U, the evolution of the system is given by the sequence of states (x^t) obtained by successive applications of the updates of U. Formally, we have: $x^{t+1} = F(x^t, u_t)$ and $x^0 = x$, with:

$$x_i^{t+1} = \begin{cases} f(x_{i-1}^t, x_i^t, x_{i+1}^t) & \text{if } i = u_t \\ x_i^t & \text{otherwise.} \end{cases}$$

This evolution can be represented in the form of *a state transition diagram*. For example, Fig. 1 shows a partial state transition diagram of rule 110 with state $x = 1010$ and update pattern $U = (2, 1, 4, 3, 1, 3, \ldots)$. The index of the cell that is updated is noted over the arrows.

Definition 2. *A state x is* reachable *if it has at least one predecessor, that is, if there exists a CA state $y \in \mathcal{E}_n$ and an update position $u \in \mathcal{L}$ such that $F(y, u) = x$; otherwise the state is* non-reachable *(or a garden-of-Eden state).*

For instance, for ECA 110, the state 1110 is reachable as it has 1010 as a predecessor (see Fig. 1). By contrast, for ECA 87, 1111 is non-reachable (see Fig. 2). Indeed, if it had a predecessor, it would necessarily be equal to 1101, up to shifts, as only one cell can change at a time. But as the transition 101 (RMT 5) is passive, the last 0 can not disappear. Remark that a system may contain both types of states, reachable and non-reachable.

A state x is converted to an RMT sequence $\tilde{x}$ with: $\tilde{x}_i = R(x_{i-1}, x_i, x_{i+1})$ for all $i \in \mathcal{L}$. For example, the state $x = 001010$ is associated to the RMT sequence $\tilde{x} = 012524$. RMT sequences will be used to establish the proofs of recurrence or irreversibility of the ECA rules.

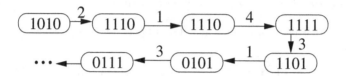

Fig. 1. Partial state transition diagram of rule 110 with $n = 4$. The cells updated during evolution are noted over arrows (convention kept).

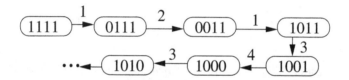

Fig. 2. Partial state transition diagram of ECA 87 with $n = 4$

3   (Ir) Reversibility of ACA

The issue of reversibility of CA has given rise to the use of various terms to name the same properties; for instance, the term "invertible" has been used as a synonymous of "reversible" [12]. This variety of terms comes from the proximity between the physical notion of reversibility and its equivalent in discrete dynamical systems. We emphasise that, in the CA context, reversibility informally denotes the possibility to "invert" the evolution of a cellular automaton, by using potentially *another* cellular automaton and *not* the fact that the evolution of the system is similar when it is run "backwards". The term *time-symmetric* has been recently used to qualify the rules whose evolution is similar if the arrow of time is "inverted" [13]. As there are multiple views on reversible CA, we note that in the deterministic synchronous case, the following statements are equivalent:

1. Each CA state has exactly one predecessor.
2. There exists no CA state that is non-reachable.
3. Each CA state lies on a cycle.
4. Each CA state is returned back in the course of dynamic evolution.

However, these definitions can not be transposed in a straightforward way to asynchronous cellular automata and in that case, the *classical* definition of reversibility needs to be revisited. One solution was that proposed consisted in associating the notion of reversibility with a given update pattern, that is, to a sequence of updates decided in advance [9]. However, in the case where cells are updated *randomly*, new difficulties arise. For instance, in the ACA case, Statement 4 also implies Statement 2 and Statement 3, but does not imply Statement 1. This leads us to search for another definition of reversibility for an ACA. Here, we choose to start from Statement 4 for defining the reversibility of ACA: we require that in an asynchronous reversible CA, each state has to be returned back almost surely during the evolution of the system.

As we use the fully asynchronous updating, the evolution of our ACA is described by a Markov chain over the space of CA states Q^L. We thus define the reversibility properties using the classical tools from Markov chain theory, which leads to identify reversibility and recurrence.

Definition 3. *For a couple of states $x, y \in Q^L$, we say that y is reachable from x if there is a sequence of updates that leads from x to y, that is:*

$$\exists k \in N^*, U = (u_0, \ldots, u_{k-1}), x^0 = x, x^k = y$$

and

$$x^{i+1} = F(x^i, u_i) \text{ for all } i \in \{0, \ldots, k-1\}.$$

We now introduce the main tool of our study :

Definition 4. *A state $x \in Q^L$ is recurrent if for every state y that is reachable for x, x is also reachable from y. A state that is not recurrent is transient.*

Intuitively, a transient state is such that a particular sequence of updates may bring into a particular state from which it will never be possible to return back to the initial state. More formally, if y is reachable from x and x is reachable from y, we say that x and y *communicate*. By convention, all states communicate with themselves. Clearly, the relationship "communicate" is an equivalence relation; this relation partitions the set of states into communication classes. In words, two major behaviours exist: for the transient states, the system remains for an almost surely finite time in the communication class, then "escapes" this class and never returns back to it. In contrast, when the system is in a recurrent state, it remains in the communication class for ever.

We can now define the (ir)reversible cellular automata:

Definition 5. *An ACA is recurrent if each CA state is recurrent, otherwise it is irreversible.*

The definitions above allow us to know if some irreversibility is present in the system but they do not say anything about the "degree of irreversibility" of the system. Indeed, it may well be that the system does possess a transient state but that the sequence of updates that leads to observe the irreversibility is never observed in practice when the updates are random. This is a difficult problem to tackle in all generality. As first step, we propose here to deal with the states where it is not possible to return back *whatever* the sequence of updates.

Definition 6. *A state x is evanescent if it is not reachable from itself. An ACA that possesses an evanescent state is strongly irreversible.*

It is interesting to remark that the set of evanescent and non-reachable states are equal. Indeed, by definition a non-reachable state is evanescent. To see why the converse is true, let us assume by contradiction that x is an evanescent state that is reachable from y. We say that the cell $i \in \mathcal{L}$ of a state $x \in Q^{\mathcal{L}}$ is *active* if the transition which applies in i is active, that is, if $f(x_{i-1}, x_i, x_{i+1}) \neq x_i$. Note that x is a *fully unstable state* (all its cells are active). It is then easy to see that y is also reachable from x (as the two states differ in only one cell) and thus, that x is reachable from itself, which contradicts the evanescence hypothesis.

As a consequence, if a rule is strongly irreversible it possesses at least one non-reachable state. However, the converse is not true: for instance, for rule 51 (the NOT rule), all states are fully unstable but the rule is reversible.

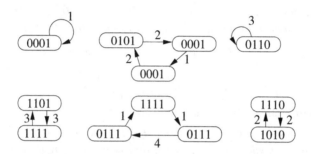

Fig. 3. Transition diagram for ECA 99 with $n = 4$

Fig. 3 depicts an example for a recurrent CA. In the state transition diagram of ECA 99, each state can be returned to with some given update of cells noted over the arrows. It can be shown (e.g., with an exhaustive search) that all the states of rule 99 ACA are recurrent.

4   Identifying the Strongly Irreversible Rules

We first present the theorem that allows us to identify strongly irreversible ECA:

Theorem 1. *An ECA is strongly irreversible if and only if one of the following conditions is verified:*

1. *RMT 0 (resp. RMT 7) is active and RMT 2 (resp. RMT 5) is passive.*
2. *RMTs 2 and 5 are active and RMTs 0 and 7 are passive.*
3. *RMTs 1, 2 and 4 (resp. RMTs 3, 6 and 5) are active, and RMTs 0, 3 and 6 (resp. 1, 4 and 7) are passive.*

Proof. First, let us prove the "if" part, that is, if one of the conditions is verified then the ECA is strongly irreversible. We proceed by examining the conditions one by one and by exhibiting for each case a non-reachable (and thus an evanescent) state.

Case 1: Let us show that **0** (with RMT 0 only) is non-reachable. Assume that y is a predecessor of **0**. First $y \neq \mathbf{0}$ as **0** is fully unstable (RMT 0 is active). The CA state y thus contains a single 1 (as the number of ones can only vary by 1 in the fully asynchronous update) and the transition from y to **0** was applied on the single 1 and with RMT 2. However, this is impossible as RMT 2 is passive. The case of RMT 5 and 7 is identical up to the 0/1 exchange.

Case 2: Let us show that **01** (with RMT 2 and 5 only) is non-reachable. First, if RMTs 2 and 5 are active, then this CA state is fully unstable. Now, assume that there is a CA state $x \neq \mathbf{01}$ and an updated cell i such that $F(x,i) = \mathbf{01}$, then, as x and **01** differ on only cell, it is easy to see that either RMT 0 or 7 produced a change of state on i, which is impossible if RMT 0 and RMT 7 are both passive.

Case 3: Let us show that **001** (with RMT 1, 2 and 4 only) is non-reachable. First, we note that this CA state is fully unstable as its RMT sequence is **124**. Again, if **001** had a predecessor $x \neq \mathbf{001}$, then the last update on x is either a 0 changed into a 1 (application of RMT 0) or a 1 changed into a 0 (application of RMT 3 or 6), which in both cases can not happen if RMTs 0, 3 and 6 are all passive. The proof for the RMTs shown into parentheses is identical up to the 0/1 exchange.

Let us now show that the three conditions above of Th. 1 are also necessary for an ECA to be strongly irreversible.

Proof. Let us consider an ECA that has a non-reachable state x. We will show that x has only four "forms" (up to the 0/1 exchange) that each brings us to the three conditions of the theorem. Let x be a non-reachable state. First, let us note that x is fully unstable and that no transition can lead to x. As a consequence, we can state an *exclusion rule* : $\tilde{x}$, the RMT sequence of x, can not contain two transitions in one of the following couples of RMTs $\{0, 2\}$, $\{1, 3\}$, $\{4, 6\}$ and $\{5, 7\}$. To see why, assume for example that RMTs 0 and 2 are both present in $\tilde{x}$, that is, $\exists i, j \in \mathcal{L}, \tilde{x}_i = 0$ and $\tilde{x}_j = 2$. As x is fully unstable, RMT 0 and 2 are both active, then, it can be remarked that two successive updates on i (or j) make the system return to x, that is, $F(F(x,i),i) = x$, which is in contradiction with the fact that x is non-reachable.

Now, let us discuss the various possibilities for x.

Case a: Let us assume that $\tilde{x}$ contains a 0. If x contains at least one 1, that is, if $x \neq \mathbf{0}$, then, we can note that x contains either the sequence 00010 or the sequence 00011, that is, $\tilde{x}$ either contains the sequence 012 or the sequence 013. However, the two cases lead to a contradiction due to the exclusion rule. The only possibility is then $x = \mathbf{0}$, which implies that RMT 0 is active and RMT 2 is passive (due to the exclusion rule). We are thus in case 1 of the theorem. The case with RMTs 5 and 7 is symmetric by 0/1 exchange.

Now, let us assume that $\tilde{x}$ does not contain RMT 0 nor RMT 7. This implies that x contains at least one 01 pattern. We need to distinguish several sub-cases.

Case b: If x does not contain the 00 or 11 pattern, that is $x = \mathbf{01}$ and $\tilde{x} = \mathbf{25}$, we can deduce that RMT 0 and 5 are active and that RMT 1 and 7 are passive (exclusion rule); we are then verifying Case 2 of the theorem.

Case c: Let us now assume that $x \neq \mathbf{01}$, and without loss of generality, that x contains the 00 pattern. Then, x necessarily contains the pattern 1001 (otherwise, it would contain the pattern 000) which means that $\tilde{x}$ thus contains the RMTs 4 and 1, and, because of the exclusion rule, does not contain RMT 3 nor 6. Two possibilities are now offered :

Case d: x contains the pattern 10011 : this is excluded because of the exclusion rule as this would imply that $\tilde{x}$ contains RMT 1 and 3.

Case e: x contains the pattern 10010 but does not contain pattern 000 (RMT 0), nor 011 (RMT 3), nor 110 (RMT 6). This means that $\tilde{x}$ contains 1, 2 and 4 (if $x = \mathbf{001}$) and possibly RMT 5 (if x contains 100101). In both cases, this means that RMT 1, 2 and 4 are active and that RMT 0, 3, 6 are passive and the last case of the theorem is proved.

The parts of the theorem presented into parentheses are symmetric to the cases discussed above by the 0/1 exchange.

As a consequence, it can be seen that the RMT sequence of a non-reachable state necessarily verifies one of the four following combinations of RMTs: (a) only RMT 0 (or only RMT 7), (b) only RMTs 2 and 5 (c) only RMTs 1, 2 and 4 (or only RMTs 3, 5 and 6), (d) only RMTs 1, 2, 4 and 5 (or only RMTs 2, 3, 5 and 6).

There are 132 such strongly irreversible rules; they are listed in Table 2.

Table 2. List of the 132 strongly irreversible rules that verify Theorem 1. Bold fonts show the *minimal representative rules* (rules with the smallest code among the group of rules that are obtained by left-right and 0-1 exchange).

0	**1**	**2**	**3**	**4**	**5**	**6**	**7**	**8**	**9**
10	**11**	**12**	**13**	**14**	**15**	16	17	**18**	**19**
20	21	**22**	**23**	24	**25**	**26**	**27**	**28**	29
30	31	**37**	39	**45**	47	53	55	61	63
64	65	66	67	68	69	70	71	**72**	**73**
74	75	**76**	**77**	**78**	79	80	81	82	83
84	85	86	87	88	89	**90**	91	92	93
94	95	101	103	109	111	117	119	**122**	125
127	133	135	141	143	149	151	157	159	**160**
161	**162**	**164**	165	167	**168**	**170**	173	175	176
178	181	183	**184**	186	189	191	197	199	205
207	213	215	218	221	223	224	226	229	231
232	234	237	239	240	242	245	247	248	250
253	255								

Example 1. *Let us consider ECA 87, a rule which satisfies the conditions of Th. 1 as RMT 7 is active and RMT 5 is passive (see Table 1). Consider the evolution of state* **1**. *If a cell is updated with RMT 7, then one 0 appears and it is easy to see that afterwards the last 0 cannot disappear as RMT 5 is passive. Hence,* **1** *is non-reachable and is evanescent which makes the rule strongly irreversible.*

5   Identification of the Recurrent Rules

This section identifies the set of rules which are irreversible and, by complementation, those which are recurrent. We proceed by identifying the rules for which particular states are transient.

Theorem 2. *A rule R is irreversible if one of the following conditions is verified:*

1. *RMT 0 (resp. RMT 7) is active and RMT 2 (resp. RMT 5) is passive or RMT 2 (resp. RMT 5) is active and RMT 0 (resp. RMT 7) is passive.*
2. *RMTs 0, 1, 2 and 4 (resp. RMTs 3, 5, 6 and 7) are passive and RMT 3 or 6 (resp. RMT 1 or 4) are active.*
3. *RMTs 0, 2, 3 and 6 (resp. RMTs 1, 4, 5 and 7) are passive and RMT 1 or 4 (resp. RMT 3 or 6) are active.*

Proof. Case 1: If RMT 0 is active and RMT 2 is passive, as shown the proof of Case 1 of Th. 1, **0** is evanescent and thus transient. If RMT 2 is active and RMT 0 is passive, let us consider the state $x = 00100$. If the third cell is updated, the system reaches **0**, which is a fixed point, and which implies that x transient.

Case 2: Now, consider $x = 001100$; its RMT sequence is $\tilde{x} = 013640$. If RMT 3 is active, $y = 000100$ can be reached by updating the third cell. This a fixed point as its RMT sequence contains only 0, 1, 2 and 4, which all correspond to passive RMTs. Similarly, if RMT 6 is active the fixed point $y = 001000$ can be reached; which shows that x is transient.

Case 3: We start with $x = 00100$; its RMT sequence is $\tilde{x} = 01240$. As RMT 1 and 4 are active, $y = 00110$ or $y' = 01100$ can be reached. However, from any CA state that contains two or more 1s, it is not possible to return to x as RMTs 2, 3 and 6 are passive. (This implies that a 1 that has at least one 0 next to it can not disappear). Hence, x is transient.

The proofs for the RMTs mentioned in the parentheses is identical by exchanging the cell states 0 and 1.

By rewriting the conditions of the theorem, it can be verified that the rules for which it does *not* apply verify the following conditions: RMTs 0 and 2 (resp. 5 and 7) are either both active or both passive, and : a) there is at least one couple of active RMTs in the following sets: $\{2,5\}$, $\{1,6\}$, $\{3,4\}$, $\{1,3\}$, $\{4,6\}$ or b) RMTs 1, 3, 4 and 6 are all passive. There are 46 rules which verify these conditions, which are listed in Table 3. Our conjecture is that all these rules are recurrent, that is, all their states are recurrent.

Table 3. List of the 46 rules that are conjectured to be recurrent

33	35	38	41	43	46	49	51	52	54
57	59	60	62	97	99	102	105	107	108
113	115	116	118	121	123	131	134	139	142
145	147	148	150	153	155	156	158	195	198
201	204	209	211	212	214				

Table 4. List of the 78 remaining rules: conjectured to be the irreversible ACA that are not strongly irreversible

32	34	36	40	42	44	48	50	56	58
96	98	100	104	106	110	112	114	120	124
126	128	129	130	132	136	137	138	140	144
146	152	154	163	166	169	171	172	174	177
179	180	182	185	187	188	190	192	193	194
196	200	202	203	206	208	210	216	217	219
220	222	225	227	228	230	233	235	236	238
241	243	244	246	249	251	252	254		

The 210 rules which satisfies at least one condition of Th. 2 are irreversible. We have already identified 132 rules (Tab. 2) as strongly irreversible. The remaining 78 irreversible ACA are listed in Table 4; we conjecture that they are not strongly irreversible, that is, they have at least one transient state but no evanescent state.

6 Conclusion

We reported a classification of the ECA space according to reversibility properties under fully asynchronous update with periodic boundary conditions. The main step now consists in completing this classification by showing that the list of recurrent rules presented are closed. This could be done analysing the communication classes of the state space of these rules. While the classifications based on the convergence time to a fixed point remain mainly open [14], achieving this result would represent an important step in the understanding of the dynamics of asynchronous CA.

As usual in the field of CA, one may ask how to extend the results to other types of asynchronism and to the CA spaces with a higher radius or higher dimension. As suggested by I. Marcovici, the classification can also be refined by considering "escaping states", that is, states where there is a possibility to stay but for which once this state is leaved, it can not be returned to.

Another question is to know if the reversibility issues presented here are similar to other views, for instance the one recently studied by Wacker and Worsch [10].

References

1. Fatès, N.: A guided tour of asynchronous cellular automata. In: Kari, J., Kutrib, M., Malcher, A. (eds.) AUTOMATA 2013. LNCS, vol. 8155, pp. 15–30. Springer, Heidelberg (2013); An extended version is available from: http://hal.inria.fr/hal-00908373
2. Amoroso, S., Patt, Y.N.: Decision procedures for surjectivity and injectivity of parallel maps for tesselation structures. Journal of Computer and System Sciences 6, 448–464 (1972)
3. Richardson, D.: Tessellations with local transformations. Journal of Computer Systems and Sciences 6, 373–388 (1972)
4. Toffoli, T.: Computation and construction universality of reversible cellular automata. Journal of Computer Systems and Sciences 15, 213–231 (1977)
5. Das, S., Sikdar, B.K.: Classification of CA rules targeting synthesis of reversible cellular automata. In: El Yacoubi, S., Chopard, B., Bandini, S. (eds.) ACRI 2006. LNCS, vol. 4173, pp. 68–77. Springer, Heidelberg (2006)
6. Das, S., Sikdar, B.K.: Characterization of 1-d periodic boundary reversible CA. Electronic Notes in Theoretical Computer Science 252, 205–227 (2009)
7. Das, S., Sarkar, A., Sikdar, B.K.: Synthesis of reversible asynchronous cellular automata for pattern generation with specific hamming distance. In: Sirakoulis, G.C., Bandini, S. (eds.) ACRI 2012. LNCS, vol. 7495, pp. 643–652. Springer, Heidelberg (2012)
8. Sarkar, A., Das, S.: On the Reversibility of 1-dimensional Asynchronous Cellular Automata. In: Local Proceedings of Automata 2011, pp. 29–40 (2011), http://hal.inria.fr/hal-00654706
9. Sarkar, A., Mukherjee, A., Das, S.: Reversibility in asynchronous cellular automata. Complex Systems 21(1), 71–84 (2012)
10. Wacker, S., Worsch, T.: On completeness and decidability of phase space invertible asynchronous cellular automata. Fundamenta Informaticae 126(2-3), 157–181 (2013)
11. Wolfram, S.: Theory and applications of cellular automata. World Scientific, Singapore (1986)
12. Toffoli, T., Margolus, N.: Invertible cellular automata: A review. Physica D 45(13), 229–253 (1990)
13. Gajardo, A., Kari, J., Moreira, A.: On time-symmetry in cellular automata. Journal of Computer and Systems Sciences 78(4), 1115–1126 (2012)
14. Fatès, N.: A note on the classification of the most simple asynchronous cellular automata. In: Kari, J., Kutrib, M., Malcher, A. (eds.) AUTOMATA 2013. LNCS, vol. 8155, pp. 31–45. Springer, Heidelberg (2013)

Finite State Incompressible Infinite Sequences*

Cristian S. Calude[1], Ludwig Staiger[2], and Frank Stephan[3]

[1] Department of Computer Science, University of Auckland,
Private Bag 92019, Auckland, New Zealand
cristian@cs.auckland.ac.nz
[2] Martin-Luther-Universität Halle-Wittenberg, Institut für Informatik,
Von-Seckendorff-Platz 1, D-06099 Halle, Germany
staiger@informatik.uni-halle.de
[3] Department of Mathematics and Department of Computer Science,
National University of Singapore, Singapore 119076, Republic of Singapore
fstephan@comp.nus.edu.sg

Abstract. In this paper we define and study finite state complexity of finite strings and infinite sequences and study connections of these complexity notions to randomness and normality. We show that the finite state complexity does not only depend on the codes for finite transducers, but also on how the codes are mapped to transducers. As a consequence we relate the finite state complexity to the plain (Kolmogorov) complexity, to the process complexity and to prefix-free complexity. Working with prefix-free sets of codes we characterise Martin-Löf random sequences in terms of finite state complexity: the weak power of finite transducers is compensated by the high complexity of enumeration of finite transducers. We also prove that every finite state incompressible sequence is normal, but the converse implication is not true. These results also show that our definition of finite state incompressibility is stronger than all other known forms of finite automata based incompressibility, in particular the notion related to finite automaton based betting systems introduced by Schnorr and Stimm [28]. The paper concludes with a discussion of open questions.

1 Introduction

Algorithmic Information Theory (AIT) [7,18,25] uses various measures of descriptional complexity to define and study various classes of "algorithmically random" finite strings or infinite sequences. The theory, based on the existence of a universal Turing machine (of various types), is very elegant and has produced many important results.

---

* This work was done in part during C. S. Calude's visits to the Martin-Luther-Universität Halle-Wittenberg in October 2012 and the National University of Singapore in November 2013, and L. Staiger's visits to the CDMTCS, University of Auckland and the National University of Singapore in March 2013. The work was supported in part by NUS grant R146-000-181-112 (PI F. Stephan).

T V Gopal et al. (Eds.): TAMC 2014, LNCS 8402, pp. 50–66, 2014.

The incomputability of all descriptional complexities was an obstacle towards more "down-to-earth" applications of AIT (e.g. for practical compression). One possibility to avoid incomputability is to restrict the resources available to the universal Turing machine and the result is resource-bounded descriptional complexity [6]. Another approach is to restrict the computational power of the machines used, for example, using context-free grammars or straight-line programs instead of Turing machines [13,20,21,27].

The first connections between finite state machine computations and randomness have been obtained for infinite sequences. Agafonov [1] proved that every subsequence selected from a (Borel) normal sequence by a regular language is also normal. Characterisations of normal infinite sequences have been obtained in terms of finite state gamblers, information lossless finite state compressors and finite state dimension: (a) a sequence is normal if and only if there is no finite state gambler that succeeds on it [28] (see also [5,15]) and (b) a sequence is normal if and only if it is incompressible by any information lossless finite state compressor [33]. Doty and Moser [16,17] used computations with finite transducers for the definition of finite state dimension of infinite sequences. The NFA-complexity of a string [13] can be defined in terms of finite transducers that are called in [13] "NFAs with advice"; the main problem with this approach is that NFAs used for compression can always be assumed to have only one state.

The definition of *finite state complexity of a finite string* x in terms of a computable enumeration of finite transducers and the input strings used by transducers which output x proposed in [9,10] is utilised to define *finite state incompressible sequences*. We show basic connections of this new notion compared to standard complexity measures in Theorem 5: It lies properly between the plain complexity as a lower bound and the prefix-free complexity as an upper bound in the case that the enumeration of transducers considered is a universal one. Furthermore, while finite state incompressibility depends on the enumeration of finite transducers, many results presented here are *independent* of the chosen enumeration. For example, we show that for every enumeration S every C_S–incompressible sequence is normal, Theorem 13. Furthermore, we can show that a sequence is Martin-Löf random iff it satisfies a strong incompressibility condition (parallel to the one for prefix-free Kolmogorov complexity) for every measure C_S based on some perfect enumeration S. One can furthermore transfer this characterisation to the measure C_S for universal enumerations S.

Our notation follows standard textbooks [4,7]:

- By $\{0,1\}^*$ we denote the set of all binary strings (words) with ε denoting the empty string; $\{0,1\}^\omega$ is the set of all (infinite) binary sequences.
- The length of $x \in X^*$ is denoted by $|x|$.
- Sequences are denoted by $\mathbf{x}, \mathbf{y}$; the prefix of length n of the sequence $\mathbf{x}$ is denoted by $\mathbf{x} \upharpoonright n$; the nth element of $\mathbf{x}$ is denoted by $\mathbf{x}(n)$.
- By $w \sqsubseteq u$ and $w \sqsubseteq \mathbf{y}$ we denote that w is a prefix of u and $\mathbf{y}$, respectively.
- If A, B are sets of strings then the concatenation is defined as $A \cdot B = \{xy : x \in A, y \in B\}$.
- A prefix-free set $A \subset X^*$ is a set with the property that for all strings $p, q \in X^*$, if $p, pq \in A$ then $p = pq$.

- By K, Km_D, and H we denote, respectively, the plain (Kolmogorov) complexity, the process complexity and the prefix-free complexity for appropriately fixed universal Turing machines.

2 Admissible Transducers and Their Enumerations

We consider transducers which try to generate prefixes of infinite binary sequences from shorter binary strings and consider hence the following transducers: An *admissible transducer* is a deterministic transducer given by a finite set of states Q with starting state q_0 and transition functions δ, μ with domain $Q \times \{0, 1\}$, and say that the transducer on state q and current input bit a transitions to $q' = \delta(q, a)$ and appends $w = \mu(q, a)$ to the output produced so far.

One can generalise inductively the functions μ and δ by stating that $\mu(q, \varepsilon) = \varepsilon$ and $\mu(q, av) = \mu(q, a) \cdot \mu(\delta(q, a), v)$ for states q and input strings av with a being one bit; similarly, $\delta(q, \varepsilon) = q$ and $\delta(q, av) = \delta(\delta(q, a), v)$. The output $T(v)$ of a transducer T on input-string v is then $\mu(q_0, v)$.

A partially computable function S with a prefix-free domain mapping binary strings to admissible transducers is called an enumeration; for a σ in the domain of S, the admissible transducer assigned by S to σ is denoted as T_σ^S.

Definition 1 (Calude, Salomaa and Roblot [9,10]). A *perfect enumeration* S of all admissible transducers is a partially computable function with a prefix-free and computable domain mapping each binary string σ in the domain to an admissible transducer T_σ^S in a one-one and onto way.

Note that partially computable one-one functions with a computable range (as considered here) have also a computable inverse. It is known that there are perfect enumerations with a regular domain and that every perfect enumeration S can be improved to a better perfect enumeration S' such that for each c there is transducer represented by σ in S and σ' in S' and these representations satisfy $|\sigma'| < |\sigma| - c$, [9,10].

Definition 2. A *universal enumeration* S of transducers is a partially computable function with prefix-free domain whose range contains all admissible transducers such that for each further enumeration S' of admissible transducers there exists a constant c such that for all σ' in the domain of S', the transducer $T_{\sigma'}^{S'}$ equals to some transducer T_σ^S where σ is in the domain of S and $|\sigma| \leq |\sigma'| + c$.

The construction of a universal enumeration S can be carried over from Kolmogorov complexity: If U is a universal machine for prefix-free Kolmogorov complexity and S' is a perfect enumeration of the admissible transducers, then the domain of S is the set of all σ such that $U(\sigma)$ is defined and in the domain of S' and T_σ^S is $T_{U(\sigma)}^{S'}$. The fact that U is a universal machine for prefix-free Kolmogorov complexity implies that also S is a universal enumeration of admissible transducers.

3 Complexity and Randomness

Recall that the plain complexity (Kolmogorov) of a string $x \in \{0,1\}^*$ w.r.t. a partially computable function $\phi : \{0,1\}^* \to \{0,1\}^*$ is $K_\phi(x) = \inf\{|p| : \phi(p) = x\}$. It is well-known that there is a universal partially computable function $U : \{0,1\}^* \to \{0,1\}^*$ such that

$$K_U(x) \leq K_\phi(x) + c_\phi$$

holds for all strings $x \in \{0,1\}^*$. Here the constant c_ϕ depends only on U and ϕ but not on the particular string $x \in \{0,1\}^*$. We will denote the complexity K_U simply by K. Furthermore, in the case that one considers only partially computable functions with prefix-free domain, there are also universal ones among them and the corresponding complexity, called *prefix complexity* is denoted with H; like K, the prefix-free complexity H depends only up to a constant on the given choice of the underlying universal machine.

Schnorr [29] considered the subclass of partially computable prefix-monotone functions (or *processes*) $\psi : \{0,1\}^* \to \{0,1\}^*$, that is, functions which satisfy the additional property that for strings $v, w \in \mathrm{dom}(\psi)$, if $v \sqsubset w$, then $\psi(v) \sqsubset \psi(w)$. For this class of functions there is also a universal partially computable prefix-monotone function $W : \{0,1\}^* \to \{0,1\}^*$ such that for every further such ψ (with the same properties) there is a constant c_ψ, depending only on W and ψ, fulfilling

$$K_W(x) \leq K_\psi(x) + c_\psi, \tag{1}$$

for all binary strings $x \in \{0,1\}^*$.

Martin-Löf [23] introduced the notion of the random sequences in terms of tests and Schnorr — as cited by Chaitin [11] — characterised them in terms of prefix-free complexity; we take this characterisation as a definition. Furthermore, Schnorr [29] showed that the same definition holds for process complexity.

Definition 3 (Martin-Löf [23]; Schnorr [11,29]). An infinite sequence $\mathbf{x} \in \{0,1\}^\omega$ is *Martin-Löf random* if there is a constant c such that $H(\mathbf{x} \upharpoonright n) \geq |n| - c$, for all $n \geq 1$. Equivalently one can say that $\mathbf{x}$ is Martin-Löf random iff there is a constant c such that $Km_D(\mathbf{x} \upharpoonright n) \geq |n| - c$, for all $n \geq 1$.

4 Complexity Based on Transducers

For a fixed admissible transducer T, one usually denotes the complexity $C_T(x)$ of a binary string x as the length of the shortest binary string y such that $T(y) = x$. This definition is now adjusted to enumerations S of admissible transducers.

Definition 4. Let S be an enumeration of the admissible transducers. For each string x, the S-complexity $C_S(x)$ is the minimum $|\sigma| + |y|$ taken over all σ in the domain of S and y in the domain of T_σ^S such that $T_\sigma^S(y) = x$.

This S-complexity is also called the *finite state complexity based on S* of a given string. Note that if S is universal and S' is any other enumeration then there is a constant c such that

$$C_S(x) \leq C_{S'}(x) + c$$

for all binary strings x. Thus the universal enumerations define an abstract finite state complexity in the same way as it is done for prefix-free and plain complexity. The next result relates the complexity C_S for universal enumerations S to the plain complexity K, the prefix-free complexity H and the process complexity Km_D.

Theorem 5. *Let S be a universal enumeration of the admissible transducers. Then there are constants c, c', c'' such that, for all binary strings x,*

$$K(x) \leq C_S(x) + c, \quad Km_D(x) \leq C_S(x) + c', \quad C_S(x) \leq H(x) + c''.$$

Furthermore, one cannot obtain equality up to constant for any of these inequalities.

Proof.   For the first inequality, note that if $T_\sigma^S(y) = x$ then σ stems from a prefix-free set and hence there is a plain Turing machine ψ which on input p first searches for a prefix σ of p which is in $\mathrm{dom}(S)$ and, in the case that such a σ is found, outputs $T_\sigma^S(y)$ for the unique y with $\sigma y = p$. Thus the mapping from all σy to $T_\sigma^S(y)$ with $\sigma \in \mathrm{dom}(S)$ and $y \in \mathrm{dom}(T_\sigma^S)$ is partially computable and well-defined. The inequality follows then from the universality of the plain Kolmogorov complexity K. One can furthermore see that ψ is also prefix-monotone and therefore also witnessing that $Km_D(x) \leq C_S(x) + c'$ for some constant c'.

   To see that the first inequality is proper, note that $K(x) \leq Km_D(x) + c'''$ but there is no constant c'''' such that $Km_D(x) \leq K(x) + c''''$ for all x [29].

   Theorem 6 below implies that the second inequality is proper.

   Let S' be a fixed perfect enumeration of all admissible transducers; it is known that S' exists [9,10]. The inequality $C_S(x) \leq H(x) + c''$ might be obtained by choosing an enumeration S which for every p in the domain of a prefix-free universal machine U assigns to $p0$ a transducer mapping ε to $U(p)$ and, in the case that $U(p) \in \mathrm{dom}(S')$, to $p1$ the transducer $T_{U(p)}^{S'}$. Clearly, if $U(p) = x$ then $T_{p0}^S(\varepsilon) = x$ and therefore $C_S(x) \leq |p| + 1$. This enumeration of transducers is universal.

   Furthermore, there is a fixed code σ for the transducer realising the identity ($T_\sigma^S(x) = x$), hence $C_S(x) \leq |x| + |\sigma|$ for all x. It is known that this bound is not matched by longer and longer prefixes of Chaitin's Ω with respect to H, hence one cannot reverse the third inequality to an equality up to constant.   □

The properness of one inequality was missing in the previous result. It follows from the following theorem.

Theorem 6. *There is a prefix-monotone partially computable function ψ such that for every enumeration S and each constant c there is a binary string x with $K_\psi(x) < C_S(x) - c$.*

Proof. Let Ω be Chaitin's random set and let Ω_s be an approximation to Ω from the left for s steps. Now define

$$\psi(x) = 0^{\min\{s:x\leq_{lex}\Omega_s\}}.$$

Note that this function is partially computable and furthermore it is monotone. It is defined on all x with $x \leq_{lex} \Omega$. Note that for $x = \Omega \restriction n$, $\psi(x)$ coincides with the convergence module $c_\Omega(n) = \min\{s : \forall m < n\,[\Omega_s(m) = \Omega(m)]\}$.

The goal of the construction is now to show that for all constants c and all enumerations S of admissible transducers, almost all prefixes $x \sqsubset \Omega$ satisfy that $\psi(x)$ is larger than the length of any value $T^S_\sigma(y)$ with $|\sigma y| \leq |x| + c$. So fix one enumeration S.

The first ingredient for this is to use that for almost all σ, if $T^S_\sigma(y)$ is longer than $\psi(\Omega \restriction |\sigma|+|y|-c)$ then y is shorter than $|\sigma|$. Assume by way of contradiction that this is not be true and that there are infinitely many n with corresponding σ, y such that $n = |\sigma| + |y| - c$ and $|T^S_\sigma(y)| \geq \psi(\Omega \restriction n) = c_\Omega(n)$ and $|\sigma| \leq n/2$. Now one can compute from σ and $|y|$ the maximum length s of an output of $T^S_\sigma(z)$ with $|z| \leq |y|$ and then take $\Omega \restriction n$ as $\Omega_s \restriction n$. Hence $H(\Omega \restriction n)$ is, up to a constant, bounded by $|\sigma| + 2\log(|y|)$ which is bounded by $n/2$ plus a constant, in contradiction to the fact that $H(\Omega \restriction n) \geq n$ for almost all n. Thus the above assumption cannot be true.

Hence, for the further proof, one has only to consider transducers whose input is at most as long as the code. The correspdonding definition would be to let, for each $\sigma \in \text{dom}(S)$, $\phi(\sigma)$ be the length of the longest output of the form $T^S_\sigma(y)$ with $y \leq |\sigma|$.

Now assume by way of contradiction that there are a constant c and infinitely many $x \sqsubset \Omega$ such that there exists a σ with $|\psi(x)| \leq \phi(\sigma)$ and $|\sigma| \leq |x| + c$. Then one can construct a prefix-free machine V with the same domain as S such that $V(\sigma)$ for all $\sigma \in \text{dom}(S)$ outputs $z = \Omega_{\phi(\sigma)} \restriction |\sigma| - c$. As $|\sigma| \leq |x| + c$ it follows that z is a prefix of x and a prefix of Ω.

The domains of V and S are the same, hence V is a partially computable function with prefix-free domain which has for infinitely many prefixes $z \sqsubset \Omega$ an input σ of length up to $|z| + 2c$ with $V(\sigma) = z$, that is, which satisfies $H_V(z) \leq |z| + 2c$ for infinitely many prefixes z of Ω. This again contradicts the fact that Ω is Martin-Löf random, hence this does not happen.

Note that $K_\psi(x) \leq Km_D(x)+c'$ for some constant c'. Now one has, for almost all n that the string $u_n = 0^{c_\Omega(n)}$ satisfies $u_n = \psi(\Omega \restriction n)$ and $K_\psi(u_n) = n$ and $Km_D(u_n) \leq n+c'$ while, for all S and c and almost all n, $C_S(u_n) > n+c$, hence $C_S(u_n) - Km_D(u_n)$ goes to ∞ for $n \to \infty$. So C_S and Km_D cannot be equal up to constant for any enumeration S of admissible transducers.   □

Furthermore, for perfect enumerations S, one can show that there is an algorithm to compute C_S.

Proposition 7. *Let S be a perfect enumeration of the admissible transducers. Then the mapping $x \mapsto C_S(x)$ is computable.*

Proof. Note that there is a fixed transducer T^S_τ such that $T^S_\tau(x) = x$ for all x. Now $C_S(x)$ is the length of the shortest σy with $\sigma \in \text{dom}(S)$, $y \in \{0,1\}^*$,

$|\sigma y| \leq |\tau x|$ and $T^S_\sigma(y) = x$. Due to the length-restriction $|\sigma y| \leq |\tau x|$, the search space is finite and due to the perfectness of the enumeration S, the search can be carried out effectively.   □

5   Complexity of Infinite Sequences

Martin-Löf randomness can be formalised using both prefix-free Kolmogorov complexity and process complexity, see Definition 3. Therefore it is natural to ask whether such a characterisation does also hold for the C_S complexity. The answer is affirmative as given in the following theorem.

Theorem 8. *The following statements are equivalent:*

(a) *The sequence* **x** *is not Martin-Löf random;*
(b) *There is a perfect enumeration* S *such that for every* $c > 0$ *and almost all* $n > 0$ *we have* $C_S(\mathbf{x} \upharpoonright n) < n - c$;
(c) *There is a perfect enumeration* S *such that for every* $c > 0$ *there exists an* $n > 0$ *with* $C_S(\mathbf{x} \upharpoonright n) < n - c$;
(d) *For every universal enumeration* S *and for every* $c > 0$ *and almost all* $n > 0$ *we have* $C_S(\mathbf{x} \upharpoonright n) < n - c$;
(e) *For every universal enumeration* S *and for every* $c > 0$ *there exists an* $n > 0$ *with* $C_S(\mathbf{x} \upharpoonright n) < n - c$.

Proof.   If **x** is Martin-Löf random then, as noted after Definition 3, $Km_D(\mathbf{x} \upharpoonright n) \geq n - c$ for some constant c and all n. It follows that, for every enumeration S, from Theorem 5 that $C_S(\mathbf{x} \upharpoonright n) \geq n - c'$ for some constant c' and all n. Hence non of the conditions (a-e) is satisfied.

Now assume that (a) is satisfied, that is, that **x** is not Martin-Löf random. Let U be a universal prefix-free machine and $H_U = H$. Using U we define the following enumeration S of finite transducers:

> For $\sigma\eta$ such that $\sigma \in \mathrm{dom}(U)$ and $\mathrm{time}(U(\sigma)) = |\eta|$, let T^S_σ be defined as the trnasucer which maps every string τ to $U(\sigma)\eta\tau$.

Here $\mathrm{time}(U(\sigma))$ denotes the time till the computation stops; S is computable and prefix-free because $\mathrm{dom}(U)$ is prefix-free.

If the sequence **x** is not Martin-Löf random, then for every $c > 0$ there exists an $n > 0$ such that $H(\mathbf{x} \upharpoonright n) < n - c$. Hence, for every $c > 0$ there exist $n > 0$, $\sigma \in \{0,1\}^*$, $s > 0$ such that $U(\sigma) = \mathbf{x} \upharpoonright n$, $|\sigma| < n - c$ and $\mathrm{time}(U(\sigma)) = s$. Consequently, for every $c > 0$ there exist $n > 0$, $\sigma \in \{0,1\}^*$, $s > 0$ and $\eta \in \{0,1\}^s$ such that $\sigma\eta \in \mathrm{dom}(S)$, $|\sigma| < n - c$, $T^S_{\sigma\eta}(\varepsilon) = \mathbf{x} \upharpoonright (n + s)$, hence for every $c > 0$ there exist $n, s > 0$ such that $C_S(\mathbf{x} \upharpoonright (n + s)) < n + s - c$. We have showed that for every $c > 0$ and almost all $m > 0$, $C_S(\mathbf{x} \upharpoonright m) < m - c$. Thus (b) holds. If S' is a universal enumeration, then $C_S(x) \leq C_{S'}(x) + c''$ for some constant c'' and all binary strings x. Hence (d) holds. Furthermore, (b) implies (c) and (d) implies (e). So (a-e) hold. Hence the conditions (a-e) are equivalent.   □

Corollary 9. *A sequence* **x** *is Martin-Löf random iff for every enumeration* S *there is a constant* c *such that for every* $n \geq 1$ *the inequality* $C_S(\mathbf{x} \upharpoonright n) \geq n - c$ *holds true.*

6   Finite State Incompressibility and Normality

In this section we define finite state incompressible sequences and prove that each such sequence is normal. Given an enumeration S of all admissible transducers, a sequence $\mathbf{x} = x_1 x_2 \cdots x_n \cdots$ is C_S–*incompressible* if $\liminf_n C_S(\mathbf{x} \upharpoonright n)/n = 1$.

Proposition 10. *Every Martin-Löf random sequence is C_S–incompressible for all enumerations S, but the converse implication is not true.*

Proof. If $\mathbf{x}$ is a Martin-Löf random sequence, then $\liminf_n K(\mathbf{x} \upharpoonright n)/n = 1$, so by Theorem 5, $\mathbf{x}$ is C_S–incompressible. Next we take a Martin-Löf random sequence $\mathbf{x}$ and modify it to be not random: define $\mathbf{x}'(n) = 0$ whenever n is a power of 2 and $\mathbf{x}'(n) = \mathbf{x}(n)$, otherwise. Clearly, $\mathbf{x}'$ is not Martin-Löf random, but $\liminf_n K(\mathbf{x} \upharpoonright n)/n = 1$, so $\mathbf{x}$ is C_S–incompressible for every enumeration S of all admissible transducers. $\square$

A sequence is *normal* if all digits are equally likely, all pairs of digits are equally likely, all triplets of digits equally likely, etc. This means that the sequence $\mathbf{x} = x_1 x_2 \cdots x_n \cdots$ is normal if the frequency of every string y in $\mathbf{x}$ is $2^{-|y|}$, where $|y|$ is the length of y.

Lemma 11. *If the sequence $\mathbf{x}$ is not normal, then there exist a transducer T_σ^S and a constant α with $0 < \alpha < 1$ (depending on $\mathbf{x}, \sigma, S$) such that for infinitely many integers $n > 0$ we have $C_{T_\sigma^S}(\mathbf{x} \upharpoonright n) < \alpha \cdot n$.*

Proof.   According to [16,17,28], if the sequence $\mathbf{x}$ is not normal, then there exist a transducer T_σ^S, a sequence $\mathbf{y}$, and a real $\alpha \in (0,1)$ such that $\lim_{m \to \infty} T_\sigma^S(\mathbf{y} \upharpoonright m) = \mathbf{x}$ and for infinitely many $m > 0$

$$T_\sigma^S(\mathbf{y} \upharpoonright m) \sqsubset \mathbf{x} \text{ and } m < \alpha \cdot |T_\sigma^S(\mathbf{y} \upharpoonright m)|.$$

Consequently, for infinitely many $m > 0$

$$C_{T_\sigma^S}(T_\sigma^S(\mathbf{y} \upharpoonright m)) \leq m < \alpha \cdot |T_\sigma^S(\mathbf{y} \upharpoonright m)|,$$

hence $C_{T_\sigma^S}(\mathbf{x} \upharpoonright n) < \alpha \cdot n$ for infinitely many $n > 0$ because $T_\sigma^S(\mathbf{y} \upharpoonright m) \sqsubset \mathbf{x}$ for infinitely many $m > 0$. $\square$

Example 12. Ambos-Spies and Busse [2,3] as well as Tadaki [31] investigated infinite sequences $\mathbf{x}$ which can be predicted by finite automata in a certain way. The formalisations result in the following equivalent characterisations for a sequence $\mathbf{x}$ to be finite state predictable:

- The sequence $\mathbf{x}$ can be predicted by a finite automaton in the sense that every state is either passing or has a prediction on the next bit and when reading $\mathbf{x}$ the finite automaton makes infinitely often a correct prediction and passes in those cases where it does not make a correct prediction, that is, it never predicts wrongly.

- There is a finite automaton which has in every state a label from $\{0,1\}^*$ such that, whenever the automaton is in a state with a non-empty label w then some of the next bits of $\mathbf{x}$ are different from the corresponding ones in w.
- $\mathbf{x}$ fails to contain some string w as a substring.
- There is a finite connected automaton with binary input alphabet such that not all states of it are visited when reading $\mathbf{x}$.
- The sequence $\mathbf{x}$ is the image $T(\mathbf{y})$ for some binary sequence $\mathbf{y}$ and a finite transducer T which has only labels of the form (a, aw) with $a \in \{0,1\}$ and $w \in \{0,1\}^*$ and where in the translation from $\mathbf{y}$ into $\mathbf{x}$ infinitely often a label (a, aw) with $w \neq \varepsilon$ is used.

Finite state predictable sequences are not normal and, by the work of Schnorr and Stimm [28], there is a finite-automaton martingale which succeeds on such a sequence. Furthermore, there are sequences which are not normal but also not finite-state predictable. An example can be obtained by translating the decimal Champernowne sequence $\mathbf{y}$ [12] into a binary sequence $\mathbf{x}$ such that $\mathbf{x}(k) = 1$ iff $\mathbf{y}(k) \in \{1, 2, \dots, 9\}$ and $\mathbf{x}(k) = 0$ iff $\mathbf{y}(k) = 0$; now the resulting $\mathbf{x}$ is not normal; however, $\mathbf{x}$ contains every substring as a substring and is thus also not finite-state predictable.

Theorem 13. *Every C_S-incompressible sequence is normal.*

Proof.   Assume that the sequence $\mathbf{x}$ is not normal. According to Lemma 11 there exist $\alpha \in (0, 1)$ and $\sigma \in \text{dom}(S)$ such that for infinitely many integers $n > 0$ we have $C_{T_\sigma^S}(\mathbf{x} \restriction n)) < \alpha \cdot n$. For these n it also holds that $C_S(\mathbf{x} \restriction n) < \alpha \cdot n + |\sigma|$. Since $\alpha < 1$, $\mathbf{x}$ is not C_S-incompressible.   □

7   How Large Is the Set of Incompressible Sequences?

It is natural to ask whether the converse of Theorem 13 is true. The results in [1,5,28,33] discussed in Introduction might suggest a positive answer. In fact, the answer is *negative*.

To prove this result we will use binary *de Bruijn strings* of order $r \geq 1$ which are strings w of length $2^r + r - 1$ over alphabet $\{0,1\}$ such that any binary string of length r occurs as a substring of w (exactly once). It is well-known that de Bruijn strings of any order exist, and have an explicit construction [14,32]. For example, 00110 and 0001011100 are de Bruijn strings of orders 2 and 3 respectively.

Note that de Bruijn strings are derived in a circular way, hence their prefix of length $r - 1$ coincides with the suffix of length $r - 1$. Denote by $B(r)$ the prefix of length 2^r of a de Bruijn string of order r. The examples of de Bruijn strings of orders 2 and 3 previously presented are derived from the strings $B(2) = 0011$ and $B(3) = 00010111$, respectively. Thus the string $B(r) \cdot B'(r)$, where $B'(r)$ is the length $r - 1$ prefix of $B(r)$, contains every binary string of length string r exactly once as a substring.

In [26] it is shown that every sequence of the form

$$\mathbf{b}_f = B(1)^{f(1)} B(2)^{f(2)} \cdots B(n)^{f(n)} \cdots$$

is normal provided that the function $f : \mathbf{N} \to \mathbf{N}$ is increasing and satisfies the condition $f(i) \geq i^i$ for all $i \geq 1$. Moreover, in this case the real $0.\mathbf{b}_f$ is a Liouville number, i.e. it is a transcendental real number with the property that, for every positive integer n, there exist integers p and q with $q > 1$ and such that $0 < |0.\mathbf{b}_f - \frac{p}{q}| < q^{-n}$.

Lemma 14. *Every string w, $B(1) \sqsubseteq w \sqsubset \mathbf{b}_f$ can be represented in the form*

$$w = B(1)^{f(1)} B(2)^{f(2)} \cdots B(n-1)^{f(n-1)} B(n)^j w' \tag{2}$$

where $n \geq 1$, $1 \leq j \leq f(n)$ and $|w'| < 2^{n+1} = |B(n+1)|$.

Proof. Indeed, in the case

$$B(1)^{f(1)} B(2)^{f(2)} \cdots B(n-1)^{f(n-1)} \sqsubseteq w \sqsubset B(1)^{f(1)} B(2)^{f(2)} \cdots B(n)^{f(n)}$$

we can choose $w' \sqsubset B(n)$, and if

$$B(1)^{f(1)} B(2)^{f(2)} \cdots B(n)^{f(n)} \sqsubseteq w \sqsubset B(1)^{f(1)} B(2)^{f(2)} \cdots B(n)^{f(n)} B(n+1)$$

we can choose $w' \sqsubset B(n+1)$.   □

Next we show that there are normal sequences which are simultaneously Liouville numbers and compressible by transducers, that is, the converse of Theorem 13 is false. This also proves that C_S–incompressibility is stronger than all other known forms of finite automata based incompressibility, cf. [1,5,15,28,33].

Theorem 15. *For every enumeration S there are normal sequences $\mathbf{x}$ such that $\lim_{n \to \infty} C_S(\mathbf{x} \restriction n)/|n| = 0$, so $\mathbf{x}$ is C_S–compressible.*

Proof. Define the transducer $T_n = (\{0,1\}, \{s_1, \ldots, s_{n+1}\}, s_1, \delta_n, \mu_n)$ as follows:

$$\begin{aligned}
\delta_n(s_i, 0) &= s_i, & \mu_n(s_i, 0) &= B(i), \text{ for } i \leq n, \\
\delta_n(s_i, 1) &= s_{i+1}, & \mu_n(s_i, 1) &= B(i), \text{ for } i \leq n, \\
\delta_n(s_{n+1}, a) &= s_{n+1}, & \mu_n(s_{n+1}, a) &= a, & \text{for } a \in \{0,1\}.
\end{aligned}$$

For example, the transducer T_4 is presented in Figure 1. Let σ_n be an encoding of T_n according to S. Choose a function $f : \mathbf{N} \to \mathbf{N}$ which satisfied the following two conditions for all $n \geq 1, i > 1$:

$$f(n) \geq \max\{|\sigma_{n+1}|, n^n, 2^{n+2}\} \text{ and } f(i) \geq 2 \cdot f(i-1). \tag{3}$$

Finally, let $p_i = 0^{f(i)-1} 1$ and $p'_j = 0^{j-1} 1$. Eq. (2) shows that

$$T_n(p_1 \cdots p_{n-1} p'_j w') = B(1)^{f(1)} \cdots B(n-1)^{f(n-1)} B(n)^j w'$$

is a prefix of the normal sequence $\mathbf{x} = \mathbf{b}_f$. We then have:

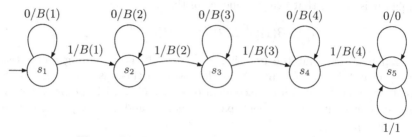

Fig. 1. Block representation of the transducer T_4

$$|T_n(p_1 \cdots p_{n-1}p'_j w')| = \sum_{i=1}^{n-1} 2^i f(i) + 2^n j + |w'|$$
$$\geq 2^{n-1} f(n-1) + 2^n j,$$

and

$$|\sigma_n| + |p_1 \cdots p_{n-1} \cdot p'_j \cdot w'|$$

$$= |\sigma_n| + \sum_{i=1}^{n-1} |p_i| + |p'_{n-1}| + |w'|$$
$$\leq f(n-1) + 2f(n-1) + j + f(n-1)$$
$$= 4f(n-1) + j.$$

This shows that for every prefix w of $\mathbf{b}_f$ presented in the form (2) as

$$w = B(1)^{f(1)} \cdots B(n-1)^{f(n-1)} \cdot B(n)^j \cdot w',$$

we have $B(1) \sqsubseteq w \sqsubseteq \mathbf{b}_f$ and (by using the inequality $\frac{a+b}{c+d} \leq \max\left\{\frac{a}{c}, \frac{b}{d}\right\}$, when $0 < a, b, c, d$):

$$\frac{C_S(w)}{|w|} \leq \frac{4f(n-1) + j}{2^{n-1} f(n-1) + 2^n j} \leq \frac{4}{2^{n-1}}.$$

This shows that $\lim_{n \to \infty} C_S(\mathbf{x} \restriction n)/|n| = 0$. $\qquad \square$

In the proof of Theorem 15 we have used an arbitrary function f satisfying (3). Of course, there exist computable and incomputable such functions.

Corollary 16. *For every perfect enumeration S there are normal and C_S-compressible computable and incomputable sequences.*

One might also consider transducers which satisfy that $|\mu(q, a)| \leq m$ for all $(q, a) \in Q \times \{0, 1\}$, that is, the output can always be at most m times as long as the input. For these one can then also consider the variant $C_S^{(m)}$ of C_S which looks at complexity using m-bounded transducers. The following result is a sample result for this area.

Theorem 17. *For every enumeration S of all 2-bounded admissible transducers, there are normal sequences* $\mathbf{x}$ *such that* $\lim_{n\to\infty} C_S^{(2)}(\mathbf{x} \restriction n)/n = 1/2$.

Proof. We start from the transducers T_n defined in the proof of Theorem 15 and we split every long output $B(i)$ of T_n into 2^{i-1} pieces of length 2. Formally, we replace the states $s_i, i \leq n$, by sub-transducers $A_i = (\{0,1\}, R_i, r_{i,1}, \delta_n^{(i)}, \mu_n^{(i)})$ where $R_i = \{r_{i,1}, \dots, r_{i,2^{i-1}}\}$,

$$\delta_n^{(i)}(r_{i,j}, a) = r_{i,j+1}, \quad \mu_n^{(i)}(r_{i,j}, a) = u_{i,j}, j < 2^i, \ a < 2,$$
$$\delta_n^{(i)}(r_{i,2^{i-1}}, 0) = r_{i,1}, \quad \mu_n^{(i)}(r_{i,2^{i-1}}, 0) = u_{i,2^{i-1}},$$
$$\delta_n^{(i)}(r_{i,2^{i-1}}, 1) = r_{i+1,1}, \ \mu_n^{(i)}(r_{i,2^{i-1}}, 1) = u_{i,2^{i-1}},$$

and $B(i) = u_{i,1} \cdots u_{i,2^{i-1}}$ with $|u_{ij}| = 2$. Observe that the transition with input 1 on state $r_{i,2^{i-1}}$ leads to the initial state of the next sub-transducer (for $i = n$ this leads to state $r_{n+2,1} = s_{n+1}$ of T_n).

Then, the new transducer is defined as follows:

$$Q_n = \bigcup_{i=1}^{n} R_i \cup \{s_{n+1}\}, q_{0n} = r_{1,1},$$

$$\delta_n' = \bigcup_{i=1}^{n} \delta_n^{(i)} \cup \{(s_{n+1}, 0, s_{n+1}), (s_{n+1}, 1, s_{n+1})\},$$

and

$$\mu_n' = \bigcup_{i=1}^{n} \mu_n^{(i)} \cup \{(s_{n+1}, 0, 0), (s_{n+1}, 1, 1)\}.$$

Again let σ_n' be an encoding of T_n' in S, and let $\bar{p}_i = (0^{2^{i-1}})^{f(i)-1}0^{2^{i-1}-1}1$ where $f : \mathbf{N} \to \mathbf{N}$, $f(n) \geq \max\{|\sigma_{n+1}'|, n^n, 2^{n+2}\}$, $f(i) \geq 2 \cdot f(i-1)$, is as in the proof of Theorem 15. Let $\bar{p}_{i,j}' = (0^{2^{i-1}})^{j-1}0^{2^{i-1}-1}1$.

Furthermore, let $B(1) \sqsubseteq w \sqsubset \mathbf{b}_f$. According to Eq. (2) we have:

$$w = B(1)^{f(1)} \cdots B(n-1)^{f(n-1)}B(n)^j w' = T_n'(\bar{p}_1 \cdots \bar{p}_{n-1}\bar{p}_j' w').$$

We then have:

$$|T_n'(\bar{p}_1 \cdots \bar{p}_{n-1}(0^{j-1})1 \cdot w')| = \sum_{i=1}^{n-1} 2^i \cdot f(i) + 2^n j + |w'|$$
$$\geq \sum_{i=1}^{n-1} 2^i \cdot f(i) + 2^n j,$$

and

$$C_S^{(m)}(w) \leq |\sigma_n'| + \sum_{i=1}^{n-1} 2^{i-1}f(i) + 2^{n-1}j + |w'|$$
$$\leq f(n-1) + \sum_{i=1}^{n-1} 2^{i-1}f(i) + 2^{n-1}j + f(n-1),$$

finally obtaining

$$\frac{C_S^{(m)}(w)}{|w|} \leq \frac{\sum_{i=1}^{n-2} 2^{i-1}f(i) + 2^{n-1}j + (2^{n-2}+2)f(n-1)}{\sum_{i=1}^{n-2} 2^i f(i) + 2^n j + 2^{n-1}f(n-1)}$$
$$\leq \frac{2^{n-2}+2}{2^{n-1}}.$$

This proves that $\lim_{t\to\infty} C_S^{(2)}(\mathbf{x} \restriction t)/t = 1/2$.   □

Theorem 17 can be easily generalised to m-bounded complexity thereby yielding the bound $\lim_{n\to\infty} C_S^{(m)}(\mathbf{x} \restriction n)/n = 1/m$. Moreover, the results of Theorems 15 and 17 can be also generalised to arbitrary (output) alphabets Y. Here the circular de Bruijn strings of order n, $CB_{|Y|}(n)$, have length $|Y|^n$.

In connection with Theorem 15, we can ask whether the finite state complexity of each sequence $\mathbf{x}$ representing a Liouville number satisfies the inequality $\limsup_{n\to\infty} C_S(\mathbf{x} \restriction n)/n < 1$. The answer is negative: Example 12 of [30] shows that there are sequences $\mathbf{x}$ representing Liouville numbers having $\limsup_{n\to\infty} K(\mathbf{x} \restriction n)/n = 1$, hence by Theorem 5, $\limsup_{n\to\infty} C_S(\mathbf{x} \restriction n)/n = 1$.

The following result complements Theorem 15: the construction is valid for every enumeration, but the degree of incompressibility is slightly smaller.

Theorem 18. *There exists an infinite, normal and computable sequence* $\mathbf{x}$ *which satisfies the condition* $\liminf_{n\to\infty} C_S(\mathbf{x} \restriction n)/n = 0$, *for all enumerations* S.

Proof.   Fix a computable enumeration $(T_m)_{m\geq 1}$ of all admissible transducers such that each T_m has at most m states and each transition in T_m from one state to another has only labels which produce outgoing strings of at most length m (that is, complicated transducers appear sufficiently late in the list).

Now define a sequence of strings α_n such that each α_n is the length-lexicographic first string longer than n such that for all transducers T_m with $1 \leq m \leq n$, for all states q of T_m and for each string γ of less than n bits, there is no string β of length below $\frac{n-1}{n} \cdot |\alpha_n|$ such that $\gamma T_m(q, \beta)$ is α_n or an extension of it. Note that these α_n must exist, as every sufficiently long prefix of the Champernowne sequence meets the above given specifications due to Champernowne sequence normality [12]. Furthermore, $\alpha_0 = 0$ as the only constraint is that α_0 is longer than 0. An easy observation shows that also $|\alpha_n| \leq |\alpha_{n+1}|$ for all n.

In what follows we will use an acceptable numbering of all partially computable functions from natural numbers to natural numbers of one variable $(\varphi_e)_{e\geq 1}$. Now let f be a computable function from natural numbers to natural numbers satisfying the following conditions:

Short: For all $t \geq 1$, $|\alpha_{f(t)}| \leq \sqrt{t}$.
Finite-to-one: For all $n \geq 1$ and almost all $t \geq 1$, $f(t) > n$.
Match: $\forall n \,\forall e < n \,\exists t \,[\varphi_e(n) < \infty \implies t > \varphi_e(n) \wedge f(t) = n \wedge f(t+1) = n \wedge \ldots \wedge f(t^2) = n]$.

In order to construct f, consider first a computable one-one enumeration $(e_0, n_0, m_0), (e_1, n_1, m_1), \ldots$ of the set of all (e, n, m) such that $e < n \wedge \varphi_e(n) = m$. The function f is now constructed in stages where the requirement "Short" is satisfied all the time, the requirement "Finite-to-one" will be a corollary of the way the function is constructed and the requirement "Match" will be satisfied for the k-th constraint (e_k, n_k, m_k) in the k-th stage.

In the k-th stage, let s_k be the first value where $f(s_k)$ was not defined in an earlier stage and let t_k be the first number such that $t_k > s_k + m_k$ and $|\alpha_{n_k}| \leq \sqrt{s_k}$. Having these properties, for u with $s_k \leq u < t_k$, let $f(u)$ be the maximal ℓ with $|\alpha_\ell| \leq \sqrt{\max\{1, u\}}$, and for u with $t_k \leq u \leq t_k^2$, let $f(u) = n_k$.

It is clear that the function f is computable. Next we verify that it satisfies the required three conditions.

Short: This condition, which is more or less hard-coded into the algorithm, directly follows from the way t_k is selected and $f(u)$ is defined in the two cases.

Finite-to-one: The inequality $f(u) \leq n$ is true only in stages k where for some u either $|\alpha_{n+1}| > \sqrt{s_k}$ or $n_k \leq n$; both conditions happen only for finitely many stages k.

Match: For each n and e with $\varphi_e(n)$ being defined, there is a stage k such that $(e_k, n_k, m_k) = (e, n, \varphi_e(n))$. The choice of t_k makes then f to be equal to n_k on $t_k, t_k + 1, \ldots, t_k^2$ and furthermore $t_k > \varphi_{e_k}(n_k)$.

Let $\mathbf{x}$ be the sequence $\alpha_{f(0)} \alpha_{f(1)} \alpha_{f(2)} \ldots$ which is obtained by concatenating all the strings $\alpha_{f(n)}$ for the n in default order. It is clear that $\mathbf{x}$ is computable.

Consider any enumeration S of transducers. Choose e such that $\varphi_e(n)$ takes the value the length of the code of that transducer T_n which has the starting state q and a further state q' and follows the following transition table:

state	input	output	new state
q	0	ε	q'
q	1	α_n	q
q'	0	0	q
q'	1	1	q

As φ_e is total, there is for each $n > e$ a t larger than the code of the transducer T_n such that $f(t), f(t+1), \ldots, f(t^2)$ are all n. Now $\sigma = \alpha_{f(0)} \ldots \alpha_{f(t^2)}$ can be generated by T_n by a code of the form $\beta = 0\sigma(0)0\sigma(1) \ldots 0\sigma(u-1)1^{t^2-t}$ where u is the length of $\alpha_{f(0)} \alpha_{f(1)} \ldots \alpha_{f(t-1)}$. The length of β is $2u + t^2 - t$. Note that $u \leq t \cdot \sqrt{t}$ by the condition "Short" and therefore $|\beta| \leq t^2 + t^{3/2} - t$ while the string σ generated from β by the transducer T_n has at least the length $(t^2 - t) \cdot |\alpha_n|$ which is at least $(t^2 - t) \cdot (n+1)$. Furthermore, the representation of T_n in S has at most length t, thus

$$C_S(\sigma)/|\sigma| \leq (t^2 + t^{3/2})/(n \cdot (t^2 - t)) \leq \frac{2}{n}.$$

It follows that $\liminf_{n \to \infty} C_S(\mathbf{x} \restriction n)/n = 0$.

Next we prove that $\mathbf{x}$ is normal. Fix a transducer T_m. Then, for every $n > m$, there is a sufficiently large t such that $(n-1) \cdot t$ of the first $n \cdot t$ values $s < n \cdot t$ satisfy $f(s) > n$. Fix such a t and let $\beta = \beta_0 \beta_1 \ldots \beta_{n \cdot t}$ be such that $\beta_0 \ldots \beta_s$ is the shortest prefix of β with T_m producing from the starting state and input

$\beta_0 \ldots \beta_s$ an extension of $\alpha_{f(0)} \ldots \alpha_{f(s)}$. Note that the image of $\beta_0 \ldots \beta_s$ is at most $m - 1$ symbols longer than $\alpha_{f(0)} \ldots \alpha_{f(s)}$. Let $\sigma = \alpha_{f(0)} \ldots \alpha_{f(t \cdot n)}$. One can prove by induction that for all s with $f(s) \geq n$ we have

$$|\beta_s| \geq \frac{n-1}{n} \cdot |\alpha_{f(s)}|,$$

and for all s where $f(s) < n$ we have

$$|\alpha_{f(s)}| \leq |\sigma|/(t \cdot n).$$

It follows that $|\beta| \geq \frac{(n-1)^2}{n^2} \cdot |\sigma|$ and therefore we have sufficiently long prefixes of **x** which are concatenations of the strings $\alpha_{f(0)} \ldots \alpha_{f(t \cdot n)}$, all having complexity relative to T_m near 1. Furthermore, the length difference between any given prefix and a prefix of such a form is smaller than the square root of the length and therefore one can conclude that the sequence is incompressible with respect to each fixed transducer T_m. Hence, by Theorem 13, it is normal. $\square$

The proof method in Theorem 18 can be adapted to obtain the following result.

Theorem 19. *There exists a perfect enumeration S and a sequence which is computable, normal and C_S-incompressible.*

Proof. The sequence of the T_n and α_n is defined as in the proof of Theorem 18; furthermore, it is assumed that the listing of the T_n is one-one. However, f has is chosen such that it satisfies the following three conditions:

Short: For all $t \geq 1$, $|\alpha_{f(t)}| \leq \sqrt{t}$.
Finite-to-one: For all $n \geq 1$ and almost all $t \geq 1$, $f(t) > n$.
Monotone: For all $t \geq 1$, $f(t) \leq f(t+1)$.

This is achieved by selecting

$$f(t) = \max\{m : |\alpha_m| \leq \sqrt{t}\}.$$

It is clear that f is computable and satisfies the conditions "Short" and "Monotone". The condition "Finite-to-one" follows from the observation that $f(t) > n$ for all t with $|\alpha_{n+1}| \leq \sqrt{t}$ and the fact that almost all t satisfy this condition.

As above one can see that whenever $f(t) > n$ and $m \leq n$ then $T_m(\beta)$ extends $\alpha_{f(0)} \alpha_{f(1)} \ldots \alpha_{f(n \cdot t)}$ only if $|\beta| \geq (n-1)^2/n^2$. Now one makes S such that the transducer T_m has the code word $0^m 1^{m^2 \cdot t_m}$ for the first t_m such that $f(t_m) > m$. It can be concluded that $C_{T_m}(\sigma)/|\sigma| \geq (m-1)^2/m^2 \cdot |\sigma|$, for all prefixes σ of **x** and that $C_{T_m}(\sigma)/|\sigma|$ goes to 1 for longer and longer prefixes of **x**. Thus the sequence **x** is normal and furthermore **x** is incompressible with respect to the here chosen S. $\square$

8   Conclusion and Open Questions

Enumerations are — in the context of this paper — computable listings of all admissible transducers and have a prefix-free domain. We have investigated two main notions of enumerations, the perfect ones (which have a decidable domain, are one-one, are surjective and have a computable inverse) and the universal ones (which optimise the codes for the transducers up to a constant for the best possible value). We have showed that Martin-Löf randomness of infinite sequences can be characterised with both types of enumerations. Furthermore, we have related the finite-state complexity based on universal enumerations with the prominent notions of algorithmic description complexity of binary strings.

The results of Sections 6 and 7 show that our definition of finite state incompressibility is stronger than all other known forms of finite automata based incompressibility, in particular the notion related to finite automaton based betting systems introduced by Schnorr [28].

There are various interesting open questions. Here are three more: Are there an enumeration S, a computable sequence $\mathbf{x}$ and a constant c such that $C_S(\sigma) > |\sigma| - c$, for all prefixes σ of $\mathbf{x}$? For which enumerations S is it true that every sequence satisfying $C_S(\mathbf{x} \restriction n) \geq n-c$ is Martin-Löf random? What is the relation between C_S–incompressible sequences and ε–random sequences, [8]? Note that some ε–random sequences can be finite-state predictable by not having a certain substring, cf. [31], hence they can be compressed by a single transducer; this is, however, not true for all ε–random sequences.

Acknowledgments. The authors would like to thank Sanjay Jain and the anonymous referees of TAMC 2014 for helpful comments.

References

1. Agafonov, V.N.: Normal sequences and finite automata. Soviet Mathematics Doklady 9, 324–325 (1968)
2. Ambos-Spies, K., Busse, E.: Automatic forcing and genericity: On the diagonalization strength of finite automata. In: Calude, C.S., Dinneen, M.J., Vajnovszki, V. (eds.) DMTCS 2003. LNCS, vol. 2731, pp. 97–108. Springer, Heidelberg (2003)
3. Ambos-Spies, K., Busse, E.: Computational aspects of disjunctive sequences. In: Fiala, J., Koubek, V., Kratochvíl, J. (eds.) MFCS 2004. LNCS, vol. 3153, pp. 711–722. Springer, Heidelberg (2004)
4. Berstel, J.: Transductions and Context-free Languages. Teubner (1979)
5. Bourke, C., Hitchcock, J.M., Vinodchandran, N.V.: Entropy rates and finite-state dimension. Theoretical Computer Science 349(3), 392–406 (2005)
6. Buhrman, H., Fortnow, L.: Resource-bounded Kolmogorov complexity revisited. In: Reischuk, R., Morvan, M. (eds.) STACS 1997. LNCS, vol. 1200, pp. 105–116. Springer, Heidelberg (1997)
7. Calude, C.S.: Information and Randomness. An Algorithmic Perspective, 2nd edn. Springer, Berlin (2002)
8. Calude, C.S., Hay, N.J., Stephan, F.: Representation of left-computable ε–random reals. Journal of Computer and System Sciences 77, 812–839 (2011)

9. Calude, C.S., Salomaa, K., Roblot, T.K.: Finite state complexity. Theoretical Computer Science 412, 5668–5677 (2011)
10. Calude, C.S., Salomaa, K., Roblot, T.K.: State-size hierarchy for FS-complexity. International Journal of Foundations of Computer Science 25(1), 37–50 (2012)
11. Chaitin, G.J.: A theory of program size formally identical to information theory. Journal of the Association for Computing Machinery 22, 329–340 (1975)
12. Champernowne, D.G.: The construction of decimals normal in the scale of ten. Journal of the London Mathematical Society 8, 254–260 (1933)
13. Charikar, M., Lehman, E., Liu, D., Panigrahy, R., Prabhakaran, M., Rasala, A., Sahai, A., Shelat, A.: Approximating the smallest grammar: Kolmogorov complexity in natural models. In: Proceedings of STOC 2002, pp. 792–801. ACM Press (2002)
14. de Bruijn, N.: A combinatorial problem. Proceedings of the Koninklijke Nederlandse Akademie van Wetenschappen 49, 758–764 (1946)
15. Dai, J.J., Lathrop, J.I., Lutz, J.H., Mayordomo, E.: Finite-state dimension. Theoretical Computer Science 310, 1–33 (2004)
16. Doty, D., Moser, P.: Finite-state dimension and lossy compressors, arxiv:cs/0609096v2 (2006)
17. Doty, D., Moser, P.: Feasible Depth. In: Cooper, S.B., Löwe, B., Sorbi, A. (eds.) CiE 2007. LNCS, vol. 4497, pp. 228–237. Springer, Heidelberg (2007)
18. Downey, R., Hirschfeldt, D.: Algorithmic Randomness and Complexity. Springer, Heidelberg (2010)
19. Katseff, H.P.: Complexity dips in random infinite binary sequences. Information and Control 38(3), 258–263 (1978)
20. Lehman, E.: Approximation Algorithms for Grammar-based Compression, PhD Thesis. MIT (2002)
21. Lehman, E., Shelat, A.: Approximation algorithms for grammar-based compression. In: Proceedings of SODA 2002, pp. 205–212. SIAM Press (2002)
22. Li, M., Vitányi, P.: An Introduction to Kolmogorov Complexity and Its Applications, 3rd edn. Springer (2007)
23. Martin-Löf, P.: The definition of random sequences. Information and Control 9, 602–619 (1966)
24. Martin-Löf, P.: Complexity oscillations in infinite binary sequences. Zeitschrift für Wahrscheinlichkeitstheorie und Verwandte Gebiete 19, 225–230 (1971)
25. Nies, A.: Computability and Randomness. Clarendon Press, Oxford (2009)
26. Nandakumar, S., Vangapelli, S.K.: Normality and finite-state dimension of Liouville numbers, arxiv:1204.4104v1 [cs.IT] (2012)
27. Rytter, W.: Application of Lempel-Ziv factorization to the approximation of grammar-based compression. Theoretical Computer Science 302, 211–222 (2002)
28. Schnorr, C.P., Stimm, H.: Endliche Automaten und Zufallsfolgen. Acta Informatica 1, 345–359 (1972)
29. Schnorr, C.P.: Process complexity and effective randomness tests. Journal of Comput. System Sciences 7, 376–388 (1973)
30. Staiger, L.: The Kolmogorov complexity of real numbers. Theoretical Computer Science 284, 455–466 (2002)
31. Tadaki, K.: Phase Transition and Strong Predictability, CDMTCS Research Report 435 (2013)
32. van Lint, J.H., Wilson, R.M.: A Course in Combinatorics. Cambridge University Press (1993)
33. Ziv, J., Lempel, A.: Compression of individual sequences via variable-rate coding. IEEE Transactions on Information Theory 24, 530–536 (1978)

Finding Optimal Strategies of Almost Acyclic Simple Stochastic Games

David Auger, Pierre Coucheney, and Yann Strozecki

PRiSM, Université de Versailles Saint-Quentin-en-Yvelines,
Versailles, France
{david.auger,pierre.coucheney,yann.strozecki}@uvsq.fr

Abstract. The optimal value computation for turned-based stochastic games with reachability objectives, also known as *simple stochastic games*, is one of the few problems in NP ∩ coNP which are not known to be in P. However, there are some cases where these games can be easily solved, as for instance when the underlying graph is acyclic. In this work, we try to extend this tractability to several classes of games that can be thought as "almost" acyclic. We give some fixed-parameter tractable or polynomial algorithms in terms of different parameters such as the number of cycles or the size of the minimal feedback vertex set.

Keywords: algorithmic game theory, stochastic games, FPT algorithms.

Introduction

A *simple stochastic game*, SSG for short, is a zero-sum, two-player, turn-based version, of the more general *stochastic games* introduced by Shapley [17]. SSGs were introduced by Condon [6] and they provide a simple framework that allows to study the algorithmic complexity issues underlying reachability objectives. A SSG is played by moving a pebble on a graph. Some vertices are divided between players MIN and MAX: if the pebble attains a vertex controlled by a player then he has to move the pebble along an arc leading to another vertex. Some other vertices are ruled by chance; typically they have two outgoing arcs and a fair coin is tossed to decide where the pebble will go. Finally, there is a special vertex named the 1-sink, such that if the pebble reaches it player MAX wins, otherwise player MIN wins.

Player MAX's objective is, given a starting vertex for the pebble, to maximize the probability of winning against any strategy of MIN. One can show that it is enough to consider stationary deterministic strategies for both players [6]. Though seemingly simple since the number of stationary deterministic strategies is finite, the task of finding the pair of optimal strategies, or equivalently, of computing the so-called *optimal values* of vertices, is not known to be in P.

SSGs are closely related to other games such as parity games or discounted payoff games to cite a few [2]. Interestingly, those games provide natural applications in model checking of the modal μ-calculus [18] or in economics. While it is

T V Gopal et al. (Eds.): TAMC 2014, LNCS 8402, pp. 67–85, 2014.

known that they can be reduced to simple stochastic games [4], hence seemingly easier to solve, so far no polynomial algorithm are known for these games either.

Nevertheless, there are some very simple restrictions for SSGs for which the problem of finding optimal strategies is tractable. Firstly, if there is only one player, the game is reduced to a Markov Decision Process (MDP) which can be solved by linear programming. In the same vein, if there is no randomness, the game can be solved in almost linear time [1].

As an extension of that fact, there is a Fixed Parameter Tractable (FPT) algorithm, where the parameter is the number of average vertices [11]. The idea is to get rid of the average vertices by sorting them according to a guessed order. Finally, when the (graph underlying the) game is a directed acyclic graph (DAG), the values can be found in linear time by computing them backwardly from sinks.

Without the previous restrictions, algorithms running in exponential time are known. Among them, the Hoffman-Karp [13] algorithm proceeds by successively playing a local best-response named switch for one player, and then a global best-response for the other player. Generalizations of this algorithm have been proposed and, though efficient in practice, they fail to run in polynomial time on a well designed example [9], even in the simple case of MDPs [8]. These variations mainly concern the choice of vertices to switch at each turn of the algorithm which is quite similar to the choice of pivoting in the simplex algorithm for linear programming. This is not so surprising since computing the values of an SSG can be seen as a generalization of solving a linear program. The best algorithm so far is a randomized sub-exponential algorithm [15] that is based on an adaptation of a pivoting rule used for the simplex.

Our Contribution

In this article, we present several graph parameters such that, when the parameter is fixed, there is a polynomial time algorithm to solve the SSG value problem. More precisely, the parameters we look at will quantify how close to a DAG is the underlying graph of the SSG, a case that is solvable in linear time. The most natural parameters that quantify the distance to a DAG would be one of the directed versions of the tree-width such as the DAG-width. Unfortunately, we are not yet able to prove a result even for SSG of bounded pathwidth. In fact, in the simpler case of parity games the best algorithms for DAG-width and clique-width are polynomials but not even FPT [16,3]. Thus we focus on restrictions on the number of cycles and the size of a minimal feedback vertex set.

First, we introduce in Section 2 a new class of games, namely *MAX-acyclic games*, which contains and generalizes the class of acyclic games. We show that the standard Hoffman-Karp algorithm, also known as strategy iteration algorithm, terminates in a linear number of steps for games in this class, yielding a polynomial algorithm to compute optimal values and strategies. It is known that, in the general case, this algorithm needs an exponential number of steps to compute optimal strategies, even in the simple case of Markov Decision Processes [8,9].

Then, we extend in Section 3 this result to games with very few cycles, by giving an FPT-algorithm where the parameter is the number of fork vertices which bounds the number of cycles. To obtain a linear dependance in the total number of vertices, we have to reduce our problem to several instances of acyclic games since we cannot even rely on computing the values in a general game.

Finally, in Section 4, we provide an original method to "eliminate" vertices in an SSG. We apply it to obtain a polynomial time algorithm for the value problem on SSGs with a feedback vertex set of bounded size (Theorem 8).

1   Definitions and Standard Results

Simple stochastic games are turn-based stochastic games with reachability objectives involving two players named MAX and MIN. In the original version of Condon [6], all vertices except sinks have outdegree exactly two, and there are only two sinks, one with value 0 and another with value 1. Here, we allow more than two sinks with general rational values, and more than an outdegree two for positional vertices.

Definition 1 (SSG). *A simple stochastic game (SSG) is defined by a directed graph $G = (V, A)$, together with a partition of the vertex set V in four parts V_{MAX}, V_{MIN}, V_{AVE} and V_{SINK}. To every $x \in V_{SINK}$ corresponds a value $Val(x)$ which is a rational number in $[0, 1]$. Moreover, vertices of V_{AVE} have outdegree exactly 2, while sink vertices have outdegree 1 consisting of a single loop on themselves.*

In the article, we denote by n_M, n_m and n_a the size of V_{MAX}, V_{MIN} and V_{AVE} respectively and by n the size of V. The set of *positional vertices*, denoted V_{POS}, is $V_{POS} = V_{MAX} \cup V_{MIN}$. We now define strategies which we restrict to be stationary and pure, which turns out to be sufficient for optimality. Such strategies specify for each vertex of a player the choice of a neighbour.

Definition 2 (Strategy). *A strategy for player MAX is a map σ from V_{MAX} to V such that*

$$\forall x \in V_{MAX}, \quad (x, \sigma(x)) \in A.$$

Strategies for player MIN are defined analogously and are usually denoted by τ. We denote Σ and T the sets of strategies for players MAX and MIN respectively.

Definition 3 (play). *A play is a sequence of vertices $x_0, x_1, x_2, \ldots$ such that for all $t \geq 0$,*

$$(x_t, x_{t+1}) \in A.$$

Such a play is consistent with strategies σ and τ, respectively for player MAX and player MIN, if for all $t \geq 0$,

$$x_t \in V_{MAX} \Rightarrow x_{t+1} = \sigma(x_t)$$

and

$$x_t \in V_{MIN} \Rightarrow x_{t+1} = \tau(x_t).$$

A couple of strategies σ, τ and an initial vertex $x_0 \in V$ define recursively a random play consistent with σ, τ by setting:

- if $x_t \in V_{MAX}$ then $x_{t+1} = \sigma(x_t)$;
- if $x_t \in V_{MIN}$ then $x_{t+1} = \tau(x_t)$;
- if $x_t \in V_{SINK}$ then $x_{t+1} = x_t$;
- if $x_t \in V_{AVE}$, then x_{t+1} is one of the two neighbours of x_t, the choice being made by a fair coin, independently of all other random choices.

Hence, two strategies σ, τ, together with an initial vertex x_0 define a measure of probability $\mathbb{P}_{\sigma,\tau}^{x_0}$ on plays consistent with σ, τ. Note that if a play contains a sink vertex x, then at every subsequent time the play stays in x. Such a play is said to *reach* sink x. To every play $x_0, x_1, \ldots$ we associate a value which is the value of the sink reached by the play if any, and 0 otherwise. This defines a random variable X once two strategies are fixed. We are interested in the expected value of this quantity, which we call the value of a vertex $x \in V$ under strategies σ, τ:

$$\mathrm{Val}_{\sigma,\tau}(x) = \mathbb{E}_{\sigma,\tau}^x(X)$$

where $\mathbb{E}_{\sigma,\tau}^x$ is the expected value under probability $\mathbb{P}_{\sigma,\tau}^x$. The goal of player MAX is to maximize this (expected) value, and the best he can ensure against a strategy τ is

$$\mathrm{Val}_\tau(x) = \max_{\sigma \in \Sigma} \mathrm{Val}_{\sigma,\tau}(x)$$

while against σ player MIN can ensure that the expected value is at most

$$\mathrm{Val}_\sigma(x) = \min_{\tau \in T} \mathrm{Val}_{\sigma,\tau}(x).$$

Finally, the value of a vertex x, is the common value

$$\mathrm{Val}(x) = \max_{\sigma \in \Sigma} \min_{\tau \in T} \mathrm{Val}_{\sigma,\tau}(x) = \min_{\tau \in T} \max_{\sigma \in \Sigma} \mathrm{Val}_{\sigma,\tau}(x). \qquad (1)$$

The fact that these two quantities are equal is nontrivial, and it can be found for instance in [6]. A pair of strategies σ^*, τ^* such that, for all vertices x,

$$\mathrm{Val}_{\sigma^*,\tau^*}(x) = \mathrm{Val}(x)$$

always exists and these strategies are said to be *optimal strategies*. It is polynomial-time equivalent to compute optimal strategies or to compute the values of all vertices in the game, since values can be obtained from strategies by solving a linear system. Conversely if values are known, optimal strategies are given by greedy choices in linear time (see [6] and Lemma 1). Hence, we shall simply write "solve the game" for these tasks.

We shall need the following notion:

Definition 4 (Stopping SSG). *A SSG is said to be* stopping *if for every couple of strategies all plays eventually reach a sink vertex with probability 1.*

Condon [6] proved that every SSG G can be reduced in polynomial time into a stopping SSG G' whose size is quadratic in the size of G, and whose values almost remain the same.

Theorem 1 (Optimality conditions, [6]). *Let G be a stopping SSG. The vector of values $(Val(x))_{x \in V}$ is the only vector w satisfying:*

- *for every $x \in V_{MAX}$, $w(x) = \max\{w(y) \mid (x, y) \in A\}$;*
- *for every $x \in V_{MIN}$, $w(x) = \min\{w(y) \mid (x, y) \in A\}$;*
- *for every $x \in V_{AVE}$ $w(x) = \frac{1}{2}w(x^1) + \frac{1}{2}w(x^2)$ where x^1 and x^2 are the two neighbours of x;*
- *for every $x \in V_{SINK}$, $w(x) = Val(x)$.*

If the underlying graph of an SSG is acyclic, then the game is stopping and the previous local optimality conditions yield a very simple way to compute values. Indeed, we can use backward propagation of values since all leaves are sinks, and the values of sinks are known. We naturally call these games *acyclic SSGs*.

Once a pair of strategies has been fixed, the previous theorem enables us to see the values as solution of a system of linear equations. This yields the following lemma, which is an improvement on a similar result in [6], where the bound is 4^n instead of $6^{\frac{n_a}{2}}$.

Lemma 1. *Let G be an SSG with sinks having rational values of common denominator q. Then under any pair of strategies σ, τ, the value $Val_{\sigma,\tau}(x)$ of any vertex x can be computed in time $O(n_a^\omega)$, where ω is the exponent of the matrix multiplication, and n_a the number of average (binary) vertices. Moreover, the value can be written as a rational number $\frac{a}{b}$, with*

$$0 \le a, b \le 6^{\frac{n_a}{2}} \times q.$$

Proof. We sketch the proof since it is standard. First, one can easily compute all vertices x such that

$$Val_{\sigma,\tau}(x) = 0.$$

Let Z be the set of these vertices. Then:

- all AVE vertices in Z have all their neighbours in Z;
- all MAX (resp. MIN) vertices x in Z are such that $\sigma(x)$ (resp. $\tau(x)$) is in Z.

To compute Z, we can start with the set Z of all vertices except sinks with positive value and iterate the following

- if Z contains an AVE vertex x with a neighbour out of Z, remove x from Z;
- if Z contains a MAX (resp. MIN) vertex x with $\sigma(x)$ (resp. $\tau(x)$) out of Z, remove x from Z.

This process will stabilize in at most n steps and compute the required set Z. Once this is done, we can replace all vertices of Z by a sink with value zero, obtaining a game G' where under σ, τ, the values of all vertices will be unchanged.

Consider now in G' two corresponding strategies σ, τ (keeping the same names to simplify) and a positional vertex x. Let x' be the first non positional vertex that can be reached from x under strategies σ, τ. Clearly, x' is well defined and

$$\text{Val}_{\sigma,\tau}(x) = \text{Val}_{\sigma,\tau}(x').$$

This shows that the possible values under σ, τ of all vertices are the values of average and sink vertices. The same is true if one average vertex has its two arcs towards the same vertex, thus we can forget those also. The value of an average vertex being equal to the average value of its two neighbours, we see that we can write a system

$$z = Az + b \tag{2}$$

where

- z is the n_a-dimensional vector containing the values of average vertices
- A is a matrix where all lines have at most two $\frac{1}{2}$ coefficients, the rest being zeros
- b is a vector whose entries are of the form 0, $\frac{p_i}{2q}$ or $\frac{p_i+p_j}{2q}$, corresponding to transitions from average vertices to sink vertices.

Since no vertices but sinks have value zero, it can be shown that this system has a unique solution, i.e. matrix $I - A$ is nonsingular, where I is the n_a-dimensional identity matrix. We refer to [6] for details, the idea being that since in $n - 1$ steps there is a small probability of transition from any vertex of G' to a sink vertex, the sum of all coefficients on a line of A^{n-1} is strictly less than one, hence the convergence of

$$\sum_{k \geq 0} A^k = (I - A)^{-1}.$$

Rewriting (2) as

$$2(I - A)z = 2b,$$

we can use Cramer's rule to obtain that the value z_v of an average vertex v is

$$z_v = \frac{\det B_v}{\det 2(I - A)}$$

where B_v is the matrix $2(I - A)$ with the column corresponding to v replaced by $2b$. Hence by expanding the determinant we see that z_v is of the form

$$\frac{1}{\det 2(I - A)} \sum_{w \in V_{AVE}} \pm 2b_w \det(2(I - A)_{v,w})$$

where $2(I - A)_{v,w}$ is the matrix $2(I - A)$ where the line corresponding to v and the column corresponding to w have been removed.

Since $2b_w$ has either value 0, $\frac{p_i}{q}$ or $\frac{p_i+p_j}{q}$ for some $1 \leq i, j \leq n_a$, we can write the value of z_v as a fraction of integers

$$\frac{\sum_{w \in v} \pm 2b_w q \cdot \det(2(I - A)_{v,w})}{\det 2(I - A) \cdot q}$$

It remains to be seen, by Hadamard's inequality, that since the nonzero entries of $2(I - A)$ on a line are a 2 and at most two -1, we have

$$\det 2(I - A) \leq 6^{\frac{n_a}{2}},$$

which concludes the proof.

The bound $6^{\frac{n_a}{2}}$ is almost optimal. Indeed a caterpillar tree of n average vertices connected to the 0 sink except the last one, which is connected to the 1 sink, has a value of $\frac{1}{2}^n$ at the root. Note that the lemma is slightly more general (rational values on sinks) and the bound a bit better ($\sqrt{6}$ instead of 4) than what is usually found in the literature.

In all this paper, the complexity of the algorithms will be given in term of number of arithmetic operations and comparisons on the values as it is customary. The numbers occurring in the algorithms are rationals of value at most exponential in the number of vertices in the game, therefore the bit complexity is increased by at most an almost linear factor.

2   MAX-acyclic SSGs

In this section we define a class of SSG that generalize acyclic SSGs and still have a polynomial-time algorithm solving the value problem.

A cycle of an SSG is an *oriented* cycle of the underlying graph.

Definition 5. *We say that an SSG is MAX-acyclic (respectively MIN-acyclic) if from any MAX vertex x (resp. MIN vertex), for all outgoing arcs a but one, all plays going through a never reach x again.*

Therefore this class contains the class of acyclic SSGs and we can see this hypothesis as being a mild form of acyclicity. From now on, we will stick to MAX-acyclic SSGs, but any result would be true for MIN-acyclic SSGs also. There is a simple characterization of MAX-acyclicity in term of the structure of the underlying graph.

Lemma 2. *An SSG is MAX-acyclic if and only if every MAX vertex has at most one outgoing arc in a cycle.*

Let us specify the following notion.

Definition 6. *We say that an SSG is strongly connected if the underlying directed graph, once sinks are removed, is strongly connected.*

Lemma 3. *Let G be a MAX-acyclic, strongly connected SSG. Then for each MAX x, all neighbours of x but one must be sinks.*

Proof. Indeed, if x has two neighbours y and z which are not sinks, then by strong connexity there are directed paths from y to x and from z to x. Hence, both arcs xy and xz are on a cycle, contradicting the assumption of MAX-acyclicity.

From now on, we will focus on computing the values of a strongly connected MAX-acyclic SSG. Indeed, it easy to reduce the general case of a MAX-acyclic SSG to strongly connected by computing the DAG of the strongly connected components in linear time. We then only need to compute the values in each of the components, beginning by the leaves.

We will show that the Hoffman-Karp algorithm [13,7], when applied to a strongly connected MAX-acyclic SSG, runs for at most a linear number of steps before reaching an optimal solution. Let us remind the notion of *switchability* in simple stochastic games. If σ is a strategy for MAX, then a MAX vertex x is *switchable* for σ if there is an neighbour y of x such that $\mathrm{Val}_\sigma(y) > \mathrm{Val}_\sigma(\sigma(x))$. *Switching* such a vertex x consists in considering the strategy σ', equal to σ but for $\sigma'(x) = y$.

For two vectors v and w, we note $v \geq w$ if the inequality holds componentwise, and $v > w$ if moreover at least one component is strictly larger.

Lemma 4 (See Lemma 3.5 in [19]). *Let σ be a strategy for MAX and S be a set of switchable vertices. Let σ' be the strategy obtained when all vertices of S are switched. Then*

$$\mathrm{Val}_{\sigma'} > \mathrm{Val}_\sigma.$$

Let us recall the Hoffman-Karp algorithm:

1. Let σ_0 be any strategy for MAX and τ_0 be a best response to σ_0
2. while (σ_t, τ_t) is not optimal:
   (a) let σ_{t+1} be obtained from σ_t by switching one (or more) switchable vertex
   (b) let τ_{t+1} be a best response to σ_{t+1}

The Hoffman-Karp algorithm computes a finite sequence $(\sigma_t)_{0 \leq t \leq T}$ of strategies for the MAX player such that

$$\forall 0 \leq t \leq T - 1, \quad \mathrm{Val}_{\sigma_{t+1}} > \mathrm{Val}_{\sigma_t}.$$

If any MAX vertex x in a strongly connected MAX-acyclic SSG has more than one sink neighbour, say $s_1, s_2, \cdots s_k$, then these can be replaced by a single sink neighbour s' whose value is

$$\mathrm{Val}(s') := \max_{i=1..k} \mathrm{Val}(s_i).$$

Hence, we can suppose that all MAX vertices in a strongly connected MAX-acyclic SSG have degree two. For such a reduced game, we shall say that a MAX vertex x is *open* for a strategy σ if $\sigma(x)$ is the sink neighbour of x and that x is *closed* otherwise.

Lemma 5. *Let G be a strongly connected, MAX-acyclic SSG, where all MAX vertices have degree 2. Then the Hoffman-Karp algorithm, starting from any strategy σ_0, halts in at most $2n_M$ steps. Moreover, starting from the strategy where every MAX vertex is open, the algorithm halts in at most n_M steps. All in all, the computation is polynomial in the size of the game.*

Proof. We just observe that if a MAX vertex x is closed at time t , then it remains so until the end of the computation. More precisely, if $s := \sigma_{t-1}(x)$ is a sink vertex, and $y := \sigma_t(x)$ is not, then since x has been switched we must have

$$\text{Val}_{\sigma_t}(y) > \text{Val}_{\sigma_t}(s).$$

For all subsequent times $t' > t$, since strategies are improving we will have

$$\text{Val}_{\sigma_{t'}}(y) \geq \text{Val}_{\sigma_t}(y) > \text{Val}_{\sigma_t}(s) = \text{Val}_{\sigma_{t'}}(s) = \text{Val}(s),$$

so that x will never be switchable again.

Thus starting from any strategy, if a MAX vertex is closed it cannot be opened and closed again, and if it is open it can only be closed once.

Each step of the Hoffman-Karp algorithm requires to compute a best-response for the MIN player. A best-response to any strategy can be simply computed with a linear program with as many variables as vertices in the SSG, hence in polynomial time. We will denote this complexity by $O(n^\eta)$; it is well known that we can have $\eta \leq 4$, for instance with Karmarkar's algorithm.

Theorem 2. *A strongly connected MAX-acyclic SSG can be solved in time $O(n_M n^\eta)$.*

Before ending this part, let us note that in the case where the game is also MIN-acyclic, one can compute directly a best response to a MAX strategy σ without linear programming: starting with a MIN strategy τ_0 where all MIN vertices are open, close all MIN vertices x such that their neighbour has a value strictly less than their sink. One obtains a strategy τ_1 such that

$$\text{Val}_{\sigma,\tau_1} < \text{Val}_{\sigma,\tau_0},$$

and the same process can be repeated. By a similar argument than in the previous proof, a closed MIN vertex will never be opened again, hence the number of steps is at most the number of MIN vertices, and each step only necessitates to compute the values, i.e. to solve a linear system (see Lemma 1).

Corollary 1. *A strongly connected MAX and MIN-acyclic SSG can be solved in time $O(n_M n_m n^\omega)$, where ω is the exponent of matrix multiplication.*

3   SSG with Few Fork Vertices

Work on this section has begun with Yannis Juglaret during his Master internship at PRiSM laboratory. Preliminary results about SSGs with one simple cycle can be found in his report [14]. We shall here obtain fixed-parameter tractable (FPT) algorithms in terms of parameters quantifying how far a graph is from being MAX-acyclic and MIN-acyclic, in the sense of section 2. These parameters are:

$$k_p = \sum_{x \in V_{POS}} (|\{y : (x, y) \in A \text{ and is in a cycle}\}| - 1)$$

and

$$k_a = \sum_{x \in V_{AVE}} (|\{y : (x, y) \in A \text{ and is in a cycle}\}| - 1).$$

We say that an SSG is POS-acyclic (for *positional* acyclic) when it is both MAX and MIN-acyclic. Clearly, parameter k_p counts the number of edges violating this condition in the game. Similarly, we say that the game is AVE-acyclic when average vertices have at most one outgoing arc in a cycle. We call *fork vertices*, those vertices that have at least two outgoing arcs in a cycle. Since averages vertices have only two neighbours, k_a is the number of fork average vertices.

Note that:

1. When $k_p = 0$ (respectively $k_a = 0$), the game is POS-acyclic (resp. AVE-acyclic).
2. When $k_a = k_p = 0$, the strongly connected components of the game are cycles. We study these games, which we call *almost acyclic*, in detail in subsection 3.1.
3. Finally, the number of simple cycles of the SSG is always less than $k_p + k_a$, therefore getting an FTP algorithm in k_p and k_a immediately gives an FTP algorithm in the number of cycles.

We obtain:

Theorem 3. *There is an algorithm which solves the value problem for SSGs in time $O(nf(k_p, k_a))$, with $f(k_p, k_a) = k_a! 4^{k_a} 2^{k_p}$.*

As a corollary, by remark 3 above we have:

Theorem 4. *There is an algorithm which solves the value problem for SSGs with k simple cycles in time $O(ng(k))$ with $g(k) = (k-1)! 4^{k-1}$.*

Note that in both cases, when parameters are fixed, the dependance in n is *linear*.

Before going further, let us explain how one could easily build on the previous part and obtain an FPT algorithm in parameter k_p, but with a much worse dependance in n.

When $k_p > 0$, one can fix partially a strategy on positional fork vertices, hence obtaining a POS-acyclic subgame that can be solved in polynomial time according to Corollary 1, using the Hoffman-Karp algorithm. Combining this with a bruteforce approach looking exhaustively through all possible local choices at positional fork vertices, we readily obtain a polynomial algorithm for the value problem when k_p is fixed:

Theorem 5. *There is an algorithm to solve the value problem of an SSG in time $O(n_M n_m n^\omega 2^{k_p})$.*

We shall conserve this brute-force approach. In the following, we give an algorithm that reduces the polynomial complexity to a linear complexity when k_a is fixed. From now on, up to applying the same bruteforce procedure, we assume $k_p = 0$ (all fork vertices are average vertices). We also consider the case of a strongly connected SSG, since otherwise the problem can be solved for each strongly connected component as done in Section 2. We begin with the baseline $k_a = 0$ and extend the algorithm to general values of k_a. Before this, we provide some preliminary lemmas and definitions that will be used in the rest of the section.

A *partial strategy* is a strategy defined on a subset of vertices controlled by the player. Let σ be such a partial strategy for player MAX, we denote by $G[\sigma]$ the subgame of G where the set of strategies of MAX is reduced to the ones that coincide with σ on its support. According to equation (1), the value of an SSG is the highest expected value that MAX can guarantee, then it decreases with the set of actions of MAX:

Lemma 6. *Let G be an SSG with value v, σ a partial strategy of MAX, and $G[\sigma]$ the subgame induced by σ with value v'. Then $v \geq v'$.*

In strongly connected POS-acyclic games, positional vertices have at least one outgoing arc to a sink. Recall that, in case the strategy chooses a sink, we say that it is open at this vertex (the strategy is said open if it is open at a vertex), and closed otherwise. We can then compare the value of an SSG with that of the subgame generated by any open strategy.

Lemma 7. *Let x be a MAX vertex of an SSG G with a sink neighbour, and σ the partial strategy open at x. If it is optimal to open x in G, then it is optimal to open it in $G[\sigma]$.*

Proof. Since it is optimal to open x in G, the value of its neighbour sink is at least that of any neighbour vertex, say y. But, in view of Lemma 6, the value of y in $G[\sigma]$ is smaller than in G, and then it is again optimal to play a strategy open at x in the subgame.

This lemma allows to reduce an almost acyclic SSG (resp. an SSG with parameter $k_a > 0$) to an acyclic SSG (resp. an SSG with parameter $k_a - 1$). Indeed, if the optimal MAX strategy is open at vertex x, then the optimal strategy of the subgame open at any MAX vertex will be open at x. A solution to find x once the subgame is solved (and then to reduce the parameter k_a) consists in testing all the open MAX vertices. But it may be the case that all MAX vertices are open which would not yield a FPT algorithm. In Lemma 8 (resp. Lemma 9), we give a restriction on the set of MAX vertices that has to be tested when $k_a = 0$ (resp. $k_a > 0$) which provides an FPT algorithm.

3.1 Almost Acyclic SSGs

We consider an SSG with $k_a = 0$. Together with the hypothesis that it is POS-acyclic and strongly connected, its graph, once sinks are removed, consists of a single cycle. A naive algorithm to compute the value of such SSG consists in looking for, if it exists, a vertex that is open in the optimal strategy, and then solve the acyclic subgame:

1. For each positional vertex x:
    (a) compute the values of the acyclic SSG $G[\sigma]$, where σ is the partial strategy open at x,
    (b) if the local optimality condition is satisfied for x in G, return the values.
2. If optimal strategies have not been found, return the value when all vertices are closed.

This naive algorithm uses the routine that computes the value of an SSG with only one cycle. When the strategies are closed, the values can be computed in linear time as for an acyclic game. Indeed, let x be an average vertex (if none, the game can be solved in linear time) and $s_1 \ldots s_\ell$ be the values of the average neighbour sinks in the order given by a walk on the cycle starting from x. Then the value of x satisfies the equation $\mathrm{Val}(x) = \frac{1}{2}s_1 + \frac{1}{2}(\frac{1}{2}s_2 + \frac{1}{2}(\cdots + \frac{1}{2}(\frac{1}{2}s_\ell + \frac{1}{2}\mathrm{Val}(x))))$, so that

$$\mathrm{Val}(x) = \frac{2^\ell}{2^\ell - 1} \sum_{i=1}^{\ell} 2^{-i} s_i, \tag{3}$$

which can be computed in time linear in the size of the cycle. The value of the other vertices can be computed by walking backward from x, again in linear time. Finally, since solving an acyclic SSG is linear, the complexity of the algorithm is $O(n^2)$ which is still better than the complexity $O(n_M n_m n^\omega)$ obtained with the Hoffman-Karp algorithm (see Theorem 5).

Remark that this algorithm can readily be extended to a SSG with k cycles with a complexity $O(n^{k+1})$. Hence it is not an FPT algorithm for the number of cycles. However, we can improve on this naive algorithm by noting that the optimal strategy belongs to one of the following subclasses of strategies:

(i) strategies closed everywhere,
(ii) strategies open at least at one MAX vertex,
(iii) strategies open at least at one MIN vertex.

The trick of the algorithm is that, knowing which of the three classes the optimal strategy belongs to, the game can be solved in linear time. Indeed:

(i) If the optimal strategy is closed at every vertex, the value can be computed in linear time as shown before.
(ii) If the optimal strategy is open at a MAX vertex (the MIN case is similar), then it suffices to solve in linear time the acyclic game $G[\sigma]$ where σ is any partial strategy open at a MAX vertex, and then use the following Lemma to find an open vertex for the optimal strategy of the initial game.

Lemma 8. *Let G be a strongly connected almost-acyclic SSG. Assume that the optimal strategy is open at a MAX vertex. For any partial strategy σ open at a MAX vertex x, let $x = x_0, x_1 \ldots x_\ell = x$ be the sequence of the ℓ open MAX vertices for the optimal strategy of $G[\sigma]$ listed in the cycle order. Then it is optimal to open x_1.*

Proof. Let $\bar{x}$ be a MAX vertex that is open when solving G. From Lemma 7, there is an index i such that $x_i = \bar{x}$ (in particular there exists an open MAX vertex when solving $G[\sigma]$). If $\ell = 1$ then $x_0 = \bar{x}$ so that the optimal strategies of $G[\sigma]$ and G coincide. Otherwise, if $i = 1$, the result is immediate. At last, if $i > 1$, x_i has the same value in G and $G[\sigma]$ (the value of its sink), and so has the vertex just before if it is different from x_0. Going backward from x_i in the cycle, all the vertices until x_0 (not included) have the same value in G and $G[\sigma]$. In particular, if $i > 1$, this is the case for x_1 whose value is then the value of its sink. So it is optimal to open x_1 in G as well.

All in all, a linear algorithm that solves a strongly connected almost-acyclic SSG G is:

1. Compute the values of the strategies closed everywhere. If optimal, return the values.
2. Else compute the optimal strategies of $G[\sigma_1]$ where σ_1 is a partial strategy open at a MAX vertex x; let y be the first open MAX vertex after x; compute the values of $G[\sigma_2]$ where σ_2 is the partial strategy open at y; if the local optimality condition is satisfied for y in G, return the values.
3. Else apply the same procedure to any MIN vertex.

Theorem 6. *There is an algorithm to solve the value problem of a strongly connected almost-acyclic SSG in time $O(n)$ with n the number of vertices.*

3.2 Fixed Number of Non Acyclic Average Vertices

Again, we assume that the SSG is strongly connected and POS-acyclic. The algorithm for almost-acyclic games can be generalized as follow.

Firstly, it is possible to compute the values of the strategies closed everywhere in polynomial time in k_a and check if this strategy is optimal. Indeed the value of each fork vertex can be expressed as an affine function of the value of all fork vertices in the spirit of Eq. (3). Then the linear system of size k_a can be solved in polynomial time, and the value of the remaining vertices computed by going backward from each fork vertex. This shows that computing the values of a game once strategies are fixed is polynomial in the number of fork average vertices, which slightly improves the complexity of Lemma 1.

Otherwise, the following Lemma allows to find a positional vertex that is open for the optimal strategy. We say that a vertex x is the last (resp. first) vertex in a set S before (resp. after) another vertex y if there is a unique simple path from x to y (resp. y to x) that does not contain any vertex in S, x and y being excluded.

Lemma 9. *Let G be a strongly connected SSG with a set $A = \{a_1, \ldots, a_\ell\}$ of fork average vertices and no positional fork vertices. Assume that the optimal strategy of G is open at a MAX vertex. Let σ be a partial strategy open at any vertex that is the last MAX vertex before a vertex in A. Let $S[\sigma]$ be the set of open MAX vertices when solving $G[\sigma]$. Then, there exists $x \in S[\sigma]$ that satisfies*

- *it is optimal to open x in G,*
- *x is the first vertex in $S[\sigma]$ after a vertex in A.*

Proof. Let $\bar{x}$ be a MAX vertex that is open when solving G, and i be such that a_i is the last vertex in A before $\bar{x}$. By Lemma 7, $\bar{x}$ is open in $G[\sigma]$ and since σ is not open at a MAX vertex between a_i and $\bar{x}$, all the vertices between the successor of a_i and $\bar{x}$ have the same value in $G[\sigma]$ and G, and then the optimal strategy of $G[\sigma]$ at these vertices is optimal for G. This property holds in particular for the first vertex in $S[\sigma]$ after a_i in the path leading to $\bar{x}$.

The Lemma is illustrated on Figure 1.

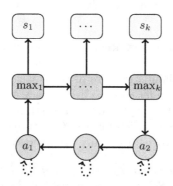

Fig. 1. Illustration of Lemma 9: a_1 and a_2 are average fork vertices. Vertices on the top are MAX vertices, labelled from 1 to k, that lead to sinks. $\max_k$ is the last MAX vertex before a fork vertex. Assume it is optimal to open at least one MAX vertex. Then, the first open MAX vertex (wrt to the labelling) of the optimal strategy of the subgame open at $\max_k$ is open as well in the optimal strategy of the initial game.

Finally, if the optimal strategy is open at some MAX vertex, then the following algorithm can be run to compute the values of G:

1. Let x be the last MAX vertex before some fork vertex, and σ_1 the partial strategy open at x. $G[\sigma_1]$ is an SSG that has $k_a - 1$ fork vertices (recall that G is strongly connected). When solved, it provides a set $S[\sigma_1]$ of open MAX vertices. There are at most $k_a + k_a - 1$ vertices that are the first in $S[\sigma_1]$ after a fork vertex. Then, from Lemma 9, it is optimal to open at least one of them in G.
2. For each y that is the first in $S[\sigma_1]$ after a fork vertex:
   (a) compute the values of $G[\sigma_2]$, σ_2 being the partial strategy open at y,
   (b) if local optimality condition is satisfied for y in G, return the values.

This algorithm computes at most $2k_a$ SSGs with $k_a - 1$ average fork vertices. In the worst case, the same algorithm must be run for the MIN vertices. Using theorem 6 for the case $k_p = k_a = 0$, we obtain Theorem 3 and its corollary.

4   Feedback Vertex Set

A feedback vertex set is a set of vertices in a directed graph such that removing them yields a DAG. Computing a minimal vertex set is an NP-hard problem [10], but it can be solved with a FPT algorithm [5]. Assume the size of the minimal vertex set is fixed, we prove in this section that we can find the optimal strategies in polynomial time. Remark that, to prove such a theorem, we cannot use the result on bounded number of cycles since a DAG plus one vertex may have an exponential number of cycles. Moreover a DAG plus one vertex may have a large number of positional vertices with several arcs in a cycle, thus we cannot use the algorithm to solve MAX-acyclic plus a few non acyclic MAX vertices.

The method we present works by transforming k vertices into sinks and could thus be used for other classes of SSGs. For instance, it could solve in polynomial time the value problem for games which are MAX-acyclic when k vertices are removed.

4.1   The Dichotomy Method

We assume from now on that all SSGs are stopping. In this subsection, we explain how to solve an SSG by solving it several times but with one vertex less.

First we remark that turning any vertex into a sink of its own value in the original game does not change any value.

Lemma 10. *Let G be an SSG and x one of its vertex. Let G' be the same SSG as G except that x has been turned into a sink vertex of value $Val_G(x)$. For all vertices y, $Val_G(y) = Val_{G'}(y)$.*

Proof. The optimality condition of Theorem 1 are exactly the same in G and G'. Since the game is stopping, there is one and only one solution to these equations and thus the values of the vertices are identical in both games.

The values in an SSG are monotone with regards to the values of the sinks, as proved in the next lemma.

Lemma 11. *Let G be an SSG and s one of its sink vertex. Let G' be the same SSG as G except that the value of s has been increased. For all vertices x, $Val_G(x) \leq Val_{G'}(x)$.*

Proof. Let fix a pair of strategy (σ, τ) and a vertex x. We have:

$$\mathrm{Val}_{(\sigma,\tau),G}(x) = \sum_{y \in V_{SINK}} \mathbb{P}(x \rightsquigarrow y)\, \mathrm{Val}_G(y)$$

$$\text{Val}_{(\sigma,\tau),G}(x) \le \sum_{y \in V_{SINK}} \mathbb{P}(x \rightsquigarrow y)\,\text{Val}_{G'}(y) = \text{Val}_{(\sigma,\tau),G'}(x)$$

because $\text{Val}_G(x) = \text{Val}_{G'}(x)$ except when $x = s$, $\text{Val}_G(s) \le \text{Val}_{G'}(s)$. Since the inequality is true for every pair of strategies and every vertex, the lemma is proved.

Let x be an arbitrary vertex of G and let $G[v]$ be the same SSG, except that x becomes a SINK vertex of value v. We define the function f by:

$$\begin{cases} \text{if x is a MAX vertex,} & f(v) = \max\{\text{Val}_{G[v]}(y) : (x,y) \in A\} \\ \text{if x is a MIN vertex,} & f(v) = \min\{\text{Val}_{G[v]}(y) : (x,y) \in A\} \\ \text{if x is an AVE vertex,} & f(v) = \frac{1}{2}\text{Val}_{G[v]}(x^1) + \text{Val}_{G[v]}(x^2) \end{cases}$$

Lemma 12. *There is a unique v_0 such that $f(v_0) = v_0$ which is $v_0 = \text{Val}_G(x)$. Moreover, for all $v > v_0$, $f(v_0) < v_0$ and for all $v < v_0$, $f(v_0) > v_0$.*

Proof. The local optimality conditions given in Theorem 1 are the same in G and $G[v]$ except the equation $f(\text{Val}_G(x)) = \text{Val}_G(x)$. Therefore, when $f(v_0) = v_0$, the values of $G[v]$ satisfy all the local optimality conditions of G. Thus v_0 is the value of s in G. Since the game is stopping there is at most one such value.

Conversely, let v_0 be the value of s in G. By Lemma 10, the values in $G[v_0]$ are the same as in G for all vertices. Therefore the local optimality conditions in G contains the equation $f(v_0) = v_0$.

We have seen that $f(v) = v$ is true for exactly one value of v. Since the function f is increasing by Lemma 11 and because $f(0) \ge 0$ and $f(1) \le 1$, we have for all $v > v_0$, $f(v_0) < v_0$ and for all $v < v_0$, $f(v_0) > v_0$.

The previous lemma allows to determine the value of x in G by a dichotomic search by the following algorithm. We keep refining an interval $[min, max]$ which contains the value of x, with starting values $min = 0$ and $max = 1$.

1. While $max - min \le 6^{-n_a}$ do:
   (a) $v = (min + max)/2$
   (b) Compute the values of $G[v]$
   (c) If $f(v) > v$ then $min = v$
   (d) If $f(v) < v$ then $max = v$
2. Return the unique rational in $[min, max]$ of denominator less than $6^{-\frac{n_a}{2}}$

Theorem 7. *Let G be an SSG with n vertices and x one of its vertex. Denote by $C(n)$ the complexity to solve $G[v]$, then we can compute the values of G in time $O(nC(n))$. In particular an SSG which can be turned into a DAG by removing one vertex can be solved in time $O(n^2)$.*

Proof. Let v_0 be the value of x in G, which exists since the game is stopping. By Lemma 12 it is clear that the previous algorithm is such that v_0 is in the interval $[min, max]$ at any time. Moreover, by Lemma 1 we know that $v_0 = \frac{a}{b}$ where $b \le 6^{\frac{n_a}{2}}$. At the end of the algorithm, $max - min \le 6^{-n_a}$ therefore there is at most one rational of denominator less than $6^{\frac{n_a}{2}}$ in this interval. It can be found exactly with $O(n_a)$ arithmetic operations by doing a binary search in the Stern-Brocot tree (see for instance [12]).

One last call to $G[v_0]$ gives us all the exact values of G. Since the algorithm stops when $max - min \leq 6^{-n_a}$, we have at most $O(n_a)$ calls to the algorithm solving $G[v]$. All in all the complexity is $O(n_a C(n) + n_a)$ that is $O(n_a C(n))$.

In the case where $G[v]$ is an acyclic graph, we can solve it in linear time which gives us the stated complexity.

4.2   Feedback Vertex Set of Fixed Size

Let G be an SSG such that X is one of its minimal vertex feedback set. Let $k = |X|$. The game is assumed to be stopping. Since the classical transformation [6] into a stopping game does not change the size of a minimal vertex feedback set, it will not change the polynomiality of the described algorithm. However the transformation produces an SSG which is quadratically larger, thus a good way to improve the algorithm we present would be to relax the stopping assumption.

In this subsection we will consider games whose sinks have dyadic values, since they come from the dichotomy of the last subsection. The gcd of the values of the sinks will thus be the maximum of the denominators. The idea to solve G is to get rid of X, one vertex at a time by the previous technique. The only thing we have to be careful about is the precision up to which we have to do the dichotomy, since each step adds a new sink whose value has a larger denominator.

Theorem 8. *There is an algorithm which solves any stopping SSG in time $O(n^{k+1})$ where n is the number of vertices and k the size of the minimal feedback vertex set.*

Proof. First recall that we can find a minimal vertex with an FPT algorithm. You can also check every set of size k and test in linear time whether it is a feedback vertex set. Thus the complexity of finding such a set, that we denote by $X = \{x_1, \ldots, x_k\}$, is at worst $O(n^{k+1})$. Let denote by G_i the game G where x_1 to x_i has been turned into sinks of some values. If we want to make these values explicit we write $G_i[v_1, \ldots, v_i]$ where v_1 to v_i are the values of the sinks.

We now use the algorithm of Theorem 7 recursively, that is we apply it to reduce the problem of solving $G_i[v_1, \ldots, v_i]$ to the problem of solving $G_{i+1}[v_1, \ldots, v_i, v_{i+1}]$ for several values of v_{i+1}. Since G_k is acyclic, it can be solved in linear time. Therefore the only thing we have to evaluate is the number of calls to this last step. To do that we have to explain how precise should be the dichotomy to solve G_i, which will give us the number of calls to solve G_i in function of the number of calls to solve G_{i+1}.

We prove by induction on i that the algorithm, to solve G_i, makes $\log(p_i)$ calls to solve G_{i+1}, where the value v_{i+1} is a dyadic number of numerator bounded by $p_i = 6^{(2^{i+1}-1)n_a}$. Theorem 7 proves the case $i = 0$. Assume the property is proved for $i - 1$, we prove it for i. By induction hypothesis, all the denominators of $v_1, \ldots, v_i$ are power of two and their gcd is bounded by p_i. By Lemma 1, the value of x_i is a rational of the form $\frac{a}{b}$ where $b \leq p_i 6^{\frac{n_a}{2}}$. We have to do the dichotomy up to the square of $p_i 6^{\frac{n_a}{2}}$ to recover the exact value of x_i in the game $G_i(v_1, \ldots, v_{i-1})$. Thus the bound on the denominator of v_{i+1} is $p_{i+1} = p_i^2 6^{n_a}$. That is $p_{i+1} = 6^{2(2^{i+1}-1)n_a} 6^{n_a} = 6^{(2^{i+2}-1)n_a}$, which proves the induction hypothesis. Since we do a dichotomy up to a precision p_{i+1}, the number of calls is clearly $\log(p_{i+1})$.

In conclusion, the number of calls to G_k is

$$\prod_{i=0}^{k-1} \log(6^{(2^{i+1}-1)n_a}) \leq 2^{k^2} \log(6)^k n_a^k.$$

Since solving a game G_k can be done in linear time the total complexity is in $O(n^{k+1})$.

Acknowledgements. This research was supported by grant ANR 12 MONU-0019 (project MARMOTE). Thanks to Yannis Juglaret for being so motivated to learn about SSGs and to Luca de Feo for insights on rationals and their representations.

References

1. Andersson, D., Hansen, K.A., Miltersen, P.B., Sørensen, T.B.: Deterministic graphical games revisited. In: Beckmann, A., Dimitracopoulos, C., Löwe, B. (eds.) CiE 2008. LNCS, vol. 5028, pp. 1–10. Springer, Heidelberg (2008)
2. Andersson, D., Miltersen, P.B.: The complexity of solving stochastic games on graphs. In: Dong, Y., Du, D.-Z., Ibarra, O. (eds.) ISAAC 2009. LNCS, vol. 5878, pp. 112–121. Springer, Heidelberg (2009)
3. Berwanger, D., Dawar, A., Hunter, P., Kreutzer, S., Obdržálek, J.: The dag-width of directed graphs. Journal of Combinatorial Theory, Series B 102(4), 900–923 (2012)
4. Chatterjee, K., Fijalkow, N.: A reduction from parity games to simple stochastic games. In: GandALF, pp. 74–86 (2011)
5. Chen, J., Liu, Y., Lu, S., O'sullivan, B., Razgon, I.: A fixed-parameter algorithm for the directed feedback vertex set problem. Journal of the ACM (JACM) 55(5), 21 (2008)
6. Condon, A.: The complexity of stochastic games. Information and Computation 96(2), 203–224 (1992)
7. Condon, A.: On algorithms for simple stochastic games. Advances in Computational Complexity Theory 13, 51–73 (1993)
8. Fearnley, J.: Exponential lower bounds for policy iteration. In: Abramsky, S., Gavoille, C., Kirchner, C., Meyer auf der Heide, F., Spirakis, P.G. (eds.) ICALP 2010. LNCS, vol. 6199, pp. 551–562. Springer, Heidelberg (2010)
9. Friedmann, O.: An exponential lower bound for the parity game strategy improvement algorithm as we know it. In: 24th Annual IEEE Symposium on Logic In Computer Science, LICS 2009, pp. 145–156. IEEE (2009)

10. Gary, M.R., Johnson, D.S.: Computers and intractability: A guide to the theory of NP-completeness (1979)
11. Gimbert, H., Horn, F.: Simple stochastic games with few random vertices are easy to solve. In: Amadio, R.M. (ed.) FOSSACS 2008. LNCS, vol. 4962, pp. 5–19. Springer, Heidelberg (2008)
12. Graham, K., Knuth, D.E.: Patashnik, concrete mathematics. In: A Foundation for Computer Science (1989)
13. Hoffman, A.J., Karp, R.M.: On nonterminating stochastic games. Management Science 12(5), 359–370 (1966)
14. Juglaret, Y.: Étude des simple stochastic games
15. Ludwig, W.: A subexponential randomized algorithm for the simple stochastic game problem. Information and Computation 117(1), 151–155 (1995)
16. Obdržálek, J.: Clique-width and parity games. In: Duparc, J., Henzinger, T.A. (eds.) CSL 2007. LNCS, vol. 4646, pp. 54–68. Springer, Heidelberg (2007)
17. Shapley, L.S.: Stochastic games. Proceedings of the National Academy of Sciences of the United States of America 39(10), 1095 (1953)
18. Stirling, C.: Bisimulation, modal logic and model checking games. Logic Journal of IGPL 7(1), 103–124 (1999)
19. Tripathi, R., Valkanova, E., Anil Kumar, V.S.: On strategy improvement algorithms for simple stochastic games. Journal of Discrete Algorithms 9(3), 263–278 (2011)

The Parameterized Complexity
of Domination-Type Problems
and Application to Linear Codes

David Cattanéo[1] and Simon Perdrix[2]

[1] LIG UMR 5217, University of Grenoble, France
David.Cattaneo@imag.fr
[2] CNRS, LORIA UMR 7503, Nancy, France
Simon.Perdrix@loria.fr

Abstract. We study the parameterized complexity of domination-type problems. (σ, ρ)-domination is a general and unifying framework introduced by Telle: given $\sigma, \rho \subseteq \mathbb{N}$, a set D of vertices of a graph G is (σ, ρ)-dominating if for any $v \in D$, $|N(v) \cap D| \in \sigma$ and for any $v \notin D, |N(v) \cap D| \in \rho$. Our main result is that for any σ and ρ recursive sets, deciding whether there exists a (σ, ρ)-dominating set of size k, or of size at most k, are both in W[2]. This general statement is optimal in the sense that several particular instances of (σ, ρ)-domination are W[2]-complete (e.g. DOMINATING SET). We prove the W[2]-membership for the *dual* parameterization too, i.e. deciding whether there exists a (σ, ρ)-dominating set of size $n - k$ (or at least $n - k$) is in W[2], where n is the order of the input graph. We extend this result to a class of domination-type problems which do not fall into the (σ, ρ)-domination framework, including CONNECTED DOMINATING SET. We also consider problems of coding theory which are related to domination-type problems with parity constraints. In particular, we prove that the problem of the minimal distance of a linear code over $\mathbb{F}_q$ is in W[2] when q is a power of prime, for both standard and dual parameterizations, and W[1]-hard for the dual parameterization.

To prove the W[2]-membership of the domination-type problems we extend the Turing-way to parameterized complexity by introducing a new kind of non-deterministic Turing machine with the ability to perform 'blind' transitions, i.e. transitions which do not depend on the content of the tapes. We prove that the corresponding problem SHORT BLIND MULTI-TAPE NON-DETERMINISTIC TURING MACHINE is W[2]-complete. We believe that this new machine can be used to prove W[2]-membership of other problems, not necessarily related to domination.

1 Introduction

Domination-Type Problems. Domination problems are central in graph theory. Telle [20] introduced the notion of (σ, ρ)-domination as a unifying framework for many problems of domination: for any two sets of integers σ and ρ, a set D of vertices of a graph G is (σ, ρ)-*dominating* if for any vertex $v \in D, |N(v) \cap D| \in \sigma$ and

T V Gopal et al. (Eds.): TAMC 2014, LNCS 8402, pp. 86–103, 2014.
© Springer International Publishing Switzerland 2014

for any vertex $v \notin D$, $|N(v) \cap D| \in \rho$. Among others, dominating sets, independent sets, and perfect codes are some particular instances of (σ, ρ)-domination. When $\sigma, \rho \in \{\text{ODD}, \text{EVEN}\}$ (where $\text{EVEN} := \{2n, n \in \mathbb{N}\}$ and $\text{ODD} := \mathbb{N} \setminus \text{EVEN}$), (σ, ρ)-domination is strongly related to problems in coding theory such as finding the minimal distance of a linear code [17]. Despite its generality, the (σ, ρ)-domination framework does not capture all the variants of domination. For instance, connected dominating set (i.e. a dominating set which induces a connected subgraph) does not fall into the (σ, ρ)-domination framework.

Parameterized Complexity of Domination-Type Problems. Most of the domination-type problems are NP-hard [20], though some of them are fixed-parameter tractable. We assume the reader is familiar with parameterized complexity and the W-hierarchy, otherwise we refer to [10,12]. The parameterized complexity of domination-type problems has been intensively studied [14,18,19] since the seminal paper by Downey and Fellows [7]. For instance, DOMINATING SET is known to be W[2]-complete [7], whereas INDEPENDENT SET and PERFECT CODE are W[1]-complete [7,3] (see Figure 1 for a list of domination-type problems with their parameterized complexity). Another example is TOTAL DOMINATING SET which is known to be W[2]-hard [1]. Parameterized complexity of domination-type problems with parity constraints – and as a consequence the parameterized complexity of the corresponding problems in coding theory – has been studied in [9]: ODDSET and WEIGHT DISTRIBUTION are W[1]-hard and in W[2], whereas EVENSET and MINIMAL DISTANCE are in W[2]. Additionally to these particular cases of domination-type problems, general results reveal how the parameterized complexity of (σ, ρ)-domination depends on the choice of σ and ρ. For instance, Golovach et al. [14] proved that when $\sigma \subseteq \mathbb{N}$ and $\rho \subseteq \mathbb{N}^+$ are non-empty finite sets, the problem of deciding whether a graph has a (σ, ρ)-dominating set of size greater than a fixed-parameter k is W[1]-complete.

In parameterized complexity, the choice of the parameter is decisive. For all the problems mentioned above the standard parameterization is considered, i.e. the parameter is the size of the solution, i.e. the (σ, ρ)-dominating set. Domination-type problems have also been studied according to the dual parameterization, i.e. the parameter is the size of the (σ, ρ)-*dominated* set. With the dual parameterization, the problem associated with (σ, ρ)-domination is FPT when σ and ρ are either finite or cofinite [14]. As a consequence, INDEPENDENT SET, DOMINATING SET and PERFECT CODE are FPT for the dual parameterization. With parity constraints (i.e. $\sigma, \rho \in \{\text{ODD}, \text{EVEN}\}$), the problem associated with (σ, ρ)-domination has been proved to be W[1]-hard [14] for the dual parameterization. Attention was also paid to the parameterized complexity of (σ, ρ)-domination when parameterized by the tree-width of the graph [6,21].

Our Results. The main result of the paper is that for any σ and ρ recursive sets, (σ, ρ)-domination belongs to W[2] for the *standard* parameterization i.e. (σ, ρ)-dominating set of size k (and at most k).

This general statement is optimal in the sense that problems of (σ, ρ)-domination are known to be W[2]-hard for some particular instances of σ and ρ (e.g.

DOMINATING SET). We also prove that for any σ and ρ recursive sets, (σ, ρ)-domination belongs to W[2] for the *dual* parameterization i.e. (σ, ρ)-dominating set of size $n - k$ (and at least $n - k$). For several particular instances of σ and ρ, the W[2]-membership was unknown: the standard parameterization of TOTAL DOMINATING SET was not known to belong to W[2], and neither did the dual parameterization of (σ, ρ)-domination for $\sigma, \rho \in \{\text{ODD}, \text{EVEN}\}$.

Moreover, we prove that STRONG STABLE SET (known to be in W[1] [14]) is W[1]-complete for the standard parameterization.

We also consider more general problems that do not fall into the (σ, ρ)-domination framework. For any property P and any set ρ of integers, D is a (P, ρ)-dominating set in a graph G if (i) the subgraph induced by D satisfies the property P and (ii) for any vertex $v \notin D$, $|N(v) \cap D| \in \rho$. A connected dominating set corresponds to $\rho = \mathbb{N}^+$ and P being the property that the graph is connected. We prove that the standard parameterization of (P, ρ)-domination is in W[2] i.e. (P, ρ)-dominating set of size k (and at most k) for any P and ρ recursive. As a consequence, CONNECTED DOMINATING SET is W[2]-complete. We also prove that another domination problem, DIGRAPH KERNEL, is W[2]-complete.

Finally, regarding problems in linear coding theory, we show that the dual parameterization of WEIGHT DISTRIBUTION and MINIMAL DISTANCE are both in W[2]. We also consider extensions of these two problems from the field $\mathbb{F}_2$ to $\mathbb{F}_q$ for any power of prime q, and show that WEIGHT DISTRIBUTION OVER $\mathbb{F}_q$ is W[1]-hard and in W[2] for both standard and dual parameterizations; and that MINIMAL DISTANCE OVER $\mathbb{F}_q$ is in W[2] for the standard parameterization, and W[1]-hard and in W[2] for the dual parameterization.

Our contributions are summarized in Figure 1.

Our Approach: Extending the Turing Way to Parameterized Complexity. The Turing way to parameterized complexity [4] consists in solving a problem with a particular kind of Turing machine to prove that the problem belongs to some class of the W-hierarchy. For instance, if a problem can be solved by a single-tape non-deterministic Turing machine in a number of steps which only depends on the parameter, then the problem is in W[1]. The W[1]-membership of PERFECT CODE has been proved using such a Turing machine [3]. When the problem is solved by a multi-tape non-deterministic machine in a number of steps which only depends on the parameter, it proves that the problem is in W[2]. To prove the W[2]-membership of (σ, ρ)-domination for any σ and ρ when parameterized by the size of the solution, we introduce an extension of the multi-tape non-deterministic Turing machine by allowing 'blind' transitions, i.e. transitions which do not depend on the symbols pointed out by the heads. We show that the extra capability of doing blind transitions does not change the computational power of the machine in terms of parameterized complexity by proving that the problem SHORT BLIND MULTI-TAPE TURING MACHINE is W[2]-complete. Blindness of the transitions makes the design of the Turing machine far more easier; moreover it seems that there is no simple and efficient simulation of the blind transitions using the standard Turing machine, even though a (not necessarily simple) efficient simulation exists because of the

W[2]-completeness of SHORT MULTI-TAPE TURING MACHINE. For these reasons, we believe that the blind Turing machine can be used to prove W[2]-membership of other problems, not necessarily related to domination-type problems.

The Paper is Organized as Follows: the next section is dedicated to the introduction of the blind multi-tape Turing machine and the proof that the corresponding parameterized problem is W[2]-complete. In Section 3, several results on the parameterized complexity of (σ, ρ)-domination are given. In Section 4, the parameterized complexity of domination-type problems which do not fall in the (σ, ρ)-domination framework are given. Finally, Section 5 is dedicated to problems from coding theory which are related to domination-type problems with parity conditions.

2 Blind Multi-tape Non-deterministic Turing Machine

A *blind* Turing machine is a Turing Machine able to do 'blind' transitions, i.e. transitions which do not depend on the symbol under the head. Blind transitions are of interest in the multi-tape case when the size of the Turing machine (i.e. the number of defined transitions) matters, since a single blind transition can be seen as a shortcut for up to $|\Gamma|^m$ transitions, where Γ is the alphabet and m the number of tapes. For the description of the transitions of a blind m-tape Turing Machine $M = (Q, \Gamma, \Delta, \Sigma, b, q_I, Q_A)$, we introduce a neutral symbol '$\lrcorner$' and define the transitions as: $\Delta \subseteq \underline{\Gamma}^m \times Q \times \underline{\Gamma}^m \times Q \times \{(-1), 0, (+1)\}^m$, where $\underline{\Gamma} = \Gamma \cup \{\lrcorner\}$. A neutral symbol on the left part means that the transition can be applied whatever the symbol of the alphabet on the corresponding tape is, and a neutral symbol on the right part means that the symbol on the tape is kept. For instance $\langle \lrcorner\ \lrcorner, q, aa, q', 00 \rangle$ is a blind transition of a 2-tape machine which, whatever the symbols under the heads are, changes the internal state q into q' and writes 'a' on both tapes. $\langle \lrcorner^m, q, \lrcorner^m, q, 1^m \rangle$ (where σ^m stands for $\sigma, \ldots, \sigma$, m times) is a blind transition of a m-tape machine which moves all the m heads to the right without modifying the content of the tapes.

The parameterized problem associated with the Blind Multi-Tape Non-Deterministic Turing Machines is defined as:

SHORT BLIND MULTI-TAPE NON-DETERMINISTIC TURING MACHINE COMPUTATION
Input: A blind m-tape non-deterministic Turing Machine M, a word w on the alphabet Σ, an integer k.
Parameter: k.
Question: Is there a computation of M on w that reaches an accepting state in at most k steps?

---

[1] When parameterized by the size of the dominating set, the parameterized complexity of (σ, ODD)-domination (resp. (σ, EVEN)-domination) for $\sigma \in \{\text{ODD}, \text{EVEN}\}$ can be derived from the parameterized complexity of ODDSET (resp. EVENSET) which has been proved in [14].

(σ, ρ)-Domination			
Name	(σ, ρ) Formulation	Standard	Dual
DOMINATING SET	$(\mathbb{N}, \mathbb{N}^+)$	W[2]-complete [7]	FPT [14]
INDEPENDENT SET	$(\{0\}, \mathbb{N})$	W[1]-complete [8]	FPT [14]
PERFECT CODE	$(\{0\}, \{1\})$	W[1]-complete [8,3]	FPT [14]
STRONG STABLE SET	$(\{0\}, \{0,1\})$	**W[1]-complete** (W[1] [14])	FPT [14]
TOTAL DOMINATING SET	$(\mathbb{N}^+, \mathbb{N}^+)$	**W[2]-complete** (W[2]-hard [1])	FPT [14]
(σ,ODD)-DOMINATING SET, $\sigma \in \{\text{ODD}, \text{EVEN}\}$		W[1]-hard, **W[2]**[1]	W[1]-hard [14], **W[2]**
(σ, EVEN)-DOMINATING SET, $\sigma \in \{\text{ODD}, \text{EVEN}\}$		**W[2]**[1]	W[1]-hard [14], **W[2]**
(σ, ρ)-DOMINATING SET, when σ, ρ recursive		**W[2]**	**W[2]**
Other Domination Problems			
CONNECTED DOMINATING SET (Dual: MAXIMAL LEAF SPANNING TREE)		**W[2]-complete** (W[2]-hard [11])	FPT [13]
DIGRAPH KERNEL		**W[2]-complete** (W[2]-hard [16])	Unknown
Problems in Coding Theory			
WEIGHT DISTRIBUTION		W[1]-hard,**W[2]** [9]	W[1]-hard [14], **W[2]**
MINIMUM DISTANCE		**W[2]** [9]	W[1]-hard [14], **W[2]**
WEIGHT DISTRIBUTION OVER $\mathbb{F}_q$, (q power of prime)		**W[1]-hard,W[2]**	**W[1]-hard**, **W[2]**
MINIMUM DISTANCE OVER $\mathbb{F}_q$, (q power of prime)		**W[2]**	**W[1]-hard**, **W[2]**

Fig. 1. Overview of the parameterized complexity of domination-type problems and some problems from coding theory. The 'Standard' column corresponds to a parameterization by the size of the (σ, ρ)-dominating set (or the Hamming weight for the problems in coding theory). In this column we consider the problem of (σ, ρ)-dominating set of size k and at most k except for INDEPENDENT SET and STRONG STABLE SET which are considered for the equality case only. The 'Dual' column corresponds to the dual parameterization, e.g. parameterized by the size of the (σ, ρ)-dominated set for domination-type problems. Our contributions, depicted in bold font, improve the results indicated in parenthesis.

Theorem 1. SHORT BLIND MULTI-TAPE NON-DETERMINISTIC TURING MACHINE COMPUTATION *is complete for* W[2].

Proof. The hardness for W[2] comes from the non-blind case which has been proven to be complete for W[2] [5]. The proof of the W[2]-membership is similar to the non-blind case [4] and consists in a reduction to WEIGHTED WEFT-2 CIRCUIT SATISFIABILITY. This problem consists in deciding whether a weft-2 mixed-type boolean circuit of depth bounded by a function of the parameter k, accepts some input of Hamming weight k. A mixed type circuit is composed of 'small' gates of fan-in ≤ 2 and 'large' AND and OR gates of unbounded fan-in. The weft of the circuit is the maximum number of unbounded fan-in gates on an input/output path.

First, we transform M into a machine which accepts its input in (exactly) k steps iff M accepts its input in at most k steps. To this end, all accepting states of M are merged into a fresh accepting state q_A and the blind transition $\langle \_^m, q_A, \_^m, q_A, 0^m \rangle$ is added.

In the following, a weft-2 mixed circuit C is constructed in such a way that the accepted inputs correspond to the sequences of k transitions of a machine M from the initial state to the accepting state. The set Δ of the transitions of M are indexed by $j \in [1, |\Delta|]$. The symbols of Γ are indexed by $s \in [0, |\Gamma|]$, where 0 is the index of the blank symbol and $|\Gamma|$ is the index of the neutral symbol '$\_$'. Let $x[i, j]$ for $i \in [1, k], j \in [1, |\Delta|]$ and $x[-1, -1]$ be the input wires of the circuit. For $i \in [1, k], j \in [1, |\Delta|]$, $x[i, j]$ is true if and only if the ith transition of the sequence is the transition indexed by j, $x[-1, -1]$ represents the constant 0. The following gates encode some information about the transitions of M: $\forall i \in [1, k], \forall q \in [1, |Q|], \forall s \in [0, |\Gamma|], \forall t \in [1, m], \forall d \in \{-1, 0, 1\}$,

- $\tau_o(i, q)$ outputs true iff the initial state on the ith transition is q:

$$\tau_o(i, q) := \bigvee_{j \in J_q} x[i, j]$$

  where $J_q = \Delta \cap (\underline{\Gamma}^m \times \{q\} \times \underline{\Gamma}^m \times Q \times \{-1, 0, 1\}^m)$
- $\tau_n(i, q)$ outputs true iff the final state on the ith transition is q:

$$\tau_n(i, q) := \bigvee_{j \in J'_q} x[i, j]$$

  where $J'_q = \Delta \cap (\underline{\Gamma}^m \times Q \times \underline{\Gamma}^m \times \{q\} \times \{-1, 0, 1\}^m)$
- $\sigma_o(i, s, t)$ outputs true iff either the symbol read by the ith transition on tape t is s, or the transition does not read the symbol on tape t in the 'blind' case $s = |\Gamma|$:

$$\sigma_o(i, s, t) := \bigvee_{j \in J_{s,t}} x[i, j]$$

  where $J_{s,t} = \Delta \cap (\underline{\Gamma}^{t-1} \times \{s\} \times \underline{\Gamma}^{m-t} \times Q \times \underline{\Gamma}^m \times Q \times \{-1, 0, 1\}^m)$

– $\sigma_n(i, s, t)$ outputs true iff either the symbol written by the ith transition on tape t is s, or the transition does not write any symbol on tape t in the 'blind' case $s = |\Gamma|$:

$$\sigma_n(i, s, t) := \bigvee_{j \in J'_{s,t}} x[i, j]$$

where $J'_{s,t} = \Delta \cap (\underline{\Gamma}^m \times Q \times \underline{\Gamma}^{t-1} \times \{s\} \times \underline{\Gamma}^{m-t} \times Q \times \{\text{-}1, 0, 1\}^m)$

– $\mu(i, d, t)$ outputs true iff the head of t has a movement d on the ith transition:

$$\mu(i, d, t) := \bigvee_{j \in J_{d,t}} x[i, j]$$

where $J_{d,t} = \Delta \cap (\underline{\Gamma}^m \times Q \times \underline{\Gamma}^m \times Q \times \{\text{-}1, 0, 1\}^{t-1} \times \{d\} \times \{\text{-}1, 0, 1\}^{m-t})$

Notice that most of these gates require unbounded fan-in OR gates in general.

The following gates encode the position of the heads and all the symbols in every cell of the tapes. These gates guarantee the correctness of the transition sequence. $\forall i \in [1, k], \forall l \in [\text{-}k, k], \forall t \in [1, m], \forall s \in [0, |\Gamma| - 1]$,

– $\beta(i, l, t)$ outputs true iff the head of tape t is at position l before step i. Since the transition sequence is of length k, l is in the interval $[\text{-}k, k]$. The gate is defined as:

$$\beta(0, l, t) := \begin{cases} 1 & \text{if } l = 0 \\ 0 & \text{otherwise} \end{cases}$$

$$\begin{aligned} \beta(i, l, t) := &(\beta(i{-}1, l, t) \wedge \mu(i{-}1, 0, t)) \\ &\vee (\beta(i{-}1, l{-}1, t) \wedge \mu(i{-}1, 1, t)) \\ &\vee (\beta(i{-}1, l{+}1, t) \wedge \mu(i{-}1, \text{-}1, t)) \end{aligned}$$

– $\sigma(i, l, s, t)$ outputs true iff the cell l of tape t contains the symbol s before step i. Let w be the input word of the machine, located on tape 1.

$$\sigma(0, l, s, t) := \begin{cases} 1 & \text{if } ((s \text{ is the index of } w[l]) \wedge (t = 1) \wedge (0 \leq l < |w|)) \\ 1 & \text{if } ((s = 0) \wedge (t \neq 1 \vee l < 0 \vee l \geq |w|)) \\ 0 & \text{otherwise} \end{cases}$$

$$\begin{aligned} \sigma(i, l, s, t) := &(\neg\beta(i - 1, l, t) \wedge \sigma(i - 1, l, s, t)) \\ &\vee (\beta(i - 1, l, t) \wedge \sigma_n(i - 1, s, t)) \\ &\vee (\beta(i - 1, l, t) \wedge \sigma_n(i - 1, |\Gamma|, t) \wedge \sigma(i - 1, l, s, t)) \end{aligned}$$

One can see in the definition of $\sigma(i, l, s, t)$ for $i > 0$ that there are three different cases: either the head was not pointing at the cell l, so the symbol remains unchanged; or the head was pointing at the cell l, and the symbol has been written in the previous step; or the head was on the cell but the transition was blind, so the symbol was already s.

Notice that these gates have a bounded fan-in, and that the recursion is on the number of transitions, so their depth is bounded by the parameter k. Notice also that there is a polynomial number of such gates since there are $k \cdot 2k \cdot m$, β gates and $k \cdot 2k \cdot |\Gamma| \cdot m$, σ gates.

All the information about the computation path has been encoded so the remaining gates check the validity of this transition sequence:

- $E := E_0 \wedge E_1 \wedge E_2 \wedge E_3 \wedge E_4$ is the final gate of the circuit. As a consequence, for any input accepted by the circuit, the following conditions $E_0, \ldots, E_4$ must be satisfied.
- $E_0 := \neg x[-1,-1]$ ensures that $x[-1,-1]$ is the constant 0, so $\neg x[-1,-1]$ is the constant 1 used by the other gates.
- E_1 ensures that for every i, at most one wire among the block $x[i,1], \ldots, x[i,|\Delta|]$ is true, which means that at each step at most one transition is performed. E_1 is defined as:

$$E_1 := \bigwedge_{i=1}^{k} \bigwedge_{j=1}^{|\Delta|} \bigwedge_{j'=1, j' \neq j}^{|\Delta|} (\neg x[i,j] \vee \neg x[i,j'])$$

- E_2 ensures that the initial state of each step is equal to the final state of the previous step. E_2 is defined as:

$$E_2 := \bigwedge_{i=2}^{k} \bigwedge_{q=1}^{|Q|} (\neg \tau_n(i-1,q) \vee \tau_o(i,q))$$

Notice that this formula encodes: $\forall i \in [2,k], \forall q \in [1,|Q|], \tau_n(i-1,q) \Rightarrow \tau_o(i,q)$.
- E_3 ensures that the symbol read by a transition on a tape is either the one pointed out by the head or any symbol when the transition is blind.

$$E_3 := \bigwedge_{i=1}^{k} \bigwedge_{t=1}^{m} \bigwedge_{l=-k}^{k} \bigwedge_{s=0}^{|\Gamma|-1} (\neg \beta(i,l,t) \vee \neg \sigma(i,l,s,t) \vee \sigma_o(i,s,t) \vee \sigma_o(i,|\Gamma|,t))$$

Notice that this formula encodes: $\forall i \in [1,k], \forall l \in [-k,k], \forall s \in [0,|\Gamma|], \forall t \in [1,m], (\beta(i,l,t) \wedge \sigma(i,l,s,t)) \Rightarrow (\sigma_o(i,s,t) \vee \sigma_o(i,|\Gamma|,t))$.
- E_4 ensures that the initial state on the first step is q_0, the initial state of M of index 0, and that the last state is the accepting state q_A of index $|Q|-1$. So E_4 is defined as:

$$E_4 := \tau_o(0,0) \wedge \tau_n(k-1,|Q|-1)$$

All the $E_i, i \in [0,4]$ gates are independent, so every input-output path goes through at most one of these unbounded fan-in gates. Since it is also the case for the gates encoding the transitions, and that the σ and β gates are bounded fan-in gates, the weft of this circuit is 2. Since the only recursive gates have a depth bounded by the parameter, the depth of this circuit is bounded by the parameter. Notice also that the number of gates is polynomial in $|M|$. This circuit outputs true if and only if M has an accepting computation path of length k on the word w, i.e. if and only if M has an accepting computation path of length at most k on w. Therefore, SHORT BLIND MULTI-TAPE NON-DETERMINISTIC TURING MACHINE COMPUTATION belongs to W[2]. □

What is interesting is that although the blindness of the transition does not change the computational power of short multi-tape non-deterministic Turing machines, there is no simple way to simulate the blind machine with the original one. Indeed, intuitively, a blind transition on m tapes is a short-cut for up to $|\Gamma|^m$ transitions, so a machine with no blind transition may have an exponentially larger size. A tape-by-tape (sequentialization) simulation would avoid this blow up of the number of transitions, but will not reach an accepting state within a number of steps depending only on the parameter.

3   Parameterized Complexity of (σ, ρ)-Domination

In this section, we prove the central result of the paper: for any recursive sets σ and ρ, (σ, ρ)-domination belongs to W[2] for both *standard* and *dual* parameterizations, i.e. the four problems which consists in deciding whether a graph has a (σ, ρ)-dominating set of size k; of size $n - k$; of size at most k; and of size at least $n - k$ are in W[2] with respect to k. To this end, we show that for any σ, ρ recursive sets, these problems of (σ, ρ)-domination can be decided using a blind multi-tape Turing machine. The only assumption on σ and ρ is that they are recursive, i.e. there exists a Turing machine which decides whether a given integer j belongs to σ (resp. ρ).

(σ, ρ)-DOMINATING SET OF SIZE AT MOST k:
Input: A graph $G = (V, E)$, an integer k.
Parameter: k.
Question: Is there a (σ, ρ)-dominating set $D \subseteq V$ such that $|D| \le k$?

(σ, ρ)-DOMINATING SET OF SIZE k, (σ, ρ)-DOMINATING SET OF SIZE $n-k$, and (σ, ρ)-DOMINATING SET OF SIZE AT LEAST $n - k$ are defined likewise.

Theorem 2. *For any recursive sets of integers σ and ρ, (σ, ρ)-DOMINATING SET OF SIZE AT MOST k and (σ, ρ)-DOMINATING SET OF SIZE k belong to W[2].*

Proof. We prove that (σ, ρ)-DOMINATING SET OF SIZE AT MOST k is in W[2], the proof that (σ, ρ)-DOMINATING SET OF SIZE k belongs to W[2] is similar. Given two recursive sets $\sigma, \rho \subseteq \mathbb{N}$, an integer k, and a graph $G = (\{v_1, \ldots, v_n\}, E)$, we consider the following $(n+1)$-tape Turing machine M, which decides whether G has a (σ, ρ)-dominating set of size at most k. M works in 3 phases (see an example in Figure 2): (1) a subset D of size at most k is non-deterministically chosen and written on the first tape. Moreover, the first $k + 1$ cells of the following n tapes – one tape for each vertex of the graph – are filled with 0s and 1s such that the ith cell of each tape is 1 iff $i \in \rho$; (2) The content of the tapes associated with the vertices in D is removed and replaced by the characteristic vector of σ, i.e. the ith cell is 1 iff $i \in \sigma$. At the end of this second phase, all heads are located on the leftmost non-blank symbol; (3) For each vertex v in D, the heads of all the tapes associated with a neighbor of v move to the right. At the end of this third phase, for every $v \in D$ (resp. $v \in \overline{D}$), the head of the tape associated with v reads 1 iff $|N(v) \cap D| \in \sigma$ (resp. $|N(v) \cap D| \in \rho$), so D is a (σ, ρ)-dominating set iff all heads but the first one read a symbol 1.

The actual description of the blind $(n+1)$-tape non-deterministic Turing machine is as follows: $M = (Q, \Gamma, \Delta, \Sigma, b, q_I, Q_A)$, where $\Gamma = \{\square, 0, 1, v_1, \ldots, v_n\}$, $b = \square$, $\Sigma = \varnothing$, $Q = \{q_{r,s} \mid r \in [1, n+1], s \in [0, k]\} \cup \{q_s^{\text{ret}} \mid s \in [1, k+1]\} \cup \{q_{i,s}^{\text{sig}} \mid i \in [1, n], s \in [0, k]\} \cup \{q^{\text{sig}}, q_\rho^{\text{end}}, q_\sigma^{\text{end}}, q^{\text{read}}, q_A\}$, $q_I = q_{1,0}$ and $Q_A = \{q_A\}$. Given an integer set A, $\overline{A}$ is the complementary set of A, it is defined as the only set such that $A \cap \overline{A} = \varnothing$ and $A \cup \overline{A} = \mathbb{N}$. The initial word w is the empty word, so every cell initially contains the blank symbol $\square$. The transitions are:

Phase 1 – Initialization of D and ρ:

$$\langle \square\square^n, q_{r,s}, v_i 1^n, q_{i+1,s+1}, (+1)(+1)^n \rangle \quad r \in [1, n], s \in \rho \cap [0, k-1], i \in [r, n]$$
$$\langle \square\square^n, q_{r,s}, v_i 0^n, q_{i+1,s+1}, (+1)(+1)^n \rangle \quad r \in [1, n], s \in \overline{\rho} \cap [0, k-1], i \in [r, n]$$
$$\langle \square\square^n, q_{r,s}, \square 1^n, q_{r,s+1}, 0(+1)^n \rangle \quad r \in [1, n+1], s \in \rho \cap [0, k-1]$$
$$\langle \square\square^n, q_{r,s}, \square 0^n, q_{r,s+1}, 0(+1)^n \rangle \quad r \in [1, n+1], s \in \overline{\rho} \cap [0, k-1]$$
$$\langle \square\square^n, q_{r,k}, \square 1^n, q_\rho^{\text{end}}, (-1)(-1)^n \rangle \quad r \in [1, n+1], \text{ if } k \in \rho$$
$$\langle \square\square^n, q_{r,k}, \square 0^n, q_\rho^{\text{end}}, (-1)(-1)^n \rangle \quad r \in [1, n+1], \text{ if } k \in \overline{\rho}$$
$$\langle v_i \_^n, q_\rho^{\text{end}}, v_i \_^n, q_\rho^{\text{end}}, (-1)(-1)^n \rangle \quad i \in [1, n]$$
$$\langle \square 1^n, q_\rho^{\text{end}}, \square 1^n, q_\rho^{\text{end}}, 0(-1)^n \rangle$$
$$\langle \square 0^n, q_\rho^{\text{end}}, \square 0^n, q_\rho^{\text{end}}, 0(-1)^n \rangle$$
$$\langle \square\square^n, q_\rho^{\text{end}}, \square\square^n, q^{\text{sig}}, (+1)(+1)^n \rangle$$

The state $q_{r,s}$ means that $s-1$ vertices among $v_1, \ldots, v_{r-1}$ have already been written on the first tape, q_ρ^{end} that the initializations of D and ρ are done and that the heads are going back to the leftmost non blank cell on every tape.

Phase 2 – Initialization of σ:

$$\langle v_i \_^n, q^{\text{sig}}, v_i \_^n, q_{i,0}^{\text{sig}}, 00^n \rangle \quad i \in [1, n]$$
$$\langle v_i \_^n, q_{i,s}^{\text{sig}}, v_i \_^{i-1} 1 \_^{n-i}, q_{i,s+1}^{\text{sig}}, 0(+1)^n \rangle \quad i \in [1, n], s \in \sigma \cap [0, k-1]$$
$$\langle v_i \_^n, q_{i,s}^{\text{sig}}, v_i \_^{i-1} 0 \_^{n-i}, q_{i,s+1}^{\text{sig}}, 0(+1)^n \rangle \quad i \in [1, n], s \in \overline{\sigma} \cap [0, k-1]$$
$$\langle v_i \_^n, q_{i,k}^{\text{sig}}, v_i \_^{i-1} 1 \_^{n-i}, q_1^{\text{ret}}, (+1)0^n \rangle \quad i \in [1, n], \text{ if } k \in \sigma$$
$$\langle v_i \_^n, q_{i,k}^{\text{sig}}, v_i \_^{i-1} 0 \_^{n-i}, q_1^{\text{ret}}, (+1)0^n \rangle \quad i \in [1, n], \text{ if } k \in \overline{\sigma}$$
$$\langle v_i \_^n, q_s^{\text{ret}}, v_i \_^n, q_{s+1}^{\text{ret}}, 0(-1)^n \rangle \quad i \in [1, n], s \in [1, k]$$
$$\langle v_i \_^n, q_{k+1}^{\text{ret}}, v_i \_^n, q_{i,0}^{\text{sig}}, 00^n \rangle \quad i \in [1, n]$$
$$\langle \square \_^n, q_1^{\text{ret}}, \square \_^n, q_\sigma^{\text{end}}, (-1)0^n \rangle$$
$$\langle v_i \_^n, q_\sigma^{\text{end}}, v_i \_^n, q_\sigma^{\text{end}}, (-1)0^n \rangle \quad i \in [1, n]$$
$$\langle \square \_^n, q_\sigma^{\text{end}}, \square \_^n, q^{\text{read}}, (+1)0^n \rangle$$

The state $q_{s,i}^{\text{sig}}$ means that the first s symbols of the characteristic vector of σ have been written on the tape associated with the vertex v_i.

Phase 3: Neighborhood Checking

$$\langle v_i\_^n, q^{\text{read}}, v_i\_^n, q^{\text{read}}, (+1)d_1 \dots d_n \rangle \qquad i \in [1,n], \text{ where } d_t = \begin{cases} +1 & \text{if } v_t \in N(v_i) \\ 0 & \text{otherwise} \end{cases}$$

$$\langle \Box 1^n, q^{\text{read}}, \Box 1^n, q_A, 00^n \rangle$$

Since σ and ρ are recursive, their characteristic vector of length k can be computed and written on the tapes in time $f(k)$ for some fixed function f. In the first phase D and ρ of size k are written on the tapes and then the heads comes back so there are $2(k+1)$ steps. In the second phase σ of size k is written sequentially on at most k tapes corresponding to the elements of D and then the heads come back so there are at most $k(2k)$ steps. Finally the third phase goes through D and moves the heads on the tapes of the neighbours so there are at most k steps. The number of transitions is polynomial in $|G|$ and the acceptance is made in at most $2(k+1) + k(2k+2)$ steps if a (σ, ρ)-dominating set of size at most k exists. As a consequence, (σ, ρ)-DOMINATING SET OF SIZE AT MOST k belongs to W[2]. Notice that the use of blind transitions in the third phase is crucial. Indeed, a naive simulation of any of these blind transitions uses 2^n non-blind transitions since the transition should be applicable for any of the 2^n possible configurations read by the heads of the machine.   □

Theorem 3. *For any recursive sets of integers σ and ρ, (σ, ρ)-DOMINATING SET OF SIZE AT LEAST $n - k$ and (σ, ρ)-DOMINATING SET OF SIZE $n - k$ belong to* W[2].

Proof. We prove that (σ, ρ)-DOMINATING SET OF SIZE AT LEAST $n - k$ is in W[2], the proof that (σ, ρ)-DOMINATING SET OF SIZE $n - k$ belongs to W[2] is similar. To decide whether a given graph G has a (σ, ρ)-dominating set of size at least $n-k$, we slightly modify the blind Turing machine used in the proof of Theorem 2 in such a way that at the end of phase (2), the first tape contains the description of a set D of size at most k, and for any $v \in D$ (resp. $v \notin D$), the ith cell of the tape associated with v is 1 if $\delta(v)-i \in \rho$ (resp. $\delta(v)-i \in \sigma$) and 0 otherwise, where $\delta(v)$ is the degree of v. Therefore, the machine reaches the accepting state if there exists a set D of size at most k such that $\forall v \in D$, $\delta(v) - |N(v) \cap D| \in \rho$ and $\forall v \in V \setminus D$, $\delta(v) - |N(v) \cap D| \in \sigma$. Since for any $v \in V$, $|N(v) \cap (V \setminus D)| = \delta(v) - |N(v) \cap D|$, $V \setminus D$ is a (σ, ρ)-dominating set of size at least $n-k$.   □

For any recursive sets σ and ρ, (σ, ρ)-domination problems are in W[2], but for some particular instances of σ and ρ this general result can be refined. In particular, we show that when $\sigma = \{0\}$ and $\rho = \{0, 1\}$, the problem is W[1]-complete:

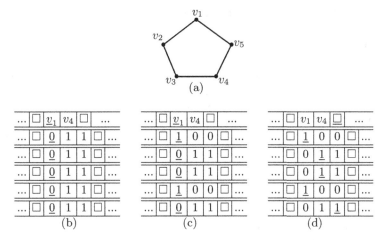

Fig. 2. Computation of $(\{0\}, \mathbb{N}^+)$-DOMINATING SET OF SIZE AT MOST k on a blind multitape Turing machine whith $k = 2$ on C_5 (see proof of Theorem 2). (a) Input graph; (b) State of the machine at the end of phase (1). The candidate set D is on the first tape, the other tapes are initialized according to ρ; (c) End of phase (2): the tapes associated with vertices in D are now initialized according to σ; (d) End of phase (3): all heads (underlined symbols) read 1, so $\{v_1, v_4\}$ is a $(\{0\}, \mathbb{N}^+)$-dominating set.

STRONG STABLE SET $((\{0\}, \{0, 1\})$-DOMINATION):
Input: A graph $G = (V, E)$, an integer k.
Parameter: k.
Question: Is there an independent set $S \subseteq V$ of size k such that $\forall v \in V \setminus S, |N(v) \cap S| \leq 1$?

Theorem 4. STRONG STABLE SET *is complete for* $W[1]$.

Proof. The $W[1]$-membership is an application of Theorem 8 in [14]. We prove the hardness by a reduction from INDEPENDENT SET which is complete for $W[1]$ [7]. Given an instance $(G = (V, E), k)$ of INDEPENDENT SET, we consider the instance (G', k) of STRONG STABLE SET where $G' = (V', E')$ with $V' = V \cup E$ and $E' = \{(u, e) \mid e$ incident to u in $G\} \cup (E \times E)$. By construction, G' consists of a stable set V and a clique E, the edges between these two sets representing the edges of G. Let S be an independent set in G, then by construction, S is a strong stable set in G'. Let S' be a strong stable set of size k in G'. Since E is a clique, $|S' \cap E| \in \{0, 1\}$. If $|S' \cap E| = 0$, then $S' \subseteq V$ and for any $u, v \in S'$, they have no common neighbor in G', so there is no edge between u and v in G, so S' is an independent set in G. Otherwise, if $|S' \cap E| = 1$ then every $u \in S' \cap V$ is isolated in G', so there are at least $k - 1$ isolated vertices in G. Since E is not empty there also exist non isolated vertices and we can take at least one of them to form together with the $k - 1$ isolated vertices, an independent set of size k in G. $\square$

4   Other Domination Problems

Some natural domination problems cannot be described in terms of (σ, ρ)-domination such as CONNECTED DOMINATING SET. In this section, we show that the proof of the (σ, ρ)-domination W[2]-membership (Theorem 2) can be generalized to (P, ρ)-domination, where P is no longer a domination constraint but any recursive property. It implies that CONNECTED DOMINATING SET, known to be hard for W[2], is actually complete for W[2]. We also show that this technique can be applied to digraph problems with the example of DIGRAPH KERNEL.

(P, ρ)-DOMINATING SET OF SIZE AT MOST k:
Input: A graph $G = (V, E)$, an integer k.
Parameter: k.
Question: Is there a subset $D \subseteq V$ such that $|D| \leq k$ and:
– the sub-graph of G induced by D satisfies the property P;
– $\forall v \in V \setminus D, |N(v) \cap D| \in \rho$?

Theorem 5. *If ρ is a recursive set of integers and P is a recursive property, then (P, ρ)-DOMINATING SET OF SIZE AT MOST k belongs to W[2].*

Proof. We use the blind multitape Turing machine of Theorem 2 with $\sigma = \mathbb{N}$, which outputs a $(\mathbb{N}, \rho)$-dominating set D if it exists, then we compose this machine with another one which decides whether such a set D induces a subgraph satisfying the property P. Since the subgraph is of size $O(k^2)$ and P is recursive, the computation time of the second machine is $f(k)$ for some function f.   □

DIGRAPH KERNEL:
Input: A directed graph $G = (V, A)$, an integer k.
Parameter: k.
Question: Is there a kernel of D of size at most k? A *kernel* is an independent set S (there exists no $u, v \in S$ such that uv or vu is in A) such that for every vertex $x \in V \setminus S$, there exists $y \in S$ such that $xy \in A$.

Theorem 6. DIGRAPH KERNEL *is complete for* W[2].

Proof. The hardness for W[2] is proved in [16]. The proof of the membership is very similar to the W[2] membership of (σ, ρ)-DOMINATING SET (Theorem 2). The machine and the initialization are the same, with $\sigma = \{0\}$ and $\rho = \mathbb{N}^+$. In phase (3), only the heads of the tapes associated with *incoming neighbors* move to the right.   □

5   Problems from Coding Theory

Parameterized complexity of problems from coding theory, in particular MINIMAL DISTANCE and WEIGHT DISTRIBUTION, have been studied in [9]. We prove that the dual parameterizations of these problems are in W[2]. Moreover, we consider extensions of these problems to linear codes over $\mathbb{F}_q$ for any q power of prime.

MINIMAL DISTANCE OVER $\mathbb{F}_q$:
Input: q a power of prime, k an integer, an $m \times n$ matrix H with entries in $\mathbb{F}_q$.
Parameters: k, q.
Question: Is there a linear combination of at least one and at most k columns of H which is equal to the all-zero vector?

WEIGHT DISTRIBUTION OVER $\mathbb{F}_q$:
Input: q a power of prime, k an integer, an $m \times n$ matrix H with entries in $\mathbb{F}_q$.
Parameters: k, q.
Question: Is there a linear combination of exactly k columns of H which is equal to the all-zero vector?

Theorem 7. WEIGHT DISTRIBUTION OVER $\mathbb{F}_q$ *is hard for* W[1] *and belongs to* W[2], *and* MINIMAL DISTANCE OVER $\mathbb{F}_q$ *belongs to* W[2].

Proof. Since WEIGHT DISTRIBUTION is a particular case of WEIGHT DISTRIBUTION OVER $\mathbb{F}_q$, with $q = 2$, WEIGHT DISTRIBUTION OVER $\mathbb{F}_q$ is hard for W[1] [9]. For the W[2] membership, let $\psi : [0, q) \to \mathbb{F}_q$ be an arbitrary indexing of the elements of $\mathbb{F}_q$ s.t. $\psi(0) = 0$. There exist a prime p and an integer c such that $q = p^c$, and there is an isomorphism $\varphi : \mathbb{F}_q \to F_p[X]/P(X)$, where $F_p[X]/P(X)$ is the set of polynomials in X with coefficients in $\mathbb{F}_p$ modulo $P(X)$. Let H' be a $mc \times (n(q-1))$-matrix over $\mathbb{F}_p$ such that $\forall i, j, \ell \in [0, m) \times [0, n) \times [1, q)$, $\sum_{u=0}^{c-1} H'_{it,j\ell} X^t = \varphi(\psi(\ell) \cdot H_{i,j})$. Intuitively, each of the $n(q - 1)$ columns of H' corresponds to one column of H multiplied by a non-zero element of $\mathbb{F}_q$.

Moreover any element $a \in \mathbb{F}_q$ is encoded using a $c \times 1$-block $\begin{pmatrix} r_0 \\ \vdots \\ r_{c-1} \end{pmatrix}$ such that $\varphi(a) = \sum_{t=0}^{c-1} r_t X^t$. It leads to the $mc \times (n(q-1))$-matrix H' which can be computed in time $m.n.f(q)$ for some function f.

Notice that there exists a linear combination of k columns of H which is equal to 0 if and only if there exist $0 \le i_1 < i_2 < \ldots < i_k < m(q - 1)$ such that the corresponding columns of H' sums to 0 (i.e. $\forall j \in [0, mc), \sum_{r=1}^{k} H'_{j,i_r} = 0$) and $\forall r \in [1, k), \lfloor \frac{i_r}{m} \rfloor \ne \lfloor \frac{i_{r+1}}{m} \rfloor$. The last condition guarantees that the k chosen columns in H' correspond to actually k distinct columns in H.

To decide whether such $i_1, \ldots, i_k$ exists we use the following blind $(mc + 1)$-tape Turing Machine $M = (Q, \Gamma, \Delta, \Sigma, b, q_I, Q_A)$. The first tape is associated with the set of columns of H' and each of the remaining tape is associated with a row of H'. The alphabet is $\Gamma = \{\square, 0, 1\} \cup \{h_i | i \in [1, n]\}$ and the states are $Q = \{q_{i,s} \mid i \in [1, n(q-1)+1], s \in [0, k \cdot p]\} \cup \{q_s^{ret} \mid s \in [1, k \cdot p + 1]\} \cup \{q_{i,s}^{av} \mid i \in [1, n], s \in [0, p-1]\} \cup \{q^{read}, q_A\}$, with $q_I = q_{1,0}$, $b = \square$, $\Sigma = \varnothing$ and $Q_A = \{q_A\}$. The transitions are separated in two phases:

Phase 1 - Initialization: First, k columns of H' are non-deterministically chosen on the first tape, while of the other tapes is initialized with k times the pattern 10^{p-1} (i.e. 1 followed by $p-1$ times 0), such that the ith cell is 1 iff $i \equiv 0 \mod p$. In order to avoid choosing two columns of H' corresponding to the same column of H but with a different factor, we go strait to the next block of columns, i.e. when a column j is chosen, the next column is chosen in among the columns indexed from ℓ to $n(q-1)$ with $\ell > j$ and $\ell \equiv 0 \bmod (q-1)$:

$\langle \square\square^{m\cdot c}, q_{i,s}, h_j 1^{m\cdot c}, q_{\ell,s+1}, (+1)(+1)^{m\cdot c}\rangle$
$\quad i\in[1, n(q-1)], s\in[0, k-1], j\in[i, n(q-1)],$ if $s\equiv 0 (mod\ p)$
$\quad \ell$ is the smallest integer such that $\ell > j$ and $\ell\equiv 0(mod\ q-1)$

$\langle \square\square^{m\cdot c}, q_{i,s}, h_j 0^{m\cdot c}, q_{\ell,s+1}, (+1)(+1)^{m\cdot c}\rangle$
$\quad i\in[1, n(q-1)], s\in[0, k-1], j\in[i, n],$ if $s\not\equiv 0 (mod\ p)$
$\quad \ell$ is the smallest integer such that $\ell > j$ and $\ell\equiv 0(mod\ q-1)$

$\langle \square\square^{m\cdot c}, q_{i,s}, \square 1^{m\cdot c}, q_{i,s+1}, 0(+1)^{m\cdot c}\rangle$
$\quad i\in[1, n(q-1)+1], s\in[k, kp),$ if $s\equiv 0 (mod\ p)$

$\langle \square\square^{m\cdot c}, q_{i,s}, \square 0^{m\cdot c}, q_{i,s+1}, 0(+1)^{m\cdot c}\rangle$
$\quad i\in[1, n+1], s\in[k, kp),$ if $s\not\equiv 0 (mod\ p)$

$\langle \square\square^{m\cdot c}, q_{i,k\cdot p}, \square 1^{m\cdot c}, q_1^{ret}, (-1)(-1)^{m\cdot c}\rangle \quad i\in[1, n(q-1)+1]$

$\langle \llcorner\lrcorner^{m\cdot c}, q_s^{ret}, \llcorner\lrcorner^{m\cdot c}, q_{s+1}^{ret}, (-1)(-1)^{m\cdot c}\rangle \quad s\in[1, k]$

$\langle \llcorner\lrcorner^{m\cdot c}, q_s^{ret}, \llcorner\lrcorner^{m\cdot c}, q_{s+1}^{ret}, 0(-1)^{m\cdot c}\rangle \qquad s\in[k+1, kp+1]$

$\langle \llcorner\lrcorner^{m\cdot c}, q_{k\cdot p+1}^{ret}, \llcorner\lrcorner^{m\cdot c}, q^{read}, 00^{m\cdot c}\rangle$

Phase 2 - Recognition: In order to check that the sum of those columns is the all-zero vector on $\mathbb{F}_p$, for any column h_i in the chosen set, the head of each tape j moves to the right $H'_{i,j}$ times using blind transitions.

$\langle h_i \llcorner^{mc}, q^{read}, h_i \llcorner^{mc}, q_{i,1}^{av}, (+1)0^{mc}\rangle \qquad i\in[1, n]$

$\langle \llcorner\lrcorner^{mc}, q_{i,s}^{av}, \llcorner\lrcorner^{mc}, q_{i,s+1}^{av}, 0d_1\ldots d_{mc}\rangle \quad i\in[1, n], s\in[0, p-2]$

$$\text{with } \forall j\in[1, mc], d_j = \begin{cases} 1 & \text{if } H'_{i,j} > s \\ 0 & \text{otherwise} \end{cases}$$

$\langle \llcorner\lrcorner^m, q_{i,p-1}^{av}, \llcorner\lrcorner^{mc}, q^{read}, 00^{mc}\rangle \qquad i\in[1, n]$

$\langle \square 1^{mc}, q^{read}, \square 1^{mc}, q_A, 00^{mc}\rangle$

In the first phase a set D of columns is non-deterministically chosen on the first tape and on each of the remaining tapes, kp cells are filled with 0 or 1 depending on the rest modulo p of their position. Then all the heads move back to leftmost non blanc symbol. Notice that the columns in D are chosen to guarantee that $\forall i \neq i' \in D$, $\lfloor \frac{i}{m} \rfloor \neq \lfloor \frac{i'}{m} \rfloor$. In the second phase, the sum of the columns in D is computed by moving the heads of the tapes to the right. The machine accepts iff at the end all the heads (but the first one) point out a symbol 0, i.e. the sum of all the columns in D of H' is the zero vector. Regarding the number of transitions, in the first phase there are $2kp$ transitions and at most kp in the second phase. Moreover the size of the machine is polynomial in n, m, q and k. As a consequence WEIGHT DISTRIBUTION is in W[2].

The proof of W[2]-membership for MINIMAL DISTANCE is the similar, except that D is chosen of size at most k. □

DUAL MINIMAL DISTANCE OVER $\mathbb{F}_q$:
Input: q a power of prime, k an integer, an $m \times n$ matrix H with entries in $\mathbb{F}_q$.
Parameters: k, q.
Question: Is there a linear combination of at least $n - k$ columns of H equal to the all-zero vector?

DUAL WEIGHT DISTRIBUTION OVER $\mathbb{F}_q$:
Input: q a power of prime, k an integer, an $m \times n$ matrix H with entries in $\mathbb{F}_q$.
Parameters: k, q.
Question: Is there a linear combination of exactly $n - k$ columns of H equal to the all-zero vector?

Theorem 8. DUAL MINIMUM DISTANCE OVER $\mathbb{F}_q$ *and* DUAL WEIGHT DISTRIBUTION OVER $\mathbb{F}_q$ *are in* W[2].

Proof. First we execute the same FPT preprocessing as in standard parameterization (Theorem 7) to get the matrix H' over $\mathbb{F}_p$ where p is the characteristic of $\mathbb{F}_q$. Let the vector v be the sum of all the columns of H', and notice that there is set D of at $n - k$ columns that sum to the zero vector iff the sum of all the columns but those in D sum to v. To this end we consider the matrix $\tilde{H} = (-v|H')$ and we slightly modify the machine used in Theorem 7 to decide whether the exists a set of at most $k+1$ columns which includes the first column and which sum to 0. So the phase 1 is modified to force the set of chosen columns to include the first column of $\tilde{H}$. The proof that DUAL WEIGHT DISTRIBUTION OVER $\mathbb{F}_q$ is in W[2] is the same except that the size of S' is fixed to k. □

Theorem 8 shows that the problem MINIMAL DISTANCE OVER $\mathbb{F}_q$ with q power of prime, which consists in deciding whether there exists a subset of at most k columns of a matrix H with entries in $\mathbb{F}_q$ that sum to the all-zero vector is in W[2]. We can prove similarly the W[2]-membership of the problem which consists in deciding whether there exists a set of at most k columns that sum to a given vector. However it is not clear whether the problem which consists in deciding the existence of a set of at most k columns that sum to a vector with no zero entry (or equivalently to a vector of maximal Hamming weight). To be more precise when q is prime one can use the same machine as in proof of Theorem 7 and change the last transition to check that non of the entries is 0, but this technique fails when q is not prime (say $q = p^2$).

6   Conclusion and Perspectives

We have demonstrated several results on the parameterized complexity of domination-type problems, including that for any (recursive) σ and ρ, (σ, ρ)-domination is in W[2] for both standard and dual parameterizations i.e. (σ, ρ)-dominating set of size k (and at most k) and (σ, ρ)-dominating set of size $n - k$ (and at least $n - k$). To this end, we have extended the Turing way to parameterized complexity with a new way to prove W[2]-membership using 'blind' Turing machines. We believe that this machine can be used to prove W[2]-membership of other problems, not necessarily related to domination.

Several questions remain open. First, the long-standing question regarding the W[1]-hardness of MINIMAL DISTANCE remains open [9]. Moreover, several problems related to domination with parity constraints, such as WEIGHT DISTRIBUTION, are W[1]-hard and in W[2], are they complete for one of these two classes, or intermediate? This question is particularly interesting since these problems have been proved to form an equivalence class with other problems from quantum computing [15,2].

It is interesting to notice that, for the dual parameterization, the difference between MINIMAL DISTANCE and WEIGHT DISTRIBUTION seems to vanish in the sense that both problems are W[1]-hard, while the completeness for W[1] or W[2] remains open. In fact, no problem of (σ, ρ)-domination is known to be W[2]-complete for the dual parameterization, thus one can wonder if such a problem exists or if for any σ and ρ, (σ, ρ)-domination is in W[1] for the dual parameterization? It would be interesting to examine σ and ρ not ultimately periodic since they are among the few known cases of hardness when the problem is parameterized by the tree-width [6].

Acknowledgments. We would like to thank Sylvain Gravier and Mehdi Mhalla for several helpful discussions, and anonymous referees for fruitful comments. This work has been partially funded by the ANR-10-JCJC-0208 CausaQ grant and by the Rhône-Alpes region.

References

1. Bodlaender, H.L., Kratsch, D.: A note on fixed parameter intractability of some domination-related problems (1994) (unpublished)
2. Cattanéo, D., Perdrix, S.: Parameterized complexity of weak odd domination problems. In: Gąsieniec, L., Wolter, F. (eds.) FCT 2013. LNCS, vol. 8070, pp. 107–120. Springer, Heidelberg (2013)
3. Cesati, M.: Perfect Code is W[1]-complete. Inf. Proc. Let. 81, 163–168 (2002)
4. Cesati, M.: The Turing way to parameterized complexity. Journal of Computer and System Sciences 67, 654–685 (2003)
5. Cesati, M., Di Ianni, M.: Computation models for parametrerized complexity. MLQ 43, 179–202 (1997)
6. Chapelle, M.: Parameterized Complexity of Generalized Domination Problems on Bounded Tree-Width Graphs. Computing Research Repository, abs/1004.2 (2010)

7. Downey, R.G., Fellows, M.R.: Fixed-Parameter Tractability and Completeness I: Basic Results. SIAM Journal on Computing 24, 873–921 (1995)
8. Downey, R.G., Fellows, M.R.: Fixed-Parameter Tractability and Completeness II: On Completeness for W[1]. Theoretical Computer Science 141, 109–131 (1995)
9. Downey, R.G., Fellows, M.R., Vardy, A., Whittle, G.: The Parametrized Complexity of Some Fundamental Problems in Coding Theory. SIAM Journal on Computing 29, 545–570 (1999)
10. Downey, R.G., Fellows, M.R.: Parameterized Complexity. Springer (1999)
11. Fellows, M.R.: Blow-ups, win/win's, and crown rules: Some new directions in *FPT*. In: Bodlaender, H.L. (ed.) WG 2003. LNCS, vol. 2880, pp. 1–12. Springer, Heidelberg (2003)
12. Flum, J., Grohe, M.: Parameterized Complexity Theory. In: Texts in theoretical computer science. Springer (2006)
13. Galbiati, G., Maffioli, F., Morzenti, A.: A short note on the approximability of the maximum leaves spanning tree problem. Information Processing Letters 52(1), 45–49 (1994)
14. Golovach, P.A., Kratochvìl, J., Suchỳ, O.: Parameterized complexity of generalized domination problems. Discrete Applied Mathematics 160(6), 780–792 (2009)
15. Gravier, S., Javelle, J., Mhalla, M., Perdrix, S.: Quantum secret sharing with graph states. In: Kučera, A., Henzinger, T.A., Nešetřil, J., Vojnar, T., Antoš, D. (eds.) MEMICS 2012. LNCS, vol. 7721, pp. 15–31. Springer, Heidelberg (2013)
16. Gutin, G., Kloks, T., Lee, C.-M., Yeo, A.: Kernels in planar digraphs. Journal of Computer and System Sciences 71(2), 174–184 (2005)
17. Halldórsson, M.M., Kratochvíl, J., Telle, J.A.: Mod-2 independence and domination in graphs. In: Widmayer, P., Neyer, G., Eidenbenz, S. (eds.) WG 1999. LNCS, vol. 1665, pp. 101–109. Springer, Heidelberg (1999)
18. Kloks, T., Cai, L.: Parameterized tractability of some (efficient) y-domination variants for planar graphs and t-degenerate graphs. In: International Computer Symposium, ICS (2000)
19. Moser, H., Thilikos, D.M.: Parameterized complexity of finding regular induced subgraphs. Journal of Discrete Algorithms 7, 181–190 (2009)
20. Telle, J.A.: Complexity of Domination-Type Problems in Graphs. Nordic Journal of Computing 1, 157–171 (1994)
21. Telle, J.A., Proskurowski, A.: Algorithms for Vertex Partitioning Problems on Partial k-Trees. SIAM Journal on Discrete Mathematics 10, 529–550 (1997)

On Representations of Abstract Systems with Partial Inputs and Outputs

Ievgen Ivanov

Université Paul Sabatier
118 route de Narbonne, Toulouse, France
ivanov.eugen@gmail.com

Abstract. We consider a class of mathematical models called blocks which generalize some input-output models which appear in mathematical systems theory, control theory, signal processing. A block maps partial functions of time to nonempty sets of partial functions of time. A class of strongly nonanticipative blocks can be considered as an analog of the class of causal time systems studied by M. Mesarovic and Y. Takahara. The behavior of a strongly nonanticipative block can be represented using an abstract dynamical system called Nondeterministic Complete Markovian System (NCMS) which is close to the notion of a solution system by O. Hájek. We show that conversely, each initial input-output NCMS (i.e. NCMS with inputs and outputs) is a representation of a strongly nonanticipative block. This result generalizes a link between causality and the existence of state-space representations that exists in several variants of mathematical systems theory to models with partial inputs and outputs.

Keywords: input-output system, signal, time, partial function, state space, dynamical system, representation.

1 Introduction

Many computing systems used today act as agents interacting with physical processes. Such systems are now frequently called cyber-physical systems [1, 2]. Examples include autonomous automotive systems, robots, medical devices, energy conservation systems, etc. The development and specification languages that have been recently applied to cyber-physical systems include Simulink, AADL, SysML, and others. Such languages frequently employ block diagram notations in which blocks denote system components and links are associated with time varying quantities (signals) which are shared between components.

From the theoretical perspective, abstract models of systems that interact with other systems or the environment using signals or time varying quantities have been studied for a long time in several variants of the mathematical systems theory, including the works by L. Zadeh [3, 4], R. Kalman [5], M. Arbib [6], G. Klir [7], W. Wymore [8], M. Mesarovic [9], B.P. Zeigler [10], V.M. Matrosov [11], and others [12–14]. Many of such studies were influenced by the General Systems Theory by L. Bertalanffy, Cybernetics by N. Wiener, information theory

T V Gopal et al. (Eds.): TAMC 2014, LNCS 8402, pp. 104–123, 2014.

introduced by C. Shannon, control theory, automata theory, and circuit theory in electrical engineering [15, 7]. Although among these works there is no generally accepted formal notion or model of a system, many approaches on some level of abstraction describe the observable (external) behavior a system as an input-output relation on time-varying quantities, e.g. a general time system [9, Section 2.5], an external behavior of a dynamical system [5, Section 1.1], an oriented abstract object [3, Chapter 1, Paragraph 4], an I/O observational frame [10, Section 5.3], etc. The most basic example of this view is a Mesarovic time system [9] which is defined as a binary relation $S \subseteq I \times O$, where I and O are sets of input and output functions on a time domain T, i.e. $I \subseteq A^T$ and $O \subseteq B^T$. It should be noted that most attention in many approaches is given to models in which either both inputs and outputs are total functions of time, or the domains of corresponding instances of inputs and outputs are equal. The case when a system is considered on the abstract level as a "black box" with inputs and outputs that are partial functions of time with possibly different domains did not receive much attention. There are certain works in this direction, e.g. [16, 17], but they impose additional assumptions (e.g. determinism) which are not assumed by default in the theories which deal with total inputs and outputs.

On the other hand, the aspect of partiality is important in many concrete mathematical models of system behavior. In particular, the models described by differential equations and hybrid automata admit situations when the inputs of a system (e.g. input control signals), if there are any, are defined on the entire time domain, but the system's behavior (solution, trajectory, execution) and its outputs are not defined on the entire time domain. Examples of such situations are finite time blow-ups in differential equations [18–20] and Zeno behaviors of hybrid systems [21–23]. They can indicate real phenomena (e.g. termination or destruction of a real system) or inadequacy of a mathematical model [18]. If such a situation is present in a particular concrete model, this model cannot be adequately abstracted to a "black box" model that admits only total inputs and outputs, like the Mesarovic time system. This discrepancy between the abilities of concrete and abstract models makes the problem of investigation of abstract models of system behavior which admit partial inputs and outputs relevant.

In the previous works [24, 25] we introduced a class of input-output abstractions of real-world systems which we called a class of *blocks*. A block can be thought of as a generalization of a Mesarovic time system which takes into account partiality of inputs and outputs as functions of time. Basically, a block is a multifunction which maps a collection of input signals (an *input signal bunch*, or simply an *input*) to a non-empty set of collections of output signals (*output signal bunches*, or simply *outputs*), and a signal is a partial function on the continuous time domain. The operation of a block can be alternatively described by a set of input-output pairs (I/O pairs) (i, o), where i is an input signal bunch and o is a corresponding output signal bunch.

The main aspects of blocks are partiality of inputs and outputs, continuous time, nondeterminism (multiple output signal bunches may correspond to one input signal bunch). The main theoretical goals of investigation of blocks are:

1) to generalize or reinterpret well-known results about abstract systems with total inputs and outputs or systems which admit only inputs and outputs with equal time domains;

2) to gain understanding of the phenomena which are inherently associated with partiality.

With regard to the first goal, in [24] we introduced and studied the notions of causality (nonanticipation), refinement, and composition for blocks and in [25] we considered the question of the existence of representations of blocks in the form of dynamical systems. With regard to the second goal, in [25] we considered the problem of the existence of total input-output pairs (i.e. pairs (i, o) such that both i and o are total) as members of the set of all input-output pairs of a block.

In this paper we continue to investigate properties of blocks and consider the following question (in the scope of first goal). Various variants of the mathematical systems theory show a link between the property of causality (nonanticipation) and the existence of a representation of an abstract input-output model of a system in the form of a dynamical system of a certain kind. For example, in the theory by M. Mesarovic and Y. Takahara [9] it can be shown that a time system is causal if and only if it has a state space representation [9, Proposition 2.8]. The goal of this paper is to find an analogy to this fact in the case of blocks.

In [25] we considered a class of abstract dynamical systems called *Nondeterministic Complete Markovian Systems* (*NCMS*) which is based on the notion of a solution system by O. Hájek [26] and showed that each *strongly nonanticipative block* [25] has a representation in the form of a so-called *initial input-output* (*I/O*) *NCMS* (a kind of a dynamical system with inputs and outputs). A strongly nonanticipative block is an adaptation of the notion of a causal time system [9] (non-anticipatory system [27]) to blocks. In this work we show that conversely, each initial I/O NCMS is a representation of a strongly nonanticipative block.

To make the paper self-contained, in Section 2 we give all necessary definitions and facts about blocks, nonanticipation, NCMS. Also, we give an overview of the problem of the existence of total input-output pairs of strongly nonanticipative blocks which is connected with NCMS representations, and describe the proposed solution and potential applications. In Section 3 we prove the main result.

2   Preliminaries

2.1   Notation

We use the following notation: $\mathbb{N} = \{1, 2, 3, ...\}$, $\mathbb{N}_0 = \mathbb{N} \cup \{0\}$, $\mathbb{R}_+$ is the set of nonnegative real numbers, $f : A \to B$ is a total function from A to B, $f : A \rightarrow B$ is a partial function from A to B, 2^A is the power set of a set A, $f|_X$ is the restriction of a function f to a set X. If A, B are sets, then B^A denotes the set of all total functions from A to B and $^A B$ denotes the set of all partial function from A to B. For a function $f : A \rightarrow B$ the symbol $f(x) \downarrow (f(x) \uparrow)$ means that $f(x)$ is defined (respectively undefined) on the argument x. We do not distinguish formally the notion of a function and a functional binary relation. When we write that a function $f : A \rightarrow B$ is total or surjective, we mean that f is total on A

specifically (i.e. $f(x) \downarrow$ for all $x \in A$), or, respectively, is onto B (i.e. for each $y \in B$ there exists $x \in A$ such that $y = f(x)$). We denote the domain and range of a function as $dom(f) = \{x \mid f(x) \downarrow\}$ and $range(f) = \{y \mid \exists x\, f(x) \downarrow \wedge y = f(x)\}$ respectively. For partial functions f, g and an argument x, $f(x) \cong g(x)$ denotes the strong equality: $f(x) \downarrow$ if and only if $g(x) \downarrow$, and $f(x) \downarrow$ implies $f(x) = g(x)$. By $f \circ g$ we denote the functional composition: $(f \circ g)(x) \cong f(g(x))$. The notation $X \mapsto y$, where X is a given set and y is a given value means a constant function on X which takes the value y.

By T we denote the non-negative real time scale $[0, +\infty)$, equipped with a topology induced by the standard topology on $\mathbb{R}$. We also define the following class of intervals: $\mathcal{T}_0 = \{\emptyset, T\} \cup \{[0, t) \mid t \in T \backslash \{0\}\} \cup \{[0, t] \mid t \in T\}$.

The symbols $\neg$, $\vee$, $\wedge$, $\Rightarrow$, $\Leftrightarrow$ denote the logical operations of negation, disjunction, conjunction, implication, and equivalence respectively.

2.2 Multi-valued Functions

A multi-valued function [28] assigns one or more resulting values to each argument value. An application of a multi-valued function to an argument is interpreted as a nondeterministic choice of a result.

Definition 1. *[28] A (total) multi-valued function from a set A to a set B (denoted as $f : A \overset{tm}{\longrightarrow} B$) is a function $f : A \to 2^B \backslash \{\emptyset\}$.*

2.3 Named Sets

We will use a simple notion of a named set [28] to formalize an assignment of values to variable names.

Definition 2. *[28] A named set is a function $f : V \tilde{\to} W$ from a non-empty set of names V to a set of values W.*

We use a special notation for the set of named sets: $^V W$ denotes the set of all named sets $f : V \tilde{\to} W$ (this notation emphasizes that V is a set of names).

An expression of the form $[n_1 \mapsto a_1, n_2 \mapsto a_2, ...]$ (where $n_1, n_2, ...$ are distinct names) denotes a named set d such that the graph of d is $\{(n_1, a_1), (n_2, a_2), ...\}$.

The *empty named set*, denoted as $[]$, is a named set with empty domain.

2.4 Blocks

Informally, a block is an abstract model of a system which receives input signals and produces output signals (Fig. 1). The input signals can be considered as time-varying characteristics (attributes) of the external environment of the system which are relevant for (the operation of) this system. Each instance of an input signal has an associated time domain on which it is defined (present). An input signal bunch can be considered as a collection of instances of input signals of the system. Each input signal bunch i has an associated domain of existence

$(dom(i))$ which is a superset of the union of the domains of signals contained in i. The domain of an input signal bunch can be thought of as a time span of the existence of the external environment of the system. The output signals can be considered as effects (results) of the system's operation. An output signal bunch, or simply an output of the block, can be considered as a collection of instances of output signals of the system. The output signals have associated domains of definition (presence) and each output signal bunch o has a domain of existence $(dom(o))$ which is a superset of the union of the domains of signals contained in o. The domain of an output signal bunch can be thought of as a time span during which the system operates. It is assumed that for an output signal bunch o which corresponds to a given input signal bunch i the inclusion $dom(o) \subseteq dom(i)$ holds (i.e. the system does not operate when the environment does not exist). In the general case, the presence of a given input signal at a given time moment does not imply the presence of a certain output signal at the same or any other time moment. A block can be nondeterministic, i.e. for one input signal bunch it may choose an output signal bunch from a set of possible variants. But for each input signal bunch there is at least one corresponding output signal bunch (although the values of all signals in it may be absent at all times).

Normally a block processes the whole input signal bunch and does or does not produce output values. But in certain cases a block may not process the whole input signal bunch and may terminate at a time which precedes the right endpoint of the domain of the input signal bunch. This situation is interpreted as an abnormal termination of a block.

Consider formal definitions [24]. Let W be a fixed non-empty set of values.

Definition 3. *(1) A signal is a partial function from T to W $(f : T \xrightarrow{\cdot} W)$.*
(2) A V-signal bunch (where V is a set of names) is a function $sb : T \xrightarrow{\cdot} {}^V W$ such that $dom(sb) \in \mathcal{T}_0$. The set of all V-signal bunches is denoted as $Sb(V, W)$.
(3) A signal bunch is a V-signal bunch for some V.
(4) A signal bunch sb is trivial, if $dom(sb) = \emptyset$ and is total, if $dom(sb) = T$. The unique trivial signal bunch is denoted as $\perp$.
(5) For a given signal bunch sb, a signal corresponding to a name x is a partial function $t \mapsto sb(t)(x)$. This signal is denoted as $sb[x]$.
(6) A signal bunch sb_1 is a prefix of a signal bunch sb_2 (denoted as $sb_1 \preceq sb_2$), if $sb_1 = sb_2|_A$ for some $A \in \mathcal{T}_0$.

The relation $\preceq$ on V-signal bunches is a partial order (for an arbitrary V). It can be generalized to pairs as follows: for any signal bunches sb_1, sb_2, sb_1', sb_2' denote $(sb_1, sb_2) \preceq^2 (sb_1', sb_2')$ if and only if there exists $A \in \mathcal{T}_0$ such that $sb_1 = sb_1'|_A$ and $sb_2 = sb_2'|_A$. The relation $\preceq^2$ is a partial order (this is not a product order).

A block has a syntactic aspect (e.g. a description in a specification language) and a semantic aspect – a partial multi-valued function on signal bunches.

Definition 4. *(1) A block is an object B (syntactic aspect) with an associated set of input names $In(B)$, a set of output names $Out(B)$, and a total multi-valued function $Op(B) : Sb(In(B), W) \xrightarrow{tm} Sb(Out(B), W)$ (operation, semantic aspect) such that $o \in Op(B)(i)$ implies $dom(o) \subseteq dom(i)$.*

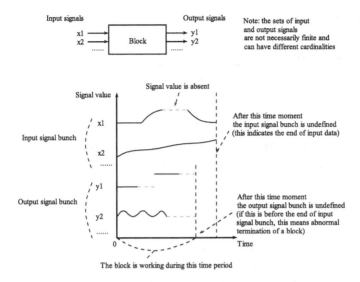

Fig. 1. An illustration of a block with the input signals x_1, x_2, ... and the output signals y_1, y_2, ... Solid curves represent (present) signal values. Dashed horizontal segments denote absence of a signal value. Dashed vertical lines indicate the right boundaries of the domains of signal bunches.

(2) *Blocks B_1, B_2 are semantically identical, if $In(B_1) = In(B_2)$, $Out(B_1) = Out(B_2)$, and $Op(B_1) = Op(B_2)$.*

(3) *An I/O pair of a block B is a pair of signal bunches (i, o) such that $o \in Op(B)(i)$. The set of all I/O pairs of B is denoted as $IO(B)$ and is called the input-output (I/O) relation of B.*

An inclusion $o \in Op(B)(i)$ or $(i, o) \in IO(B)$ means that o is a possible output of a block B on the input i.

Definition 5. *An I/O pair (i, o) of a block B is called*

(1) *trivial, if $(i, o) = (\bot, \bot)$;*
(2) *normal, $dom(i) = dom(o)$;*
(3) *total, if $dom(i) = dom(o) = T$;*
(4) *abnormal, if $dom(o) \subset dom(i)$.*

An abnormal I/O pair (i, o) can be interpreted as an unrecoverable error during the operation of a block on the input i. The next lemma supports this.

Lemma 1. *Let (i, o) be an abnormal I/O pair of some block and i', o' be signal bunches such that $(i, o) \preceq^2 (i', o')$. Then $o = o'$.*

Proof. Follows immediately from the definitions. □

Definition 6. *([25]) A block B is deterministic, if $Op(B)(i)$ is a singleton set for each $In(B)$-signal bunch i.*

Definition 7. *([25]) A deterministic block B is causal, if for all signal bunches i_1, i_2 and $A \in \mathcal{T}_0$, if $i_1|_A = i_2|_A$ and $o_j \in Op(B)(i_j)$, $j = 1, 2$, then $o_1|_A = o_2|_A$.*

Definition 8. *([25]) A block B is a sub-block of a block B' (denoted as $B \trianglelefteq B'$), if $In(B) = In(B')$, $Out(B) = Out(B')$, and $IO(B) \subseteq IO(B')$.*

Definition 9. *([24]) A block B is (weakly) nonanticipative, if for each $A \in \mathcal{T}_0$ and i_1, i_2, if $i_1|_A = i_2|_A$, then $\{o|_A \mid o \in Op(B)(i_1)\} = \{o|_A \mid o \in Op(B)(i_2)\}$.*

Definition 10. *([25]) A block B is strongly nonanticipative, if for each $(i, o) \in IO(B)$ there is a deterministic causal sub-block $B' \trianglelefteq B$ (a deterministic response strategy) such that $(i, o) \in IO(B')$.*

Weakly nonanticipative blocks generalize pre-causal time systems in the sense of M. Mesarovic and Y. Takahara [9]. Similarly, the notion of a strongly nonanticipative block generalizes the notion of a causal time system [9, 27].

Lemma 2. *For any deterministic block B the following holds: B is causal if and only if B is strongly nonanticipative if and only if B is weakly nonanticipative.*

Proof. Follows immediately from the definitions. $\qquad\square$

Lemma 3. *(1) Each strongly nonanticipative block is weakly nonanticipative.*
(2) There is a weakly nonanticipative block which is not strongly nonanticipative.

Proof. Follows from [24, Theorem 2]. $\qquad\square$

2.5 Nondeterministic Complete Markovian Systems (NCMS)

The notion of a NCMS was introduced in [29] for the purpose of studying the relation between the existence of global and local trajectories of dynamical systems. This notion is close to the notion of a *solution system* introduced by O. Hájek in [26], but there are some differences. A comparison was given in [25].

Let $T = \mathbb{R}_+$ be the non-negative real time scale. Denote by $\mathfrak{T}$ the set of all intervals (connected subsets) in T which have cardinality greater than one.

Let Q be a set (a state space) and Tr be some set of functions of the form $s : A \to Q$, where $A \in \mathfrak{T}$. Let us call its elements trajectories.

Definition 11. *([29, 25]) A set of trajectories Tr is closed under proper restrictions (CPR), if $s|_A \in Tr$ for each $s \in Tr$ and $A \in \mathfrak{T}$ such that $A \subseteq dom(s)$.*

Definition 12. *([29, 25])*

(1) A trajectory $s_1 \in Tr$ is a subtrajectory of $s_2 \in Tr$ (denoted as $s_1 \sqsubseteq s_2$), if $dom(s_1) \subseteq dom(s_2)$ and $s_1 = s_2|_{dom(s_1)}$.
(2) A trajectory $s_1 \in Tr$ is a proper subtrajectory of $s_2 \in Tr$ (denoted as $s_1 \sqsubset s_2$), if $s_1 \sqsubseteq s_2$ and $s_1 \neq s_2$.

The set $(Tr, \sqsubseteq)$ is a (possibly empty) partially ordered set (poset).

Definition 13. *([29, 25]) A CPR set of trajectories Tr is*

(1) Markovian (Fig. 2), if for each $s_1, s_2 \in Tr$ and $t \in T$ such that $t = \sup dom(s_1) = \inf dom(s_2)$, $s_1(t) \downarrow$, $s_2(t) \downarrow$, and $s_1(t) = s_2(t)$, the following function s belongs to Tr:
$s(t) = s_1(t)$, if $t \in dom(s_1)$, and $s(t) = s_2(t)$, if $t \in dom(s_2)$.
(2) complete, if each non-empty chain in $(Tr, \sqsubseteq)$ has a supremum.

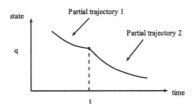

Fig. 2. Markovian property of NCMS. If one trajectory ends and another begins in a state q at time t, then their concatenation is a trajectory.

Definition 14. *([29]) A nondeterministic complete Markovian system (NCMS) is a triple (T, Q, Tr), where Q is a set (state space) and Tr (trajectories) is a set of functions $s : T \tilde{\rightarrow} Q$ such that $dom(s) \in \mathfrak{T}$, which is CPR, complete, and Markovian.*

This is an *intensional* definition. An alternative *extensional* definition (or an overview of the class of all NCMS) can be given using the notion of an *LR representation* of NCMS which is described below.

Definition 15. *Let $s_1, s_2 : T \tilde{\rightarrow} Q$. Then s_1 and s_2 coincide:*

(1) on $A \subseteq T$ (denoted as $s_1 \doteq_A s_2$), if $s_1|_A = s_2|_A$ and $A \subseteq dom(s_1) \cap dom(s_2)$;
(2) in a left neighborhood of $t \in T$, if $t > 0$ and there exists $t' \in [0, t)$ such that $s_1 \doteq_{(t',t]} s_2$ (this is denoted as $s_1 \doteq_{t-} s_2$);
(3) in a right neighborhood of $t \in T$, if there exists $t' > t$, such that $s_1 \doteq_{[t,t')} s_2$ (this is denoted as $s_1 \doteq_{t+} s_2$).

Let Q be a set. Denote by $ST(Q)$ the set of pairs (s, t) where $s : A \rightarrow Q$ for some $A \in \mathfrak{T}$ and $t \in A$.

Definition 16. *([29, 25]) A predicate $p : ST(Q) \rightarrow Bool$ is called*

(1) left-local, if $p(s_1, t) \Leftrightarrow p(s_2, t)$ whenever $(s_1, t), (s_2, t) \in ST(Q)$ and $s_1 \doteq_{t-} s_2$, and, moreover, $p(s, t)$ whenever t is the least element of $dom(s)$;

(2) right-local, if $p(s_1, t) \Leftrightarrow p(s_2, t)$ whenever $(s_1, t), (s_2, t) \in ST(Q)$, $s_1 \doteq_{t+} s_2$, and, moreover, $p(s, t)$ whenever t is the greatest element of $dom(s)$.

Denote by $LR(Q)$ the set of all pairs (l, r), where $l : ST(Q) \to Bool$ is a left-local predicate and $r : ST(Q) \to Bool$ is a right-local predicate.

Definition 17. *([25]) A pair $(l, r) \in LR(Q)$ is called a LR representation of a NCMS $\Sigma = (T, Q, Tr)$, if $Tr = \{s : A \to Q \mid A \in \mathfrak{T} \wedge (\forall t \in A\ l(s, t) \wedge r(s, t))\}$.*

Theorem 1. *([25, Theorem 1])*

(1) Each pair $(l, r) \in LR(Q)$ is a LR representation of a NCMS with the set of states Q.
(2) Each NCMS has a LR representation.

2.6   NCMS Representation of a Strongly Nonanticipative Block

As was shown in [25], NCMS can be used to represent strongly nonanticipative blocks. Let W denote a fixed non-empty set of values.

Definition 18. *([25]) An input-output (I/O) NCMS is a NCMS (T, Q, Tr) such that Q has a form $^I W \times X \times {}^O W$ for some sets I (set of input names), $X \neq \emptyset$ (set of internal states), and O (set of output names). The set $^I W$ is called an input data set and $^O W$ is called an output data set.*

Informally, an I/O NCMS describes possible evolutions of triples (d_{in}, x, d_{out}) of input data $(d_{in} \in {}^I W)$, internal state $(x \in X)$, and output data $(d_{out} \in {}^O W)$.

Each I/O NCMS $\Sigma = (T, Q, Tr)$ has unique sets of input names, internal states, and output names, denoted as $In(\Sigma)$, $IState(\Sigma)$, $Out(\Sigma)$ respectively.

For any state $q \in Q$ of an I/O NCMS the symbols $in(q)$, $istate(q)$, $out(q)$ will denote the projections of q on the first, second, and third coordinate respectively.

For any trajectory $s \in Tr$ the symbols $in \circ s$, $istate \circ s$, $out \circ s$ will denote the compositions of the respective projection maps with s.

For each $i \in Sb(In(\Sigma), W)$ denote

- $S(\Sigma, i) = \{s \in Tr \mid dom(s) \in \mathcal{T}_0 \ \wedge in \circ s \preceq i\}$;
- $S_{max}(\Sigma, i)$ is the set of all $\sqsubseteq$-maximal trajectories from $S(\Sigma, i)$;
- $S_{init}(\Sigma, i) = \{s(0) \mid s \in S(\Sigma, i)\}$;
- $S_{init}(\Sigma) = \{s(0) \mid s \in Tr \wedge dom(s) \in \mathcal{T}_0\}$.

For each $Q' \subseteq Q$ denote $Sel_{1,2}(Q', d, x) = \{q \in Q' \mid \exists d'\ q = (d, x, d')\}$, i.e. the states from Q' selected by the value of the first and second component.

For each $Q' \subseteq Q$ and $i \in Sb(In(\Sigma), W)$ denote

$$
o_{all}(\Sigma, Q', i) = \begin{cases} \{\bot\}, & \text{if } Q' = \emptyset \text{ or } i = \bot; \\ \{\{0\} \mapsto out(q) \mid q \in Q'\}, & \text{if } Q' \neq \emptyset \text{ and} \\ & dom(i) = \{0\}; \\ \{out \circ s \mid s \in S_{max}(\Sigma, i) \wedge s(0) \in Q'\} \cup \\ \quad \cup \{\{0\} \mapsto out(q) \mid q \in Q' \backslash S_{init}(\Sigma, i)\}, & \text{otherwise}; \end{cases}
$$

(where $\{0\} \mapsto out(q)$ is a constant function on $\{0\}$). For each $Q_0 \subseteq Q$ denote

$$O_{all}(\Sigma, Q_0, i) = \begin{cases} \{\bot\}, & dom(i) = \emptyset; \\ \bigcup_{x \in IState(\Sigma)} O_{all}(\Sigma, Sel_{1,2}(Q_0, i(0), x), i), & dom(i) \neq \emptyset. \end{cases}$$

Definition 19. *([25]) An initial I/O NCMS is a pair (Σ, Q_0), where $\Sigma = (T, Q, Tr)$ is an I/O NCMS and Q_0 is a set (admissible initial states) such that $S_{init}(\Sigma) \subseteq Q_0 \subseteq Q$.*

Definition 20. *([25]) A NCMS representation of a block B is an initial I/O NCMS (Σ, Q_0) such that*

(1) $In(B) = In(\Sigma)$ and $Out(B) = Out(\Sigma)$;
(2) $Op(B)(i) = O_{all}(\Sigma, Q_0, i)$ for all $i \in Sb(In(B), W)$.

Informally, the operation of a block B represented by an initial I/O NCMS (Σ, Q_0) on an input signal bunch i can be described as follows [25]:

(1) If $i(0)$ is undefined, then B stops (the output signal bunch is $\bot$).
(2) Otherwise, B chooses an arbitrary internal state $x \in IState(\Sigma)$.
(3) If there is no admissible initial state $q \in Q_0$ with $in(q) = i(0)$ and $istate(q) = x$ (i.e. $Sel_{1,2}(Q_0, i(0), x) = \emptyset$), then B stops.
(4) Otherwise, B chooses $q \in Q_0$ such that $in(q) = i(0)$ and $istate(q) = x$.
(5) If $dom(i) = \{0\}$ or there is no trajectory s which starts in q and is defined on some interval from $\mathcal{T}_0$, then B outputs $out(q)$ at time 0 and stops.
(6) Otherwise, B chooses a maximal trajectory s defined on an interval from $\mathcal{T}_0$ such that $s(0) = q$ and $in \circ s \preceq i$ and outputs the signal bunch $out \circ s$.

Theorem 2. *([25, Theorem 2]) Each strongly nonanticipative block has a NCMS representation.*

2.7 Existence of Total I/O Pairs of Strongly Nonanticipative Blocks

As we have noted above, a block can admit input-output pairs (i, o) in which signal bunches have different domains $(dom(o) \subset dom(i))$. This is interpreted as an abnormal termination. In particular, a block can output a non-total output signal bunch for each total input signal bunch, i.e. can have no total I/O pairs.

The problem of the existence of total I/O pairs of strongly nonanticipative blocks was investigated in [25], where the following questions were considered:

(A) How can one prove that a given strongly nonanticipative block B has a total I/O pair (if B indeed has a total I/O pair) ?
(B) How can one prove that for a given input signal bunch $i \in Sb(In(B), W)$, where $dom(i) = T$, there exists $o \in Op(B)(i)$ with $dom(o) = T$?

From the practical perspective, (A) and (B) can be linked with the problems of analysis of specifications of cyber-physical systems, real-time information processing systems, and other similar systems. If the semantics of such a specification is a strongly nonanticipative block and partiality of an I/O pair is interpreted as an indication of an error (problematic situation) in its behavior, the methods that give an answer to (A) and (B) can be used for checking that the specification admits a correct behavior (total I/O pair) and may be applied in such domains as viability theory, control synthesis, real-time software verification, etc.

The following results obtained in [25] show that for strongly nonanticipative blocks represented by NCMS the questions (A) and (B) can be reduced to the problem of the existence of global trajectories of NCMS.

Definition 21. *A global trajectory of a NCMS* $\Sigma = (T, Q, Tr)$ *is a trajectory* $s \in Tr$ *such that* $dom(s) = T$.

Theorem 3. *([25, Theorem 3]) Let B be a strongly nonanticipative block and* (Σ, Q_0) *be its NCMS representation. Then B has a total I/O pair if and only if* Σ *has a global trajectory.*

Theorem 4. *([25, Theorem 4]) Let B be a strongly nonanticipative block and* (Σ, Q_0) *be its NCMS representation, where* $\Sigma = (T, Q, Tr)$.
Let $i \in Sb(In(B), W)$ *and* $dom(i) = T$. *Let* (l, r) *be a LR representation of* Σ *and* $l' : ST(Q) \to Bool$ *and* $r' : ST(Q) \to Bool$ *be predicates such that*

$$l'(s,t) \Leftrightarrow l(s,t) \wedge (\min dom(s) \downarrow = t \vee in(s(t)) = i(t));$$

$$r'(s,t) \Leftrightarrow r(s,t) \wedge (\max dom(s) \downarrow = t \vee in(s(t)) = i(t)).$$

Then:

(1) $(l', r') \in LR(Q)$;
(2) If (l', r') *is a LR representation of a NCMS* $\Sigma' = (T, Q, Tr')$, *then* $\{o \in Op(B)(i) \mid dom(o) = T\} \neq \emptyset$ *if and only if* Σ' *has a global trajectory.*

The latter problem of the existence of global trajectories of NCMS was considered in [29, 25] and was reduced to the problem of the existence of certain *locally* defined trajectories of NCMS (which is usually more tractable).

Both Theorem 3 and Theorem 4 explicitly use a representation of a block in the form of an initial I/O NCMS (which always exists by Theorem 2) and also give a motivation to investigate whether any initial I/O NCMS defines a strongly nonanticipative block (which is a topic of this paper).

2.8   Existence of Global Trajectories of NCMS

To give a better overview of the context of our work, below we briefly state the main results about the existence of global trajectories of NCMS from [25].

Let $\Sigma = (T, Q, Tr)$ be a fixed NCMS.

Definition 22. *([25])* Σ *satisfies*

(1) local forward extensibility (LFE) property, if for each $s \in Tr$ *of the form* $s : [a, b] \to Q$ $(a < b)$ *there exists a trajectory* $s' : [a, b'] \to Q$ *such that* $s' \in Tr$, $s \sqsubseteq s'$ *and* $b' > b$.

(2) global forward extensibility (GFE) property, if for each $s \in Tr$ of the form $s : [a,b] \to Q$ there exists a trajectory $s' : [a,+\infty) \to Q$ such that $s \sqsubseteq s'$.

Definition 23. *([29, 25]) A right dead-end path (in Σ) is a trajectory $s : [a,b) \to Q$, where $a,b \in T$, $a < b$, such that there is no $s' : [a,b] \to Q$, $s \in Tr$ such that $s \sqsubset s'$ (i.e. s cannot be extended to a trajectory on $[a,b]$).*

Definition 24. *([29, 25]) An escape from a right dead-end path $s : [a,b) \to Q$ (in Σ) is a trajectory $s' : [c,d) \to Q$ $(d \in T \cup \{+\infty\})$ or $s' : [c,d] \to Q$ $(d \in T)$ such that $c \in (a,b)$, $d > b$, $s(c) = s'(c)$. An escape s' is called infinite, if $d = +\infty$.*

Definition 25. *([29, 25]) A right dead-end path $s : [a,b) \to Q$ in Σ is called strongly escapable, if there exists an infinite escape from s.*

Definition 26. *(1) A right extensibility measure is a function $f^+ : \mathbb{R} \times \mathbb{R} \dashrightarrow \mathbb{R}$ such that $A = \{(x,y) \in T \times T \mid x \leq y\} \subseteq dom(f^+)$, $f(x,y) \geq 0$ for all $(x,y) \in A$, $f^+|_A$ is strictly decreasing in the first argument and strictly increasing in the second argument, and for each $x \geq 0$, $f^+(x,x) = x$, and $\lim_{y \to +\infty} f^+(x,y) = +\infty$.*
(2) A right extensibility measure f^+ is called normal, if f^+ is continuous on $\{(x,y) \in T \times T \mid x \leq y\}$ and there exists a function α of class K_∞ (i.e. the function $\alpha : [0,+\infty) \to [0,+\infty)$ is continuous, strictly increasing, and $\alpha(0) = 0$, $\lim_{x \to +\infty} \alpha(x) = +\infty$) such that $\alpha(y) < y$ for all $y > 0$ and the function $y \mapsto f^+(\alpha(y), y)$ is of class K_∞.

Let us fix a right extensibility measure f^+.

Definition 27. *A right dead-end path $s : [a,b) \to Q$ is called f^+-escapable (Fig. 3), if there exists an escape $s' : [c,d] \to Q$ from s such that $d \geq f^+(c,b)$.*

An example of a right extensibility measure is $f^+(x,y) = 2y - x$. In this case, s is f^+-escapable, if there is an escape $s' : [c,d] \to Q$ from s with $d - b \geq b - c$.

Theorem 5. *([25, Theorem 6]) Assume that f^+ is a normal right extensibility measure and Σ satisfies LFE. Then each right dead-end path is strongly escapable if and only if each right dead-end path is f^+-escapable.*

Lemma 4. *([25, Lemma 14]) Σ satisfies GFE if and only if Σ satisfies LFE and each right dead-end path is strongly escapable.*

Theorem 6. *([25, Theorem 5]) Let (l,r) be a LR representation of Σ. Then Σ has a global trajectory if and only if there exists a pair $(l',r') \in LR(Q)$ such that*

(1) $l'(s,t) \Rightarrow l(s,t)$ and $r'(s,t) \Rightarrow r(s,t)$ for all $(s,t) \in ST(Q)$;
(2) $\forall t \in [0,\epsilon]$ $l'(s,t) \wedge r'(s,t)$ holds for some $\epsilon > 0$ and a function $s : [0,\epsilon] \to Q$;
(3) if (l',r') is a LR representation of a NCMS Σ', then Σ' satisfies GFE.

Theorem 6, Lemma 4, and Theorem 6 give a method for proving the existence of global trajectories of NCMS. Informally, this method consists of finding/guessing

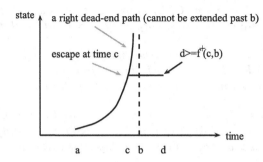

Fig. 3. An f^+-escapable right dead-end path $s : [a, b) \to Q$ (a curve) and a corresponding escape $s' : [c, d] \to Q$ (a horizontal segment) with $d \geq f^+(c, b)$

some region (subset of trajectories) which presumably contains a global trajectory and is described by the (left-/right-local) predicates l', r' and then proving that this region indeed contains a global trajectory. After choosing a suitable right extensibility measure f^+, the latter step can be done by finding certain locally defined trajectories (f^+-escapes) in a neighborhood of each time moment (the right endpoint of the domain of any potential right dead-end path).

2.9 Potential Applications

The mentioned method of proving the existence of global trajectories of NCMS can be illustrated on the following example problem: propose sufficient conditions under which for a given a set $D \subseteq \mathbb{R}^n$, a compact set $U \subset \mathbb{R}^m$ and a continuous and bounded function $f : T \times \mathbb{R}^n \times U \to \mathbb{R}^n$ which is Lipschitz-continuous in the second argument, there exists a Lebesgue-measurable function $u : T \to U$ and a corresponding Caratheodory solution $x : T \to \mathbb{R}^n$ of the differential equation $dx(t)/dt = f(t, x(t), u(t))$ (which is absolutely continuous on each compact segment $[a, b] \subset T$) such that $x(t) \neq \mathbf{0}$ for all $t \in T$ and the function u is constant on each time interval $I \subset T$ such that $x(t) \notin D$ for all $t \in I$. This question can be interpreted as a question about sufficient conditions under which a controlled system which travels in accordance with a known law of motion (x describes the system's trajectory, u is the input control) can avoid hitting the origin ($x(t) \neq \mathbf{0}$) under the assumption that the system lacks maneuverability outside a known set D. Some related problems were considered e.g. in [30] and were studied using control-theoretic methods, but one can also use NCMS.

One can show that if the problem is properly formalized in terms of NCMS or strongly nonanticipative blocks, one can derive the following easily verifiable sufficient conditions for the existence of the required x and u from the results described above: (1) for each $t \in T$ there exist $u_1, u_2 \in U$ such that $f(t, \mathbf{0}, u_1)$ and $f(t, \mathbf{0}, u_2)$ are noncollinear; (2) $\{\mathbf{0}\}$ is a path-component of $\{\mathbf{0}\} \cup (\mathbb{R}^n \backslash D)$.

3   Main Result

Theorem 7. *Each initial I/O NCMS Σ is a NCMS representation of a strongly nonanticipative block. Moreover, if Σ is a NCMS representation of each of two blocks B_1, B_2, then B_1 and B_2 are semantically identical.*

Together with Theorem 2, this theorem gives a generalization of the results about the existence of state-space representations of time systems in the sense of M. Mesarovic and Y. Takahara [9] to blocks. Note that unlike a state-space representation of a time system, a NCMS admits a nondeterministic behavior, i.e. a situation when several trajectories start in one state at a certain time moment, so in some sense Theorem 7 describes a more general case than the statement about causality of time systems which have a state space representation.

In the rest of the section we give a proof of Theorem 7 in the form of a series of lemmas. A detailed proof of each lemma cannot be given in this paper due to space limitations, so we will give proofs of the most important lemmas and sketch proofs of more straightforward lemmas. A detailed proof of Theorem 7 will be given in [31].

Lemma 5. *([25, Lemma 4]) Let Σ be an I/O NCMS, $i \in Sb(In(\Sigma), W)$, and $s \in S(\Sigma, i)$. Then there exists $s' \in S_{max}(\Sigma, i)$ such that $s \sqsubseteq s'$.*

Lemma 6. *([25, Lemma 6]) Each initial I/O NCMS is a NCMS representation of a unique (up to semantic identity) block.*

The proofs of the next two lemmas are straightforward, so we omit them.

Lemma 7. *Let B be a block with a NCMS representation, $(i, o) \in IO(B)$, and $(i', o') \preceq (i, o)$. Then $(i', o') \in IO(B)$.*

Lemma 8. *Assume that a block B has a NCMS representation, $o \in Op(B)(i)$, and $i \widetilde{\preceq} i'$. Then there exists $o' \in Op(B)(i')$ such that $(i, o) \preceq (i', o')$.*

Lemma 9. *If a block B has a NCMS representation, then it is weakly nonanticipative.*

Proof. Follows from Lemma 7, Lemma 8, and Theorem 3.

The next two lemmas deal with degenerate cases. We omit the proofs.

Lemma 10. *Assume that a block B is weakly nonanticipative and $dom(o) \subseteq \{0\}$ for each $(i, o) \subseteq IO(B)$. Then B is strongly nonanticipative.*

Lemma 11. *Assume that a block B has a NCMS representation, $(i_*, o_*) \in IO(B)$, $\{0\} \subset dom(i_*)$, and $dom(o_*) = \{0\}$. Then there exists a sub-block $B' \trianglelefteq B$ such that B' has a NCMS representation and $Op(B')(i_*) = \{o_*\}$.*

Lemma 12. *Assume that a block B has a NCMS representation, $(i_*, o_*) \in IO(B)$, and $\{0\} \subset dom(o_*)$. Then there exists a deterministic block B' such that B' has a NCMS representation, $In(B') = In(B)$, $Out(B') = Out(B)$, and $Op(B')(i) \subseteq Op(B)(i)$ for each $i \in Sb(In(B), W)$ such that $i(0) \downarrow= i_*(0)$, and $(i_*, o_*) \in IO(B')$.*

Proof (Sketch). Let (Σ, Q_0) be a NCMS representation of B and $\Sigma = (T, Q, Tr)$. Then $o_* \in O_{all}(\Sigma, Q_0, i_*)$. Then $i_* \neq \bot$, because $\{0\} \subset dom(o_*) \subseteq dom(i_*)$. Then there exists $x_* \in IState(\Sigma)$ such that $o_* \in o_{all}(\Sigma, Sel_{1,2}(Q_0, i_*(0), x_*), i_*)$. Then because $\{0\} \subset dom(o_*)$, there exists $s_* \in S_{max}(\Sigma, i_*)$ such that $s_*(0) \in Sel_{1,2}(Q_0, i_*(0), x_*)$ and $o_* = out \circ s_*$. Then $s_* \in Tr$.

Let $\mathcal{X}$ be the set of all sets $X \subseteq Tr$ such that

a) $s_* \in X$;
b) $0 \in dom(s)$ and $s(0) = s_*(0)$ for each $s \in X$;
c) for each $s \in X$ and $t \in T\backslash\{0\}$, $s|_{[0,t)} \in X$ and $s|_{[0,t]} \in X$;
d) for each $s_1, s_2 \in X$, if $in \circ s_1 = in \circ s_2$, then $s_1 = s_2$.

It follows immediately that

$$\{s_*|_{[0,t)} | t \in T\backslash\{0\}\} \cup \{s_*|_{[0,t]} | t \in T\backslash\{0\}\} \cup \{s_*\} \in \mathcal{X},$$

and $\bigcup c \in \mathcal{X}$ for each non-empty $\subseteq$-chain $c \subseteq \mathcal{X}$. Then Zorn's lemma implies that $\mathcal{X}$ has some $\subseteq$-maximal element X^*.

Let us show that each non-empty $\sqsubseteq$-chain in X^* has a supremum in X^*. Let $C \subseteq X^*$ be a non-empty $\sqsubseteq$-chain. Let s_0 be a function (the graph of) which is a union of (graphs of) elements of C (s_0 is indeed a function, because C is a $\sqsubseteq$-chain). Then $s_0 \in Tr$ by the completeness and CPR properties of the NCMS Σ. Besides, $0 \in dom(s_0)$. It is easy to show that $s_0 \in X^*$. It follows immediately that s_0 is a $\sqsubseteq$-supremum of C. Because C is arbitrary, it follows that each non-empty $\sqsubseteq$-chain in X^* has a supremum in X^*.

Let $Y = IState(\Sigma) \times X^*$ and $Q' = {}^{In(B)}W \times Y \times {}^{Out(B)}W$. Then $Y \neq \emptyset$, because $X^* \neq \emptyset$.

For each $s \in X^*$ and $y \in Y$ let $f_s^y : dom(s) \to Q'$ be a function such that $f_s^y(0) = (in(s(0)), y, out(s(0)))$ and

$$f_s^y(t) = \big(in(s(t)), (istate(s(t)), s|_{[0,t]}), out(s(t))\big)$$

for each $t \in dom(s)\backslash\{0\}$.

Note that because $X^* \in \mathcal{X}$, for each $s \in X^*$ we have $0 \in dom(s)$, $s \in Tr$, and $s|_{[0,t]} \in X^*$ for all $t > 0$. This implies that f_s^y indeed takes values in Q'.

Let us define the following set:

$$Tr' = \{\tilde{s} \mid \exists s \in X^* \exists y \in Y \exists A \in T \ A \subseteq dom(s) \wedge \tilde{s} = f_s^y|_A\}.$$

Because $dom(f_s^y) = dom(s)$ and $range(f_s^y) \subseteq Q'$ for all $s \in X^*$ and $y \in Y$, we have that Tr' is the set of all functions of the form $\tilde{s} : A \to Q'$, where $A \in T$, such that there exist $s \in X^*$ and $y \in Y$ such that $\tilde{s} \sqsubseteq f_s^y$.

Let $\Sigma' = (T, Q', Tr')$. It is not difficult to show that Σ' is a NCMS. The definition of Q' implies that Σ' is an I/O NCMS.

For each $d_{in} \in Sb(In(B), W)$ denote

$$O_0(d_{in}) = \{d_{out} \mid \exists x \in IState(\Sigma)(d_{in}, x, d_{out}) \in Q_0)\};$$
$$D_0 = \{d_{in} \in Sb(In(B), W) \mid O_0(d_{in}) \neq \emptyset\}.$$

Note that because $s_* \in Tr$, $s_*(0) \downarrow$, and (Σ, Q_0) is an initial I/O NCMS,

$$(in(s_*(0)), istate(s_*(0)), out(s_*(0))) = s_*(0) \in S_{init}(\Sigma),$$

whence $out(s_*(0)) \in O_0(in(s_*(0)))$ and $in(s_*(0)) \in D_0$.

Then there exists a function $\eta : D_0 \to Sb(Out(B), W)$ (selector) such that $\eta(in(s_*(0))) = out(s_*(0))$ and $\eta(d_{in}) \in O_0(d_{in})$ for each $d_{in} \in D_0$.

Let us define

$$Q_0' = \{(d_{in}, y, d_{out}) | d_{in} \in D_0 \wedge y \in Y \wedge d_{out} = \eta(d_{in})\}.$$

Obviously, we have $Q_0' \subseteq Q'$. It is easy to check that $in(q) = in(s_*(0))$ and $out(q) = out(s_*(0))$ for each $q \in S_{init}(\Sigma')$. Then for each $q \in S_{init}(\Sigma')$, $\eta(in(q)) \downarrow= \eta(in(s_*(0))) = out(s_*(0)) = out(q)$, whence $S_{init}(\Sigma') \subseteq Q_0'$. Thus $S_{init}(\Sigma') \subseteq Q_0' \subseteq Q'$. Then (Σ', Q_0') is an initial I/O NCMS.

Then by Lemma 6, (Σ', Q_0') is a NCMS representation of some block B'. Then $In(B') = In(\Sigma') = In(B)$, $Out(B') = Out(\Sigma') = Out(B)$, and $Op(B')(i) = O_{all}(\Sigma', Q_0', i)$ for all $i \in Sb(In(B), W)$.

It is not difficult to show that the block B' is deterministic and that $Op(B')(i) \subseteq Op(B)(i)$ for each $i \in Sb(In(B), W)$ such that $i(0) \downarrow= i_*(0)$.

Let us show that $(i_*, o_*) \in IO(B')$. We have $s_* \in X^*$, $s_* \in S_{max}(\Sigma, i_*)$, and $o_* = out \circ s_*$. Let $y \in Y$ be an arbitrary element. Because $s_* \in X^*$, we have $f_{s_*}^y \in Tr'$. Denote $\tilde{s} = f_{s_*}^y$. Then $dom(\tilde{s}) = dom(s_*)$ and $in \circ \tilde{s} = in \circ f_{s_*}^y = in \circ s_* \preceq i_*$, so $\tilde{s} \in S(\Sigma', i_*)$. By Lemma 5, there exists $\tilde{s}' \in S_{max}(\Sigma', i_*)$ such that $\tilde{s} \sqsubseteq \tilde{s}'$. Because $\tilde{s}' \in Tr'$, there exists $s \in X^*$ and $y' \in Y$ such that $\tilde{s}' \sqsubseteq f_s^{y'}$. Then $f_{s_*}^y \sqsubseteq f_s^{y'}$. This implies that $dom(s_*) \subseteq dom(s)$, $y = y'$, and $s_*(t) = s(t)$ for all $t \in dom(s_*) \backslash \{0\}$. Moreover, $s(0) = s_*(0)$, because $s \in X^*$. Then $s_* \sqsubseteq s$. Denote $A = dom(\tilde{s}')$. Then $dom(s_*) = dom(\tilde{s}) \subseteq dom(\tilde{s}') = A$, so $s_* \sqsubseteq s|_A$. Besides, $s|_A \in X^*$, because $s \in X^*$. Then we have $f_{s|_A}^y \sqsubseteq f_s^y$. Moreover, $dom(f_{s|_A}^y) = dom(s|_A) = A$, because $A = dom(\tilde{s}') \subseteq dom(f_s^y) = dom(s)$. Then because $\tilde{s}' \sqsubseteq f_s^{y'} = f_s^y$ and $A = dom(\tilde{s}')$, we have $f_{s|_A}^y = \tilde{s}'$. Then $in \circ (s|_A) = in \circ (f_{s|_A}^y) = in \circ \tilde{s}' \preceq i_*$. Besides, $s|_A \in X^* \subseteq Tr$, whence $s|_A \in S(\Sigma, i_*)$. Because $s_* \sqsubseteq s|_A$ and $s_* \in S_{max}(\Sigma, i_*)$, we have $s_* = s|_A$. Then $\tilde{s} = f_{s_*}^y = f_{s|_A}^y = \tilde{s}' \in S_{max}(\Sigma', i_*)$. We have $\tilde{s}(0) \in S_{init}(\Sigma') \subseteq Q_0'$, because (Σ', Q_0') is an initial I/O NCMS. Moreover, $in(\tilde{s}(0)) = in(s_*(0)) = i_*(0)$ and $istate(\tilde{s}(0)) = istate(f_{s_*}^y(0)) = y \in IState(\Sigma')$, whence $\tilde{s}(0) \in Sel_{1,2}(Q_0', i_*(0), y)$. Moreover, $out \circ \tilde{s} = out \circ f_{s_*}^y = out \circ s_* = o_*$. Then

$$o_* \in o_{all}(\Sigma', Sel_{1,2}(Q_0', i_*(0), y), i_*) \subseteq O_{all}(\Sigma', Q_0', i_*).$$

Thus $(i_*, o_*) \in IO(B')$.

Lemma 13. *Assume that a block B has a NCMS representation, $(i_*, o_*) \in IO(B)$, and $i_*(0) \downarrow$. Then there exists a deterministic causal block B' such that $In(B') = In(B)$, $Out(B') = Out(B)$, $Op(B')(i) \subseteq Op(B)(i)$ for each $i \in Sb(In(B), W)$ such that $i(0) \downarrow= i_*(0)$, and $(i_*, o_*) \in IO(B')$.*

Proof. Consider the following cases.

1) Either $dom(i_*) = \{0\}$, or $o_* = \bot$, and also the inclusion $dom(o) \subseteq \{0\}$ holds for each $(i, o) \in IO(B)$ such that $(i_*, o_*) \preceq (i, o)$.

Let us define a function $O : Sb(In(B), W) \to 2^{Sb(Out(B), W)}$ as follows: $O(\bot) = \{\bot\}$ and $O(i) = \{o_*\}$, if $i \neq \bot$. Then $O(i)$ is a singleton set for each i. Moreover, $dom(o_*) \subseteq \{0\}$, so $dom(o) \subseteq dom(i)$ holds for all i, o such that $o \in O(i)$. Then there exists a deterministic block B' such that $In(B) = In(B')$, $Out(B) = Out(B')$, and $Op(B') = O$. If $o_1 \in Op(B')(i_1)$ and $o_2 \in Op(B')(i_2)$ for some i_1, i_2 such that $i_1|_{[0,t]} = i_2|_{[0,t]}$ for some $t \in T$, then $i_1 = \bot$ if and only if $i_2 = \bot$, so $o_1 = o_2$, whence $o_1|_{[0,t]} = o_2|_{[0,t]}$. Thus B' is causal.

Moreover, $o_* \in O(i_*) = Op(B')(i_*)$, because $i_* \neq \bot$. Then $(i_*, o_*) \in IO(B')$.

Let $i \in Sb(In(B), W)$ and $i(0) \downarrow = i_*(0)$.

Consider the case when $o_* = \bot$. Then because $(i_*, o_*) \in IO(B)$ and $(i_*|_{\{0\}}, \bot) \preceq (i_*, o_*)$, we have $(i_*|_{\{0\}}, \bot) \in IO(B)$ by Lemma 7. Then because $i_*|_{\{0\}} \preceq i$, by Lemma 8 there exists $o' \in Op(B)(i)$ such that $(i_*|_{\{0\}}, \bot) \preceq (i, o')$. Then because $i_*(0) \downarrow$, we have $o' = \bot$. Then $\{\bot\} = \{o_*\} = Op(B')(i) \subseteq Op(B)(i)$.

Consider the case when $o_* \neq \bot$. Then $dom(i_*) = \{0\}$ and because $(i_*, o_*) \in IO(B)$ and $i_* = i_*|_{\{0\}} \preceq i$, by Lemma 8 there exists $o' \in Op(B)(i)$ such that $(i_*, o_*) \preceq (i, o')$. Then $dom(o') \subseteq \{0\}$ and $dom(o_*) = \{0\}$, so $o' = o_*$ and $\{o_*\} = Op(B')(i) \subseteq Op(B)(i)$.

Thus B' satisfies the statement of the lemma.

2) $\{0\} \subset dom(i_*)$, $o_* \neq \bot$, and the inclusion $dom(o) \subseteq \{0\}$ holds for each $(i, o) \in IO(B)$ such that $(i_*, o_*) \preceq (i, o)$.

Then $\{0\} \subset dom(i_*)$ and $dom(o_*) = \{0\}$, so by Lemma 11 there exists a subblock $B' \trianglelefteq B$ such that B' has a NCMS representation and $Op(B')(i_*) = \{o_*\}$.

By Lemma 9, B' is weakly nonanticipative. Consider the following cases.

2.1) There exists $(i_0, o_0) \in IO(B')$ such that $i_0(0) \downarrow = i_*(0)$ and $\{0\} \subset dom(o_0)$. Then by Lemma 12 (applied to B'), there exists a deterministic block B'' which has a NCMS representation, such that $In(B'') = In(B') = In(B)$, $Out(B'') = Out(B') = Out(B)$, $Op(B'')(i) \subseteq Op(B')(i) \subseteq Op(B)(i)$ for each $i \in Sb(In(B'), W)$ such that $i(0) \downarrow = i_0(0) = i_*(0)$, and $(i_0, o_0) \in IO(B'')$. Then $Op(B'')(i_*) \subseteq Op(B')(i_*) = \{o_*\}$, so $(i_*, o_*) \in IO(B'')$. Besides, B'' is causal by Lemma 9 and Lemma 2. Then B'' satisfies the statement of the lemma.

2.2) For each $(i, o) \subseteq IO(B')$, if $i(0) \downarrow = i_*(0)$, then $\{0\} \subset dom(o)$ is not satisfied (which is implies the inclusion $dom(o) \subseteq \{0\}$).

Let B_0' be a block such that $In(B_0') = In(B)$, $Out(B_0') = Out(B)$, and $Op(B_0')(i) = Op(B')(i)$, if $i(0) \downarrow = i_*(0)$, and $Op(B_0')(i) = \{\bot\}$, otherwise. Obviously, B_0' is indeed correctly defined as a block.

Let us show that B_0' is weakly nonanticipative. Let $A \in \mathcal{T}_0$, $i_1, i_2 \in Sb(In(B_0'), W)$, and $i_1|_A = i_2|_A$. If $A = \emptyset$ or $i_1 = \bot$ or $i_2 = \bot$, then

$$\{o|_A | o \in Op(B_0')(i_1)\} = \{\bot\} = \{o|_A | o \in Op(B_0')(i_2)\}.$$

Assume that $0 \in A \cap dom(i_1) \cap dom(i_2)$. If $i_1(0) = i_*(0)$, then $i_2(0) = i_*(0)$, whence $Op(B_0')(i_j) = Op(B')(i_j)$ for $j = 1, 2$, and so

$$\{o|_A | o \in Op(B_0')(i_1)\} = \{o|_A | o \in Op(B_0')(i_2)\}$$

because B' is weakly nonanticipative. Otherwise, $i_2(0) = i_1(0) \neq i_*(0)$, whence $Op(B'_0)(i_1) = Op(B'_0)(i_2) = \{\bot\}$. Then

$$\{o|_A | o \in Op(B'_0)(i_1)\} = \{\bot\} = \{o|_A | o \in Op(B'_0)(i_2)\}.$$

We conclude that B'_0 is weakly nonanticipative. Moreover, $dom(o) \subseteq \{0\}$ for each $(i, o) \in IO(B'_0)$. Then B'_0 is strongly nonanticipative by Lemma 10, so it has some deterministic causal sub-block $B'' \trianglelefteq B'_0$ (because $IO(B'_0) \neq \emptyset$). Then $Op(B'')(i_*) \subseteq Op(B'_0)(i_*) = Op(B')(i_*) = \{o_*\}$, whence $(i_*, o_*) \in IO(B'')$. Besides, $In(B'') = In(B)$, $Out(B'') = Out(B)$, and for each $i \in Sb(In(B), W)$ such that $i(0) \downarrow= i_*(0)$ we have

$$Op(B'')(i) \subseteq Op(B'_0)(i) = Op(B')(i) \subseteq Op(B)(i).$$

Then B'' satisfies the statement of the lemma.

3) There exists $(i_0, o_0) \in IO(B)$ such that $(i_*, o_*) \preceq (i_0, o_0)$ and $dom(o_0) \subseteq \{0\}$ does not hold. Then $\{0\} \subset dom(o_0)$, $i_0(0) \downarrow= i_*(0)$, and by Lemma 12 there exists a deterministic block B' which has a NCMS representation, such that $In(B') = In(B)$, $Out(B') = Out(B)$, $Op(B')(i) \subseteq Op(B)(i)$ for each $i \in Sb(In(B), W)$ such that $i(0) \downarrow= i_0(0) = i_*(0)$, and $(i_0, o_0) \in IO(B')$. Then B' is weakly nonanticipative by Lemma 9, so it is causal by Lemma 2. Because $(i_0, o_0) \in IO(B')$ and $(i_*, o_*) \preceq (i_0, o_0)$, we have $(i_*, o_*) \in IO(B')$ by Theorem 3. Then B' satisfies the statement of the lemma.

Lemma 14. *Assume that a block B has a NCMS representation. Then B is strongly nonanticipative.*

Proof. Let us fix an arbitrary $(i_0, o_0) \in IO(B)$.

If $i_0 = \bot$, then let $i_* = \{0\} \mapsto []$ and o_* be an arbitrary member of $Op(B)(i_*)$. Otherwise, i.e. if $i_0 \neq \bot$, then let $i_* = i_0$ and $o_* = o_0$. In both cases we have defined a pair (i_*, o_*) such that $(i_*, o_*) \in IO(B)$ and $i_* \neq \bot$.

Denote $D = ^{In(B)} W$. For each each $d \in D$ let $i_d = \{0\} \mapsto d$, if $d \neq i_*(0)$ and $i_d = i_*$, if $d = i_*(0)$. Then $i_d(0) \downarrow= d$ and $Op(B)(i_d) \neq \emptyset$ for each $d \in D$ and $o_* \in Op(B)(i_{i_*(0)})$. Then there exists a (selector) function $f : D \to Sb(Out(B), W)$ such that $f(d) \in Op(B)(i_d)$ for each $d \in D$ and $f(i_*(0)) = o_*$.

Then by Lemma 13, for each $d \in D$ let us choose a deterministic causal block B_d such that $In(B_d) = In(B)$, $Out(B_d) = Out(B)$, $Op(B_d)(i) \subseteq Op(B)(i)$ for each $i \in Sb(In(B), W)$ such that $i(0) \downarrow= i_d(0)$, and $(i_d, f(d)) \in IO(B_d)$.

Let $O : Sb(In(B), W) \to 2^{Sb(Out(B), W)}$ be a function such that $O(i) = Op(B_{i(0)})(i)$, if $i \neq \bot$ and $O(\bot) = \{\bot\}$.

Then $O(i) \neq \emptyset$ for all i and $dom(o) \subseteq dom(i)$ whenever $o \in O(i)$. Then there exists a block B' such that $In(B') = In(B)$, $Out(B') = Out(B)$, $Op(B') = O$.;

Because for each $d \in D$ the block B_d is deterministic, B' is deterministic.

Let us show that $B' \trianglelefteq B$. Let $(i, o) \in IO(B')$. If $i = \bot$, then $(i, o) = (\bot, \bot) \in IO(B)$. Otherwise, $o \in O(i) = Op(B_{i(0)})(i) \subseteq Op(B)(i)$, because $i(0) = i_{i(0)}(0)$, whence $(i, o) \in IO(B)$. Thus $B' \trianglelefteq B$.

Let us show that B' is causal. Let $i, i' \in Sb(In(B'), W)$, $t \in T$, $i|_{[0,t]} = i'|_{[0,t]}$, $o \in Op(B')(i)$, and $o' \in Op(B')(i')$. If $i = \bot$ or $i' = \bot$, then $i = i' = o = o' = \bot$, so $o|_{[0,t]} = o'|_{[0,t]}$. Consider the case when $i \neq \perp$ and $i' \neq \perp$. Then $i(0) \downarrow$, $i'(0) \downarrow$, and $i(0) = i'(0)$. Denote $d = i(0)$. Then $o \in Op(B')(i) = Op(B_d)(i)$ and $o' \in Op(B')(i') = Op(B_d)(i')$,whence $o|_{[0,t]} = o'|_{[0,t]}$, because B_d is causal.

We conclude that B' is causal. Moreover, $Op(B')(i_*) = Op(B_{i_*(0)})(i_*) = Op(B_{i_*(0)})(i_{i_*(0)}) = \{o_*\}$. Then $(i_*, o_*) \in IO(B')$. If $i_0 \neq \perp$, this implies that $(i_0, o_0) = (i_*, o_*) \in IO(B')$. Otherwise, i.e. if $i_0 = \perp$, then $(i_0, o_0) \in IO(B')$.

We conclude that for each $(i_0, o_0) \in IO(B)$ there exists a deterministic causal sub-block $B' \unlhd B$ such that $(i_0, o_0) \in IO(B')$. So B is strongly nonanticipative.

Now we can prove Theorem 7.

Proof (of Theorem 7). Let (Σ, Q_0) be an initial I/O NCMS. Then by Lemma 6 and Lemma 14, it is a NCMS representation of a strongly nonanticipative block.

Now assume that (Σ, Q_0) is a NCMS representation of each of the blocks B_1 and B_2. Then by Lemma 6, B_1 and B_2 are semantically identical.   □

Acknowledgments. I would like to thank Dr. Martin Strecker and Prof. Louis Féraud of Institut de Recherche en Informatique de Toulouse (IRIT), France and Prof. Mykola Nikitchenko of Taras Shevchenko National University of Kyiv, Ukraine for their comments and advices regarding this work.

References

1. Baheti, R., Gill, H.: Cyber-physical systems. The Impact of Control Technology, 161–166 (2011)
2. Lee, E.A., Seshia, S.A.: Introduction to embedded systems: A cyber-physical systems approach. Lulu.com (2013)
3. Zadeh, L.A., Desoer, C.A.: Linear System Theory: The State Space Approach. McGraw-Hill (1963)
4. Zadeh, L.A.: The concepts of system, aggregate, and state in system theory (1969)
5. Kalman, R.E., Falb, P.L., Arbib, M.A.: Topics in Mathematical System Theory (Pure & Applied Mathematics S.). McGraw-Hill Education (1969)
6. Padulo, L., Arbib, M.: System theory: a unified state-space approach to continuous and discrete systems. W.B. Saunders Company (1974)
7. Klir, G.J.: Facets of Systems Science (IFSR International Series on Systems Science and Engineering). Springer (2001)
8. Wymore, A.W.: A mathematical theory of systems engineering: the elements. Wiley (1967)
9. Mesarovic, M.D., Takahara, Y.: Abstract Systems Theory. LNCIS, vol. 116. Springer, Heidelberg (1989)
10. Zeigler, B.P., Praehofer, H., Kim, T.G.: Theory of modeling and simulation: integrating discrete event and continuous complex dynamic systems. Academic Press (2000)
11. Matrosov, V.M., Anapolskiy, L., Vasilyev, S.: The method of comparison in mathematical systems theory. Nauka, Novosibirsk (1980) (in Russian)
12. Willems, J.C.: Paradigms and puzzles in the theory of dynamical systems 36(3), 259–294 (1991)

13. Polderman, J.W., Willems, J.C.: Introduction to mathematical systems theory: a behavioral approach. Springer, Berlin (1997)
14. Lin, Y.: General systems theory: A mathematical approach. Springer (1999)
15. Seising, R.: Cybernetics, system(s) theory, information theory and fuzzy sets and systems in the 1950s and 1960s. Information Sciences 180(23), 4459–4476 (2010)
16. Liu, X., Matsikoudis, E., Lee, E.A.: Modeling timed concurrent systems. In: Baier, C., Hermanns, H. (eds.) CONCUR 2006. LNCS, vol. 4137, pp. 1–15. Springer, Heidelberg (2006)
17. Matsikoudis, E., Lee, E.A.: On fixed points of strictly causal functions. Technical Report UCB/EECS-2013-27, EECS Department, University of California, Berkeley (April 2013)
18. Ball, J.: Finite time blow-up in nonlinear problems. Nonlinear Evolution Equations, 189–205 (1978)
19. Galaktionov, V., Vazquez, J.L.: The problem of blow-up in nonlinear parabolic equations. Discrete and Continuous Dynamical Systems 8(2), 399–433 (2002)
20. Goriely, A.: Integrability and nonintegrability of dynamical systems, vol. 19. World Scientific Publishing Company (2001)
21. Goebel, R., Sanfelice, R.G., Teel, A.: Hybrid dynamical systems 29(2), 28–93 (2009)
22. Henzinger, T.A.: The theory of hybrid automata. In: Proc. Eleventh Annual IEEE Symp. Logic in Computer Science, LICS 1996, pp. 278–292 (1996)
23. Zhang, J., Johansson, K.H., Lygeros, J., Sastry, S.: Zeno hybrid systems. International Journal of Robust and Nonlinear Control 11(5), 435–451 (2001)
24. Ivanov, I.: An abstract block formalism for engineering systems. In: Ermolayev, V., Mayr, H.C., Nikitchenko, M., Spivakovsky, A., Zholtkevych, G., Zavileysky, M., Kravtsov, H., Kobets, V., Peschanenko, V.S. (eds.) ICTERI. CEUR Workshop Proceedings, vol. 1000, pp. 448–463. CEUR-WS.org (2013)
25. Ivanov, I.: On existence of total input-output pairs of abstract time systems. In: Ermolayev, V., Mayr, H.C., Nikitchenko, M., Spivakovsky, A., Zholtkevych, G. (eds.) ICTERI 2013. CCIS, vol. 412, pp. 308–331. Springer, Heidelberg (2013)
26. Hájek, O.: Theory of processes, i. Czechoslovak Mathematical Journal 17, 159–199 (1967)
27. Windeknecht, T.: Mathematical systems theory: Causality. Mathematical Systems Theory 1(4), 279–288 (1967)
28. Nikitchenko, N.S.: A composition nominative approach to program semantics. Technical report, IT-TR 1998-020, Technical University of Denmark (1998)
29. Ivanov, I.: A criterion for existence of global-in-time trajectories of non-deterministic markovian systems. In: Ermolayev, V., Mayr, H.C., Nikitchenko, M., Spivakovsky, A., Zholtkevych, G. (eds.) ICTERI 2012. CCIS, vol. 347, pp. 111–130. Springer, Heidelberg (2013)
30. Frankowska, H.: Optimal control under state constraints. In: Proceedings of the International Congress of Mathematicians, Hyderabad, India, August 19-27, pp. 2915–2942 (2010)
31. Ivanov, I.: Investigation of abstract systems with inputs and outputs as partial functions of time. PhD thesis, Université Paul Sabatier, France and Taras Shevchenko National University of Kyiv, Ukraine (to appear, 2014)

Complexity Information Flow
in a Multi-threaded Imperative Language

Jean-Yves Marion and Romain Péchoux

Université de Lorraine, CNRS and INRIA
LORIA, UMR 7503
Nancy, France
{jean-yves.marion,romain.pechoux}@loria.fr

Abstract. In this paper, we propose a type system to analyze the time consumed by multi-threaded imperative programs with a shared global memory, which delineates a class of safe multi-threaded programs. We demonstrate that a safe multi-threaded program runs in polynomial time if (i) it is strongly terminating wrt a non-deterministic scheduling policy or (ii) it terminates wrt a deterministic and quiet scheduling policy. As a consequence, we also characterize the set of polynomial time functions. The type system presented is based on the fundamental notion of data tiering, which is central in implicit computational complexity. It regulates the information flow in a computation. This aspect is interesting in that the type system bears a resemblance to typed based information flow analysis and notions of non-interference. As far as we know, this is the first characterization by a type system of polynomial time multi-threaded programs.

Keywords: Implicit computational complexity, Ptime, multi-threaded imperative language, non-interference, type system.

1 Introduction

The objective of this paper is to study the notion of complexity flow analysis introduced in [20,18] in the setting of concurrency. Our model of concurrency is a simple multi-threaded imperative programming language where threads communicate through global shared variables. The measure of time complexity that we consider for multi-threaded programs is the processing time. That is the total time for all threads to complete their tasks. As a result, the time measure gives an upper bound on the number of scheduling rounds. The first outcome of this paper is a novel type system, which guarantees that each strongly terminating safe multi-threaded program runs in polynomial time (See Section 3.2 and Theorem 6). Moreover, the runtime upper bound holds for all thread interactions. As a simple example, consider the two-thread program:

$$x:\ \mathtt{while}(X^1 == Y^1)\{\mathtt{skip}\} \quad y:\ \mathtt{while}(X^1 \neq Y^1)\{\mathtt{skip}\}$$
$$C; \qquad\qquad\qquad C';$$
$$X^1 := \neg X^1 \qquad\qquad\quad Y^1 := \neg Y^1$$

T V Gopal et al. (Eds.): TAMC 2014, LNCS 8402, pp. 124–140, 2014.
© Springer International Publishing Switzerland 2014

This example illustrates a simple synchronization protocol between two threads x and y. Commands C and C' are critical sections, which are assumed not to modify X and Y. The operator $\neg$ denotes boolean negation. Both threads are safe if commands C and C' are safe with respect to the same typing environment. Our first result establishes that this two-thread program runs in polynomial time (in the size of the initial shared variable values) if it is strongly terminating and safe.

Then, we consider a class of deterministic schedulers, that we call quiet (see Section 9). The class of deterministic and quiet schedulers contains all deterministic scheduling policies which depend only on threads. A typical example is a round-robin scheduler. The last outcome of this paper is that a safe multi-threaded program which is terminating wrt a deterministic and quiet scheduler, runs in polynomial time. Despite the fact that it is not strongly terminating, the two-thread program (see below) terminates under a round-robin scheduler, if C and C' terminate.

$$x : \; \texttt{while}(X^1 > 0) \qquad y : \; \texttt{while}(Z^1 > 0)$$
$$\{C; \qquad\qquad\qquad\quad \{C';$$
$$Z^1 \texttt{:=} 0 : 1\} : 1 \qquad\qquad X^1 \texttt{:=} 0 : 1\} : 1$$

As a result, if commands C and C' are safe, then this two-thread program runs in polynomial time wrt to a round-robin scheduler. The last outcome consists in that if we just consider one-thread programs, then we characterize exactly FPtime, which is the class of polynomial time functions. (See Theorem 7).

The first rationale behind our type system comes from data-ramification concept of Bellantoni and Cook [5] and Leivant [17]. The type system has two atomic types **0** and **1** that we call tiers. The type system precludes that values flow from tier **0** to tier **1** variables. Therefore, it prevents circular algorithmic definitions, which may possibly lead to an exponential length computation. More precisely, explicit flow from **0** to **1** is forbidden by requiring that the type level of the assigned variable is less or equal to the type level of the source expression. Implicit flow is prevented by requiring that (i) branches of a conditional are of the same type and (ii) guard and body of while loops are of tier **1**. If we compare with the data-ramification concept of [5,17], tier **1** parameters correspond to variables on which a ramified recursion is performed whereas tier **0** parameters correspond to variables on which recursion is forbidden.

The second rationale behind our type system comes from secure flow analysis. In order to have an overview on information flow analysis, see Sabelfeld and Myers survey [24]. In [26] for sequential imperative programs and in [25] for multi-threaded imperative programming language, Irvine, Smith and Volpano define a type system to certify confidentiality policies. Types are based on security levels say H (High) and L (Low). The type system prevents the leak of information from level H to level L. This is similar to the principle governing our type system: **0** (resp. **1**) corresponds to H (resp. L). In fact, our approach rather coincides with an integrity policy [6] (i.e the rule "no read down") than with a confidentiality one [4]. A key notion is non-interference. We establish a first non-interference result which states that values stored in tier **1** variables are independent from tier 0 variables. See Section 5 for a precise statement. Then, we demonstrate a temporal non-interference property which expresses that the length of while loops only depends on tier 1 variables, see Section 6. The temporal non-interference property is the crucial point to establish complexity bounds. **The main contributions** of this paper are the following:

1. We demonstrate the ability of the complexity flow analysis for the implicit computational complexity (ICC) field. It is worth noticing that compared to ICC's results, the expressivity is quite interesting. This point is illustrated by several examples.
2. To our knowledge, this is the first type system which provides a polynomial time upper bound on a class of multi-threaded programs.

Beside complexity analysis, we may think to applications to security. Indeed, Nowak and Zhang [23] formalize cryptographic proofs thanks to a type system in order to guarantee polynomial-time computations. Here, this work could provide a smoother framework where we can deal with multiple adversaries in security proofs.

Related Works. Implicit Computational Complexity (ICC) was an important source of inspiration for this paper. Besides the works of Bellantoni, Cook and Leivant already cited, there are other works on light logics [10,3], on linear types [12], and interpretation methods [7,21], just to mention a few. There are also works on resource control of imperative language like [13,14,22]. Only a few studies based on ICC methods are related to resource control of concurrent computational models. In [2], a bound on the resource needed by synchronous cooperative threads in a functional framework is computed. The paper [1] provides a static analysis for ensuring feasible reactivity in a synchronous π-calculus. In [19] an elementary affine logic is introduced to tame the complexity of a modal call-by-value lambda calculus with multi-threading and side effects. There are also works on the termination of multi-threaded imperative languages [9]. In this paper, we separate complexity analysis from termination analysis but the tools on termination can be combined with our results. In the complexity flow analysis framework, the main focus of [18] is sequential computation over evolving structure. In [11], we present a characterization of polynomial space computation. Finally, the type system of [20] for a sequential imperative language is similar to [26] while this work is closer to [25] for multi-threaded imperative programming language.

2   A Complexity Flow Type System

2.1   A Multi-threaded Programming Language

We introduce a multi-threaded imperative programming language similar to the language of [25,8] and which is an extension of the simple while-imperative programming language of [15]. A multi-threaded program consists in a finite set of threads where each thread is a while-program. Threads run concurrently on a common shared memory. A thread interacts with other threads by reading and writing on the shared memory.

Commands and expressions are built from a set $\mathbb{V}$ of variables, and a set $\mathbb{O}$ of operators of fixed arity including constants (operators of arity 0) as follows:

$$
\begin{array}{lll}
\textit{Expressions} & E_1, \ldots, E_n ::= X \mid op(E_1, \ldots, E_n) & X \in \mathbb{V}, op \in \mathbb{O} \\
\textit{Commands} & C, C' \quad ::= X := E \mid C \; ; \; C' \mid \texttt{skip} \\
& \quad \mid \texttt{if } E \texttt{ then } C \texttt{ else } C' \\
& \quad \mid \texttt{while}(E)\{C\}
\end{array}
$$

A multi-threaded program M (or just a program when there is no ambiguity) is a finite map from thread identifiers $x, y, \ldots$ to commands. We write $dom(M)$ to denote the set of thread identifiers. Note also that we do not consider the ability of generating new threads. Let $\mathcal{V}(I)$ be the set of variables occurring in I, where I is an expression, a command or a multi-threaded program.

2.2 Semantics, Termination and Time Usage

We give a standard small step operational semantics for multi-threaded programs. Let $\mathbb{W}$ be the set of words[1] over a finite alphabet Σ including two words tt and ff that denote true and false. The length of a word d is denoted $|d|$. A store μ is a finite mapping from $\mathbb{V}$ to $\mathbb{W}$. We write $\mu[X_1 \leftarrow d_1, \ldots, X_n \leftarrow d_n]$ to mean the store μ' where X_i is updated to d_i.

The evaluation rules for expressions and commands are given in Figure 1. Each operator of arity n is interpreted by a total function $[\![op]\!] : \mathbb{W}^n \mapsto \mathbb{W}$. The judgment $\mu \vDash E \xrightarrow{e} d$ means that the expression E is evaluated to $d \in \mathbb{W}$ wrt μ. A configuration c is either a pair $\mu \vDash C$ composed of a store and of a command, or a store μ. The judgment $\mu \vDash C \xrightarrow{s} \mu'$ expresses that C terminates and outputs the store μ'. $\mu \vDash C \xrightarrow{s} \mu' \vDash C'$ means that the evaluation of C is still in progress: the command has evolved to C' and the store has been updated to μ'.

For a multi-threaded program M, the store μ plays the role of a global memory shared by all threads. The store μ is the only way for threads to communicate. The definition of the global relation $\xrightarrow{g}$ is given in Figure 1, where $M - x$ is the restriction of M to $dom(M) - \{x\}$ and $M[x := C_1]$ is the map M where the command assigned to x is updated to C_1. At each step, a thread x is chosen non-deterministically. Then, one step of x is performed and the control returns to the upper level. Note that the rule *(Stop)* halts the computation of a thread. In what follows, let $\emptyset$ be a notation for the (empty) multi-threaded program (i.e. all threads have terminated). We will discuss about deterministic scheduling policies in the last section.

A multi-threaded program M is strongly terminating, noted $M \Downarrow$, if for any store, all reduction sequences starting from M are finite. Let $\xrightarrow{h}{}^{t}$ be the t-fold self composition and $\xrightarrow{h}{}^{*}$ be the reflexive and transitive closure of the relation

---

[1] Our result could be generalized to other domains such as binary trees or lists. We have restricted this study to words to have a concise presentation.

$$\frac{}{\mu \vDash X \overset{\text{e}}{\to} \mu(X)} \qquad \frac{\mu \vDash E_1 \overset{\text{e}}{\to} d_1 \quad \dots \quad \mu \vDash E_n \overset{\text{e}}{\to} d_n}{\mu \vDash op(E_1, \dots, E_n) \overset{\text{e}}{\to} \llbracket op \rrbracket(d_1, \dots, d_n)}$$

$$\frac{}{\mu \vDash \texttt{skip} \overset{\text{s}}{\to} \mu} \qquad \frac{\mu \vDash E \overset{\text{e}}{\to} d}{\mu \vDash X := E \overset{\text{s}}{\to} \mu[X \leftarrow d]} \qquad \frac{\mu \vDash C_1 \overset{\text{s}}{\to} \mu_1}{\mu \vDash C_1 \; ; \; C_2 \overset{\text{s}}{\to} \mu_1 \vDash C_2}$$

$$\frac{\mu \vDash C_1 \overset{\text{s}}{\to} \mu_1 \vDash C_1'}{\mu \vDash C_1 \; ; \; C_2 \overset{\text{s}}{\to} \mu_1 \vDash C_1'; C_2} \qquad \frac{\mu \vDash E \overset{\text{e}}{\to} w, \; w \in \{\texttt{tt}, \texttt{ff}\}}{\mu \vDash \texttt{if } E \texttt{ then } C_{\texttt{tt}} \texttt{ else } C_{\texttt{ff}} \overset{\text{s}}{\to} \mu \vDash C_w}$$

$$\frac{\mu \vDash E \overset{\text{e}}{\to} \texttt{ff}}{\mu \vDash \texttt{while}(E)\{C\} \overset{\text{s}}{\to} \mu} \qquad \frac{\mu \vDash E \overset{\text{e}}{\to} \texttt{tt}}{\mu \vDash \texttt{while}(E)\{C\} \overset{\text{s}}{\to} \mu \vDash C; \texttt{while}(E)\{C\}} (W_{\texttt{tt}})$$

$$\frac{M(x) = C \quad \mu \vDash C \overset{\text{s}}{\to} \mu_1}{\mu \vDash M \overset{\text{g}}{\to} \mu_1 \vDash M - x} (Stop) \qquad \frac{M(x) = C \quad \mu \vDash C \overset{\text{s}}{\to} \mu_1 \vDash C_1}{\mu \vDash M \overset{\text{g}}{\to} \mu_1 \vDash M[x := C_1]} (Step)$$

Fig. 1. Small step semantics of expressions, commands and multi-threaded programs

$\overset{\text{h}}{\to}$, $\text{h} \in \{\texttt{s}, \texttt{g}\}$. The running time of a strongly terminating program M is the function $Time_M$ from $\mathbb{W}^n$ to $\mathbb{N}$ defined by:

$$Time_M(\boldsymbol{d}) = \max\{t \mid \mu_0[X_i \leftarrow d_i] \vDash M \overset{\text{g}^t}{\to} \mu \vDash \emptyset\}$$

where μ_0 is the empty store that maps each variable to the empty word $\epsilon \in \mathbb{W}$. Throughout, $\mu_0[X_1 \leftarrow d_1, \dots, X_n \leftarrow d_n]$ is called the *initial store*.

A strongly terminating multi-threaded program M is running in polynomial time if for all $d_1, \dots, d_n \in \mathbb{W}$,

$$Time_M(d_1, \dots, d_n) \leq Q(\max_{i=1,n} |d_i|)$$

for some polynomial Q. Observe that, in the above definition, the time consumption of an operator is considered as constant, which is fair if operators are supposed to be computable in polynomial time.

2.3   The Type System

Atomic types are elements of the boolean lattice $(\{\mathbf{0}, \mathbf{1}\}, \preceq, \mathbf{0}, \vee, \wedge)$ where $\mathbf{0} \preceq \mathbf{1}$. We call them *tiers* accordingly to the data ramification principle of [16]. We use $\alpha, \beta, \dots$ for tiers. A variable typing environment Γ is a finite mapping from $\mathbb{V}$ to $\{\mathbf{0}, \mathbf{1}\}$, which assigns a single tier to each variable. An operator typing environment Δ is a mapping that associates to each operator op a set of operator types $\Delta(op)$, where the operator types corresponding to an operator of arity n are of the shape $\alpha_1 \to \dots \alpha_n \to \alpha$ with $\alpha_i, \alpha \in \{\mathbf{0}, \mathbf{1}\}$ using implicit right associativity of $\to$. We write $dom(\Gamma)$ (resp. $dom(\Delta)$) to denote the set of variables typed by Γ (resp. the set of operators typed by Δ). Figure 2 gives the type system for expressions and commands. Given a multi-threaded program M, a variable typing environment Γ and an operator typing environment Δ, M is *well-typed* if for every $x \in dom(M)$, $\Gamma, \Delta \vdash M(x) : \alpha$ for some tier α.

$$\frac{\Gamma(X) = \alpha}{\Gamma, \Delta \vdash X : \alpha} \qquad \frac{\Gamma, \Delta \vdash X : \beta \qquad \Gamma, \Delta \vdash E : \alpha}{\Gamma, \Delta \vdash X := E : \beta} \beta \preceq \alpha$$

$$\frac{\Gamma, \Delta \vdash E_1 : \alpha_1 \ldots \Gamma, \Delta \vdash E_n : \alpha_n \qquad \alpha_1 \to \ldots \to \alpha_n \to \alpha \in \Delta(op)}{\Gamma, \Delta \vdash op(E_1, \ldots, E_n) : \alpha}$$

$$\frac{\Gamma, \Delta \vdash E : 1 \qquad \Gamma, \Delta \vdash C : \alpha}{\Gamma, \Delta \vdash \mathtt{while}(E)\{C\} : 1} \qquad \frac{\Gamma, \Delta \vdash C : \alpha \qquad \Gamma, \Delta \vdash C' : \beta}{\Gamma, \Delta \vdash C \; ; \; C' : \alpha \vee \beta}$$

$$\frac{}{\Gamma, \Delta \vdash \mathtt{skip} : \alpha} \qquad \frac{\Gamma, \Delta \vdash E : \alpha \qquad \Gamma, \Delta \vdash C : \alpha \qquad \Gamma, \Delta \vdash C' : \alpha}{\Gamma, \Delta \vdash \mathtt{if}\ E\ \mathtt{then}\ C\ \mathtt{else}\ C' : \alpha}$$

Fig. 2. Type system for expressions and commands

Notice that the subject reduction property is not valid, because we don't explicitly have any subtyping rule. However, a weak subject reduction property holds: If $\Gamma, \Delta \vdash C : \alpha$ and $\mu \vDash C \xrightarrow{\mathtt{s}} \mu' \vDash C'$ then $\Gamma, \Delta \vdash C' : \beta$ where $\beta \preceq \alpha$.

3   Safe Multi-threaded Programs

3.1   Neutral and Positive Operators

As in [20], we define two classes of operators called neutral and positive. For this, let $\trianglelefteq$ be the sub-word relation over $\mathbb{W}$, which is defined by $v \trianglelefteq w$, iff there are u and u' such that $w = u.v.u'$, where . is the concatenation.
An operator op is *neutral* if:

1. either $[\![op]\!] : \mathbb{W} \to \{\mathtt{tt}, \mathtt{ff}\}$ is a predicate;
2. or for all $d_1, \ldots, d_n \in \mathbb{W}$,
   $\exists i \in \{1, \ldots, n\}, [\![op]\!](d_1, \ldots, d_n) \trianglelefteq d_i$.

An operator op is *positive* if there is a constant c_{op} such that:

$$|[\![op]\!](d_1, \ldots, d_n)| \leq \max_i |d_i| + c_{op}$$

A neutral operator is always a positive operator but the converse is not true. In the remainder, we assume that operators are all neutral or positive[2].

---

[2] Actually, a more general condition would be that the initial segment, generated by the closure of neutral operators, is polynomial in the size of the greatest element. Again, we skip this part because of the lack of space and because it is not essential.

3.2   Safe Environments and Safe Multi-threaded Programs

An operator typing environment Δ is *safe* if for each $op \in dom(\Delta)$ of arity n and for each $\alpha_1 \to \ldots \to \alpha_n \to \alpha \in \Delta(op)$, we have $\alpha \preceq \wedge_{i=1,n}\alpha_i$, and if the operator op is positive, but not neutral, then $\alpha = \mathbf{0}$. Given Γ a variable typing environment and Δ a operator typing environment, we say that M is a *safe multi-threaded program* if M is well-typed and Δ is safe.

Intuitively, a tier $\mathbf{0}$ argument is unsafe. This means that it cannot be used as a loop guard. So for "loop-safety" reasons, if an operator has a tier $\mathbf{0}$ argument then the result is necessarily of tier $\mathbf{0}$. In return, a positive operator can increase the size of its arguments. On the other hand, a neutral operator does not increase the size of its arguments. So, we can apply it safely everywhere. The combination of the type system, which guarantees some safety properties on the information flow, and of aforementioned operator specificities is crucial to bound the runtime.

4   Examples

In what follows, let E^α, respectively $C : \alpha$, be a notation meaning that the expression E, respectively command C, is of type α under the considered typing environments.

Example 1. Given a word d, the operator eq_d tests whether or not its argument begins with the prefix d and *pred* computes the predecessor:

$$[\![eq_d]\!](u) = \begin{cases} = \mathtt{tt} & \text{if } u = d.w \\ = \mathtt{ff} & \text{otherwise} \end{cases}$$

$$[\![pred]\!](u) = \begin{cases} = \epsilon & \text{if } u = \epsilon \\ = w & \text{if } u = \ell.w, \ \ell \in \Sigma \end{cases}$$

Both operators are neutral. This means that their types satisfy $\Delta(pred), \Delta(eq_u) \subseteq \{\mathbf{0} \to \mathbf{0}, \mathbf{1} \to \mathbf{1}, \mathbf{1} \to \mathbf{0}\}$ wrt to a safe environment Δ. The operator suc_d adds a prefix d. It is positive, but not neutral. So, $\Delta(suc_d) \subseteq \{\mathbf{1} \to \mathbf{0}, \mathbf{0} \to \mathbf{0}\}$:

$$[\![suc_d]\!](u) = d.u$$

Example 2. Consider the sequential programs add_Y and mul_Z that compute respectively addition and multiplication on unary words using the positive successor operator $+1$, in infix notation, and two neutral operators, -1 and a unary predicate > 0, both in infix notation. Both programs are safe by checking that their main commands are well-typed wrt the safe operator typing environment Δ defined by $\Delta(+1) = \{\mathbf{0} \to \mathbf{0}\}$ and $\Delta(-1) = \Delta(> 0) = \{\mathbf{1} \to \mathbf{1}\}$.

add_Y :
```
while(X¹ > 0)¹{
    X¹:=X¹ − 1; : 1
    Y⁰:=Y⁰ + 1 : 0
```

mul_Z :
```
Z⁰:=0⁰; : 0
while(X¹ > 0)¹{
    X¹:=X¹ − 1; : 1
```

$\}:1$

$$U^1:=Y^1;:1$$
$$\texttt{while}(Y^1>0)^1\{$$
$$Y^1:=Y^1-1;:1$$
$$Z^0:=Z^0+1:0$$
$$\};:1$$
$$Y^1:=U^1:1$$
$$\}:1$$

Example 3. Consider the following multi-thread M composed of two threads x and y computing on unary numbers:

$x:$ **while** $(X^1>0)^1\{$ $y:$ **while** $(Y^1>0)^1\{$
     $Z^0:=Z^0+1;:0$ $Z^0=0;:0$
     $X^1:=X^1-1;:1$ $Y^1:=Y^1-1;:1$
$\}:1$ $\}:1$

This program is strongly terminating. Moreover, given a store μ such that $\mu(X)=n$ and $\mu(Z)=0$, if $\mu\vDash M\overset{g}{\rightarrow}^k\mu'\vDash\emptyset$ then $\mu'(Z)\in[0,n]$. M is safe using an operator typing environment Δ such that $\Delta(-1)=\Delta(>0)=\{1\rightarrow1\}$ and $\Delta(+1)=\{0\rightarrow0\}$ and $M\Downarrow$. Consequently, by Theorem 5, there is a polynomial T such that for each store μ, $k\leq T(\|\mu\|_1)$.

Example 4. Consider the following multi-thread M that shuffles two strings given as inputs:

$x:$ **while** $(\neg eq_\epsilon(X^1))^1\{$
     $Z^0:=concat(head(X^1),Z^0);:0$
          $X^1:=pred(X^1);:1$
$\}:1$

$y:$ **while** $(\neg eq_\epsilon(Y^1))^1\{$
     $Z^0:=concat(head(Y^1),Z^0);:0$
          $Y^1:=pred(Y^1);:1$
$\}:1$

The negation operator $\neg$ and eq_ϵ are unary predicates and consequently can be typed by $1\rightarrow1$. The operator $head$ returns the first symbol of a string given as input and can be typed by $1\rightarrow0$ since it is neutral. The $pred$ operator can typed by $1\rightarrow1$ since its computation is a subterm of the input. Finally, the concat operator that performs the concatenation of the symbol given as first argument with the second argument can be typed by $0\rightarrow0\rightarrow0$ since $|[\![concat]\!](u,v)|=|v|+1$. This program is safe and strongly terminating consequently it also terminates in polynomial time.

Example 5. Consider the following multi-thread M:

$x:$ **while** $(X^1>0)^1\{$ $y:$ **while** $(Y^1>0)^1\{$
     $Y^1:=X^1;:1$ $Z^0:=Z^0+1;:0$
     $X^1:=X^1-1;:1$ $Y^1:=Y^1-1;:1$
$\}:1$ $\}:1$

Observe that, contrarily to previous examples, the guard of y depends on information flowing from X to Y. Given a store μ such that $\mu(X) = n$, $\mu(Y) = \mu(Z) = 0$, if $\mu \vDash M \overset{g^k}{\to} \mu' \vDash \emptyset$ then $\mu'(Z) \in [0, n \times (n + 1)/2]$. This multi-thread is safe with respect to a safe typing operator environment Δ such that $\Delta(-1) = \Delta(> 0) = \{1 \to 1\}$ and $\Delta(+1) = \{0 \to 0\}$. Moreover it strongly terminates. Consequently, it also terminates in polynomial time.

Example 6. The following program computes the exponential:

```
expY(X¹, Y⁰) :
    while(X¹ > 0){
        U?:=Y⁰;: ?
        while(U? > 0){
            Y⁰:=Y⁰ + 1;: 0
            U?:=U? − 1 : ?
        };: 1
        X¹:=X¹ − 1 : 1
    };: 1
```

It is not typable in our formalism. Indeed, suppose that it is typable. The command $Y:=Y+1$ enforces Y to be of tier $\mathbf{0}$ since $+1$ is positive. Consequently, the command $U:=Y$ enforces U to be of tier $\mathbf{0}$ because of typing discipline for assignments. However, the innermost while loop enforces $U > 0$ to be of tier $\mathbf{1}$, so that U has to be of tier $\mathbf{1}$ (because $\mathbf{0} \to \mathbf{1}$ is not permitted for a safe operator typing environment) and we obtain a contradiction.

Example 7. As another counter-example, consider now the addition *badd* on binary words:

```
baddY :
    while(X? > 0)?{
        X?:=X? − 1;: ?
        Y⁰:=Y⁰ + 1 : 0
    } : 1
```

Contrarily to Example 2, the above program is not typable because the operator -1 has now type $\Delta(-1) = \{0 \to 0\}$. Indeed it cannot be neutral since binary predecessor is not a subterm operator. Consequently, -1 is positive and the assignment $X:=X-1$ enforces X to be of type $\mathbf{0}$ whereas the loop guard enforces X to be of tier $\mathbf{1}$. Note that this counter-example is not that surprising in the sense that a binary word of size n may lead to a loop of length 2^n using the -1 operator. Of course this does not imply that the considered typing discipline rejects computations on binary words, it only means that this type system rejects exponential time programs. Consequently, "natural" binary addition algorithms are captured as illustrated by the following program that computes the binary addition on reversed binary words of equal size:

$binary\_add_Z$:
```
while(¬eq_ε(X¹))¹{
    R⁰:=result(bit(X¹), bit(Y¹), bit(C¹)); : 0
    C¹:=carry(bit(X¹), bit(Y¹), bit(C¹)); : 1
    Z⁰:=concat(R⁰, Z⁰); : 0
    X¹:=pred(X¹); : 1
    Y¹:=pred(Y¹); : 1
} : 1
```

As usual, $pred$ is typed by $\mathbf{1} \to \mathbf{1}$. The negation operator $\neg$ and eq_ϵ are predicates and, consequently, can be typed by $\mathbf{1} \to \mathbf{1}$, since they are neutral. The operator bit returns $\mathtt{tt}$ or $\mathtt{ff}$ depending on whether the word given as input has first digit 1 or 0, respectively. Consequently, it can be typed by $\mathbf{1} \to \mathbf{1}$. The operators $carry$ and $result$, that compute the carry and the result of bit addition, can be typed by $\mathbf{1} \to \mathbf{1} \to \mathbf{1} \to \mathbf{1}$ since they are neutral. Finally, the operator $concat(x, y)$ defined by if $[\![bit]\!](x) = i, i \in \{0, 1\}$ then $[\![concat]\!](x, y) = i.y$ is typed by $\mathbf{0} \to \mathbf{0} \to \mathbf{0}$. Indeed it is a positive operator since $|[\![concat]\!](x, y)| = |y| + 1$.

5 Sequential and Concurrent Non-interferences

In this section, we demonstrate that classical non-interference results are obtained through the use of the type system under consideration. For that purpose, we introduce some intermediate lemmata. The confinement Lemma expresses the fact that no tier $\mathbf{1}$ variables are modified by a command of tier $\mathbf{0}$.

Lemma 1 (Confinement). *Let Γ be a variable typing environment and Δ be a safe operator typing environment. If $\Gamma, \Delta \vdash C : \mathbf{0}$, then every variable assigned to in C is of type $\mathbf{0}$, and C does not contain while loops.*

Proof. By induction on the structure of C.   □

The following lemma, called simple security, says that only variables at level $\mathbf{1}$ will have their content read in order to evaluate an expression E of type $\mathbf{1}$.

Lemma 2 (Simple security). *Let Γ be a variable typing environment and Δ be a safe operator typing environment. If $\Gamma, \Delta \vdash E : \mathbf{1}$, then for every $X \in \mathcal{V}(E)$, we have $\Gamma(X) = \mathbf{1}$. Moreover, all operators in E are neutral.*

Proof. By induction on E, and using the fact that E is necessarily only composed of operators of type $\mathbf{1} \to \dots \to \mathbf{1} \to \mathbf{1}$, because the environment is safe.   □

Definition 1. *Let Γ be a variable typing environment and Δ be an operator typing environment.*

- *The equivalence relation $\approx_{\Gamma,\Delta}$ on stores is defined as follows:*
  $\mu \approx_{\Gamma,\Delta} \sigma$ *iff for every $X \in dom(\Gamma)$ s.t. $\Gamma(X) = \mathbf{1}$ we have $\mu(X) = \sigma(X)$*
- *The relation $\approx_{\Gamma,\Delta}$ is extended to commands as follows:*
  1. *If $C = C'$ then $C \approx_{\Gamma,\Delta} C'$*
  2. *If $\Gamma, \Delta \vdash C : \mathbf{0}$ and $\Gamma, \Delta \vdash C' : \mathbf{0}$ then $C \approx_{\Gamma,\Delta} C'$*
  3. *If $C \approx_{\Gamma,\Delta} C'$ and $D \approx_{\Gamma,\Delta} D'$ then $C; D \approx_{\Gamma,\Delta} C'; D'$*

- *Finally, it is extended to configurations as follows:*
  If $C \approx_{\Gamma,\Delta} D$ *and* $\mu \approx_{\Gamma,\Delta} \sigma$ *then* $\mu \vDash C \approx_{\Gamma,\Delta} \sigma \vDash D$

Remark 1. A consequence of Lemma 2 is that if $\mu \approx_{\Gamma,\Delta} \sigma$ and if $\Gamma, \Delta \vdash E : \mathbf{1}$, then computations of E are identical under the stores μ and σ, that is $\mu \vDash E \overset{e}{\to} d$ and $\sigma \vDash E \overset{e}{\to} d$.

We now establish a sequential non-interference Theorem which states that if X is variable of tier $\mathbf{1}$ then the value stored in X is independent from variables of tier $\mathbf{0}$.

Theorem 1 (Sequential non-interference). *Assume that Γ is a variable typing environment and Δ is a safe operator typing environment s.t. $\Gamma, \Delta \vdash C : \alpha$ and $\Gamma, \Delta \vdash D : \alpha$. Assume also that $\mu \vDash C \approx_{\Gamma,\Delta} \sigma \vDash D$. Then, we have:*

- *if $\mu \vDash C \overset{s}{\to} \mu' \vDash C'$ then there are σ' and D' such that*
  $\sigma \vDash D \overset{s}{\to}^{*} \sigma' \vDash D'$ *and* $\mu' \vDash C' \approx_{\Gamma,\Delta} \sigma' \vDash D'$
- *if $\mu \vDash C \overset{a}{\to} \mu'$ then there exists σ' such that*
  $\sigma \vDash D \overset{s}{\to}^{*} \sigma'$ *and* $\mu' \approx_{\Gamma,\Delta} \sigma'$

Proof. First suppose that $\alpha = \mathbf{0}$. Confinement Lemma 1 implies that $\mu' \approx_{\Gamma,\Delta} \sigma'$ since no tier $\mathbf{1}$ variable is changed. Thus, $\mu' \vDash C' \approx_{\Gamma,\Delta} \sigma' \vDash D'$. Second suppose that $\alpha = \mathbf{1}$. We proceed by induction on C. Suppose that C is $\mathtt{while}(E)\{C_1\}$ and the evaluation under μ is:

$$\frac{\mu \vDash E \overset{e}{\to} \mathtt{tt}}{\mu \vDash \mathtt{while}(E)\{C_1\} \overset{s}{\to} \mu \vDash C_1; \mathtt{while}(E)\{C_1\}} \quad (W_{\mathtt{tt}})$$

By Remark 1, the evaluation of E under σ is necessarily $\mathtt{tt}$. Since C is an atomic command, $C \approx_{\Gamma,\Delta} D$ implies $C = D$. As a result, $\sigma \vDash \mathtt{while}(E)\{C_1\} \overset{s}{\to} \sigma \vDash C_1; \mathtt{while}(E)\{C_1\}$. We have $\mu' \approx_{\Gamma,\Delta} \sigma'$ because $\mu = \mu'$ and $\sigma = \sigma'$. We conclude that both configurations are equivalent, that is $\mu' \vDash C' \approx_{\Gamma,\Delta} \sigma' \vDash D'$. The other cases are treated similarly. □

Sequential non-interference can be adapted to multi-threaded programs. For this, we extend $\approx_{\Gamma,\Delta}$ to multi-threaded programs and to configurations:

- If $dom(M) = dom(M')$ and $\forall x \in dom(M)$,
  $M(x) \approx_{\Gamma,\Delta} M'(x)$ then $M \approx_{\Gamma,\Delta} M'$
- If $M \approx_{\Gamma,\Delta} M'$ and $\mu \approx_{\Gamma,\Delta} \sigma$ then $\mu \vDash M \approx_{\Gamma,\Delta} \sigma \vDash M'$

Theorem 2 (Concurrent Non-interference). *Assume that Γ is a variable typing environment, that Δ is a safe operator typing environment such that M is well-typed. Assume also that $\mu \vDash M_1 \approx_{\Gamma,\Delta} \sigma \vDash M_2$. Then, if $\mu \vDash M_1 \overset{g}{\to} \mu' \vDash M_1'$ then there are σ' and M_2' s.t. $\sigma \vDash M_2 \overset{g}{\to}^{*} \sigma' \vDash M_2'$ and $\mu' \vDash M_1 \approx_{\Gamma,\Delta} \sigma' \vDash M_2$.*

Proof. Consequence of Theorem 1. □

$$\frac{\mu \vDash E \overset{e}{\to} d}{\mu \vDash_t X := E \overset{s}{\to} \mu[X \leftarrow d]} \qquad \mu \vDash_t \mathtt{skip} \overset{s}{\to} \mu \qquad \frac{\mu \vDash_0 C_1 \overset{s}{\to} \mu_1}{\mu \vDash_t C_1 \; ; \; C_2 \overset{s}{\to} \mu_1 \vDash_t C_2}$$

$$\frac{\mu \vDash_t C_1 \overset{s}{\to} \mu_1 \vDash_{t'} C_1'}{\mu \vDash_t C_1 \; ; \; C_2 \overset{s}{\to} \mu_1 \vDash_{t'} C_1'; C_2} \qquad \frac{\mu \vDash E \overset{e}{\to} w, \; w \in \{\mathtt{tt}, \mathtt{ff}\}}{\mu \vDash_t \mathtt{if} \; E \; \mathtt{then} \; C_{\mathtt{tt}} \; \mathtt{else} \; C_{\mathtt{ff}} \; \overset{s}{\to} \; \mu \vDash_t C_w}$$

$$\frac{\mu \vDash E \overset{e}{\to} \mathtt{ff}}{\mu \vDash_t \mathtt{while}(E)\{C\} \; \overset{s}{\to} \; \mu} \qquad \frac{\mu \vDash E \overset{e}{\to} \mathtt{tt}}{\mu \vDash_t \mathtt{while}(E)\{C\} \; \overset{s}{\to} \; \mu \vDash_{t+1} C; \mathtt{while}(E)\{C\}} \quad (TW_{tt})$$

$$\frac{M(x) = C \quad \mu \vDash_0 C \overset{s}{\to} \mu'}{\mu \vDash_t M \overset{g}{\to} \mu' \vDash_t M - x} \qquad \frac{M(x) = C \quad \mu \vDash_t C \overset{s}{\to} \mu' \vDash_{t'} C'}{\mu \vDash_t M \overset{g}{\to} \mu' \vDash_{t'} M[x := C']}$$

Fig. 3. Loop length measure for commands and multi-threaded programs

6   Sequential and Concurrent Temporal Non-interferences

Now we establish a property named temporal non-interference. This property ensures that the length of while-loops does not depend on variables of tier $\mathbf{0}$, but depends only on tier $\mathbf{1}$ variables. Consequently, a change in the value of a variable of tier $\mathbf{0}$ does not affect loop lengths.

For this, we define a loop length measure in Figure 3 based on the small step semantics of Figure 1. $\sigma \vDash_0 C \overset{s}{\to}^* \sigma' \vDash_t C'$ holds if t is the number of while-loops, which are unfolded to reach $\sigma' \vDash C'$ from $\sigma \vDash C$, that is t is the number of applications of the Figure 3's rule (TW_{tt}) in a computation. It is convenient to define the relation $\Rightarrow_t$ by $\sigma \vDash C \Rightarrow_t \sigma' \vDash C'$ iff $\sigma \vDash_0 C \overset{s}{\to}^* \sigma' \vDash_t C'$.

Remark 2. If $\Gamma, \Delta \vdash C : \mathbf{0}$ and $\sigma \vDash C \overset{s}{\to}^* \sigma' \vDash C'$ then $\sigma \vDash C \Rightarrow_0 \sigma' \vDash C'$ since there is no while loop inside C, by Lemma 1. Moreover, if $\sigma \vDash C \Rightarrow_t \sigma' \vDash C'$, then for every $k \leq t$ there are σ'' and C'' such that $\sigma \vDash C \Rightarrow_k \sigma'' \vDash C'' \Rightarrow_{t-k} \sigma' \vDash C'$.

Theorem 3 (Temporal non-interference). *Assume that Γ is a variable typing environment and Δ is a safe operator typing environment s.t. $\Gamma, \Delta \vdash C : \alpha$ and $\Gamma, \Delta \vdash D : \alpha$. Assume also that $\mu \vDash C \approx_{\Gamma, \Delta} \sigma \vDash D$. Then, if $\mu \vDash C \Rightarrow_t \mu' \vDash C'$ then there are σ' and D' s.t. $\sigma \vDash D \Rightarrow_t \sigma' \vDash D'$ and $\mu' \vDash C' \approx_{\Gamma, \Delta} \sigma' \vDash D'$.*

Proof. The proof goes by induction on t. Suppose that $t = 0$. This means that no rule (TW_{tt}) has been fired. The conclusion is a consequence of the sequential non-interference Theorem 1.

Next, suppose that $\mu \vDash C \Rightarrow_{t+1} \mu' \vDash C'$. This means that a rule (TW_{tt}) has been applied. So suppose that $C = \mathtt{while}(E)\{C_1\}$ and that $\mu \vDash E \overset{e}{\to} \mathtt{tt}$. First, from the hypothesis $\mu \approx_{\Gamma, \Delta} \sigma$ and Lemma 2, we deduce $\sigma \vDash E \overset{e}{\to} \mathtt{tt}$. Second, since $C \approx_{\Gamma, \Delta} D$, we have $C = D$, by definition of $\approx_{\Gamma, \Delta}$. Since $C' = C_1; C$, we also have $D' = C_1; C$. Thus, $C' \approx_{\Gamma, \Delta} D'$ and $\sigma \vDash D \Rightarrow_{t+1} \sigma' \vDash D'$ hold. Since D runs a while, we have $\mu' = \mu$ and $\sigma' = \sigma$. Thus $\mu' \approx_{\Gamma, \Delta} \sigma'$ holds. So, $\mu' \vDash C' \approx_{\Gamma, \Delta} \sigma' \vDash D'$. The other cases are similar. $\square$

We extend the relation $\Rightarrow_t$ as follows: $\mu \vDash M \Rightarrow_t \mu' \vDash M'$ if and only if $\mu \vDash_0 M \overset{\S}{\to}^* \mu' \vDash_t M'$. As a corollary, we obtain a temporal non-interference result for multi-threaded programs.

Theorem 4 (Concurrent temporal non-interference). *Assume Γ is a variable typing environment and Δ is a safe operator typing environment s.t. M and N are well typed. Assume that $\mu \vDash M \approx_{\Gamma,\Delta} \sigma \vDash N$. Then, if $\mu \vDash M \Rightarrow_t \mu' \vDash M'$ then there are σ' and N' s.t. $\sigma \vDash N \Rightarrow_t \sigma' \vDash N'$ and $\mu' \vDash M' \approx_{\Gamma,\Delta} \sigma' \vDash N'$.*

Proof. Consequence of Theorem 3.   □

7   Analysis of Multi-threaded Program Running Time

An important point is that the number of reachable tier 1 configurations in a computation is polynomially bounded in the size of tier 1 initial values.

Lemma 3. *Let M be a safe multi-threaded program wrt environments Γ and Δ. If $\mu \vDash M \Rightarrow_t \mu' \vDash M'$ then $\forall X \in \mathcal{V}(M)$ such that $\Gamma(X) = 1$ either $\mu'(X) \in \{\text{tt}, \text{ff}\}$ or $\exists Y \in \mathcal{V}(M)$ such that $\Gamma(Y) = 1$ and $\mu'(X) \trianglelefteq \mu(Y)$.*

Proof. Take one global computational step $\mu \vDash M \overset{g}{\to} \mu' \vDash M'$. Let X be a variable assigned to in $M(x)$, for some thread identifier x, such that $\Gamma(X) = 1$. X can only be assigned to an expression E of tier 1. By simple security lemma 2, E only contains neutral operators. It means that either $\mu'(X)$ is a truth value (corresponding to the computation of a predicate) or a subterm of a tier 1 variable value.   □

In the case where a multi-threaded program strongly terminates (i.e. $M \Downarrow$), we now establish that for all thread interactions, the maximal length of while-loops is polynomially bounded in the size of tier 1 values of the initial store. This is a consequence of the temporal non-interference property. Define $\mu \downarrow 1$ as the restriction of the store μ to tier 1 variables. Notice that $\mu \approx_{\Gamma,\Delta} \mu'$ iff $\mu \downarrow 1 = \mu' \downarrow 1$. It is convenient to also define $\| - \|_1$ by $\|\mu\|_1 = \max_{\Gamma(X)=1} |\mu(X)|$.

Theorem 5. *Let M be a safe multi-threaded program such that $M \Downarrow$. There is a polynomial T such that for all stores μ, if $\mu \vDash M \Rightarrow_t \mu' \vDash M'$ then $t \leq T(\|\mu\|_1)$.*

Proof. By Theorem 4, the length of while-loops depends only on variables of tier 1. It implies that if M enters twice into a configuration with the same thread command, and the same values of tier 1, then M is non-terminating. Indeed, it is possible to repeat again the same transition up to infinity. So, a computation never reaches two configurations $\sigma \vDash C$ and $\sigma' \vDash C$ such that $\sigma \approx_{\Gamma,\Delta} \sigma'$. Let μ be the initial store $\mu_0[X_1 \leftarrow d_1, \ldots, X_n \leftarrow d_n]$. For each thread x, define $Config_x = \{(\sigma \downarrow 1, N(x)) \mid \mu \vDash M(x) \overset{s}{\to}^* \sigma \vDash N(x)\}$. The total length of loops is bounded by the cardinality of $\cup_x Config_x$. Since the number of sub-words of a word of size n is bounded by n^2, Lemma 3 implies the number of distinct stores σ reachable from μ is bounded polynomially in $\|\mu\|_1$. As a result, the cardinality of $Config_x$ is also bounded by a polynomial in $\|\mu\|_1$. Thus, there is a polynomial T such that the length of each terminating multi-threaded computation starting from μ is bounded by $T(\|\mu\|_1)$. Finally, we have that $t \leq T(\|\mu\|_1)$. $\qquad\square$

We can now state our first main result:

Theorem 6. *Assume that M is a safe multi-threaded program. Moreover suppose that M strongly terminates. There is a polynomial Q such that:*

$$\forall d_1, \ldots, d_n \in \mathbb{W}, \ Time_M(d_1, \ldots, d_n) \leq Q(\max_{i=1,n}(|d_i|))$$

Proof. We conclude by Theorem 5 and by setting $Q(X) = r.T(X) + r$ for some r, which depends on M. $\qquad\square$

8   A Characterization of Polynomial Time Functions

We now come to a characterization of the set of functions computable in polynomial time. A sequential program M consists in a single thread program (i.e. $dom(M) = \{x\}$) and an output variable, say Y. The partial function $[\![M]\!]$ computed by M is then defined by:

$[\![M]\!](d_1, \ldots, d_n) = w$ iff $\mu_0[X_1 \leftarrow d_1, \ldots, X_n \leftarrow d_n] \models M \overset{g}{\twoheadrightarrow}^* \mu \models \emptyset$ and $\mu(Y) = w$

Theorem 7. *The set of functions computed by strongly terminating and safe sequential programs whose operators compute polynomial time functions is exactly FPtime, which is the set of polynomial time computable functions.*

Proof. By Theorem 6, the execution time of a safe and strongly terminating sequential program is bounded by a polynomial in the size of the initial values. In the other direction, we show that every polynomial time function over the set of words $\mathbb{W}$ can be computed by a safe and terminating program. Consider a Turing Machine TM, with one tape and one head, which computes within n^k steps for some constant k and where n is the input size. The tape of TM is represented by two variables `Left` and `Right` which contain respectively the reversed left side of the tape and the right side of the tape. States are encoded by constant words and the current state is stored in the variable `State`. We assign to each of these three variables that hold a configuration of TM the tier $\mathbf{0}$. A one step transition is simulated by a finite cascade of if-commands of the form:

```
if eq_a(Right⁰)⁰
    then
        if eq_s(State⁰)⁰
        then
            State⁰:=s'⁰;; 0
            Left⁰:=suc_b(Left⁰);; 0
            Right⁰:=pred(Right⁰) : 0
        else ... : 0
    ...
```

The above command expresses that if the current read letter is a and the state is s, then the next state is s', the head moves to the right and the read letter is replaced by b. Since each variable inside the above command is of type $\mathbf{0}$, the type of the if-command is also $\mathbf{0}$. Moreover, since suc_b is a positive operator, its type is forced to be $\mathbf{0} \to \mathbf{0}$. eq_a, eq_s and $pred$ being neutral operators, they can also be typed by $\mathbf{0} \to \mathbf{0}$.

Finally, it just remains to show that every polynomial can be simulated by a safe program of tier $\mathbf{1}$. For this, we construct k nested loop by using the multiplication template of Example 2. The guards of loops correspond to input size which are of tier $\mathbf{1}$ wlog.

9   On Deterministic Scheduling

Actually, we can extend our results to a class of deterministic schedulers. Till now, we have considered a non-deterministic scheduling policy but in return we require that multi-threaded programs strongly terminate. Now, say that a deterministic scheduler $\mathcal{S}$ is *quiet* if the scheduling policy depends only on the current multi-threaded program M and on the tier $\mathbf{1}$ restriction $\mu \downarrow 1$. For example, a deterministic scheduler whose policy just depends on running threads, is quiet. Next, we replace the non-deterministic global transition of Figure 1 by:

$$\frac{\mathcal{S}(M, \mu \downarrow 1) = x \quad \mu \models M(x) \overset{\mathbf{s}}{\to} \mu'}{\mu \models M \overset{\mathbf{g}}{\to} \mu' \models M - x}$$

$$\frac{\mathcal{S}(M, \mu \downarrow 1) = x \quad \mu \models M(x) \overset{\mathbf{s}}{\to} \mu' \models C'}{\mu \models M \overset{\mathbf{g}}{\to} \mu' \models M[x := C']}$$

Theorem 8. *Let M be a safe multi-threaded program s.t. M is terminating wrt a deterministic and quiet scheduler $\mathcal{S}$. There is a polynomial Q such that:*

$$\forall d_1, \ldots, d_n \in \mathbb{W}, \ Time_M(d_1, \ldots, d_n) \leq Q(\max_{i=1,n}(|d_i|))$$

Proof. The proof follows the outline of proofs of theorems 5 and 6 by taking into account all command thread interactions.

Example 8. We consider a simple mutual exclusion algorithm. The noncritical sections are represented by commands C and C', which are assumed not to modify X or Y. Under a fair, deterministic and quiet scheduler, this algorithm is terminating, if both C and C' are terminating.

```
x :  X¹:=tt;: 1
        while(Y¹ = tt)¹{skip} : 1
        C
        X¹:=ff;: 1
```

$y:\ Y^1{:}{=}\mathtt{tt};:1$
$\quad\quad\mathtt{while}(X^1=\mathtt{tt})^1\{$
$\quad\quad Y^1=\mathtt{ff};:1$
$\quad\quad\mathtt{while}(X^1=\mathtt{tt})^1\{\mathtt{skip}\}:1$
$\quad\quad\quad\quad Y^1{:}{=}\mathtt{tt};:1\}$
$\quad\quad C';$
$\quad\quad Y{:}{=}\mathtt{ff};:1$

Now, if C and C' are safe, then the runtime is bounded by a polynomial in the size of tier 1 inputs.

References

1. Amadio, R.M., Dabrowski, F.: Feasible reactivity in a synchronous pi-calculus. In: PPDP, pp. 221–230 (2007)
2. Amadio, R.M., Dal Zilio, S.: Resource control for synchronous cooperative threads. In: Gardner, P., Yoshida, N. (eds.) CONCUR 2004. LNCS, vol. 3170, pp. 68–82. Springer, Heidelberg (2004)
3. Baillot, P., Terui, K.: Light types for polynomial time computation in lambda-calculus. In: LICS, pp. 266–275. IEEE Computer Society Press (2004)
4. Bell, D.E., La Padula, L.J.: Secure computer system: unified exposition and multics interpretation. Technical report, Mitre corp Rep. (1976)
5. Bellantoni, S., Cook, S.: A new recursion-theoretic characterization of the poly-time functions. Computational Complexity 2, 97–110 (1992)
6. Biba, K.: Integrity considerations for secure computer systems. Technical report, Mitre corp. Rep. (1977)
7. Bonfante, G., Marion, J.Y., Moyen, J.Y.: Quasi-interpretations a way to control resources. Theo. Comput. Sci. (2011)
8. Boudol, G., Castellani, I.: Noninterference for concurrent programs. In: Orejas, F., Spirakis, P.G., van Leeuwen, J. (eds.) ICALP 2001. LNCS, vol. 2076, pp. 382–395. Springer, Heidelberg (2001)
9. Cook, B., Podelski, A., Rybalchenko, A.: Proving thread termination. In: PLDI, pp. 320–330 (2007)
10. Girard, J.-Y.: Light linear logic. Inf. Comput. 143(2), 175–204 (1998)
11. Hainry, E., Marion, J.-Y., Péchoux, R.: Type-based complexity analysis for fork processes. In: Pfenning, F. (ed.) FOSSACS 2013. LNCS, vol. 7794, pp. 305–320. Springer, Heidelberg (2013)
12. Hofmann, M.: Linear types and non-size-increasing polynomial time computation. Inf. Comput. 183(1), 57–85 (2003)
13. Jones, N.: The expressive power of higher-order types or, life without cons. J. Funct. Program. 11(1), 5–94 (2001)
14. Jones, N., Kristiansen, L.: A flow calculus of wp-bounds for complexity analysis. ACM Trans. Comput. Log. 10(4) (2009)
15. Jones, N.D.: Computability and complexity, from a programming perspective. MIT Press (1997)
16. Leivant, D.: A foundational delineation of poly-time. Inf. Comput. 110(2), 391–420 (1994)
17. Leivant, D.: Predicative recurrence and computational complexity i: Word recurrence and poly-time. In: Clote, P., Remmel, J. (eds.) Feasible Mathematrics II (1994)

18. Leivant, D., Marion, J.-Y.: Evolving graph-structures and their implicit computational complexity. In: Fomin, F.V., Freivalds, R., Kwiatkowska, M., Peleg, D. (eds.) ICALP 2013, Part II. LNCS, vol. 7966, pp. 349–360. Springer, Heidelberg (2013)
19. Madet, A., Amadio, R.M.: An elementary affine λ-calculus with multithreading and side effects. In: Ong, L. (ed.) TLCA 2011. LNCS, vol. 6690, pp. 138–152. Springer, Heidelberg (2011)
20. Marion, J.-Y.: A type system for complexity flow analysis. In: LICS, pp. 123–132 (2011)
21. Marion, J.Y., Péchoux, R.: Sup-interpretations, a semantic method for static analysis of program resources. ACM TOCL 10(4), 27 (2009)
22. Niggl, K.-H., Wunderlich, H.: Certifying polynomial time and linear/polynomial space for imperative programs. SIAM J. Comput. 35(5), 1122–1147 (2006)
23. Nowak, D., Zhang, Y.: A calculus for game-based security proofs. IACR Cryptology ePrint Archive, 2010:230 (2010)
24. Sabelfeld, A., Myers, A.C.: Language-based information-flow security. IEEE J. Selected Areas in Communications 21(1), 5–19 (2003)
25. Smith, G., Volpano, D.: Secure information flow in a multi-threaded imperative language. In: POPL, pp. 355–364. ACM (1998)
26. Volpano, D., Irvine, C., Smith, G.: A sound type system for secure flow analysis. Journal of Computer Security 4(2/3), 167–188 (1996)

A Personalized Privacy Preserving Method for Publishing Social Network Data

Jia Jiao, Peng Liu, and Xianxian Li*

College of Computer Science and Information Technology,
Guangxi Normal University, Guilin, China
jiaojia0623@163.com, {liupeng,lixx}@mailbox.gxnu.edu.cn

Abstract. One of the most important concerns in publishing social network data for social science research and business analysis is to balance between the individual's privacy protection and data utility. Recently, researchers have developed lots of privacy models and anonymous techniques to prevent re-identifying of relevant information of nodes through structure information of social networks, but most of the existing methods did not cater for the individuals' personalized privacy requirements and did not take full advantage of distributed characteristics of the social network nodes. Motivated by this, we specify three types of privacy attributes for various individuals and develop a *personalized k-degree-l-diversity (PKDLD)* anonymity model. Furthermore, we design and implement a graph anonymization algorithm with less distortion to the properties of the original graph. Finally, we conduct experiments on some real-world datasets to evaluate the practical efficiency of our methods, and the experimental results show that our algorithm reduces the anonymous cost efficiently and improves the data utility.

Keywords: privacy preserving, personalized anonymity, social network.

1 Introduction

Nowadays, partly driven by many Web 2.0 applications, more and more social network data have been made publicly available and analyzed in one way or another [1]. The social network data has significant application value for commercial and research purposes. However, the social network data often have privacy information of individuals as well as their sensitive relationship. So privacy disclosure risks arise when the data holders publish the social network data. And it has become a major concern to balance between the individual's privacy and the utility of social net work data while publishing the social network data.

Social network data can be represented as a graph, in which nodes and edges correspond to social entities and social links between them, respectively [2]. Many approaches have been proposed to preserve the privacy of published social network data in existing research[3–8]. However, most of the existing methods do not consider the personalized privacy requirement of the individuals of social

---

* Corresponding author.

T V Gopal et al. (Eds.): TAMC 2014, LNCS 8402, pp. 141–157, 2014.
© Springer International Publishing Switzerland 2014

networks, thus provide excessive preservation to some individuals, and can not
only increase the anonymous cost, but also lower the utility of data.

In this paper we focus on protecting nodes' degree and nodes' one sensitive
label, such as salaries [3], to propose our anonymous model and algorithm. Nodes'
degree information and nodes' sensitive label can help the adversaries re-identify
individuals and their relevant information in a published social network graph.

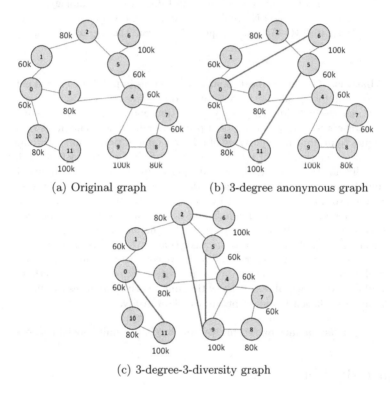

(a) Original graph   (b) 3-degree anonymous graph

(c) 3-degree-3-diversity graph

Fig. 1. A degree and label anonymous graph without considering individual privacy
requirement

For example, if an adversary knows that one person has four friends in Fig.1a,
he can immediately know that node 4 is that person and the related attributes
of node 4 are revealed. In order to solve this problem, Liu et al.[4] proposed
k-degree anonymity model to prevent nodes from degree attack. A graph is *k-
degree anonymous* if and only if for any node in this graph, there is at least *k*-1
other nodes with the same degree. Fig.1b is a *3-degree anonymous* graph, the
adversary no longer knows that node 4 is that person. When a social network
graph contain nodes with sensitive attributes, alone *k-degree anonymity* is not
sufficient to prevent the inference attack from sensitive attributes of individuals.
Fig.1b shows a graph that satisfies *3-degree anonymity* but node labels are not
considered. In the case, nodes 0, 4, 5 have the same degree 4, but they all have

the label "60k". If an attacker knows that someone has four friends in the social network, he can get that this person's salary is 60k without exactly re-identifying the node. For solving this problem, Yuan et al.[3] have proposed the proposed the *KDLD (k-degree-l-diversity) anonymity* model, which based on the *k-degree anonymity* model. A graph is *KDLD anonymous* if and only if for any node in this graph, there exist at least k-1 other nodes with the same degree and at least l distinct sensitive label values in each equivalent group. Fig.1c shows a *3-degree-3-diversity anonymous* graph.

However, both Yuan et al.[3] and Liu et al.[4] focused on a universal approach that exerted the same amount of preservation for all individuals, and did not cater for their personalized needs, which is described as Fig.1. The consequence is that it may be offer excessive privacy control to some individuals who require only lower privacy-preserving level, and damage the utility of the publishing data.

In reality, different individuals may have different personal privacy standards or preference to the same information, and in social network website, such as Pengyou and Renren, they allow users to specify whether their basic information, photo albums, blogs and friend list etc., will be accessed or not by others. Besides, both Yuan et al.[3] and Liu et al.[4] did not take full advantage of the power law distribution of the large-scale network nodes,which indicate that only a small number of nodes have high degrees and most of nodes have low degrees[9, 10]. However, the few nodes with high degree have brought a great deal of cost of both data distortion and computation in the anonymized process. Actually, the few high degree nodes in social network are the famous nodes whose corresponding individuals are often the public figures. They usually have relative lower privacy preserving requirements.

Motivated by above reasons, in this paper, we assume that an adversary with the background knowledge about the degree of some target individuals wants to re-identify an individual whose information contained in published social networks as mentioned in Yuan et al.[3], but the difference is that we consider the individual's privacy requirement. Personalized privacy requirements can be captured in the privacy information collection process and stored with the associated data by individuals.

We also assume that individuals can specify whether his own friend list and sensitive attribute can be access or not by others. We can know the node's degree and sensitive label by the responding individual's friend list's number and sensitive attributes, such as salary described in Fig.1.We classify the individual's privacy requirements into three categories which are denoted by $H(high)$, $M(middle)$, and $L(low)$ respectively:

$H(high)$ privacy: the individual specifies both his friend list and sensitive attribute not to be accessed by others, such as node 2 described in Fig.2a and Fig.2b;

$M(middle)$ privacy: the individual specifies only his friend list not to be accessed by others, such as node 1 described in Fig.2a and Fig.2b;

L(low) privacy: the individual allow his friend list and sensitive attribute to be accessed by others, such as node 0 described in Fig.2a and Fig.2b.

Our privacy preserving goal is to prevent an attacker from re-identifying an individual and finding the fact that a certain individual has a specific sensitive value, which is similar to Yuan et al.[3], and in the meanwhile every individual's privacy requirements will be satisfied.

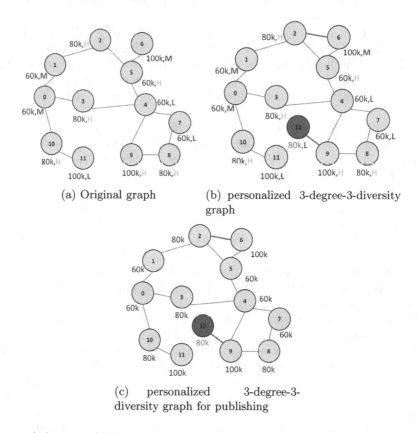

(a) Original graph

(b) personalized 3-degree-3-diversity graph

(c) personalized 3-degree-3-diversity graph for publishing

Fig. 2. A degree and label anonymous graph with considering individual privacy requirement

Our contributions are as follows: First, we propose the *PKDLD* (*personalized k-degree-l-diversity*) anonymity model based on *KDLD anonymity* model. Second, we design and implement our own graph anonymous algorithm *PKLADP* (*Personalized K-degree-L-diversity Against Degree Attack Publication*) which has less distortion to the properties of the original graph such as degrees and distances between nodes and can protect individual from degree attack. Finally, we conduct an empirical study which indicates that anonymized social network generated by our method performs well in cost and utility.

The rest of the paper is organized as follows. Section 2 defines the problems. Section 3 describes the details of the personalized anonymous algorithm implementation. We report the experimental results on some real datasets in section 4. Comparison of our work with previous proposals is given in section 5. And section 6 concludes the paper.

2   Problem Description

In this paper, we focus on the privacy preserving problem for an un-weighted undirected simple graph with a sensitive label and a privacy attribute on each node. Fig.2a shows an example of such a graph. A social network graph is defined as follows:

Definition 1. Social Network Graph. A social network graph is a four tuple $G = (V, E, La, \lambda)$, Where V is a set of nodes of the graph, $E \subseteq V \times V$ is a set of edges between nodes, La is a set of labels, and $\lambda : V \to La$ is a labeling function which maps nodes to their labels.

The label l in La is a two tuple, i.e. $l = (s, r)$, where s is the sensitive label of the corresponding node, and r is the privacy attribute of the corresponding node. The privacy attribute is one of H,M,L. For any node $v \in V$, if $r(v) = H$ or $r(v) = M$, it implies that the node v has privacy requirement, and then we should anonymize it and the nodes with attribute H should have the larger anonymous extent than that the nodes with attribute M. And $r(v) = L$ means that information of the node does not need privacy protection. For example, in Fig.2a $r(2) = H$, which means that the node 2 needs to be anonymized both the degree information and the sensitive label, and $r(1) = M$ implies that only the degree information of the node 1 requires to be protected.

Definition 2. Degree Sequence P. Degree sequence P for a graph G with n nodes is a sequence with the form: $[P[1], P[2], ..., P[n]]$, where for each $1 \leq i \leq n$, $P[i]$ is a 4-tuple (id, d, s, r), in which id identifies a node, d is the degree, s is the sensitive label and r is the privacy attribute associated with the node id. Furthermore, it satisfies that $P[1].d \geq P[2].d \geq ... \geq P[i].d \geq ... \geq P[n].d$, where $P[i].d$ is a project map from the 4-tuple $P[i]$ to its second element, e.g. $(x_1, x_2, x_3, x_4).d = x_2$. As an example, in Fig.2a, the degree sequence is $P = [(4, 4, 60k, L), (5, 3, 60k, H), (0, 3, 60k, M), (2, 2, 80k, H), (3, 2, 80k, H), (8, 2, 80k, H), (9, 2, 100k, H), (10, 2, 80k, H), (1, 2, 60k, M), (7, 2, 60k, L), (6, 1, 100k, M), (11, 1, 100k, L)]$.

Definition 3. *KDLD* Sequence. A degree sequence P is a *KDLD*(*k-degree-l-diversity*) sequence if P satisfies the following constraints. P can be divided into a group of subsequences $[[P[1], ..., P[i_1]], [P[i_1 + 1], ..., P[i_2]], ..., [P[i_m + 1], ..., P[j]]]$ such that for any subsequences $P_x = [P[i_x], ..., P[i_{x+1}]]$, P_x satisfies three constraints:(1) All the elements in P_x share the same degree ($P[i_x].d = P[i_x+1].d = ... = P[i_{x+1}].d$); (2)$P_x$ has size at least $k(i_{x+1} - i_x + 1 \geq k)$; (3)$P_x$'s label set $\{P[t].s | i_x \leq t \leq i_{x+1}\}$ have at least l distinct values [3].

Definition 4. *KD* Sequence. A degree sequence P is a $KD(k\text{-}degree)$ sequence if P satisfies the following constraints. P can be divided into a group of subsequences $[[P[1], ..., P[i_1]], [P[i_1 + 1], ..., P[i_2]], ..., [P[i_m + 1], ..., P[j]]]$ such that for any subsequences $P_y = [P[i_y], , P[i_{(y+1)}]], P_y$ satisfies two constraints: (1) All the elements in P_y share the same degree $(P[i_y].d = P[i_y + 1].d = ... = P[i_{(y+1)}].d)$; (2)$P_y$ has size at least $k(i_{y+1} - i_y + 1 \geq k)$.

Definition 5. PKDLD Sequence. A degree sequence P is a $PKDLD($ *personalized k-degree-l-diversity*)sequence if and only if for any node u, whose privacy attribute is H, i.e. $r(u) = H$, then u belongs to the subsequence of $KDLD$ sequence; for any node v, whose privacy attribute is M, i.e. $r(v) = M$, v belongs to the subsequence of $KDLD$ sequence or KD sequence.

E.g., the degree sequence P of the Fig.2b is a $P3D3D(k = 3, l = 3)$ sequence. Here $P = [(4, 4, 60k, L), (5, 3, 60k, H), (0, 3, 60k, M), (2, \mathbf{3}, 80k, H), (9, \mathbf{3}, 100k, H),$ $(3, 2, 80k, H), (8, 2, 80k, H), (10, 2, 80k, H), (1, 2, 60k, M), (6, \mathbf{2}, 100k, M), (7, 2,$ $60k, L), (11, 1, 100k, L), (12, 1, 80k, L)]$.

Definition 6. A *PKDLD* Graph. A graph is a $PKDLD$ graph if and only if its degree sequence is a $PKDLD$ sequence.

A $PKDLD$ graph preserves two aspects for the individuals who have the privacy requirement H. (1) The probability that an attacker can correctly re-identify the target individuals is at most $1/k$; (2) The sensitive label of the target individuals can be related with l different values. And a $PKDLD$ graph preserves the individuals with the privacy preserving requirement M from re-identifying correctly at most $1/k$. Besides, a $PKDLD$ graph also considers the individuals privacy preserving requirement.

Based on the above definitions, well solve the following two sub-problems.

(1) $PKDLD$ sequence generation.

Given the degree sequence P of the graph G and two integers k and l, compute a $PKDLD$ sequence which satisfies the following constraint. $L(P^{new}, P) = \sum_{\forall u} |P_u^{new}.d - P_u.d|$ has the minimum value, where $P_u.d$ is the node u's degree in the original graph, and $P_u^{new}.d$ is the node u's target degree in the anonymous graph. Clearly, smaller degree change needs fewer noise edges to implement the change [3].

(2) Graph construction Given a graph $G = (V, E, La, \lambda)$ and a degree sequence P^{new} ,construct a new graph $G' = (V', E', La, \lambda')$ with $V \subseteq V', E \subseteq E'$. The degree sequence P' of G' is a $PKDLD$ sequence and P' has all the elements in P^{new} since G' is constructed from G by adding some noise nodes. Meanwhile, $|APL_G - APL_{G'}|$ is minimized and APL is short for average shortest path length.

And here we note that we'll remove the nodes' privacy attribute for publishing after the step (2), which is described in Fig.2c.

3   Personalized Anonymity

From the problem description, there are still two sub-problems to be solved in order to achieve the personalized anonymity. In this section we'll design and implement the Algorithm 1(*P-K-L-Based Algorithm*) and Algorithm 2(*PKLADP Algorithm*) for *PKDLD* sequence generation and graph construction respectively.

3.1   PKDLD Sequence Generation

Given the degree sequence P of the original graph G, in order to generate a *PKDLD* sequence P^{new} with the minimum $L(P^{new}, P)$ value, the 4-tuples in P should be divided into groups. If the node's highest privacy attribute is H in the same group, the group should be adjusted to belong to the subsequence of *KDLD* sequence. If the node's highest privacy attribute is M in the same group, the group should be adjusted to belong to the subsequence of *KDLD* sequence or *KD* sequence.

Algorithm 1. *P-K-L-Based Algorithm*

Input: the degree sequence P of the original G, two integers k and l
Output: *PKDLD* sequence P^{new}

1. **while** (|the ungrouped nodes' distinct sensitive label values in $P| \geq l$) and (|the number of nodes in $P| \geq 2k$) **do**
2.   choose the first node v whose privacy attribute is H or M to begin our anonymity;
3.   from node v, choose the first k elements in P as a current group $g_{current}$;
4.   **if** (there is a node whose privacy attribute is H in $g_{current}$) **then**
5.   **if** (|the nodes' distinct sensitive label values in $g_{current}| < l$) **then**
6.   merging next element into $g_{current}$, until $g_{current}$ satisfies the *l-diversity* constraint;
7.   **end if**
8. $mergenum=0$;
9.   **while** (the next element's privacy attribute is H or M) and ($mergenum < k$) **do**
10.   calculate the two costs: C_{new} and C_{merge};
11.   **if** ($C_{new} < C_{merge}$) **then**
12.   merging next element into $g_{current}$; $mergenum$++;
13.   **end if**
14.   **end while**
15.   **else**
16.   **if** (in $g_{current}$, there is a node whose privacy attribute is M) **then**
17. $mergenum=0$;
18.   **end if**
19.   **while** (the next element's privacy attribute is H or M) and ($mergenum < k$) **do**
20.   repeat lines 10-13
21.   **end while**
22.   **end if**
23. **end while**

In *P-K-L-Based Algorithm*, we begin to anonymize the degree sequence P as the following steps. We choose the first node v with the privacy attribute H or M in P, if the ungrouped nodes' distinct sensitive label values is at least l and the number of nodes is at least $2k$ in P. Then we choose the first k elements in P to be a current group, and check whether there is a node whose privacy attribute is H in this group. If there is, we check the nodes' distinct sensitive label value. If the value is smaller than l, we keep on merging next element into the current group until the l-$diversity$ constraint is satisfied. After the current group satisfies k-$degree$ and l-$diversity$ constraints, we calculate two costs.

C_{new}. The cost of creating a new group for the next k elements is the total degree changes when make all the nodes in this group have the same degree.

C_{merge}. It includes the cost of merging the next element into the current group and creating a new group for the next k elements by skipping the next element.

When the next element's privacy attribute is H or M, and C_{merge} is smaller than C_{new} and $mergenum$ is smaller than k, we keep on merging the next element into the current group and continue this process.

If there is no node whose privacy attribute is H and there is a node whose privacy attribute is M in current group. We just calculate C_{new} and C_{merge} two costs. When the next element's privacy attribute is M and C_{merge} is smaller than C_{new} and $mergenum$ is smaller than k, we keep on merging the next element into the current group and continue this comparison process. After this step, we update P by deleting the elements of current group from P. And we repeat the above process when the ungrouped nodes' distinct sensitive label values are at least l and the number of nodes is at least $2k$ in P.

Since nodes are sorted by their descending degrees in P, constructing groups using the above methods is helpful to group the nodes with similar degrees together. For example, if using the this algorithm to make $P = [(4, 4, s_1, L), (5, 3, s_1, H), (0, 3, s_1, M), (2, 2, s_2, H), (3, 2, s_2, H), (8, 2, s_2, H), (9, 2, s_3, H), (10, 2, s_2, H), (1, 2, s_1, M), (7, 2, s_1, L), (6, 1, s_3, M), (11, 1, s_3, L)]$ satisfy $P3D3D$ constraint $(k = 3, l = 3)$, we get $P^{new} = [(4, 4, s_1, L), (5, 3, s_1, H), (0, 3, s_1, M), (2, \mathbf{3}, s_2, H), (9, \mathbf{3}, s_3, H), (3, 2, s_2, H), (8, 2, s_2, H), (10, 2, s_2, H), (1, 2, s_1, M), (6, \mathbf{2}, s_3, M), (7, 2, s_1, L), (11, 1, s_3, L), (12, 1, s_2, L)]$. Algorithm 1 shows the pseudo code of the P-K-L-$Based$ $Algorithm$.

P-K-L-$Based$ $Algorithm$ runs in $O(n)$ time. The process of generating the $PKDLD$ sequence needs only to scan the degree sequence P once, therefore the run time is $O(n)$. Hence the time complexity of the P-K-L-$Based$ $Algorithm$'s is $O(n)$.

All the nodes in the same group will be adjusted to have the same degree in the next graph construction process. We use the highest degree of a group as its target degree, so the degree of nodes in the same group needs to be increased to achieve its target degree. In view of this, we only consider adding edges and noise nodes to increase nodes' degree rather than deleting edges.

3.2   Graph Construction

When we get the new degree sequence P^{new} from the $PKDLD$ sequence generation, a new graph G' can be constructed based on P^{new}. The degree sequence P' of G' contains all the elements in P^{new} since G' is constructed from G by adding some noise nodes.

The *APL* has the minimal change between graphs G and G' in our graph construction algorithm *PKLADP (Personalized K-degree-L-diversity Against Degree Attack Publication)*, The algorithm is described in Algorithm 2.

First, we select the nodes in P^{new} whose degree is smaller than their target degree, and store them into V_{sdiff}, which is described in Algorithm 2(line 1). We need only scan the P^{new} once to store the nodes into V_{sdiff}, whose degree haven't achieved their target degree, so it completes in $O(n)$.

Then, if V_{sdiff} is not empty for each node u in V_{sdiff}, and if there is another node v in V_{sdiff} and the distance between u and v is 2, i.e. $d(u,v) = 2$, we link u and v to increase the degree of both u and v(See Fig.3a). The distance between u,v is changed from 2 to 1. So only the lengths of shortest paths passing though u and v change by 1. If u or v has been achieved its target degree, we erase u or v from V_{sdiff}.And the described code in Algorithm 2 is in lines 2-14. Here, the run time of *the for loop* is $O(n)$, and the run time of the *if judgment statement* of whether there is a node v satisfying constraint is also $O(n)$, so the time complexity of lines 2-14 is $O(n^2)$.

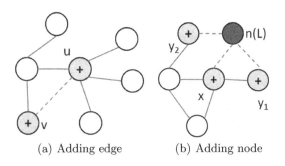

(a) Adding edge   (b) Adding node

Fig. 3. Adding edge and node

Last, through the adding edges as Fig.3a, if there still exist nodes in V_{sdiff} for each node x in V_{sdiff}, we add noise node n in which $r(n)$ is L and $s(n)$ is the same with any one of x's neighbor's. If x has been achieved its target degree, we erase x from V_{sdiff}.We link y_1 and n if y_1 and x are direct neighbor and also y_1 is in V_{sdiff} (See Fig.3b). Or we link y_2 and n if x and y_2 are two hop nodes and also y_2 is in V_{sdiff} (See Fig.3b).The distance between x, y_1 or x, y_2 has no change. And the described code in Algorithm 2 is in lines 9-16. As the same with lines 2-14, there are a *for loop* and an *if judgment statement* to judge the satisfying node y_1 or y_2 under the *for loop*, so the time complexity of lines 15-29 is also $O(n^2)$.

In conclusion, the time complexity of *PKLADP* Algorithm is $O(n^2)$.

Algorithm 2. *PKLADP Algorithm*

Input: the original graph G, the degree sequence P of G, the $PKDLD$ sequence P^{new}
Output: the $PKDLD$ graph G'

1. Select the nodes in P^{new} which for any $u \in V$, $P_u^{new}.d - P_u.d \,! = 0$,and store them in V_{diff};
2. **if** (V_{sdiff} is not empty) **then**
3.   **for all** (node u in V_{sdiff}) **do**
4.   **if** (in V_{sdiff},there is a node v, $d(u,v)=2$) **then**
5. $link(u,v);P_u.d + +;P_v.d + +;$
6.   **end if**
7.   **if** ($P_u^{new}.d - P_u.d \,= 0$) **then**
8.   erase u form V_{sdiff};
9.   **end if**
10.   **if** ($P_v^{new}.d - P_v.d \,= 0$) **then**
11.   erase v form V_{sdiff};
12.   **end if**
13.   **end for**
14. **end if**
15. **if** (V_{sdiff} is not empty) **then**
16.   **for all** (node x in V_{sdiff}) **do**
17.   add a node n which $n.r=L$ and $n.s$ is the same with any one of x's neighbor's for node x;
18. $P_x.d++;$ $n.d++;$
19.   **if** ($P_x^{new}.d - P_x.d \,= 0$) **then**
20.   erase x form V_{sdiff};
21.   **end if**
22.   **if** (in V_{sdiff},there is a node y, $d(x,y)=1$ or $d(x,y)=2$) **then**
23. $link(n,y);P_y.d++;n.d + +;$
24.   **end if**
25.   **if** ($P_y^{new}.d - P_y.d \,= 0$) **then**
26.   erase y form V_{sdiff};
27.   **end if**
28.   **end for**
29. **end if**

---

4   Experiments

In this section, we report a systematic empirical study to evaluate our anonymizing method using some real data sets. All the experiments were conducted on a *PC* computer running the *Windows 7 Ultimate operating system*, with a *3.2 GHz Intel Core i3 CPU, 4.0 GB main memory*, and a *500 GB hard disk*. The program was implemented in *C++* and was compiled using *Microsoft Visual Studio .NET 2010*.

4.1   Data Sets

We use two real data sets: *Cora dataset* and *Citation dataset*.

The *Cora Dataset* (http://www.cs.umd.edu/projects/linqs/projects/lbc/index.html) contains 2708 nodes and 5429 edges. This is similar to Yuan et al.[3]'s assumption which used the node's 7 classifications as the sensitive label.

The *Citation Dataset*(http://www.datatang.com/data/17310) is a dataset collected by an academic researcher social network ArnetMiner of Tsinghua University and published in *datatang*. It consists of 2555 nodes and 6101 edges. We use the nodes'17 publication years as the sensitive label.

4.2   Results and Analysis

To evaluate the effectiveness of our algorithm, we perform experiment on the dataset used the algorithm in Yuan et al.[3]. The algorithm proposed in [3] without considering individuals' privacy requirement is called *KLADP*, and our algorithm is called *PKLADP*. We test the *PKD5D* model (i.e. $l{=}5$) for *Cora dataset* and *Citation dataset*. In *KLADP* we focus on a universal approach that exerts the same amount of privacy preservation for all individuals. And in *PKLADP* we consider the three types of individuals' privacy protect requirement and set the proportion of the $H : M : L$ to be 3:3:4 arbitrarily.

Anonymous Cost. We compute two kinds of anonymous cost: $L(P^{new}, P)$ and $Cost(G, G')$.

$L(P^{new}, P)$ is the sum of increased degree of all nodes from degree sequence P of original G to *PKDLD* sequence P^{new} by using *PKLADP Algorithm* or *KDLD* sequence P^{new} by using *KLADP Algorithm* respectively, that is

$$L(P^{new}, P) = \sum_{\forall u} |P_u^{new}.d - P_u.d| \tag{1}$$

$Cost(G, G')$ is all the number of adding edges and nodes from original graph $G = (V, E, La, \lambda)$ to published graph $G' = (V', E', La', \lambda')$ with $V \subseteq V'$, $E \subseteq E'$, that is

$$Cost(G, G') = (|E'| - |E|) + (|V'| - |V|) \tag{2}$$

Fig.4 and Fig.5 represent $L(P^{new}, P)$ cost and $Cost(G, G')$ cost in terms of changing k values using *PKLADP Algorithm* and *KLADP Algorithm* repectively.The results show that for all the two data sets, our *PKLADP Algorithm* performs better than the *KLADP Algorithm*.The former's preserving extent is weaker than that the latter's while satisfying the privacy requirement, thus we can add less edges or nodes to satisfy anonymity, i.e. the anonymous cost is smaller. Especially when we anonymise some high degree nodes with the privacy attribute L, we can dramatically decrease the number of adding edges and nodes,so the $L(P^{new}, P)$ cost and $Cost(G, G')$ cost of our algorithm is reduced.

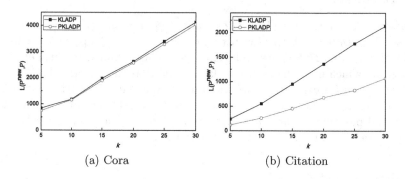

(a) Cora   (b) Citation

Fig. 4. Cora dataset and Citation dataset: $L(P^{new}, P)$ Cost for different k

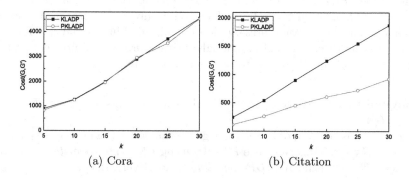

(a) Cora   (b) Citation

Fig. 5. Cora dataset and Citation dataset: $Cost(G, G')$ Cost for different k

Utility. To examine how well the published graph represents the original graph, we use two measurements the APL (average shortest path length)[3, 4] and CC (clustering coefficient)[3, 4, 6, 11].

APL is a concept in network topology that is defined as the average of distances between all pairs of nodes. A real-world data graph may be composed of several connected components $(C_1, C_2, ...C_l)$. We define a weighted APL for the whole graph as

$$APL = \frac{|V_{C_i}|}{V} \sum_{i=1}^{l} APL_{C_i} \tag{3}$$

where $V_(C_i)$ is the node number of each connected component and APL_{c_i} is the APL of each connected component.

CC represents the characteristic of its neighborhood graph. In this paper, we use the clustering coefficient for the whole graph as

$$CC' = \frac{1}{n} \sum_{i=1}^{n} CC_i \tag{4}$$

where CC_i is the local clustering coefficient of a node, CC' is the average of the local clustering coefficients of all the nodes given by Watts et al.[11].

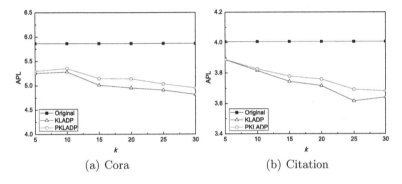

(a) Cora   (b) Citation

Fig. 6. Cora dataset and Citation dataset: APL for different k

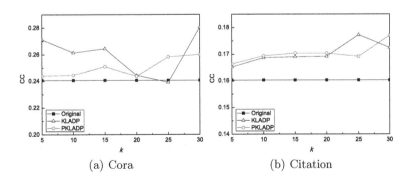

(a) Cora   (b) Citation

Fig. 7. Cora dataset and Citation dataset: CC for different k

In Fig.6, we report the values of *APL* using our algorithm and compared algorithm for the two datasets. The *APL* of the original graph is also reported in all plots. From the figures, we can see that using our algorithm's *APL* has lower differences to original graph's *APL* than that compared algorithm's.

We show the *CC* results of two datasets, respectively, with the changing k values in Fig.7. From the graphs, we can observe that the values of *CC* by using our *PKLADP Algorithm* and compared *KLADP Algorithm*, though different in varied k value, they never deviate too much from their original values. But meanwhile, our algorithm has considered the individual's privacy preserving requirement and provides personalized preserving for the individuals.

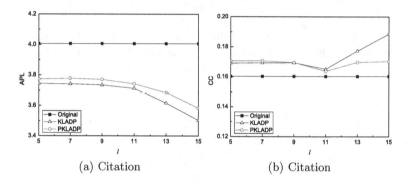

(a) Citation   (b) Citation

Fig. 8. Citation dataset: *APL* and *CC* for different l

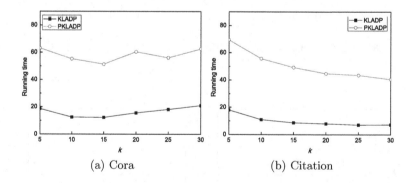

(a) Cora   (b) Citation

Fig. 9. Cora dataset and Citation dataset: Running time for different k

Since the number of sensitive labels of *Cora dataset* is not enough, in order to see the effect of l, we test *P15DLD* model (i.e. k=15) of *Citation dataset*. Fig.8a and Fig.8b show the *APL* and *CC* results with respect to different l values. We can observe the similar results, which are shown in Figs.6, 7 changing the k values, that our algorithm works better than the compared algorithm, and allows personalized protections meanwhile.

Algorithm Efficiency. We present the running time of our *PKLADP Algorithm* and compared *KLADP Algorithm* when k increases in two datasets respectively in Figs.9a, 9b. From the figures, we can observe that both our algorithm and the compared algorithm are very efficient, the largest running time is less than 70s and 20s in our algorithm and the compared algorithm respectively. The compared algorithm works much better, since when we generate the *PKDLD* sequence, it takes time to compute the nodes' highest privacy preserving requirement for each current group, this process doesn't exist in compared algorithm.

5   Related Work

The privacy preserving of microdata (tabular data) has been extensively studied recently, that relates to the privacy preserving of social network data in previous study. The *k-anonymity* was proposed by Samarati and Sweeney [12], Sweeney [13] for the sake of sensitive information's privacy preserving while publishing microdata for public use. A dataset is said to be *k-anonymous* if each record is indistinguishable from at least k-1 other records with respect to the quasi-identifier attributes in the same dataset. The privacy preserving is better with the larger k. From then on, a variety of privacy models and anonymous algorithm had been proposed based on *k-anonymity*, such as *l-diversity*[14], *t-closeness*[15], etc.

Lots of studies have been done for privacy preserving of social network data publication. In literature [16], Hay et al. proposed *naive anonymization*, which remove identifying attributes such as name or social security number and introduce synthetic identifiers to replace, but by collecting information from external sources about an individuals relationships, an adversary may be able to re-identify individuals in the graph. Based on this, Liu et al. [4] studied the *k-degree anonymity* and a graph is said to be *k-degree anonymity* if and only if for every node v, there exist at least k-1 other nodes in the graph with the same degree as v. Zhou et al. [5] identified an essential type of privacy attacks: neighborhood attacks by assuming that an adversary's background knowledge is some knowledge about the neighbors of a target victim and the relationship among the neighbors. They proposed *k-neighborhood anonymity* model to tackle the neighborhood attacks. Zou et al. [8] assumed that the adversary can know any subgraph around a certain individual and propose a *k-automorphism* protection model: a graph is *k*-automorphism if and only if for every node there exist at least k-1 other nodes do not have any structure difference with it. The difference between literatures [16, 4, 5, 8] and this paper is that the former only considered the nodes' structural attributes, didn't consider the nodes' own data attributes, especially the sensitive attributes.

In paper [6], Zhou et al. extended the *k-neighborhood* anonymity model [5] to *k-neighborhood-l-diversity* model to protect the graph data with sensitive node label. Yuan et al. [3] studied *KDLD* (*k-degree-l-diversity*) anonymity model in view of the *k-degree* anonymity model [4] as mentioned in section 1. Campan et al. [17] considered using clustering approach to anonymize social networks which used a super-node to represent some nodes in an equivalent class and a super-edge to replace the edges between two equivalent class. Like in paper [3], papers [6, 17] considered both the structural attributes and the data attributes of nodes, but they didn't take into account the individuals' personalized privacy preserving requirements.

Yuan et al.[7] allowed users to set personalized privacy assumptions based on the attacker's background knowledge. They defined three levels of attacks by gradually increasing the strength of the attacker's three levels of background knowledge and putted forward approaches and algorithms with respect to the three levels of attacks one by one. Paper [7] has proposed the personalized requirement, however, they partitioned the entities by different attackers' background knowledge, and all the nodes in the graph were anonymized with the *Anatomy* [18] method according to different levels of privacy assumptions. It is different from their work that we classify the entities according to the different attribute value of nodes which will be allowed to access by others, and then our algorithm can reduce the number of nodes that need to be anonymized, especially for the high degree nodes with lower privacy attribute value. Therefore, our method can take full advantage of the power law distribution of large social networks.

6   Conclusion

In this paper, we propose a *PKDLD* model, design and implement a graph anonymization algorithm *PKLADP* for personalized privacy preserving social network data publishing to prevent from the structural attack. By dividing nodes with different privacy requirements, we reduce the cost of data distortion and improve the utility of data for anonymizing the original data when the computation cost has little change. The algorithm is high efficient especially for the scenario that high degree nodes have relative lower privacy requirement and the network satisfies the power law distribution. By conducting experiment on real-world datasets, we show that the algorithm performs well in terms of protection it provides compared with the existing approaches.

Acknowledgments. The research is supported by the National Key Basic Research Program of China (973 Program, No. 2012CB326403), National Science Foundation of China (No. 61272535, No. 61165009), Guangxi "Bagui Scholar" Special Project Funds, and Postgraduate education innovation project of Guangxi(No.YCSZ2013042).

References

1. Zhou, B., Pei, J., Luk, W.: A brief survey on anonymization techniques for privacy preserving publishing of social network data. ACM SIGKDD Explorations Newsletter 10(2), 12–22 (2008)

2. Wasserman, S.: Social network analysis: Methods and applications, vol. 8. Cambridge University Press (1994)
3. Yuan, M., Chen, L., Yu, P., Yu, T.: Protecting sensitive labels in social network data anonymization. ACM SIGKDD Explorations Newsletter 25(3), 633–647 (2013)
4. Liu, K., Terzi, E.: Towards identity anonymization on graphs. In: Proceedings of the 2008 ACM SIGMOD International Conference on Management of Data, pp. 93–106. ACM (2008)
5. Zhou, B., Pei, J.: Preserving privacy in social networks against neighborhood attacks. In: IEEE 24th International Conference on Data Engineering, ICDE 2008, pp. 506–515. IEEE (2008)
6. Zhou, B., Pei, J.: The k-anonymity and l-diversity approaches for privacy preservation in social networks against neighborhood attacks. Knowledge and Information Systems 28(1), 47–77 (2011)
7. Yuan, M., Chen, L., Yu, P.S.: Personalized privacy protection in social networks. Proceedings of the VLDB Endowment 4(2), 141–150 (2010)
8. Zou, L., Chen, L., Özsu, M.T.: K-automorphism: A general framework for privacy preserving network publication. Proceedings of the VLDB Endowment 2(1), 946–957 (2009)
9. Faloutsos, M., Faloutsos, P., Faloutsos, C.: On power-law relationships of the internet topology. ACM SIGCOMM Computer Communication Review 29, 251–262 (1999)
10. Mislove, A., Marcon, M., Gummadi, K.P., Druschel, P., Bhattacharjee, B.: Measurement and analysis of online social networks. In: Proceedings of the 7th ACM SIGCOMM Conference on Internet Measurement, pp. 29–42. ACM (2007)
11. Watts, D.J., Strogatz, S.H.: Collective dynamics of small-world networks. Nature 393(6684), 440–442 (1998)
12. Samarati, P., Sweeney, L.: Generalizing data to provide anonymity when disclosing information. In: PODS 1998, p. 188 (1998)
13. Sweeney, L.: k-anonymity: A model for protecting privacy. International Journal of Uncertainty, Fuzziness and Knowledge-Based Systems 10(05), 557–570 (2002)
14. Machanavajjhala, A., Kifer, D., Gehrke, J., Venkitasubramaniam, M.: l-diversity: Privacy beyond k-anonymity. ACM Transactions on Knowledge Discovery from Data (TKDD) 1(1), 3 (2007)
15. Li, N., Li, T., Venkatasubramanian, S.: t-closeness: Privacy beyond k-anonymity and l-diversity. In: IEEE 23rd International Conference on Data Engineering, ICDE 2007, pp. 106–115. IEEE (2007)
16. Hay, M., Miklau, G., Jensen, D., Weis, P., Srivastava, S.: Anonymizing social networks. Computer Science Department Faculty Publication Series, p. 180 (2007)
17. Campan, A., Truta, T.M.: Data and structural k-anonymity in social networks. In: Bonchi, F., Ferrari, E., Jiang, W., Malin, B. (eds.) PinKDD 2008. LNCS, vol. 5456, pp. 33–54. Springer, Heidelberg (2009)
18. Xiao, X., Tao, Y.: Anatomy: Simple and effective privacy preservation. In: Proceedings of the 32nd International Conference on Very Large Data Bases, pp. 139–150, VLDB Endowment (2006)

A Bit-Encoding Phase Selection Strategy for Satisfiability Solvers

Jingchao Chen

School of Informatics, Donghua University
2999 North Renmin Road, Songjiang District, Shanghai 201620, P.R. China
chen-jc@dhu.edu.cn

Abstract. The phase (also called polarity) selection strategy is an important component of a SAT solver based on conflict-driven DPLL. DPLL algorithm is due to Davis, Putnam, Logemann, Loveland. It is a complete, backtracking-based search algorithm for deciding the satisfiability of propositional logic formulae. This paper studies the phase selection strategy and presents a new phase selection strategy, called bit-encoding scheme. The basic idea of this new strategy is to let the phase at each decision level correspond to a bit value of the binary representation of a counter. The counter increases in step with the increase of the number of restarts. In general, only the first 6 decision levels use this new scheme. The other levels use an existing scheme. Compared with the existing strategies, the new strategy is simple, and its cost is low. Experimental results show that the performance of the new phase strategy is good, and the new solver Glue_bit based on it can improve Glucose 2.1 which won a Gold Medal for application category at the SAT Challenge 2012. Furthermore, Glue_bit solved a few application instances that were not solved in the SAT Challenge 2012. From the results on the application SAT+UNSAT category at the SAT Competition 2013, Glue_bit was the best improved version of Glucose, and outperformed glucose 2.3 that is the latest improved version of glucose 2.1.

Keywords: SAT solver, conflict-driven DPLL, phase selection for SAT solvers.

1 Introduction

Numerous state-of-the-art SAT solvers have been developed in order to solve much more SAT problems such as computer aided design, data diagnosis, EDA, logic reasoning, cryptanalysis, planning, equivalence checking, model checking, test pattern generation etc. However, a large number of real-world SAT problems remain unsolvable yet. In general, SAT solvers are classified into Conflict Driven Clause Learning (CDCL), look-ahead and local search. Among three kinds of solvers, CDCL solvers are more practical and prevail. It is a solver based on conflict-driven DPLL. DPLL algorithm is due to Davis, Putnam, Logemann, Loveland. It is a complete, backtracking-based search algorithm, which runs by

T V Gopal et al. (Eds.): TAMC 2014, LNCS 8402, pp. 158–167, 2014.

choosing a literal, assigning a truth value to it, recursively checking if the formula is satisfiable under the current assignment. If this is the case, a solution is found. Otherwise, the same recursive check is done assuming the opposite truth value. To enhance efficiency, the DPLL algorithm adds the unit propagation and pure literal elimination mechanism. This paper focuses on this type of solvers for solving the SAT problem in conjunctive normal form. A CDCL solver consists of ingredients such as variable selection, phase (also called polarity) selection, restart, BCP (Boolean Constraint Propagation), conflict analysis, clause learning and its database maintenance. The improvement of each ingredient is significant. So far, various improvements on various ingredients have been proposed. For example, for variable selection, the most salient improvement is the discovery of VSIDS (Variable State Independent Decaying Sum) scheme [6]. To accelerate BCP, two watched-literals scheme was proposed. With respect to conflict analysis, a large amount of optimizing work has been done. For example, first UIP (unique implication points), conflict clause minimization, on-the-fly self-subsuming resolution [7], learned clause minimization [8] etc. To maintain effectively the clause learning database, in 2009, Audemard et al. introduced a Glucose-style reduction strategy [9] to remove less important learned clauses. In 2011, they presented further a freezing and reactivating policy [10] to restore the most promising learnt clauses rather than to re-compute them. To improve the restart strategy, in 2012, they proposed a postponing strategy [12]. Due to these improvements, Glucose 2.1 won a Gold Medal for application category at SAT Challenge 2012.

In this paper, we study an important ingredient of CDCL solvers: phase selection. The simplest phase selection policy is that each decision variable is always assigned to false, which is used as a default heuristic of MiniSAT. In addition to this, there are two phase selection strategies that are widely used in CDCL SAT solvers. One is the phase selection heuristic used in RSAT (RSAT heuristic for short) [1]. The other is Jeroslow-Wang heuristic [2]. The basic idea of the RSAT heuristic is to save the previous phase and assign the decision variable to the same value when it is visited once again. The basic idea of Jeroslow-Wang heuristic is to define variable polarity as the sign of a literal with the maximum weight. The weight of a literal depends on the number of clauses containing that literal and their sizes. PrecoSAT [3] used Jeroslow-Wang heuristic. Glucose adopts a phase selection policy based on MiniSAT and RSAT heuristic: it always assigns a decision variable to false if that variable was never visited, and the previous value otherwise. Such a phase selection policy is simple, but in some cases, we found it is not so efficient. In 2012, we proposed a dynamic phase selection policy [16], using the ACE (Approximation of the Combined lookahead Evaluation) weight [4,5]. This dynamic policy is an improved version of MPhaseSAT ACE phase policy [15]. It is aggressive so that the search depth is shortened. So it seems to be beneficial to UNSAT instances, but not necessarily to beneficial to SAT instances. Therefore, this policy can be applied to only a part of SAT instances.

Whether we choose a static or dynamic phase selection policy, the existing phase selection policies have a common characteristic: the phase differences between the two adjacent searches are almost zero at the first few decision levels. In the other word, the existing phase selection policies are of uniformity but short of diversity. In this paper, we want to replace uniformity of phase selection with diversity of phase selection. To attain this goal, we introduce a new phase selection policy called bit-encoding scheme. This new phase selection can ensure that the phase differences between the two adjacent searches are one or more at the first six decision levels. Its basic idea is to force the phase at each decision level to correspond to a bit value of the binary representation of a counter. The counter increases by one each time the search restarts. Clearly, such a phase selection is simple and diverse, and its computation cost is low. We show in this paper that the bit-encoding scheme is especially good for problems in hardware verification such as 11pipe_k and can enhance significantly the performance of Glucose. Furthermore, the new phase selection solved a few unsolved instances in SAT Challenge 2012. From the results on the application SAT+UNSAT category at the SAT Competition 2013 [14], Glue_bit was the best improved version of Glucose, and outperformed glucose 2.3 that is the latest improved version of glucose 2.1. We used the bit-encoding phase selection policy to develop a hack version of MiniSAT, which is called minisat_bit. In SAT competition 2013, minisat_bit solved 118 SAT application instances, while Lingeling aqw which won a Gold Medal for core sequential application SAT category solved 119 SAT application instances [14]. There was only one instance difference. This shows that the bit-encoding phase selection policy is effective and useful for some instances.

2   A Bit-Encoding Phase Selection

In modern CDCL SAT solvers, how to select the phase of a variable is an inseparable step that follows the decision variable selection, because we must assign each decision variable to a value. If we can select always correctly a phase, all satisfiable formulae will be solved in a linear number of decisions. In theory, no such heuristic for selecting always correctly a phase on any SAT problem exists unless P=NP. In practice, it is possible to develop a phase selection heuristic that significantly reduces the number of conflicts in some cases. The phase selection of Glucose are the hybrid of MiniSAT and RSAT heuristic. It always assigns a decision variable to false for the first time. No evidence shows that such a policy is always efficient. To improve the phase selection of Glucose, Here, we present a new phase selection policy called bit-encoding scheme. The basic idea of this new policy is to let the phase at each decision level correspond to a bit value of the binary representation of a counter. Here, the role of the counter is to record the current phase information of multiple decision variables. Let n denote the value of a counter, and the binary representation of n be

$$n = b_k 2^k + b_{k-1} 2^{k-1} + \cdots + b_1 2 + b_0.$$

Our new policy stipulates that during the m-th search period, the phase of a variable at the k-th decision level is equal to b_k. Every time a restart begins, the counter n increases by one. If the decision level on which this policy is applied is not limited, we cannot use the previous phase information. By the experimental result in [16], such a policy will not be a good one. On the other hand, it is possible to ignore the relevance between adjacent search spaces. Therefore, we apply bit-encoding scheme on only the first L levels. Furthermore, in order to strengthen the relevance between adjacent search spaces, we use only the first M bits of the counter n, where both L and M are a constant, and $M \leq L$. And let the phase of a variable at the k-th decision level correspond to the (k modulo M)-th bit of n, where $k < L$. When $k \geq L$, we use the phase selection policy of Glucose. Here is the C code of our phase selection.

```
// assume current decision level is k
   if(k < L ) polarity[var]=(n >>(k %M))&1;
   else polarity[var]=previous[var];
```

where previous[var] is used to save the previous phase and is initially set to false. In order to get an optimal value of L and M, we used Glucose to conduct some experiments with different values of L (search levels) and M. The instances tested were some harder ones. In total, we selected 200 application instances from SAT Challenge 2012. As shown in Table 1, the best values of L and M are 6 and 4, respectively. Therefore, in our solver Glue_bit, when the nodes of the search tree is less than or equal to 6, we use the formula polarity[var]=($n >>$(k %4))&1 to compute the phase of a variable. The last row of Table 1 presents the result of assigning randomly to a phase of each decision variable at the first L nodes of the search tree. As seen in Table 1, the performance of the random heuristic was not the best. However, its performance changes are not big as L increases.

Table 1. Number of solved instances with different values of L (search tree levels) and M (bits of the counter). Tests were conducted on 200 application instances in SAT Challenge 2012. The timeout was 1500s.

		\multicolumn{4}{c}{L}			
		4	6	8	12
M	4	179	188	179	181
	$M = L$	179	185	177	179
\multicolumn{2}{c}{random}		173	176	175	174

In theory, we can not guarantee that parameters L and M chosen here are not overfit for this experimental benchmarks. Perhaps, there may exist potential threats to generalizability. However, we believe that their variable range will be not big, because Minisat_bit based on this parameter choice performed well in SAT competition 2013.

Here we use the parameter L (search tree levels) given above to define the phase update period as 2^L. In Glue_bit, L is set to 6. That is, its phase update period is $2^6 = 64$. the value of L of a non-bit-encoding phase selection policy can be regarded as 0. Therefore, the phase update period of the other existing policies can be regarded as $2^0 = 1$. The phase update period can be considered as a metric to measure the diversity of a search procedure. If the phase update period of a solver is two or more, we call it diverse. Otherwise, it is said to be non-diverse or uniform. So far, all the known solvers are uniform, whereas our new solver is diverse.

Not all the instances are suited for a diverse phase selection policy. Therefore, we use a metric called AveMax_D (the average of maximal depths) whether to apply the bit-encoding scheme. Its definition is the following.

$$\text{AveMax\_D} = \frac{1}{8} \sum_{i=2}^{9} \max\{\text{conflict depths in } i\text{-th restart interval}\}$$

This is easy to compute. It can be done by using 8 out of the first 9 restarting intervals. The reason why we ignore the first of the first 9 restarting intervals is because on the same instance, changing the first decision variable results in different maximum conflict depth in the first restarting interval. From the second restarting interval, such a deviation becomes small. That is, different decision variable selections and different solving strategies will yield different AveMax_D. For Glucose-style solvers, using the above formula is better. When (AveMax_D $>$ 2500) or ($\#clauses > 200000$ and $\#conflicts < 300000$) or ($\#variables < 500$ and $\#conflicts > 300 \times \#restarts$), we replace the bit-encoding policy with Glucose phase policy. Except for this case, in general, we apply the bit-encoding policy above. Based on our observation, for application category at the SAT Challenge 2012, about 75% of the benchmark instances meet the conditions, and are suited for the bit-encoding policy. Minisat_bit, which is a hack version of MiniSAT, applied fully the bit-encoding phase selection policy without the above constraint conditions. In SAT competition 2013, minisat_bit solved 118 SAT application instances, while Lingeling aqw, which won a Gold Medal for core sequential application SAT category, solved 119 SAT application instances [14]. This shows that the bit-encoding phase selection policy is effective and useful.

3   Empirical Evaluation

Glucose 2.1 is the best solver of application category at SAT Challenge 2012. Lingeling aqw won a Gold Medal for core sequential application category in SAT competition 2013 [14]. Glucose 2.1 and Lingeling aqw are two different types of CDCL solvers. Lingeling aqw contains very complicated simplification procedures, while Glucose does not at all. In this paper, we focuses on the Glucose-style solver. So we do not compare with Lingeling aqw.

Glue_bit is our variant of the Glucose SAT solver using the proposed bit-encoding phase-selection algorithm. It is based on Glue_DDD, whose detail will be given more below. Glue_bit used the bit-encoding phase selection policy, while Glue_DDD did not. Except for this, they are the same. To observe the random feature of phase selection policy, we developed a random version of Glue_bit, called Glue_rand, which assigns randomly to a phase of each decision variable at the first 6 nodes of the search tree. That is, its value of L is set to 6. The only difference between Glue_rand and Glue_bit is that Glue_rand assigns to a phase of variable v with polarity$[v]$=rand()%2. All the solvers tested used SatElite as a preprocessor.

Glue_DDD is an improved version of Glucose 2.1. It improved the restart policy of Glucose 2.1, using a new postponing strategy based on a decision-depth-sensitive parameter [13]. Below we outline briefly Glue_DDD.

In addition to the restart triggering condition of Glucose 2.1, Glue_DDD embeds such additional conditions as the AveMax_D test and the DDD (its definition is given below) blocking test. Here is the C++ code of the restart triggering strategy of Glue_DDD.

```
K=AveMax_D < 250 && freeVars > 2500 ? 0.82 : 0.8;
assume learnt clause is to c;
sumLBD+= c.lbd(); conflicts++;
queueLBD.push(c.lbd());
if(queueLBD.isFull() && queueLBD.avg()*K > sumLBD/conflicts)
    if(AveMax_D < 250 || AveMax_D > 1500 || !blocked || DDD > 1) {
        queueLBD.clear();
        restart();
    }
```

Here, DDD is short for Decision Depth Decreasing, which is defined by a Longest Decreasing Subsequence (LDS). Given a sequence S, LDS(S) is the longest decreasing subsequence with the following property: (1) it contains the first term of S; (2) each term is strictly smaller than the one preceding it. In the Glue_DDD solver, S is seen as a sequence of conflict decision levels. The DDD of S is defined as the number of terms in LDS(S), that is, DDD(S)=|LDS(S)|. For the restarts that are not postponed by Glucose blocking strategy, Glue_DDD does not apply the DDD blocking strategy. For the restart postponed by Glucose, if DDD<2, even if the restart triggering condition is true, Glue_DDD continues to postpone that restart. Here is C++ code for the postponing algorithm of Glue_DDD.

```
R=AveMax_D ≥ 250 && AveMax_D ≤ 900 && conflicts < 1500000 ? 1.38 : 1.4;
if (AveMax_D ≥ 250 && freeVars > 5000 ){
    queueTrail.push(trail.size());
    if(queueLBD.isFull() && queueTrail.isFull() &&
      trail.size() > R*queueTrail.avg()) {
        queueLBD.clear();
        blocked=true;
    }
}
```

To verify the effectiveness of the bit-encoding phase selection policy, we conducted comparing experiments of four SAT solvers: Glucose 2.1, Glue_DDD, Glue_bit and Glue_rand under the following platform: Intel Core 2 Quad Q6600

CPU with speed of 2.40GHz and 2GB memory. It is a 32-bit machine, and slower than the experimental platform used by SAT Challenge 2012. On some instances, it is slower about $0.6 \sim 0.7$ times. Therefore, we decided to set the runtime limit for each solver on each instance to 1500 seconds. All the instances used in the experiments are from application category of SAT Challenge 2012.

Table 2. Runtime (in seconds) required by Glucose 2.1, Glue_DDD, Glue_bit and Glue_rand to solve 25 instances in SAT Challenge 2012. ">1500" shows that the instance cannot be solved in 1500s. The answer of an instance is denoted by either "S" or "U". They stand for SAT and UNSAT, respectively.

Instance	S/U	#var	#clause	Glucose2.1	Glue_DDD	Glue_bit	Glue_rand
rpoc_xits_08_unsat	U	1278	74789	>1500	>1500	1238.6	>1500
goldb-heqc-x1mul	U	8760	55571	>1500	1217.3	778.9	445.06
aloul-chnl11-13	U	286	1742	>1500	>1500	183.6	1089.54
ACG-20-5p0	U	324716	1390931	>1500	1466.2	1476.2	1476.2
md5_47_1	S	65604	273512	>1540	100.5	188.8	38.3
12pipe_q0_k	U	136800	4216460	>1500	>1500	1220.6	1106.7
11pipe_q0_k	U	104244	3007883	>1500	>1500	740.4	617.739
10pipe_q0_k	U	77639	2082017	>1500	>1500	464.1	426.839
IBM_04_ba_30_s_dat.k75	S	154604	642979	>1500	290.5	803.1	360.4
sha0_36_3	S	50073	210235	>1500	264.4	355.9	60.6
x1mul.miter.sat03-359	U	8756	55571	1486.45	>1500	804.8	596.7
c6288mul.miter.sat3-346	U	9540	61421	>1500	1224.7	600.9	381.0
bc57-sensor-1-k303-406	U	435701	1379987	>1500	>1500	1460.2	1445.42
schp3bc57-sensor-1-k303	U	435701	1379987	>1500	1362.2	1228.6	1208.4
aes_64_1_keyfind_2	S	596	2780	>1500	232.2	232.3	232.3
korf-17	U	6664	89966	>1500	>1500	682.1	700.18
slp-synthe-aes-bottom14	U	22886	76038	>1500	>1500	998.91	>1500
aaai10-ip5-TP-21-step11	U	99736	783991	>1500	1406.9	1486.6	1378.98
vmpc_33	S	1089	177375	>1500	>1500	843.8	1301.73
ndhf_xits_19UNKNOWN	S	4020	466486	903.4	>1500	>1500	>1500
smt-qfbv-aigs-vc149789	S	360364	1076507	816.9	>1500	>1500	>1500
simon3k2fix_gr_2pvar_w8	U	3771	270136	>1500	428.1	1169.7	>1500
simon3k2fix_gr_2pvar_w9	U	5028	307674	>1500	33.7	33.8	33.7
gldberg3:c6288mul.mter	U	9540	61421	>1500	1286.5	643.5	508.45
rbcl_xits_08_unsat	U	1278	68055	>1500	>1500	1399.5	>1500

Except for 25 instances shown in Table 2, on the other instances, the three solvers except for Glue_rand have the same number of solved instances. In other words, each of 25 instances cannot be solved by at least one solver among the four solvers. As seen in Table 2, the performance of Glue_bit is the best. The performance of Glue_rand is poorer than that of Glue_bit. The number of instances solved by Glue_bit is more than that solved by Glucose 2.1 and Glue_DDD. Glue_bit solved 23 out of 25 instances, and seems to be especially good for UNSAT problems. The ability that Glue_bit solves the pipe_q0_k family from hardware verification is also strong. It solved 3 pipe_q0_k instances over other two solvers. In SAT Challenge 2012, no solver solved rbcl_xits_08_unsat. However, as shown in the last row of Table 2, Glue_bit solved it in 1399.5 seconds.

Table 3. Performance of solvers on 600 application instances in SAT Challenge 2012 (timeout is set to 1500s)

Solver	Instances Solved	Average time (in seconds) per solved instance
Glucose 2.1	498	238.6
Glue_DDD	507	231.8
Glue_bit	518	222.8
Glue_rand	506	208.5

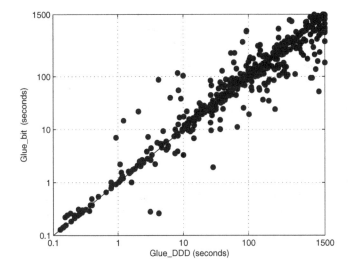

Fig. 1. Comparing the runtimes of Glue_DDD and Glue_bit on application instances from SAT Challenge 2012

On this instance, we tested also the other solvers of SAT Challenge 2012. Yet no solver solved rbcl_xits_08_unsat within 1500 seconds. So indeed the bit-encoding phase selection can enhance the performance of solvers such as Glucose.

Table 3 shows the number of solved instances and the average running time per solved instance in seconds. Glucose 2.1, Glue_DDD, Glue_bit and Glue_rand solved 498, 507, 518 and 506 out of 600 application instances, respectively. In terms of the average running time, Glue_bit was a little faster than Glue_DDD. Although Glue_rand was the fastest, the number of solved instances by it was less than Glue_DDD. That is, Glue_rand did not improve Glue_DDD, and weaken the performance of Glue_DDD. In a word, Glue_bit was the best among the four solvers. The random heuristic is not suited for the bit-encoding phase policy. Among 518 instances solved by Glue_bit, the number of instances where the new phase selection policy was actually used by Glue_bit is 397. This shows that in most cases the bit-encoding phase selection policy is useful.

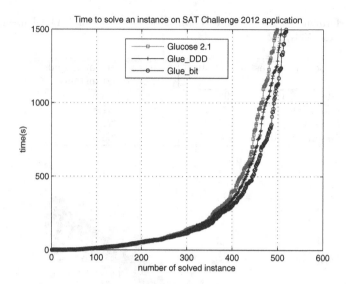

Fig. 2. The number of instances that Glucose 2.1, Glue_DDD and Glue_bit are able to solve in a given amount of time. The x-axis denotes the number of solved instances, while the y-axis denotes the running time in seconds.

Figure 1 shows a log-log scatter plot comparing the running times of Glue_bit and Glue_DDD on application instances from SAT Challenge 2012. Each point corresponds to a given instance. The climax (1500,1500) means that the instances on that point were not solved by any of the two solvers. As shown in Figure 1, some points are centralised at the nearby diagonal. This is because on those instances, both solvers did not use the bit-encoding phase selection policy. The points below the diagonal means that Glue_bit solved faster the instances denoted by them.

Figure 2 shows a cactus plot related to the performance comparison of the three solvers. Clearly, our new solver Glue_bit outperforms not only Glucose 2.1, but also Glue_DDD. In the cactus plot, the curve of Glue_bit is always below the curve of Glucose 2.1 and Glue_DDD. It implies that in a given amount of time, Glue_bit solved always more instances than Glucose 2.1 and Glue_DDD.

4   Conclusions and Future Work

In this paper, we introduced a new concept of "diversity". We used this new concept to develop a new phase selection policy called bit-encoding scheme. All the existing bit-encoding phase selections are uniform, while our bit-encoding phase selection is diverse. But more than that, the bit-encoding phase selection policy is simple, and its computation cost is very low. Furthermore, it is easily embedded into modern CDCL SAT solvers. Empirical results demonstrate that the bit-encoding phase selection policy can improve significantly the performance of solvers such as Glucose.

In our experiments, we found that different restart policies have different impacts on the phase selection policy. For a given restart, how to determine a good phase selection policy is a problem that is worth studying.

Although in theory it is impossible to select always correctly each phase every time unless P=NP, an optimal and practical phase selection policy should exist. As an open problem, how to search for such a phase selection policy is a very valuable research topic in future.

References

1. Pipatsrisawat, K., Darwiche, A.: A lightweight component caching scheme for satisfiability solvers. In: Marques-Silva, J., Sakallah, K.A. (eds.) SAT 2007. LNCS, vol. 4501, pp. 294–299. Springer, Heidelberg (2007)
2. Jeroslow, R., Wang, J.: Solving propositional satisfiability problems. Annals of Mathematics and Artificial Intelligence 1, 167–187 (1990)
3. Biere, A.: Lingeling, Plingeling, PicoSAT and PrecoSAT at SAT Race 2010 (2010), http://baldur.iti.uka.de/sat-race-2010/descriptions/solver_1+2+3+6.pdf
4. Heule, M.: March: towards a look-ahead SAT solver for general purposes, Master thesis (2004)
5. Chen, J.: Building a Hybrid SAT Solver via Conflict-driven, Look-ahead and XOR Reasoning Techniques. In: Kullmann, O. (ed.) SAT 2009. LNCS, vol. 5584, pp. 298–311. Springer, Heidelberg (2009)
6. Moskewicz, M.W., Madigan, C.F., Zhao, Y., Zhang, L.T., Malik, S.: Chaff: Engineering an Efficient SAT Solver. In: Design Automation Conference, DAC (2001)
7. Han, H., Somenzi, F.: On-the-fly clause improvement. In: Kullmann, O. (ed.) SAT 2009. LNCS, vol. 5584, pp. 209–222. Springer, Heidelberg (2009)
8. Sörensson, N., Biere, A.: Minimizing Learned Clauses. In: Kullmann, O. (ed.) SAT 2009. LNCS, vol. 5584, pp. 237–243. Springer, Heidelberg (2009)
9. Audemard, G., Simon, L.: Predicting learnt clauses quality in modern SAT solvers. In: IJCAI 2009, pp. 399–404 (2009)
10. Audemard, G., Lagniez, J.-M., Mazure, B., Saïs, L.: On Freezing and Reactivating Learnt Clauses. In: Sakallah, K.A., Simon, L. (eds.) SAT 2011. LNCS, vol. 6695, pp. 188–200. Springer, Heidelberg (2011)
11. Gomes, C.P., Selman, B., Crato, N.: Heavy-tailed distributions in combinatorial search. In: Smolka, G. (ed.) CP 1997. LNCS, vol. 1330, pp. 121–135. Springer, Heidelberg (1997)
12. Audemard, G., Simon, L.: Refining Restarts Strategies for SAT and UNSAT. In: Milano, M. (ed.) CP 2012. LNCS, vol. 7514, pp. 118–126. Springer, Heidelberg (2012)
13. Chen, J.C.: Solvers with a Bit-Encoding Phase Selection Policy and a Decision-Depth-Sensitive Restart Policy. In: Proceedings of the SAT Competition 2013, pp. 44–45 (2013)
14. SAT 2013 competition web page, http://www.satcompetition.org/2013/
15. Chen, J.C.: Phase Selection Heuristics for Satisfiability Solvers (2011), http://arxiv.org/abs/1106.1372
16. Chen, J.C.: A Dynamic Phase Selection Strategy for Satisfiability Solvers (2012), http://arxiv.org/abs/1208.1613

Generalized Finite Automata
over Real and Complex Numbers

Klaus Meer and Ameen Naif

Computer Science Institute, BTU Cottbus-Senftenberg
Platz der Deutschen Einheit 1
D-03046 Cottbus, Germany
`meer@informatik.tu-cottbus.de, naif@tu-cottbus.de`

Abstract. In a recent work, Gandhi, Khoussainov, and Liu [7] introduced and studied a generalized model of finite automata able to work over arbitrary structures. As one relevant area of research for this model the authors identify studying such automata over partciular structures such as real and algebraically closed fields.

In this paper we start investigations into this direction. We prove several structural results about sets accepted by such automata, and analyse decidability as well as complexity of several classical questions about automata in the new framework. Our results show quite a diverse picture when compared to the well known results for finite automata over finite alphabets.

1 Introduction

Finite automata represent one of the fundamental elementary models of algorithms in Computer Science. There is an elaborated theory about problems that can be solved both with and concerning finite automata which now usually is taught in basic theory courses. When dealing with algorithmic questions about finite automata like the word problem, the emptiness and finiteness problems, the equivalence problem or minimization of such automata, such questions are treated by analyzing the Turing model of computation as underlying computational model. Thus, statements like 'the equivalence problem for non-deterministic finite automata is NP-complete' are to be understood using complexity theory in the Turing model.

In recent years theoretical computer science has seen an increasing interest in alternative to the Turing machine models of computation. The reader might think of quantum computers [13], neural networks [8], analogue computers [4], several kinds of biologically inspired devices [14], and models for computations over the real and complex numbers. Models for the latter split into approaches based on the Turing machine like those followed in recursive analysis [15] and in algebraically inspired notions of algorithms [3,5].[1] One feature of such algorithm

---

[1] For all mentioned areas the given references are not thought to be exhaustive but should just serve as a starting point for readers being more interested in the corresponding models.

T V Gopal et al. (Eds.): TAMC 2014, LNCS 8402, pp. 168–187, 2014.

models is that they do not any longer exclusively work over finite alphabets as underlying structures. For example, the Blum-Shub-Smale (shortly BSS) model introduced in [3] can be and was used to define a computability notion for many different structures including $\mathbb{R}$ and $\mathbb{C}$. It is thus a reasonable question whether also the concept of a (deterministic and non-deterministic) finite automaton can be generalized to work over more general structures than just finite alphabets.

In recent work Gandhi, Khoussainov, and Liu [7] introduce such a generalized model of finite automata called $(\mathcal{S}, k)$-automata. It is able to work over an arbitrary structure $\mathcal{S}$, and here in particular over infinite alphabets like the real numbers. A structure is characterized by an alphabet (also called universe) together with a finite number of binary functions and relations over that alphabet. Now, intuitively the model processes words over the underlying alphabet componentwise. Each single step is made of finitely many test operations relying on the fixed relations as well as finitely many computational operations relying on the fixed functions. For performing the latter an $(\mathcal{S}, k)$-automaton can use a finite number k of registers. As we shall see the latter ability adds significantly power to the model in comparison to 'old-fashioned' finite automata. An automaton then moves between finitely many states and finally accepts or rejects an input.

The motivation to study such generalizations is manifold. In [7] the authors discuss different previous approaches to design finite automata over infinite alphabets and their role in program verification and database theory. One goal is to look for a generalized framework that is able to homogenize at least some of these approaches. As the authors remark, many classical automata models like pushdown automata, Petri nets, visible pushdown automata can be simulated by the new model. Another major motivation results from work on algebraic models of computation over structures like the real and complex numbers. Here, the authors suggest their model as a finite automata variant of the Blum-Shub-Smale BSS model. They then ask to analyze such automata over structures like real or algebraically closed fields.

The latter will be the focus of the present work. We restrict our attention to generalized finite automata over two special structures denoted by $\mathcal{S}_{\mathbb{R}}$ and $\mathcal{S}_{\mathbb{C}}$ and defined precisely below. These are the suitable choices for relating $(\mathcal{S}, k)$-automata to the BSS model.

The paper is structured as follows. Section 2 recalls the generalized automata model from [7] for the above two structures and gives some basic examples. It collects as well the notions from BSS computability theory necessary to relate the automata model to the latter. We shall then study a bunch of questions well known from finite automata theory. Section 3 analyzes the word problem for deterministic and non-deterministic $(\mathcal{S}, k)$-automata in comparison to certain complexity classes in the BSS model. A complexity class coming into play very naturally here is the class DNP of problems that can be verified by so called digital nondeterminism. In particular, we shall get a somehow diverse picture concerning the classes of languages accepted by (non)-deterministic $(\mathcal{S}, k)$-automata. For example, we shall see that there are easy problems in $P_{\mathbb{C}}$, the class of problems being polynomial time solvable in the complex number BSS model, that cannot be accepted by any non-deterministic automaton, whereas the complex Knapsack problem $KS_{\mathbb{C}}$ (a problem likely not located in $P_{\mathbb{C}}$) can. This problem can as well be used to show that languages accepted by non-deterministic complex $(\mathcal{S}, k)$-automata are not closed under complementation. Towards obtaining this result we discuss certain structural properties of languages accepted by complex $(\mathcal{S}, k)$-automata. We discuss as well the real case, where (partially by different arguments) the same results hold as well. A second structural result for complex automata will give a kind of weak pumping lemma.

A number of undecidability results are shown for both structures in Section 4. Among them we find the emptiness problem, the equivalence problem, several reachability questions as well as the problem to minimize an $(\mathcal{S}, k)$-automaton. For some restricted classes of automata we also give decidability results. The paper closes with a discussion concerning open future questions.

2   Generalized Finite Automata over $\mathbb{R}$ and $\mathbb{C}$

We suppose the reader to be familiar with the basics of the Blum-Shub-Smale model of computation and complexity over $\mathbb{R}$ and $\mathbb{C}$. Very roughly, algorithms in this model work over finite strings of real or complex numbers, respectively. The operations either can be computational, in which case addition, subtraction, and multiplication is allowed; without much loss of generality we do not consider divisions in this paper to avoid technical inconveniences. Or an algorithm can branch depending on the result of a binary test operation. The latter either will be an inequality test of form 'is $x \geq 0$?' over the reals or an equality test 'is $x = 0$?' when working over the complex numbers. The size of a string is the number of components it has, the cost of an algorithm is the number of operations it performs until it halts. For more details see [2]. Notions being central for this paper shall be explained in more detail below.

The generalized finite automata introduced in [7] work over structures. Here, a structure $\mathcal{S}$ consists of a universe D as well as of finite sets of (binary) functions and relations over the universe. A more precise definition concerning the structures we are interested in follows below. An automaton informally works as follows. It reads a word from the universe, i.e., a finite string of components from D and processes each component once. Reading a component the automaton can set up some tests using the relations in the structure. The tests might involve a fixed set of constants from the universe D that the automaton can use. It can as well perform in a limited way computations on the current component. Towards this aim, there is a fixed number k of registers that can store elements from D. Those registers can be changed using their current value, the current input and the functions related to $\mathcal{S}$. After having read the entire input word the automaton accepts or rejects it depending on the state in which the computation stops. These automata can both be deterministic and non-deterministic.

Since the approach allows structures to have arbitrary universes, the model in particular easily can be adapted to define generalized finite automata over structures like $\mathbb{R}$ and $\mathbb{C}$. We shall in the rest of the paper focus on these universes. Our goal is to study questions for these automata in the framework of the BSS model of computations over those structures. We thus define two structures $\mathcal{S}_\mathbb{R}$ and $\mathcal{S}_\mathbb{C}$ according to the operations used in the BSS model over $\mathbb{R}$ and $\mathbb{C}$, respectively. As in the original work [7] we equip all our structures as well with the projection operators pr_1, pr_2 which give back the first and the second component of a tuple, respectively.

Definition 1. *Let $\mathcal{S}_\mathbb{R} := (\mathbb{R}, +, -, \bullet, pr_1, pr_2, \geq, =)$ denote the structure of the reals as ring with order. Similarly, the structure of the complex numbers as ring with equality is given by $\mathcal{S}_\mathbb{C} := (\mathbb{C}, +, -, \bullet, pr_1, pr_2, =)$.*

In order to avoid technicalities for operation $-$ we allow both orders of the involved arguments, i.e., applying $-$ to two values x, v can mean $x - v$ or $v - x$. Similarly, the order test can be performed both as $x \leq v$? and $v \leq x$? As mentioned above without loss of generality we do not include division as an operation. This will not significantly change our results.

The following definition is from [7] but adapted for the special structures exclusively considered here.

Definition 2. *(Finite automata over $\mathcal{S}_\mathbb{R}$ and $\mathcal{S}_\mathbb{C}$, [7]) Let $k \in \mathbb{N}$ be fixed.*

a) *A deterministic $(\mathcal{S}_\mathbb{R}, k)$-automaton $\mathcal{A}$ consists of the following objects:*
  - *a finite state space Q and an initial state $q_0 \in Q$,*
  - *a set $F \subseteq Q$ of accepting states,*
  - *a set of ℓ registers which contain fixed given constants $c_1, \ldots, c_\ell \in \mathbb{R}$*
  - *a set of k registers which can store real numbers denoted by $v_1, \ldots, v_k$,*
  - *a transition function $\delta : Q \times \mathbb{R} \times \mathbb{R}^k \times \{0, 1\}^{k+\ell} \mapsto Q \times \mathbb{R}^k$.*

  The automaton processes elements of $\mathbb{R}^*$, i.e., words of finite length with real components. For such an $(x_1, \ldots, x_n) \in \mathbb{R}^n$ it works as follows. The computation starts in q_0 with an initial configuration for the values $v_1, \ldots, v_k \in \mathbb{R}$, say all $v_i = 0$. Then, $\mathcal{A}$ reads the input components step by step. Suppose a value x is read in state $q \in Q$. Now the next state together with an update of the values v_i is computed as follows:
    - *$\mathcal{A}$ performs the $k + \ell$ comparisons $x\sigma_1 v_1$?, $x\sigma_2 v_2$?, ..., $x\sigma_k v_k$?, $x\sigma_{k+1}c_1$?, ..., $x\sigma_{k+\ell}c_\ell$?, where $\sigma_i \in \{\geq, \leq, =\}$. This gives a vector $b \in \{0, 1\}^{k+\ell}$, where a component 0 indicates that the comparison that was tested is violated whereas 1 codes that it is valid;*
    - *depending on state q and b the automaton moves to a state $q' \in Q$ (which could again be q) and updates the v_i applying one of the operations in the structure: $v_i \leftarrow x \circ_i v_i$. Here, $\circ_i \in \{+, -, \bullet, pr_1, pr_2\}, 1 \leq i \leq k$ depends on q and b only.*

  When the final component of an input is read $\mathcal{A}$ performs the tests for this component and moves to its final state without any further computation. It accepts the input if this final state belongs to F, otherwise $\mathcal{A}$ rejects.

b) *Non-deterministic $(\mathcal{S}_\mathbb{R}, k)$-automata are defined similarly with the only difference that δ becomes a relation in the following sense: If in state q the tests result in $b \in \{0, 1\}^{k+\ell}$ the automaton can non-deterministically choose*

for the next state and the update operations one among finitely many tuples
$(q', \circ_1, \ldots, \circ_k) \in Q \times \{+, -, \bullet, pr_1, pr_2\}^k$.
*As usual, a non-deterministic automaton accepts an input if there is at least
one accepting computation.*

c) *(Non-)Deterministic automata over $\mathcal{S}_C$ are defined similarly, the only difference being that the tests all have to be equality tests.*

d) *For an automaton $\mathcal{A}$ the language of finite strings accepted by $\mathcal{A}$ is denoted
by $L(\mathcal{A})$. Clearly, $L(\mathcal{A}) \subseteq \mathbb{R}^*$ or $L(\mathcal{A}) \subseteq C^*$, depending on the structure
considered.*

Remark 1. a) A few words concerning the intuitive abilities and some technical aspects of generalized automata are appropriate here. Given k registers an
automaton can perform calculations. Depending on the problem considered this
ability can be relatively strong. It is, for example, easy to see that in appropriate
structures like $\mathcal{S}_\mathbb{R}$ and $\mathcal{S}_C$ one can count in certain situations. For example, a
language like $\{(-1)^n 1^n | n \in \mathbb{N}\}$ easily is seen to be acceptable by such an au-
tomaton if 1 is available as constant. On the other hand input components still
can be read only once. If we want to use it several times it has to be stored in
one of the k registers in order to be used again. This of course might be a severe
restriction.

b) Another technical aspect refers to the initialization of a computation. In
the original definition in [7] one can start a computation with an arbitrary assignment for the registers v_i. However, this might mean that additional constants
are introduced into calculations. Below we do not want to analyze which impact
the use of an initialization in general has with respect to the constants used.
For our results it seems not to change arguments significantly. Therefore, all our
computations start with initial values 0 for the v_i. Nevertheless, the impact of
different initializations seems an interesting problem to be analyzed further.

c) The final component of an input somehow is treated differently than all the
others by the model. The reason is that after having read the final component
only a test but no computation is performed. So in a certain sense an automaton
is working only restrictedly with it. In some cases below we circumvent this effect
by choosing a kind of dummy final component which will be the same for all
inputs of a problem. That way the crucial parts of an input all are handled the
same way by an automaton.

Since below we want to treat some elementary questions about such automata
within the framework of real and complex BSS machines, we finish this section
with the definition of a special non-deterministic complexity class in the BSS
model that turns out to be interesting for analyzing non-deterministic generalized automata.

Definition 3. *(Digital non-determinism)*
a) *In the BSS-model over $\mathbb{R}$ a problem $L \subseteq \mathbb{R}^*$ belongs to the class $\mathrm{DNP}_\mathbb{R}$
of problems verifiable in polynomial time using digital non-determinism if there
is a BSS algorithm M working as follows: M gets as its inputs tuples $(x, y) \in
\mathbb{R}^* \times \{0, 1\}^*$ and computes a result in $\{0, 1\}$ interpreted as reject or accept,*

respectively. For an $x \in L$ there has to be a $y \in \{0,1\}^*$ such that $M(x,y) = 1$, for $x \notin L$ the result $M(x,y)$ has to be 0, no matter which y is chosen.

The running time of M has to be polynomial in the (algebraic) size of x.

b) Similarly, the complex counterpart $\mathrm{DNP}_\mathbb{C}$ is defined.

Digital non-determinism is a kind of restricted non-determinism in the real and complex BSS model. Here, usually the guess y is allowed to stem from $\mathbb{R}^*$ or $\mathbb{C}^*$ having polynomial length in the size of x. Most natural generalizations of discrete NP-complete problems to the real framework lead to problems in $\mathrm{DNP}_\mathbb{R}$; the Knapsack problem treated below is a typical example. However, there are important open questions related to this class. It is easy to see that $P_\mathbb{R} \subseteq \mathrm{DNP}_\mathbb{R} \subseteq \mathrm{NP}_\mathbb{R}$, but it is currently only conjectured that both inclusions are strict. As a consequence, problems in $\mathrm{DNP}_\mathbb{R}$ are conjectured not to be $\mathrm{NP}_\mathbb{R}$-complete in the BSS model. The same is conjectured to be true for the classes $P_\mathbb{C} \subseteq \mathrm{DNP}_\mathbb{C} \subseteq \mathrm{NP}_\mathbb{C}$.

3 Basic Results, a Structural Theorem and a Weak Pumping Lemma for Complex Automata

Given the above definition of digital non-determinism our first result is immediate. The word problem for a fixed $(\mathcal{S}, k)$-automaton asks whether a given input $w \in D^*$ is accepted by the automaton. The following is easily proved.

Lemma 1. *The word problem belongs to class $P_\mathbb{R}$ for deterministic $(\mathcal{S}_\mathbb{R}, k)$-automata and to class $\mathrm{DNP}_\mathbb{R}$ for non-deterministic such automata. This holds as well if the automaton is considered part of the input. It is analogously true for complex automata and the complex BSS model.*

The previous result gives rise to several further questions which we shall treat next dealing with complex automata. Are all problems in $P_\mathbb{C}$ acceptable by a non-deterministic $(\mathcal{S}_\mathbb{C}, k)$-automaton, are there any $\mathrm{DNP}_\mathbb{C}$ problems which potentially do not belong to $P_\mathbb{C}$ but will be accepted by such an automaton, are non-deterministic automata closed under complementation?

We shall see that the answers to the above questions show a somewhat skew picture of the relation between acceptable languages in the generalized automata model and complexity classes in the BSS model.

Let us start with defining an extension of the classical NP-complete Knapsack problem.

Definition 4. *The complex Knapsack problem $KS_\mathbb{C}$ is defined as follows: Given $n \in \mathbb{N}$ and n complex numbers $x_1, \ldots, x_n$, is there a subset $S \subseteq \{1, \ldots, n\}$ such that $\sum_{i \in S} x_i = 1$?*

The real Knapsack problem $KS_\mathbb{R}$ is defined similarly.

Remark 2. Dealing with this problem using $(\mathcal{S}_\mathbb{C}, k)$-automata we consider inputs of the form $(x_1, \ldots, x_n, 1)$ of length $n + 1$. This is done in order to guarantee that the numbers $x_1, \ldots, x_n$ are treated equally, see Remark 1 c) above. This could of course be done differently.

Note that the complexities of both $KS_{\mathbb{C}}$ and $KS_{\mathbb{R}}$ in the respective BSS model are unknown. This is the case for many reasonable extensions of classically NP-complete problems to the real or complex number model. On the one hand side such problems usually fall into the classes $DNP_{\mathbb{K}}$ defined above, where $\mathbb{K} \in \{\mathbb{R}, \mathbb{C}\}$. It is not known whether $DNP_{\mathbb{K}}$ contains $NP_{\mathbb{K}}$-complete problem for one of the settings (and actually conjectured to be false [9]), so the respective generalized Knapsack problems likely will neither be $NP_{\mathbb{C}}$- nor $NP_{\mathbb{R}}$-complete. On the other side it is neither known whether problems being NP-complete in the Turing model can be solved more efficiently in the BSS model.

Example 1. $KS_{\mathbb{C}}$ can be accepted by a non-deterministic $(S_{\mathbb{C}}, 1)$-automaton. The non-deterministic automaton uses 1 as its only constant and a single register v_1. When reading a new component x_i of the input the automaton non-deterministically chooses whether x_i should participate in the final sum or not. If it should, then x_i is added to v_1, otherwise v_1 remains unchanged. When $(x_1, \ldots, x_n)$ has been processed the automaton finally checks whether the last input component x_{n+1} equals 1. If not $\mathcal{A}$ rejects, if yes (i.e., $x_{n+1} = 1$) it is also compared with v_1. The automaton accepts iff $v_1 = 1$.

3.1   A Structure Theorem for Complex Automata

Our first major result will be a structural theorem concerning the languages accepted by complex deterministic and non-deterministic automata. As an easy consequence of this theorem it follows that the class of languages accepted by non-deterministic complex automata is not closed under complementation. For a particular restricted structure over the integers the corresponding result was shown in [7]. We prove it using topological arguments. More precisely, the Knapsack problem turns out to be a counterexample here. Before doing so we recall the definition of typical paths for a BSS algorithm [2], adapted accordingly to $(S_{\mathbb{C}}, k)$-automata. Characteristic paths are important since the sets of inputs that follow those paths have a useful topological structure.

Definition 5. *a) Let $\mathcal{A}$ be a deterministic $(S_{\mathbb{C}}, k)$-automaton using ℓ constants. Let $\mathcal{P}$ be a path of this automaton, i.e., a finite sequence $(q_0, q_1, \ldots, q_s)$ of states of $\mathcal{A}$ together with a sequence $(b^{(0)}, b^{(1)}, \ldots, b^{(s-1)})$ of test results in $\{0,1\}^{k+\ell}$ such that the automaton moves from q_i to q_{i+1} when $b^{(i)}$ represents the outcome of the tests. Here, q_0 denotes the start state of $\mathcal{A}$.*

i)   The path set $V_{\mathcal{P}}$ related to a path $\mathcal{P}$ of length s is the set of points in $\mathbb{C}^s$ that are branched by $\mathcal{A}$'s computation along $\mathcal{P}$.

ii)   The characteristic path of length s of $\mathcal{A}$ is the one obtained if all $b^{(i)} = 0$, i.e., all $k + \ell$ equality tests performed at each step of the computation give result 'false'.

b)   If $\mathcal{A}$ is a non-deterministic automaton any path that corresponds to a computation where all test results give $b^{(i)} = 0$ is called a characteristic path.

Note that characteristic paths always are realizable by some computation, i.e., the corresponding path sets are non-empty. This is true because at each step of a computation only a constant number of values are stored in the registers, so there is always a next complex input component being different from all of them.

Characteristic path sets have a very special structure. This structure is made more precise in the following definition.

Definition 6. *For a set $L \subseteq \mathbb{C}^n, n \geq 1$ let $\mathcal{P}_1(L)$ denote the projection of L to the first component. Then L is called recursively co-finite, or rcf for short, if the following two conditions hold:*

i) $\mathcal{P}_1(L)$ is co-finite;
ii) if $n > 1$ for any $x^* \in \mathcal{P}_1(L)$ the set

$$\{(x_2, \ldots, x_n) \in \mathbb{C}^{n-1} | (x^*, x_2, \ldots, x_n) \in L\}$$

is recursively co-finite.

In particular, a set $L \subseteq \mathbb{C}$ is rcf if it is co-finite.

We say that L is recursively co-finite of cardinality $s \in \mathbb{N}$ if the cardinalities of all the complements of projections involved in the above definition are less than or equal to s.

Now the following structural theorem can be proven.

Theorem 1. *Let $\mathcal{A}$ be a $(\mathcal{S}_\mathbb{C}, k)$-automaton with ℓ constant registers, $L(\mathcal{A}) \subseteq \mathbb{C}^*$ the language accepted by $\mathcal{A}$. For each $n \in \mathbb{N}$ let $L_n(\mathcal{A}) := L(\mathcal{A}) \cap \mathbb{C}^n$ and $\overline{L_n(\mathcal{A})}$ its complement in $\mathbb{C}^n$.*

a) If $\mathcal{A}$ is deterministic, then for each $n \in \mathbb{N}$ exactly one of the two sets $L_n(\mathcal{A})$ and $\overline{L_n(\mathcal{A})}$ contains a rcf set of cardinality at most $s := k + \ell$. In particular, for all n the cardinalities of the respective complements are bounded by a constant that is independent of n. Which of the two sets contains the rcf set can vary with n.

b) If $\mathcal{A}$ is non-deterministic, then there is a constant M such that for each $n \in \mathbb{N}$ exactly one of the two sets $L_n(\mathcal{A})$ and $\overline{L_n(\mathcal{A})}$ contains a rcf set of cardinality at most $O(M^n)$.

Moreover, for those n where the rcf set is contained in $L_n(\mathcal{A})$ the cardinalities of the respective complements can again be bounded by $s = k + \ell$.

Thus the difference between the statements for deterministic and non-deterministic automata is the cardinality bound for the rcf sets in $\overline{L_n(\mathcal{A})}$. In the non-deterministic case in general it cannot be bounded by a constant being independent of n.

Proof. a) Suppose first that $\mathcal{A}$ is deterministic. Fix n and consider the characteristic path γ_n of length n. Its path set V_{γ_n} is rcf with cardinality at most $s = k + \ell$; in each computational step i of $\mathcal{A}$ along the characteristic path if x_i denotes the current input component all but at most s many choices for $x_i \in \mathbb{C}$ will be branched further along γ_n. Thus, V_{γ_n} is rcf of cardinality at most s. If path γ_n accepts, then $L_n(\mathcal{A})$ contains a rcf set, otherwise $\overline{L_n(\mathcal{A})}$ contains such a set.

It remains to show that only one of the two sets contains a rcf set. Towards this aim let $U \subset \mathbb{C}^n$ denote a subset of an arbitrary union of path sets of length n other than V_{γ_n}. We claim that U is not a rcf set. For any point $x \in U$ define $t(x) \in \{1, \ldots, n\}$ as index of the first component of x such that $\mathcal{A}$'s computation on x answers a test with $=$. Since by assumption $x \notin V_{\gamma_n}$ such an index exists. Now choose $x^* \in U$ with maximal value $t(x^*)$. For the point $(x_1^*, \ldots, x_{t(x^*)-1}^*)$ all tests so far have been answered by $\neq$, for component $x_{t(x^*)}^*$ an equality test is satisfied. If U would be rcf the projection of the set $\{(x_{t(x^*)}, \ldots, x_n) | (x_1^*, \ldots, x_{t(x^*)-1}^*, x_{t(x^*)}, \ldots, x_n) \in U\}$ to its first component has to be co-finite. Thus, there has to exist an $x_t \in \mathbb{C}$ such that all tests performed by $\mathcal{A}$ on input $(x_1^*, \ldots, x_{t(x^*)-1}^*, x_t)$ are answered negatively and $(x_1^*, \ldots, x_{t(x^*)-1}^*, x_t)$ can be extended to a point $\hat{x}$ in U. However, for such a point $t(\hat{x})$ would be larger than $t(x^*)$ thus contradicting our choice of x^*. It follows that U cannot be a rcf set.

b) Next, let $\mathcal{A}$ be non-deterministic. There are two cases two consider due to the fact that now several characteristic paths of a given length can exist. Let $n \in \mathbb{N}$ be fixed.

Case 1: All characteristic paths of length n reject. Then L_n has the same structure as in the deterministic case, i.e., it is the finite union of paths which are not characteristic. By the same argument as above L_n does not contain a rcf set.

Concerning the structure of $\overline{L_n}$ in Case 1 let $M \in \mathbb{N}$ denote the maximal number of non-deterministic choices automaton $\mathcal{A}$ can follow in one of its states. If φ is a characteristic path of $\mathcal{A}$ note that not necessarily $V_\varphi \subseteq \overline{L_n}$ since a point $x \in V_\varphi$ could be branched as well along an accepting non-characteristic path. Nevertheless, along a characteristic path in each step at most $k + \ell$ values are branched away from the path. There are at most M^n many characteristic paths of length n, thus at most $(k + \ell)M^n$ values for a fixed input component can be branched away from all characteristic paths (note that there might be inputs which can follow all characteristic paths of $\mathcal{A}$). It follows that $\overline{L_n}$ contains a rcf set of cardinality $O(M^n)$.

Case 2: There are characteristic paths which accept. Let φ denote one of them. Since the path accepts it follows $V_\varphi \subseteq L_n$. Clearly, V_φ is rcf of cardinality $k + \ell$. Finally, $\overline{L_n}$ does not contain a rcf set. Because if $U \subseteq \overline{L_n}$ would be rcf each point in U has to be branched away from the accepting characteristic path φ. The same argument as in the deterministic case gives a contradiction.   □

We derive two further results from the theorem. Example 1 has shown that the Knapsack problem, which is conjectured not to be efficiently solvable in the BSS model, can be accepted by a non-deterministic automaton. However, the next result shows that easier problems cannot. Thus, the class of languages acceptable by non-deterministic (S, k)-automata is contained in $\text{DNP}_\mathbb{R}$ or $\text{DNP}_\mathbb{C}$, respectively, but lies kind of skew with respect to the class of languages decidable in polynomial time in the BSS model.

Corollary 1. *There are problems in complexity class* $P_{\mathbb{C}}$ *which can not be accepted by any non-deterministic* $(\mathcal{S}_{\mathbb{C}}, k)$*-automaton. Similarly for* $P_{\mathbb{R}}$. *As a consequence, the class of languages accepted by non-deterministic real or complex automata is strictly contained in* $DNP_{\mathbb{C}}$ *and* $DNP_{\mathbb{R}}$, *respectively.*

Proof. For the complex case define a language L as

$$L := \{(a_0, \ldots, a_n, x) \in \mathbb{C}^{n+2} | n \in \mathbb{N}, \sum_{i=0}^{n} a_i \cdot \left(x^{n^n}\right)^i = 0\}$$

By repeated squaring of x and subsequent evaluation of the univariate polynomial given through the a_i membership in L can be decided in the complex BSS model in polynomial time in n, i.e., $L \in P_{\mathbb{C}}$. Now suppose a non-deterministic $(\mathcal{S}_{\mathbb{C}}, k)$-automaton $\mathcal{A}$ accepts L. As in the proof of Theorem 1 let M denote an upper bound on the number of non-deterministic choices in any state. For a canonical choice of $(a_0, \ldots, a_n) \in \mathbb{C}^{n+1}$ the polynomial $z \mapsto \sum_{i=0}^{n} a_i z^i$ has n different complex roots and each of them has n^n different n^n-th roots. Thus, there are $n \cdot n^n$ choices for x such that $(a_0, \ldots, a_n, x) \in L$. Now on the one hand side $L \cap \mathbb{C}^{n+2}$ does not contain a rcf set since given $(a_0, \ldots, a_n) \neq 0$ there are always only finitely many choices for x yielding a point in L. On the other hand, for large enough n we have $n \cdot n^n > const \cdot M^n$ which contradicts as well part b) of Theorem 1. It follows that $\mathcal{A}$ cannot exist.

The proof for the real number is similar and postponed to the full paper. The last claim now follows from Lemma 1 and the containment of $P_{\mathbb{R}}$ in $DNP_{\mathbb{R}}$ and of $P_{\mathbb{C}}$ in $DNP_{\mathbb{C}}$. $\qquad\square$

The above proof for the real case uses a certain structural property of accepted sets similar to rcf sets but seemingly weaker with respect to deriving interesting structural results. Therefore, we did not formulate it separately.

The following corollary shows another application of Theorem 1. Note that part a) follows from b), however we add the simple argument based on the previous theorem.

Corollary 2. *a)   For all* $k \in \mathbb{N}$ *there is no deterministic* $(\mathcal{S}_{\mathbb{C}}, k)$*-automaton accepting* $KS_{\mathbb{C}}$.

b)   For all $k \in \mathbb{N}$ *there is no non-deterministic* $(\mathcal{S}_{\mathbb{C}}, k)$*-automaton accepting the complement of* $KS_{\mathbb{C}}$, *i.e., the set*

$$\overline{KS_{\mathbb{C}}} = \{(x_1, \ldots, x_n, x_{n+1}) | n \in \mathbb{N} \text{ and either } x_{n+1} \neq 1 \text{ or} \\ \forall S \subseteq \{1, \ldots, n\} \sum_{i \in S} x_i \neq 1\}.$$

c)   The class of languages accepted by non-deterministic complex automata is stritcly larger than the class of languages accepted by deterministic such automata.

Proof. a) Our definition of $KS_{\mathbb{C}}$ requires as positive instances $n+1$-dimensional vectors whose final component is 1. Thus no rcf set can be a subset of $KS_{\mathbb{C}}$.

According to the theorem the only characteristic path of a potential deterministic automaton has to contain an rcf set. Consider an input $(x_1, \ldots, x_{n-1}) \in \mathbb{C}^{n-1}$ such that all 2^{n-1} possible sums of components give a different result and such that the automaton follows the characteristic path when reading $(x_1, \ldots, x_{n-1})$. Clearly, such a sequence exists. Then there are 2^{n-1} many choices for x_n such that $(x_1, \ldots, x_n, 1) \in KS_\mathbb{C}$, but for component x_n the characteristic path can only branch away a constant number s of values for x_n. Thus, such an automaton cannot exist.

b) Suppose a non-deterministic automaton $\mathcal{A}$ accepts $\overline{KS_\mathbb{C}}$. The theorem implies that precisely one of the sets $KS_\mathbb{C}$ and $\overline{KS_\mathbb{C}}$ contains an rcf set. For $KS_\mathbb{C}$ this is not possible since the final component of an input in $KS_\mathbb{C}$ is forced to equal 1. For deriving as well a contradiction in the remaining case we need an additional argument. As in a) choose n large enough and $(x_1, \ldots, x_{n-1}) \in \mathbb{C}^{n-1}$ such that all 2^{n-1} possible sums of components give a different result, this time also different from 1. In addition we require that $\mathcal{A}$ follows for an infinite number of choices for x_n an accepting characteristic path when reading $(x_1, \ldots, x_{n-1})$. Note that fixing $(x_1, \ldots, x_{n-1})$ as above there is an infinite number of x_n such that $(x_1, \ldots, x_n, 1) \notin KS_\mathbb{C}$ and all such inputs must be accepted by $\mathcal{A}$. Thus the existence of such a path φ is guaranteed because there are only finitely many paths of a given length. Consider the two final computational steps of $\mathcal{A}$ when reading x_n and $x_{n+1} = 1$. Since $\mathcal{A}$ accepts for infinitely many choices of x_n along φ, for the final tests with $x_{n+1} = 1$ only those register values v_i have an influence that depend on the choice of x_n. Each of them, however, can only branch a single value away from φ. Since there are $2^{n-1} > k$ choices for x_n such that $(x_1, \ldots, x_n, 1) \in KS_\mathbb{C}$ the automaton can still branch most of them along φ and accept, thus leading to a contradiction.

Finally, claim c) directly follows since the class of languages accepted by a deterministic automaton clearly is closed under complementation. Example 1 together with parts a), b) imply the statement.   □

Though the notion of characteristic path(s) makes sense for real automata as well it is not clear how to use it to obtain a meaningful structural result like Theorem 1. This is discussed a bit further in the final section. With respect to the real Knapsack problem, however, we can prove the same statement by applying well known results from algebraic complexity theory [1,12]. Due to space limitations a proof will be postponed to the full version.

Note that a much more general result for $KS_\mathbb{R}$ in the realm of the real BSS model has been shown in [6].

Proposition 1. *For all $k \in \mathbb{N}$ there is no non-deterministic $(\mathcal{S}_\mathbb{R}, k)$-automaton accepting the complement of $KS_\mathbb{R}$. Thus, the class of languages accepted by non-deterministic real automata is not closed under complementation. The real Knapsack problem is not accepted by a deterministic $(\mathcal{S}_\mathbb{R}, k)$-automaton.*

3.2 A Weak Pumping Lemma

One major structural tool for establishing a language not to be regular in the classical finite automata framework is the pumping lemma. It is thus natural to ask whether a similar property holds for our generalized automata. However, a short consideration immediately implies that - if at all - such a statement has to be more involved. Consider the language $L := \{(x_1, 1, x_2, 1, \ldots, 1, x_n) \in \mathbb{C}^n | n \in \mathbb{N}, x_1 = 1, x_{i+1} = x_i + 1, 1 \leq i \leq n-1\}$, i.e., $(x_1, \ldots, x_n)$ represent initial segments of $\mathbb{N}$. L clearly is acceptable by a deterministic $(\mathcal{S}_\mathbb{C}, 1)$-automaton.[2] Now, if in a word $w \in L$ we pump any of its substrings the structure of the defining recursion formula for the components clearly is destroyed.

One major obstacle for obtaining a kind of pumping lemma is the ability to perform computations. Even if an automaton runs through a loop with respect to its state set it is by no means clear whether the loop is realizable even only once more repeating the same subsequence of input components. The reason is that in most cases the assignments of registers will change. And a different assignment clearly can result in a different computation path when reading the same part of an input repeatedly.

For complex automata and some loops it turns out that we can say a bit more. Here, once again the characteristic path of a deterministic automaton is helpful because many inputs follow it. We shall now show that a weak kind of pumping is possible. As drawback two features of the classical pumping lemma are lost. First, the pumping might not be possible for words in the language but for rejected words; and secondly, it cannot be guaranteed to hold for all words of a certain length. Nevertheless, we shall see that the statement can be used to show certain problems not being acceptable by deterministic complex automata.

Theorem 2. *Let $L \subseteq \mathbb{C}^*$ be accepted by a deterministic $(\mathcal{S}_\mathbb{C}, k)$-automaton $\mathcal{A}$. Then there is a word $w := uz \in \mathbb{C}^*$ such that either all $uz^t, t \in \mathbb{N}_0$ belong to L or they all belong to $\mathbb{C}^* \setminus L$. Moreover, u and z have an algebraic length of at most K and $2K$, respectively, where K denotes the number of states of $\mathcal{A}$.*

Proof. Before going into detail we outline the main idea of the proof. We are looking for inputs that follow the characteristic path of $\mathcal{A}$ when the input dimension becomes larger. As mentioned earlier this path is realizable. For example, we could take a sequence of algebraically independent numbers. So there is a loop that can be realized as many times as we want. Let q denote the starting and final state of the first such loop; here, by first we mean first time the loop is completed. We fix this loop as the one we are interested in for the rest of the proof. Then there is a u of length at most K such that when reading u as its first input components $\mathcal{A}$ follows the characteristic path and enters q for a first time. The length of the loop is some $s \leq K$.

The problem, however, is that the above easy argument implies realizability of the characteristic path only when the input components can be changed all

---

[2] The intermediate 1's are used to avoid including the operation $+1$ in the structure; they could be removed if the operation is available.

the time. Ad hoc there is no guarantee that we can follow the loop any given number of runs always taking the *same* complex vector $z \in \mathbb{C}^s$. The main task in the proof is to establish the existence of such a z. This will be done as follows: First, we show the existence of an open set $X \subseteq \mathbb{C}^s$ such that for each $x \in X$ automaton $\mathcal{A}$ on input ux follows the loop once. For this purpose we can use a sequence ux with the set of components being algebraically independent. In that case, no equality test will be answered positively, so ux follows the characteristic path. Since the test functions are continuous in the input components there is an open set X containing x such that for all $y \in X$ the input uy follows the characteristic path as well.

The main part of the proof now shows that for each additional run through the loop only a reasonably small set of points from X have to be removed because they might not be branched along the characteristic path when passing another time the loop.

Now towards the details. In the proof we restrict ourselves to an automaton $\mathcal{A}$ that uses no contants and a single register only. However, after it has been given it should be obvious that this is no restriction at all. We add a comment on this at the end.

Let u and $X \subseteq \mathbb{C}^s$ be as above. When $\mathcal{A}$ has read u it is in state q; let v^* denote the value of the register at that moment. For all $x \in X$ the computation on ux follows the characteristic path. Fix u and $x^* := (x_1^*, \ldots, x_s^*) \in X$ such that all components are algebraically independent. Our goal is to find a $z \in X$ such that uz^t for all $t \in \mathbb{N}_0$ follows the characteristic path. To do so it must be guaranteed that for each run of $\mathcal{A}$ through the loop the current input component z_j never equals the current value in the register, for all $1 \leq j \leq s$. The latter of course can change with each new run through the loop. We thus have to analyze how the register value evolves.

Let us begin with some easy cases. First, if all computations performed during the loop are the projection pr_2 onto v, then v does not change. Since u and x^* have independent components all tests $x_j^* = v$? are answered negatively and we are done. Secondly, suppose there is an operation $pr_1(x_j^*, v)$ performed and this is the only one that changes v, i.e., all other operations are pr_2. Then from this step on $v = x_j^*$ and in the next run through the loop the corresponding equality test is positive, so the computation leaves the characteristic path. This can easily be resolved replacing x^* by $x^* \tilde{x}^* \in \mathbb{C}^{2s}$ with all components independent (and thus different) and running twice through the loop. Now if $v = x_j^*$ in the next run $v = \tilde{x}_j^*$? will be tested with negative outcome; the projection pr_1 changes v's value into $\tilde{x}_j^*$. We then consider two consecutive runs through the loop as a new loop of double length. The only price to pay for this is the length of z in the theorem's statement which changes from at most K to at most $2K$.

Thirdly, if all operations are projections but different from the first two cases the statement trivially is correct. Finally, the case that projections occur but not exclusively is covered by the arguments that follow below. We therefore without loss of generality assume that during each step $j, 1 \leq j \leq s$ along the loop an arithmetic operation $v \circ_j x_j$ is performed. For sake of notational simplicity we only consider $\circ_j \in \{+, *\}$; subtractions do not change the arguments.

The way how v's value evolves during one sweep through the loop starting from initial value $v_t, t \geq 1$ can be described as follows:

$$v_{t+1} = [((v_t + a_1) * m_1 + a_2) * m_2 + \ldots + a_{s-1}] * m_{s-1} \circ_s x_s$$

Here, we have $a_i = x_i, m_i = 1$ in case $\circ_i = +$ and $a_i = 0, m_i = x_i$ if $\circ_i = *$. Moreover note that (1) includes s updates, one for each move along the states constituting the loop. The structure of the register value at intermediate steps can be easily extracted from the above formula, this will be used below.

Each additional run through the loop formally gives the same update starting from the respective value v_t. We now have to show that there is a point $z \in X$ such that all updates given by (1) are different from the respective components of z, no matter how often the loop is passed.

For all intermediate updates leading from $v_1 = v^*$ to v_2 this is true for all points in X. We now show for each $t \geq 1$ the following

Claim: Suppose $X^{(t)}$ is the subset of points $x \in X$ such that $\mathcal{A}$ for each input $ux^j, 0 \leq j \leq t$ follows the characteristic path and thus ends in state q. Then $X^{(t+1)}$ is obtained from $X^{(t)}$ by removing a set $R^{(t)}$ of points whose final component x_s belongs to a *finite* set.

The claim implies the theorem: Since X is open, if at each loop such a set $R^{(t)}$ has to be removed, then $X \setminus \bigcup_{t \geq 1} R^{(t)}$ has a non-empty projection onto the s-th component. This is true since with respect to this component an at most countable set is removed from an interval. It follows that X contains a point z that follows the characteristic path for any given number of loops.

Proof of the claim: Suppose $x^* = (x_1^*, \ldots, x_{s-1}^*, x_s^*)$ is chosen from the open set X as explained before. Let us fix the first $s - 1$ components and analyze for which values of x_s the input $u(x_1^*, \ldots, x_{s-1}^*, x_s)^t$ is branched along the loop for $t = 1, 2, 3, \ldots$ times. If $t = 1$ this is the case at least for x_s belonging to the open interval we obtain when projecting X to its final component.

Case 1: $\mathcal{A}$'s operation when reading x_s along the loop is an addition, i.e., $\circ_s = +$ and $v_{t+1} = f(v_t, x_1^*, \ldots, x_{s-1}^*) + x_s$ with f the appropriate function extracted from (1). In order to make sure that the computation follows the characteristic path all intermediate results given implicitly by (1) must be different from the respective component x_j^* and from x_s at the final step. This restricts the possible choices for x_s. The first condition when entering the loop for the next sweep is that $v \neq x_1^*$. This implies that only one value for x_s has to be avoided, namely $x_1^* - f(v_t, x_1^*, \ldots, x_{s-1}^*)$. By expanding the representation in (1) each intermediate result for the register value can easily be seen to be a degree one polynomial in x_s as variable. The coefficient of x_s is of the form $m_1^{\alpha_1} m_2^{\alpha_2} \ldots m_{s-1}^{\alpha_{s-1}}$ with some $\alpha_i = t - 1$ and the other $\alpha_i = t$, depending on where in the loop the computation currently resides. Thus the coefficient always is a product of some x_i^* with certain powers. The choice of the x_i^* as algebraically independent numbers guarantees this product to be always non-zero and different from 1. This implies that a comparison between the current value of v and the actual component

$x_j^*, 1 \leq j \leq s-1$ always gives a negative result except for one assignment of x_s. This 'bad' value is the unique complex solution of a linear equation in x_s. The same holds for the final step in the loop and the comparison with x_s. As consequence, the computation continues to stay for one more step on the characteristic path.

Case 2: $\circ_s = *$ and $v_{t+1} = f(v_t, x_1^*, \ldots, x_{s-1}^*) * x_s$. A similar reasoning as before shows that if the computation runs for the t-th time through the loop ($t \geq 1$) the current register value is expressible as a polynomial of degree $t-1$ in x_s. More precisely, the highest coefficient, i.e., the coefficient of x_s^{t-1} has the form $f(v_t, x_1^*, \ldots, x_{s-1}^*) \cdot m_1^{\alpha_1} m_2^{\alpha_2} \ldots m_{s-1}^{\alpha_{s-1}}$. Here, again some $\alpha_i = t-1$ and the other $\alpha_i = t$. Due to the choice of u and x^* the value $f(v_t, x_1^*, \ldots, x_{s-1}^*) \neq 0$ because f is a polynomial and there is no algebraic relation between the components. It follows that the comparison between the register value and one of the x_j^* or x_s only is positive for at most $t-1$ many choices of x_s. These choices have to be excluded in order to stay on the charatteristic path.

The above reasoning shows that for each run through the loop all but a finite number of assignments to x_s are suitable in order to guarantee that the point $ux_1^* \ldots x_{s-1}^* x_s$ is branched along the characteristic path of $\mathcal{A}$. Each such point is a suitable choice for uz. The Claim and thus the theorem follow.

Two final remarks are appropriate: If the automaton has k registers and ℓ constants the arguments apply in precisely the same way. Once again, only a finite number of values have to be forbidden for one of the variables x_i with respect to each register and each sweep through the loop. Moreover, for several registers it might be the case that instead of x_s another component has to be taken into account, for example, when one register value does not depend on x_s. Once again, this does not harm the above proof.   □

We end this section with an easy example showing how the weak pumping lemma can be applied. We are confident that other interesting examples can be treated that way as well.

Example 2. Consider the following modification of the Subset Sum problem. Define the language $L \subset \mathbb{C}^*$ to consist of all points $(x_1, \ldots, x_n) \in \mathbb{C}^n$ such that there are two disjoint and non-empty sets $S_1, S_2 \subset \{1, \ldots, n\}$ satisfying $\sum_{i \in S_1} x_i = \sum_{i \in S_2} x_i$. Then L cannot be accepted by a deterministic complex $(\mathcal{S}_\mathbb{C}, k)$-automaton. The proof of the weak pumping lemma implies that we can choose all components of u, z algebraically independent. Consequently, the input uz must be rejected since validity of the defining property for L implies an algebraic relation between the input components. But uz^2 clearly is an input in L since we can choose S_1 to cover the first occurence of z and S_2 its second. This is not possible since it would imply the starting state of the loop to be at the same time accepting and rejecting. L thus cannot be accepted.

4   Undecidability Results

We now turn to a bunch of undecidability results for the generalized automata model dealing with classical problems from finite automata theory. The basic of all these results is the following well known fundamental undecidability result for the BSS model, see [3].

Proposition 2. *The set $\mathbb{Q}^+$ of positive rational numbers is neither decidable in the real nor in the complex BSS model.*[3]

The undecidability results below are obtained by embedding the decidability question for the rationals into the problems under consideration. In all cases this will be done using in one or the other way a fundamental automaton that is described in the next result.

Proposition 3. *There is a deterministic $(\mathcal{S}_{\mathbb{R}}, 3)$-automaton $\mathcal{A}$ that accepts the language $L \subseteq \mathbb{R}^*$, defined as*

$$L := \{(r, x_1, x_2, \ldots, x_n, 0, t, s) | n, t, s \in \mathbb{N}, x_i \in \{-1, 1\},$$
$$s = \sum_{i, x_i = 1} x_i, t = \sum_{i, x_i = -1} |x_i|, r = \tfrac{s}{t}\}.$$

The automaton uses three constants $-1, 0, 1$.

L as well can be accepted by a deterministic $(\mathcal{S}_{\mathbb{C}}, 3)$-automaton when considered as language in $\mathbb{C}^*$.

Proof. Before describing $\mathcal{A}$ in more detail its way of functioning is outlined briefly. A tuple accepted by $\mathcal{A}$ as its first component must have a positive rational number r of form $\frac{s}{t}$. The correct values for s and t are determined by means of the intermediate components x_i which are used as counters: a value $x_i = 1$ is used by $\mathcal{A}$ to increase a counter for s by 1, $x_i = -1$ similarly is used for t. Those counters are realized in two of the registers of the automaton.

Now towards the details. From the following description it should be obvious how the automaton formally can be devised, so we do not specify each possible transition in detail. The automaton uses three registers v_1, v_2, v_3 that are initialized with 0. It uses as its constants $-1, 0, 1$ (this is not intended to be the minimal number possible to achieve the all-over goal). Any input that does not respect the formal constraints given in the definition of L is branched into a sink state. More precisely, the automaton checks all x_i to belong to $\{-1, 1\}$ by comparing a current x_i with the two constant registers storing $-1, 1$. Similarly, $\mathcal{A}$ expects the sequence of x_i's to terminate reading a 0 component followed by two additional non-zero components s and t.

Let us then assume that an input satisfies these formal requirements (which of course can only be guaranteed after having read the entire input). $\mathcal{A}$ copies the first component r into register v_1. Now each time $\mathcal{A}$ reads a component $x_i = 1$ it adds the value 1 to register v_2, i.e., $v_2 \leftarrow v_2 + x_i$. Registers v_1 and v_3 are not changed in this case. Similarly, reading $x_i = -1$ register v_3 is increased by 1 using the operation $v_3 \leftarrow v_3 - x_i$ and v_1, v_2 remain unchanged. The first 0 read indicates that the automaton enters a new phase of its algorithm. Notice

---

[3] We work with $\mathbb{Q}^+$ instead of $\mathbb{Q}$ for sake of simplicity below, not because of any particular importance using positive rationals only.

that if already $x_1 = 0$ the computation should end in a sink as well. In the next phase the automaton checks whether the numbers constructed so far in registers v_2, v_3 constitute a representation of r as fraction, thus yielding r to be a positive rational number. First $\mathcal{A}$ checks by a corresponding test whether $v_2 = t$. If not it moves into a sink state; otherwise t is a potential candidate for the correct denominator and the automaton performs the operation $v_1 \leftarrow v_1 \cdot t$. Then, it reads s and compares it to both v_3 and the updated value of v_1. Only if both these equality tests are satisfied the automaton runs into its unique accepting state, otherwise it moves again into a sink.

It is then obvious that $\mathcal{A}$ only accepts tuples of the corresponding form for which r is a positive rational and s, t represent a valid fraction for r. Since the automaton does not use inequality branches the algorithm works exactly the same in the complex model. □

The proposition immediately implies several undecidability results. Since the theorem deals with deterministic automata the corresponding problems are as well undecidable for non-deterministic automata. The size of an automaton can be taken as sum of its number of registers and number of states.

Theorem 3. *The following problems on $(S_\mathbb{R}, 3)$-automata are undecidable in the real number BSS model. The analogue statements hold for complex automata and the complex BSS model; all $\mathcal{A}$ used below (except in part d) are deterministic $(S_\mathbb{R}, 3)$-automata, q_0 denotes their respective initial state.*

a) EMPTINESS PROBLEM: *Given $\mathcal{A}$, is $L(\mathcal{A}) = \emptyset$?*
b) EQUIVALENCE PROBLEM: *Given two automata, do they accept the same language?*
c) REACHABILITY PROBLEM I: *Given $\mathcal{A}$ and a state p of $\mathcal{A}$, is there a computation of $\mathcal{A}$ that starts in q_0 and reaches p?*
d) REACHABILITY PROBLEM II: *There is an $(S_\mathbb{R}, 4)$-automaton $\mathcal{A}$ (not part of the input) such that the following problem is undecidable: Given a state p of $\mathcal{A}$ and an assignment $v \in \mathbb{R}^k$ of the 4 registers of $\mathcal{A}$, is there a computation of $\mathcal{A}$ starting in its initial state with initialization $0 \in \mathbb{R}^4$ and leading to p attaining register values v?*
e) MINIMIZATION PROBLEM: *Given $\mathcal{A}$, is it state minimal among all deterministic automata accepting $L(\mathcal{A})$?*
   As consequence, there is no BSS algorithm minimizing any given generalized automaton.

Proof. All statements are implied by using suitable variants of the automaton constructed in Proposition 3.

For the emptiness problem consider as input an automaton $\mathcal{A}$ that uses in addition to constants $-1, 0, 1$ a constant $c \in \mathbb{R}$. This constant thus is part of the input and can be used to relate the emptiness problem with deciding the positive rationals. This can be done by modifying the automaton in Proposition 3 in such a way that in its first step it compares the first input component r with constant c. Only if $r = c$ the automaton continues to work as described in the proposition, otherwise it moves into a sink state. Now this $\mathcal{A}$ will accept a word if and only if c is rational. Thus deciding whether a given real number is a positive rational number can be reduced to deciding whether $L(\mathcal{A}) \neq \emptyset$. The latter problem is undecidable.

Claim b) is a direct consequence since one easily can construct an automaton that accepts no word from $\mathbb{R}^*$. Taking this automaton together with the one from a) as input the emptiness problem reduces to the equivalence problem.

Reachability problem I is easily seen to be undecidable as well using part a) since the automaton reaches its only accepting state iff the constant c is a positive rational.

Reachability problem II needs another modification of our standard automaton. It is necessary because now the automaton should be fixed, so we cannot code the rationals as decision problem by varying the automaton using different constants. Instead we code the rationals in the final desired register assignment as follows. First, recall that automaton $\mathcal{A}$ from Proposition 3 finishes an accepting computation on a tuple $(r, x_1, \ldots, x_n, 0, t, s)$ in its unique accepting state, say p, with register assignment $(r \cdot t, t, s)$. In that case $r = \frac{s}{t}$ is rational. However, we do not know in advance how s, t look like and whether they exist, so they cannot be used as the desired assignment for an instance of Reachability Problem II. Therefore, $\mathcal{A}$ is modified as follows giving a new $(\mathcal{S}_{\mathbb{R}}, 4)$-automaton $\mathcal{A}'$. This automaton uses one additional register in order to store twice the first component r read. The second copy is stored in register v_4 and this register will not be changed any more during the rest of the computation of $\mathcal{A}'$. If $\mathcal{A}$ has reached its final state p, then $\mathcal{A}'$ continues its computation requiring one additional 0-component as remaining input and using the projection operation to set registers $v_1 = v_2 = v_3 = 0$. The only accepting state of $\mathcal{A}'$ is a new state p' and it can only be reached from p in the above described way. If this is the case, then the four registers of $\mathcal{A}'$ have the assignment $(0, 0, 0, r)$, where r is the rational leading as first component of an input to the above final configuration. Thus, for the fixed automaton $\mathcal{A}'$ there is a computation leading from q_0 to state p' and resulting in a register assignment $(0, 0, 0, r)$ iff $r \in \mathbb{Q}^+$. It follows that the second version of the reachability problem is undecidable as well.

Finally, the minimization problem clearly cannot be computable for deterministic generalized automata; if it were one could decide the emptiness problem since a minimal automaton for the empty set has one state only.   □

4.1 Conclusion and Open Questions

In this paper we have studied the generalized model of finite automata introduced in [7] in the framework of BSS computability and complexity. The focus has been on real and complex number computability. Our results show that a lot of classical questions about finite automata in the generalized framework have different answers. Among them we find both different complexity and computability results. In addition, they lead to a lot of further open questions, a few of which are outlined below.

Another kind of reachability problem than those of Theorem 3 was studied in [7]. There, the question is whether given a (non-deterministic) automaton and a computation path there is an input such that the automaton follows the given path with its computation. One of the main results in [7] is that this problem is decidable. Since the path is part of the instance there is a finite number of steps to be performed, i.e., the dimension of a suitable input $x \in \mathbb{R}^*$ for the automaton's computation to realize the path is given. Then the problem translates into an existential formula in first order logic over the reals. The formula just asks for the existence of an input realizing the required computational steps. Thus the problem is decidable by quantifier elimination. The same holds over $\mathbb{C}$. The difference with the above Reachability problem II is that we do not know in advance (a bound for) the length of a potential accepting path. The problem thus looks a bit similar to the real Halting Problem [3]. It would then be interesting to analyze whether reachability problems can be of the same degree of undecidability than the real Halting Problem. Note, however, that the rationals are known to be of a weaker degree of undecidability [11]. This question seems interesting also from the BSS side since not many problems are known that are of the same difficulty of the Halting Problem, see [10] for one such.

Theorem 3 gives as well rise to investigate the limits of the respective undecidability results, i.e., for which kind of restricted automata some of the problems might turn out to be decidable. Here, restrictions for example can apply to the number of registers and/or the number of constants used by the automaton. One easy result into this direction is the following.

Lemma 2. *For $(\mathcal{S}_{\mathbb{C}}, 1)$-automata and $(\mathcal{S}_{\mathbb{R}}, 1)$-automata that use no constants the emptiness problem is decidable in polynomial time in the size of the automaton over the corresponding structure.*

Note that the above problem is purely discrete if the initial configuration contains no complex data. This holds as well if the automaton has k registers but no constants. Since all purely discrete problems are decidable in the BSS model [3] (though not necessarily in polynomial time) restrictions of the problems treated in Theorem 3 only become interesting if either constants are present or the initial configuration is part of the input as well. This of course does not apply to the second reachability problem since here the final register values are part of an instance. In general, we could also wonder about the impact arbitrary initial assignments to the registers have, for example, with respect to the interplay with the set of constants used.

Other open questions relate to the weak pumping lemma and further structural properties of languages accepted. Is there a similar result for real automata? One can easliy define characteristic paths in the real setting as well. Instead of requiring all tests to give a negative answer one could demand that the tests establish the current input component to be larger (or smaller) than all values it is compared to. This conditions even could be mixed with changing states. However, it is not clear to us whether a meaningful statement about the evolvement of register values could be deducted for computations along such real characteristic paths. Another problem related of course would be a stronger pumping lemma, i.e., one dealing with *accepting* computations. Once again, a main difficulty here seems to be to control the register values. And even more ambitious: What's about a Myhill-Nerode like characterization of languages accepted by (S, k)-automata? Though Theorem 3 indicates that such a result would likely look very different from the classical one, since it might not result in computable properties like state minimization, it would certainly be interesting to find such characterizations.

References

1. Ben-Or, M.: Lower bounds for algebraic decision trees. In: Proc. 15th ACM STOC, pp. 80–86 (1983)
2. Blum, L., Cucker, F., Shub, M., Smale, S.: Complexity and Real Computation. Springer (1998)
3. Blum, L., Shub, M., Smale, S.: On a theory of computation and complexity over the real numbers: NP-completeness, recursive functions and universal machines. Bull. Amer. Math. Soc. 21, 1–46 (1989)
4. Bournez, O., Campagnolo, M.L.: A Survey on Continuous Time Computations. In: Cooper, B., Löwe, B., Sorbi, A. (eds.) New Computational Paradigms, Changing Conceptions of What is Computable, pp. 383–423. Springer, New York (2008)
5. Bürgisser, P., Clausen, M., Shokrollahi, M.A.: Algebraic Complexity Theory. Grundlehren, vol. 315. Springer (1997)
6. Cucker, F., Shub, M.: Generalized Knapsack problems and fixed degree separation. Theoretical Computer Science 161, 301–306 (1996)
7. Gandhi, A., Khoussainov, B., Liu, J.: Finite Automata over Structures. In: Agrawal, M., Cooper, S.B., Li, A. (eds.) TAMC 2012. LNCS, vol. 7287, pp. 373–384. Springer, Heidelberg (2012)
8. Haykin, S.: Neural Networks - A Comprehensive Foundation, 2nd edn. Prentice Hall (1999)
9. Meer, K.: On the complexity of Quadratic Programming in real number models of computation. Theoretical Computer Science 133(1), 85–94 (1994)
10. Meer, K., Ziegler, M.: Real Computational Universality: The word problem for a class of groups with infinite presentation. Foundations of Computational Mathematics 9(5), 599–609 (2009)
11. Meer, K., Ziegler, M.: An explicit solution to Post's problem over the reals. Journal of Complexity 24(1), 3–15 (2008)
12. Meyer auf der Heide, F.: Lower bounds for solving linear diophantine equations on random access machines. Journal ACM 32(4), 929–937 (1985)
13. Nielsen, M.A., Chuang, I.L.: Quantum Computation and Quantum Information. Cambridge University Press (2000)
14. Paun, G.: Membrane Computing: An Introduction. Springer (2002)
15. Weihrauch, K.: Computable Analysis: An Introduction. Springer (2000)

An Incremental Algorithm for Computing Prime Implicates in Modal Logic

Manoj K. Raut

Dhirubhai Ambani Institute of Information and Communication Technology,
Gandhinagar, Gujarat
manoj_raut@daiict.ac.in
http://intranet.daiict.ac.in/~manoj_raut/

Abstract. The algorithm to compute prime implicates and prime implicants in modal logic $\mathcal{K}$ has been suggested in [1]. In this paper we suggest an incremental algorithm to compute the prime implicates of a knowledge base KB and a new knowledge base F from $\Pi(KB) \wedge F$ in modal logic $\mathcal{K}$, where $\Pi(KB)$ is the set of prime implicates of KB and we also prove the correctness of the algorithm.

Keywords: modal logic, prime implicates, knowledge compilation.

1 Introduction

Propositional entailment is a central issue in artificial intelligence due to its high complexity. Determining the logical entailment of a given query from a knowledge base is intractable [3] in general as all known algorithms run in time exponential in the size of the given knowledge base. To overcome such computational intractability, the propositional entailment problem is split into two phases such as *off-line* and *on-line*. In the off-line phase the original knowledge base KB is transformed into another knowledge base KB' and the queries are answered in the on-line phase from the new knowledge base in polynomial time in the size of KB'. In such type of compilation most of the computational overhead shifted into the off-line phase, is amortized over on-line query answering. The off-line computation is known as *knowledge compilation*.

Several algorithms for knowledge compilation have been suggested so far, for example, [4–17]. In these approaches of knowledge compilation, a knowledge base KB is compiled off-line into another *equivalent* knowledge base $\Pi(KB)$, i.e, the set of prime implicates of KB, so that queries can be answered from $\Pi(KB)$ in polynomial time. Most of the work in knowledge compilation have been restricted to propositional logic and first order logic in spite of increasing intrest in modal logic. Due to lack of expressive power in propositional logic and the undecidability of first order logic, modal logic is required as a knowledge representation language in many problems. Modal logic gives a trade-off between expressivity and complexity as they are more expressive than propositional logic and better behaved computationally than first order logic. An algorithm to compute the set

T V Gopal et al. (Eds.): TAMC 2014, LNCS 8402, pp. 188–202, 2014.

of prime implicates of modal logic $\mathcal{K}$ and $\mathcal{K}_n$ has been proposed in [1] and [2] respectively. Prime implicates have been proved useful to other areas of AI such as belief revision [18] and non-monotonic reasoning [19].

As a knowledge base is not static, new clauses are added to the existing knowledge base. It will be inefficient to compute the set of prime implicates of the new knowledge base from the scratch. On the other hand properties of $\Pi(KB)$ can be utilized for computing the prime implicates of the new knowledge base. Many incremental algorithm for computing prime implicates and prime implicants in propositional logic and first order logic has been suggested so far in the literature [8] [10] [11] [22] [23]. In this paper, we suggest an incremental method to compute the set of prime implicates in modal logic $\mathcal{K}$ of the new knowledge base from the prime implicates of the old knowledge base based on the algorithm [1].

The paper is organised as follows. In section 2 we present the syntax and semantics of modal logic briefly. In section 3 we present the incremental algorithm to compute prime implicates in modal logic. Section 4 concludes the paper.

2   Preliminaries

We briefly introduce modal logic $\mathcal{K}$ from [20] [21]. Formulae in $\mathcal{K}$ are formed using a set of propositional variables V, the standard logical connectives $\neg, \wedge$, and $\vee$ and the modal operators $\Box$ and $\Diamond$. We call a formula of the form $\Box\phi$ a $\Box$-formula and a $\Diamond\phi$ a $\Diamond$-formula. The length of a formula ϕ, denoted by $|\phi|$, is the number of occurrences of propositional variables, logical connectives, and modal operators in ϕ.

A formula is said to be in nagation normal form if the the nagation appears just before the propositional variable in the formula. Every formula in $\mathcal{K}$ can be transformed to a formula in NNF in linear time using the equivalences $\neg(\phi\wedge\psi) \equiv \neg\phi\vee\neg\psi$, $\neg(\phi\vee\psi) \equiv \neg\phi\wedge\neg\psi$, $\neg\neg\phi \equiv \phi$, $\neg\Box\phi \equiv \Diamond\neg\phi$, $\neg\Diamond\phi \equiv \Box\neg\phi$. For example, applying these equivalences to the formula $\neg\Box(\Diamond a \wedge \Box b)$ results in the nagation normal form $\Diamond(\Box\neg a \vee \Diamond\neg b)$. As we apply equivalence preserving operations to ϕ to obtain $NNF(\phi)$, so $\phi \equiv NNF(\phi)$.

Definition 1. *A relational structure for $\mathcal{K}$ (also called possible worlds model, Kripke model, or a modal model) is a triple $M = \langle W, R, v \rangle$, where W is a nonempty set (elements of W are called states), R is a binary relation on W (formally, $R \subseteq W \times W$), and v is a valuation function assigning truth values $v(p, w)$ to atomic propositions p at state w (formally, $v : V \times W \to \{ \text{ true }, \text{ false } \}$ where V is the set of propositional letters).*

Definition 2. *Truth of a modal formula ϕ at a world w in a relational structure $M = \langle W, R, v \rangle$ denoted by $M, w \models \phi$ is defined inductively as follows:*

- $M, w \models p$ *iff* $v(p, w) = \text{true}$ *(where $p \in V$)*
- $M, w \models \text{true}$ *and* $M, w \not\models \text{false}$
- $M, w \models \neg\phi$ *iff* $M, w \not\models \phi$

- $M, w \models \phi \wedge \psi$ iff $M, w \models \phi$ and $M, w \models \psi$
- $M, w \models \phi \vee \psi$ iff $M, w \models \phi$ or $M, w \models \psi$
- $M, w \models \Box\phi$ iff for all $w' \in W$ with wRw' we have $M, w' \models \phi$
- $M, w \models \Diamond\phi$ iff for some $w' \in W$ with wRw' we have $M, w' \models \phi$

A formula ϕ is satisfiable if there exists some model M and and some world w such that $M, w \models \phi$. A formula ϕ is said to be valid written as $\models \phi$ if $M, w \models \phi$ for all M and w. A formula ϕ is said to be unsatisfiable if there exists no M and w for which $M, w \models \phi$. A formula ψ is a logical consequence of a formula ϕ written as $\phi \models \psi$ if $M, w \models \phi$ implies $M, w \models \psi$ for every model M and world $w \in W$.

There are two types of logical consequences in modal logic which are:

1. a formula ψ is a global consequence of ϕ if whenever $M, w \models \phi$ for every world w of a model M, then $M, w \models \psi$ for every world w of M.
2. a formula ψ is a local consequence of ϕ if $M, w \models \phi$ implies $M, w \models \psi$ for every model M and world w.

The concept of local and global consequences is not there in propositional logic as each model contains a single possible world. Eventhough both consequences exist in first order logic, local consequence is only studied there. In this paper we will only study local consequences and whenever $\phi \models \psi$ we mean ψ is a local consequence of ϕ. We now present the following definition from [12].

Definition 3. *Let Y be a set of clauses. The residue of subsumption of Y, denoted by $Res(Y)$ is a subset of Y such that for every clause $C \in Y$, there is a clause $D \in Res(Y)$ where $D \models C$; and no clause in $Res(Y)$ entails any other clause in $Res(Y)$.*

With the above definition of Residue operation, the clauses which are entailed by other clauses can be deleted from a set. The definition of prime implicates and prime implicants are same as defined in propositional logic.

We now present some basic properties of logical consequences and equivalences of $\mathcal{K}$ from [1] which will be used in proofs later.

Theorem 1. *[Bienvenu] Let $\psi_1, \ldots, \psi_m, \chi, \chi_1, \ldots, \chi_n$ be formulae in $\mathcal{K}$, and let γ be a propositional formula. Then*

1. $\psi \models \chi \Leftrightarrow \models \neg\psi \vee \chi \Leftrightarrow \psi \wedge \neg\chi \models \bot$
2. $\psi \models \chi \Leftrightarrow \Diamond\psi \models \Diamond\chi \Leftrightarrow \Box\psi \models \Box\chi$
3. $\gamma \wedge \Diamond\psi_1 \wedge \ldots \wedge \Diamond\psi_m \wedge \Box\chi_1 \wedge \ldots \wedge \Box\chi_n \models \bot \Leftrightarrow (\gamma \models \bot$ *or* $\psi_i \wedge \chi_1 \wedge \ldots \wedge \chi_n \models \bot$ *for some* i)

The definitions of literals, clauses, terms and formulas in modal logic $\mathcal{K}$ known as definition $D4$ in [1] are given below.

Definition 4. *The literals L, clauses C, terms T, and formulas F are defined as follows:*

$$L ::= a \mid \neg a \mid \Box F \mid \Diamond F$$
$$C ::= L \mid C \vee C$$
$$T ::= L \mid T \wedge T$$
$$F ::= a \mid \neg a \mid F \wedge F \mid F \vee F \mid \Box F \mid \Diamond F$$

A formula is said to be in conjunctive normal form (CNF) if it is a conjunction of clauses and it is in disjunctive normal form (DNF) if it is a disjunction of terms. Any formula can be converted to a formula in NNF in linear time but the transformation to CNF or DNF is exponential in both time and space. Now we give the definitions of prime implicates and prime implicants of a knowledge base X.

Definition 5. *A clause C is said to be an implicate of a formula X if $X \models C$. A clause C is a prime implicate of X if C is an implicate of X and there is no other implicate C' of X such that $C' \models C$. The set of prime implicates of X is denoted by $\Pi(X)$.*

Definition 6. *A term C is said to be an implicant of a formula X if $C \models X$. A term C is said to be a prime implicant of X if C is an implicant of X and and there is no other implicant C' of X such that $C \models C'$.*

In the rest of the paper we compute the prime implicates of modal formulae incrementally with respect to the above definition of literals, clauses, terms and formulas.

3   Incremental Algorithm

We now consider the computational aspects of prime implicates incrementally. Given a set of prime implicates $\Pi(X) = \{\pi_1, \ldots, \pi_n\}$ of a knowledge base X and a new knowledge base F, (i.e, a formula F) we want to compute the prime implicates of $X \wedge F$ from $\Pi(X) \wedge F$. Note that we can compute the prime implicates of $X \wedge F$ from $\Pi(X \wedge F)$ using Bienvenu's algorithm [1] but the following theorem says there is no need to compute from the scratch as we can compute the same prime implicates from $\Pi(X) \wedge F$ efficiently and avoid unnecessary computation of large number of clauses.

Lemma 1. $\Pi(X \wedge F) = \Pi(\Pi(X) \wedge F)$.

Proof. Let $C_1 \in \Pi(X \wedge F)$. This implies $X \wedge F \models C_1$ and there does not exists any implicate C_2 of $X \wedge F$ such that $C_2 \models C_1$. As the notion of prime implicates induced by definition 4 satisfy **Equivalence**(refer [1]), so $\Pi(X) \equiv X$. Hence $\Pi(X) \wedge F \models C_1$ and there does not exists any implicate C_2 of $\Pi(X) \wedge F$ such that $C_2 \models C_1$. This imples $C_1 \in \Pi(\Pi(X) \wedge F)$. Hence $\Pi(X \wedge F) \subseteq \Pi(\Pi(X) \wedge F)$. Similarly the other part $\Pi(\Pi(X) \wedge F) \subseteq \Pi(X \wedge F)$ can be proved. Hence the result follows.

The following algorithm IN_MODPI computes a set of conjunctions of literals for a prime implicate π_i of X and a given formula F, and $MODPI$ computes a set of sets of conjunctions of literals for the set of all prime implicates Π of X and the formula F.

Algorithm   MODPI(NNF($\Pi(X)$),F)
Input: The set of prime implicates $\Pi(X) = \{\pi_1, \ldots, \pi_n\}$ and a formula F
Output: A set of sets of terms of $\Pi(X) \wedge F$
begin
    for i= 1 to n{
        Compute $\mathcal{T}_i = IN\_MODPI(\pi_i, \mathtt{NNF}(F))$ for each $\pi_i \in \mathtt{NNF}(\Pi(X))$
    }
    $\mathcal{T} = \cup_{i=1}^{n} \mathcal{T}_i$
end

Algorithm   IN_MODPI(π_i, R)
Input: A prime implicate π_i of X and a set R of formulas in NNF
Output: A set of terms
begin
  If $\pi_i \cup R = \pi_i \cup (\phi \wedge \psi) \cup R'$
    do $IN\_MODPI(\pi_i, \{\phi\} \cup \{\psi\} \cup R')$
  elseif $\pi_i \cup R = \pi_i \cup (\phi \vee \psi) \cup R'$
    do $IN\_MODPI(\pi_i, \phi \cup R')$ then do $IN\_MODPI(\pi_i, \psi \cup R')$
  elseif $\pi_i \cup R = (\phi \vee \psi) \cup R$
    do $IN\_MODPI(\{\phi\}, R)$ then do $IN\_MODPI(\{\psi\}, R)$
  else output $\wedge_{\sigma \in \pi_i \cup R} \sigma$
end

From the above algorithms we can assume that $\mathcal{T} = MODPI(\Pi(X), F)$. So $\mathcal{T}_i = IN\_MODPI(\pi_i, NNF(F))$.

Example 1. Consider a formula $X = a \wedge ((\Diamond(b \wedge c) \wedge \Diamond b) \vee (\Diamond b \wedge \Diamond(c \vee d) \wedge \Box e \wedge \Box f))$. The set of prime implicates of X as computed in [1] is $\Pi(X) = \{a \vee a, \Diamond(b \wedge c) \vee \Box(e \wedge f), \Diamond(b \wedge c) \vee \Diamond(b \wedge e \wedge f), \Diamond(b \wedge c) \vee \Diamond((c \vee d) \wedge e \wedge f)\} = \{\pi_1, \pi_2, \pi_3, \pi_4\}$. Let $F = (\Diamond b \vee \Box e) \wedge \neg(\Box a \vee \Diamond c)$ be a formula. We will see how to compute $\mathcal{T}$, $\mathcal{T}_1, \mathcal{T}_2, \mathcal{T}_3, \mathcal{T}_4$.

1. To compute $\mathcal{T}_1$, first $NNF(F)$ is computed which is equivalent to $R_1 = (\Diamond b \vee \Box e) \wedge \Diamond \neg a \wedge \Box \neg c$. Then we run IN_MODPI on $\pi_1 = a \vee a$ and on the singleton set $R_1 = \{(\Diamond b \vee \Box e) \wedge \Diamond \neg a \wedge \Box \neg c\}$. Since $\pi_1 = a \vee a$ and $R_1 = \{(\Diamond b \vee \Box e) \wedge \Diamond \neg a \wedge \Box \neg c\}$, then by the first else-if case of IN_MODPI, we call IN_MODPI on $\pi_1 = a \vee a$ and $R_2 = \{(\Diamond b \vee \Box e), \Diamond \neg a \wedge \Box \neg c\}$. As $\pi_1 = a \vee a$ and $R_2 = \{(\Diamond b \vee \Box e), \Diamond \neg a \wedge \Box \neg c\}$, again by first else-if case of IN_MODPI, we call IN_MODPI on $\pi_1 = a \vee a$ and $R_3 = \{(\Diamond b \vee \Box e), \Diamond \neg a, \Box \neg c\}$. As $R_3 = \{(\Diamond b \vee \Box e), \Diamond \neg a, \Box \neg c\}$, we make two recursive calls by the second else-if case.

   (a) The first subcall will be on $\pi_1 = a \vee a$ and on the set $R_4 = \{\Diamond b, \Diamond \neg a, \Box \neg c\}$. As $\pi_1 = a \vee a \equiv a$ and there are no $\wedge$ and $\vee$ operators outside the modal operators in R_4, so by the else case of IN_MODPI, the output is the conjunction of the elements which is $a \wedge \Diamond b \wedge \Diamond \neg a \wedge \Box \neg c$.

(b) The second subcall will be on $\pi_1 = a \vee a$ and on $R_5 = \{\Box e, \Diamond \neg a, \Box \neg c\}$. As $\pi_1 = a \vee a \equiv a$ and there are no $\wedge$ and $\vee$ operators outside the modal operators in R_5, so by the else case of IN_MODPI, the output is the conjunction of the elements which is $a \wedge \Box e \wedge \Diamond \neg a \wedge \neg c$

So $\mathcal{T}_1 = (a \wedge \Diamond b \wedge \Diamond \neg a \wedge \Box \neg c) \vee (a \wedge \Box e \wedge \Diamond \neg a \wedge \neg c)$

2. To compute $\mathcal{T}_2$ we run IN_MODPI on $\pi_2 = \Diamond(b \wedge c) \vee \Box(e \wedge f)$ and on the singleton set $NNF(F) = R_1 = \{(\Diamond b \vee \Box e) \wedge \Diamond \neg a \wedge \Box \neg c\}$. Then by the first else-if case we run IN_MODPI on $\pi_2 = \Diamond(b \wedge c) \vee \Box(e \wedge f)$ and on the set $R_6 = \{(\Diamond b \vee \Box e), \Diamond \neg a \wedge \Box \neg c\}$. Then again by the first elfe-if case we run IN_MODPI on $\pi_2 = \Diamond(b \wedge c) \vee \Box(e \wedge f)$ and on the set $R_7 = \{(\Diamond b \vee \Box e), \Diamond \neg a, \Box \neg c\}$. As $R_7 = \{(\Diamond b \vee \Box e), \Diamond \neg a, \Box \neg c\}$, we make two recursive calls by second else-if case of IN_MODPI.

(a) The first subcall will be on $\pi_2 = \Diamond(b \wedge c) \vee \Box(e \wedge f)$ and on $R_8 = \{\Diamond b, \Diamond \neg a, \Box \neg c\}$. As $\pi_2 = \Diamond(b \wedge c) \vee \Box(e \wedge f)$ there will be two recursive subcalls by the second else-if case of IN_MODPI.

   i. The first subcall on $\Diamond(b \wedge c)$ and on $R_8 = \{\Diamond b, \Diamond \neg a, \Box \neg c\}$. As there are no $\wedge$ and $\vee$ symbols outside the modal operators, so by the else case of IN_MODPI we get the conjunction of elements as $\Diamond(b \wedge c) \wedge \Diamond b \wedge \Diamond \neg a \wedge \Box \neg c$.

   ii. The second subcall on $\Box(e \wedge f)$ and on $R_8 = \{\Diamond b, \Diamond \neg a, \Box \neg c\}$. As there are no $\wedge$ and $\vee$ symbols outside the modal operators, so by the else case of IN_MODPI we get the conjunction of elements as $\Box(e \wedge f) \wedge \Diamond b \wedge \Diamond \neg a \wedge \Box \neg c$.

(b) The second subcall will be on $\pi_2 = \Diamond(b \wedge c) \vee \Box(e \wedge f)$ and on $R_9 = \{\Box e, \Diamond \neg a, \Box \neg c\}$. As $\pi_2 = \Diamond(b \wedge c) \vee \Box(e \wedge f)$ there will be two recursive subcalls by the second else-if case of IN_MODPI.

   i. The first subcall on $\Diamond(b \wedge c)$ and on $R_9 = \{\Box e, \Diamond \neg a, \Box \neg c\}$. As there are no $\wedge$ and $\vee$ symbols outside the modal operators, so by the else case of IN_MODPI we get the conjunction of elements as $\Diamond(b \wedge c) \wedge \Box e \wedge \Diamond \neg a \wedge \Box \neg c$.

   ii. The second subcall on $\Box(e \wedge f)$ and on $R_9 = \{\Box e, \Diamond \neg a, \Box \neg c\}$. As there are no $\wedge$ and $\vee$ symbols outside the modal operators, so by the else case of IN_MODPI we get the conjunction of elements as $\Box(e \wedge f) \wedge \Box e \wedge \Diamond \neg a \wedge \Box \neg c$.

So $\mathcal{T}_2 = (\Diamond(b \wedge c) \wedge \Diamond b \wedge \Diamond \neg a \wedge \Box \neg c) \vee (\Box(e \wedge f) \wedge \Diamond b \wedge \Diamond \neg a \wedge \Box \neg c) \vee (\Diamond(b \wedge c) \wedge \Box e \wedge \Diamond \neg a \wedge \Box \neg c) \vee (\Box(e \wedge f) \wedge \Box e \wedge \Diamond \neg a \wedge \Box \neg c)$

Similarly we can compute,

3. $\mathcal{T}_3 = (\Diamond(b \wedge c) \wedge \Diamond b \wedge \Diamond \neg a \wedge \Box \neg c) \vee (\Diamond(b \wedge e \wedge f) \wedge \Diamond b \wedge \Diamond \neg a \wedge \Box \neg c) \vee (\Diamond(b \wedge c) \wedge \Box e \wedge \Diamond \neg a \wedge \Box \neg c) \vee (\Diamond(b \wedge e \wedge f) \wedge \Box e \wedge \Diamond \neg a \wedge \Box \neg c)$

4. $\mathcal{T}_4 = (\Diamond(b \wedge c) \wedge \Diamond b \wedge \Diamond \neg a \wedge \Box \neg c) \vee (\Diamond((c \vee d) \wedge e \wedge f) \wedge \Diamond b \wedge \Diamond \neg a \wedge \Box \neg c) \vee (\Diamond(b \wedge c) \wedge \Box e \wedge \Diamond \neg a \wedge \Box \neg c) \vee (\Diamond((c \vee d) \wedge e \wedge f) \wedge \Box e \wedge \Diamond \neg a \wedge \Box \neg c)$

So $\mathcal{T} = \mathcal{T}_1 \cup \mathcal{T}_2 \cup \mathcal{T}_3 \cup \mathcal{T}_4 = [(a \wedge \Diamond b \wedge \Diamond \neg a \wedge \Box \neg c) \vee (a \wedge \Box e \wedge \Diamond \neg a \wedge \neg c)] \wedge [(\Diamond(b \wedge c) \wedge \Diamond b \wedge \Diamond \neg a \wedge \Box \neg c) \vee (\Box(e \wedge f) \wedge \Diamond b \wedge \Diamond \neg a \wedge \Box \neg c) \vee (\Diamond(b \wedge c) \wedge \Box e \wedge \Diamond \neg a \wedge$

$\Box\neg c) \vee (\Box(e \wedge f) \wedge \Box e \wedge \Diamond\neg a \wedge \Box\neg c)] \wedge [(\Diamond(b \wedge c) \wedge \Diamond b \wedge \Diamond\neg a \wedge \Box\neg c) \vee (\Diamond(b \wedge e \wedge f) \wedge \Diamond b \wedge \Diamond\neg a \wedge \Box\neg c) \vee (\Diamond(b \wedge c) \wedge \Box e \wedge \Diamond\neg a \wedge \Box\neg c) \vee (\Diamond(b \wedge e \wedge f) \wedge \Box e \wedge \Diamond\neg a \wedge \Box\neg c)] \wedge [(\Diamond(b \wedge c) \wedge \Diamond b \wedge \Diamond\neg a \wedge \Box\neg c) \vee (\Diamond((c \vee d) \wedge e \wedge f) \wedge \Diamond b \wedge \Diamond\neg a \wedge \Box\neg c) \vee (\Diamond(b \wedge c) \wedge \Box e \wedge \Diamond\neg a \wedge \Box\neg c) \vee (\Diamond((c \vee d) \wedge e \wedge f) \wedge \Box e \wedge \Diamond\neg a \wedge \Box\neg c)].$

Theorem 2. *IN_MODPI always terminates and every formula returned by IN_MODPI is a conjunctive formula. The disjunction of the formulae returned by IN_MODPI, i.e, $\mathcal{T}_i$, is equivalent to $\pi_i \cup R$, i.e, equivalent to $\pi_i \cup F$. The conjunction of the disjunction of the formulae returned by MODPI, i.e, $\mathcal{T}$, is equivalent to $\wedge_{i=1}^n \pi_i \cup R (= \Pi \wedge F)$, that is, $(\wedge_{i=1}^n(\vee_{T \in \mathcal{T}_i} T)) \equiv \Pi \wedge F$ where each $T \in \mathcal{T}_i$ is a term.*

Proof. It is obvious that IN_MODPI outputs only conjunctive formulae due to its recursive structure and the recursion stops when the number of occurrences of $\wedge$ and $\vee$ outside the scope of modal operators reaches zero.

To prove the second part we will apply induction on the number of occurences of $\wedge$ and $\vee$ which are outside the scope of modal operators in the set of formulae $\pi_i \cup R$. If the number of occurences of $\wedge$ and $\vee$ which are outside the scope of modal operators is zero then there is a single formula which is the conjunction of elements in $\pi_i \cup R$. Let the result holds for size n. Let the number of occurences of $\wedge$ and $\vee$ in $\pi_i \cup R$ be $n+1$. If $\pi_i \cup R$ contains a formula of the form $\pi_i \cup (\phi \wedge \psi)$, then we call IN_MODPI on π_i and $(R \setminus \{\phi \wedge \psi\}) \cup \{\phi\} \cup \{\psi\}$. As the number of occurrences of $\wedge$ and $\vee$ is n in the resulting set, so the disjunction of the output formula is equivalent to $\pi_i \cup (R \setminus \{\phi \wedge \psi\}) \cup \{\phi\} \cup \{\psi\}$ and hence equivalent to $\pi_i \cup R$. The next case when $\pi_i \cup R$ contains a formula of the form $\pi_i \cup (\phi \vee \psi)$ then we call IN_MODPI on π_i and $(R \setminus \{\phi \vee \psi\}) \cup \{\phi\}$ and also on π_i and $(R \setminus \{\phi \vee \psi\}) \cup \{\psi\}$. As the number of occurrences of $\wedge$ and $\vee$ in both these formulae is n so the disjunction of the output formulae is equivalent to $\pi_i \cup R$. The third case when $\pi_i \cup R$ contains a formula of the form $(\phi \vee \psi) \cup R$ then we call IN_MODPI on $(\pi_i \setminus \{\phi \vee \psi\}) \cup \{\phi\}$ and R and also on $(\pi_i \setminus \{\phi \vee \psi\}) \cup \{\psi\}$ and R. As the number of occurrences of $\wedge$ and $\vee$ in both these formulae is n so the disjunction of the output formulae is equivalent to $\pi_i \cup R$. Note that as $R = NNF(F)$, so $R \equiv F$ which implies that all the cases equivalent to $\pi_i \cup F$.

To prove the third part $(\wedge_{i=1}^n(\vee_{T \in \mathcal{T}_i} T)) \equiv \wedge_{i=1}^n \pi_i \cup R(= \Pi \wedge F)$, we use mathematical induction on i. In fact, we have to prove $((\vee_{T \in \mathcal{T}_1} T) \wedge (\vee_{T \in \mathcal{T}_2} T) \wedge \ldots \wedge (\vee_{T \in \mathcal{T}_n} T)) \equiv \wedge_{i=1}^n \pi_i \cup R$, where each $T \in \mathcal{T}_i$ is a term. For $i = 1$, we have to prove $(\vee_{T \in \mathcal{T}_1} T) \equiv \pi_1 \cup R$, i.e, we have to prove $(T_1 \vee T_2 \vee \ldots \vee T_n) \equiv \pi_1 \cup R$. This holds by the second part of the proof above. Assume the statement is true for n. We have to prove $((\vee_{T \in \mathcal{T}_1} T) \wedge (\vee_{T \in \mathcal{T}_2} T) \wedge \ldots \wedge (\vee_{T \in \mathcal{T}_n} T) \wedge (\vee_{T \in \mathcal{T}_{n+1}} T)) \equiv \wedge_{i=1}^{n+1} \pi_i \cup R$. By inductive hypothesis $((\vee_{T \in \mathcal{T}_1} T) \wedge (\vee_{T \in \mathcal{T}_2} T) \wedge \ldots \wedge (\vee_{T \in \mathcal{T}_n} T) \wedge (\vee_{T \in \mathcal{T}_{n+1}} T)) \equiv (\wedge_{i=1}^n \pi_i \cup R) \wedge (\vee_{T \in \mathcal{T}_{n+1}} T)$. In turn this is equivalent to $(\wedge_{i=1}^n \pi_i \cup R) \wedge (\pi_{n+1} \cup R)$ by the basis step. Hence this is equivalent to $\wedge_{i=1}^{n+1} \pi_i \cup R$ which is equivalent to $\wedge_{i=1}^{n+1} \pi_i \wedge F$, which is nothing but $\Pi \wedge F$. Hence proved.

Theorem 3. *The length of the formulas $\pi_i \cup NNF(F)$ and $\Pi \cup NNF(F)$ does not exceed $|\pi_i| + 2|F|$ and $|\Pi| + 2|F|$ respectively, where $\pi_i \in \Pi$ for $i = 1, 2, \ldots, n$.*

Proof. While applying NNF to a formula F the number of binary operators, the number of propositional variables and the number of modal operators do not change. But the number of negation symbols may increase because of $\neg(\Box\phi_1 \wedge \Diamond\phi_2) \equiv \Diamond\neg\phi_1 \vee \Box\neg\phi_2$. As the number of negation symbols increase with the number of propositional variables in $NNF(F)$ so the total number of symbols will not exceed $2|F|$. As π_i and Π are already in NNF so the total number of symbols in $\pi_i \cup NNF(F)$ and $\Pi \cup NNF(F)$ will not exceed $|\pi_i| + 2|F|$ and $|\Pi| + 2|F|$ respectively.

Theorem 4. *The number of formulas output by IN_MODPI does not exceed $3^{|\pi_i \cup NNF(F)|}$ on inputing $\pi_i \cup NNF(F)$ for $i = 1, 2, \ldots, n$. The number of formulas output by $MODPI$ does not exceed $n3^{|\Pi \wedge F|}$ on inputing $\Pi \wedge F$.*

Proof. We know that for every call to IN_MODPI the number of recursive calls made is at most three. Let N be the number of occurrences of $\wedge$ and $\vee$ which lie outside the scope of modal operators in the set $\pi_i \cup NNF(F)$. So the total number of recursive calls made is N on input $\pi_i \cup NNF(F)$ as with each call the value of N reduces by 1 and the recursion continues till N becomes 0. This implies, during the execution of IN_MODPI the number of terminating subcalls will not exceed 3^N. We know that each call can produce exactly one formula. As N is bounded above by $|\pi_i \cup NNF(F)|$, so there will be $3^{|\pi_i \cup NNF(F)|}$ formulas output by IN_MODPI.

As $NNF(F)$ has same N value as F and $|\pi_i \wedge NNF(F)|$ is bounded above by $|\Pi \wedge F|$, so $3^{|\Pi \wedge F|}$ formulas will be output from $MODPI$ in each iteration. When $MODPI$ takes $\Pi \wedge F$ as input, it runs IN_MODPI on π_i and $NNF(F)$ for n times, so the total number of formulas output by $MODPI$ will not exceed $n3^{|\Pi \wedge F|}$.

Theorem 5. *The length of each formula output by $MODPI$ will not exceed $|\Pi| + 2|F|$.*

Proof. When we make a call to IN_MODPI on input π_i and $R(= NNF(F))$ then IN_MODPI makes a call to π_i and R' where $|R'| < |R|$. Then the length of $\pi_i \cup R'$ is always less than $\pi_i \cup R$. So the length of any formula output by IN_MODPI on input π_i and R is always less than the length of $\pi_i \cup R$. By Theorem 3, the length of $\pi_i \cup NNF(F)$ does not exceed $|\pi_i| + 2|F|$. So the length of the formula output by IN_MODPI on input π_i and F does not exceed $|\pi_i| + 2|F|$. When we call $MODPI$ on $\Pi \wedge F$ then $MODPI$ calls IN_MODPI on input π_i and $NNF(F)$. As $|\pi_i|$ is bounded by $|\Pi|$ so the length of the formula output by $MODPI$ will not exceed $|\Pi| + 2|F|$.

Let $\mathcal{T} = MODPI(\Pi(X), F)$. So $\mathcal{T}_i = IN\_MODPI(\pi_i, NNF(F))$. Now we define $\Delta(T)$ according to [1]. For each $T \in \mathcal{T}_i$, let A_T be the set of propositional literals in T and B_T is the set of formulae ξ such that $\Diamond\xi$ is in T. If there are no literals of the form $\Box\zeta$ in T then $\Delta(T) = A_T \cup \{\Diamond\xi \mid \xi \in B_T\}$. Otherwise compute $\Delta(T) = A_T \cup \{\Box C_T\} \cup \{\Diamond(\xi \wedge C_T) \mid \xi \in B_T\}$ where C_T is the conjunction of formulae ζ such that $\Box\zeta$ is in T.

Now we present the main algorithm for computing prime implicates of $\Pi \wedge F$ which is a modification of the algorithm given in [1].

Algorithm   ITER_MODPI($\Pi(X) \wedge F$)

Input: A set prime implicates $\Pi(X) = \{\pi_1, \ldots, \pi_n\}$ and a formula F
Output: A set of clauses,i.e, a set of prime implicates of $\Pi(X) \wedge F$
begin
    If $\Pi(X) \wedge F$ is unsatisfiable then
    return $\{\Diamond(a \wedge \neg a)\}$
    else
    for i=1 to n$\{$
        Compute $\mathcal{T}_i$
        Compute $\Delta(T)$ for each $T \in \mathcal{T}_i$
        Compute $CAND_i = \vee_{T \in \mathcal{T}_i} \Delta(T)$
    $\}$
    Compute $CANDID = \cup_i^n CAND_i$
    Compute $CANDIDATES = Res(CANDID)$
    Return CANDIDATES
end

The correctness of the above algorithm is proved by the following theorems.

Theorem 6. *Every prime implicate of a term T is equivalent to some element in $\Delta(T)$.*

Proof. The following proof is a modification of the proof given in [1] and as the proof given in [1] has omitted almost all the steps so we give the full proof here. Let $T = \alpha_1 \wedge \ldots \wedge \alpha_l \wedge \Diamond\beta_1 \wedge \ldots \wedge \Diamond\beta_m \wedge \Box\gamma_1 \wedge \ldots \wedge \Box\gamma_n$ be a term. Let $\pi = \psi_1 \vee \ldots \vee \psi_p \vee \Diamond\phi_1 \vee \ldots \vee \Diamond\phi_q \vee \Box\xi_1 \vee \ldots \vee \Box\xi_r$ be a prime implicate of T. As π is an implicate of T so $T \models \pi$. This implies $\alpha_1 \wedge \ldots \wedge \alpha_l \wedge \Diamond\beta_1 \wedge \ldots \wedge \Diamond\beta_m \wedge \Box\gamma_1 \wedge \ldots \wedge \Box\gamma_n \wedge \neg\psi_1 \wedge \ldots \wedge \neg\psi_p \wedge \Box\neg\phi_1 \wedge \ldots \wedge \Box\neg\phi_q \wedge \Diamond\neg\xi_1 \wedge \ldots \wedge \Diamond\neg\xi_r$ is unsatisfiable. Then by Theorem 1, one of the following must hold.

1. $\alpha_1 \wedge \ldots \wedge \alpha_l \wedge \neg\psi_1 \wedge \ldots \wedge \neg\psi_p \models \bot$
2. $\beta_k \wedge \gamma_1 \wedge \ldots \wedge \gamma_n \wedge \neg\phi_1 \wedge \ldots \wedge \neg\phi_q \models \bot$ for some k.
3. $\neg\xi_k \wedge \gamma_1 \wedge \ldots \wedge \gamma_n \wedge \neg\phi_1 \wedge \ldots \wedge \neg\phi_q \models \bot$ for some k.

If (1) holds then there must be i and j such that $\alpha_i = \psi_j$. This implies $\alpha_i \models \pi$. As α_i is an implicate, so π must be equivalent to α_i, i.e, $\pi \equiv \alpha_i$ otherwise we are getting a stronger implicate α_i than π contradicting our assumption that π is a prime implicate. Since $\alpha_i \in \Delta(T)$, the result holds, i.e, an element of $\Delta(T)$ is equivalent to a prime implicate of T.

If (2) holds, then $\beta_k \wedge \gamma_1 \wedge \ldots \wedge \gamma_n \models \phi_1 \vee \ldots \vee \phi_q$. Then $\Diamond(\beta_k \wedge \gamma_1 \wedge \ldots \wedge \gamma_n) \models \Diamond(\phi_1 \vee \ldots \vee \phi_q)$. Also $\models \Diamond(\phi_1 \vee \ldots \vee \phi_q) \leftrightarrow (\Diamond\phi_1 \vee \ldots \vee \Diamond\phi_q)$. This implies $\Diamond(\beta_k \wedge \gamma_1 \wedge \ldots \wedge \gamma_n) \models \pi$. Now we have to prove $\Diamond(\beta_k \wedge \gamma_1 \wedge \ldots \wedge \gamma_n)$ is an implicate of T, i.e, we have to prove $T \models \Diamond(\beta_k \wedge \gamma_1 \wedge \ldots \wedge \gamma_n)$, i.e, we have to prove $\alpha_1 \wedge \ldots \wedge \alpha_l \wedge \Diamond\beta_1 \wedge \ldots \wedge \Diamond\beta_m \wedge \Box\gamma_1 \wedge \ldots \wedge \Box\gamma_n \models \Diamond(\beta_k \wedge \gamma_1 \wedge \ldots \wedge \gamma_n)$. This is true iff $\alpha_1 \wedge \ldots \wedge \alpha_l \wedge \Diamond\beta_1 \wedge \ldots \wedge \Diamond\beta_m \wedge \Box\gamma_1 \wedge \ldots \wedge \Box\gamma_n \wedge \Box\neg(\beta_k \wedge \gamma_1 \wedge \ldots \wedge \gamma_n)$

is unsatisfiable. Then by Theorem 1, one of the following must hold. So either $\alpha_1 \wedge \ldots \wedge \alpha_l \models \bot$ or $\beta_k \wedge \gamma_1 \wedge \ldots \wedge \gamma_n \wedge \neg(\beta_k \wedge \gamma_1 \wedge \ldots \wedge \gamma_n) \models \bot$. As second part is true so $\Diamond(\beta_k \wedge \gamma_1 \wedge \ldots \wedge \gamma_n)$ is an implicate of T. As π is a prime implicate so $\Diamond(\beta_k \wedge \gamma_1 \wedge \ldots \wedge \gamma_n)$ must be equivalent to π, otherewise we get a stronger implicate $\Diamond(\beta_k \wedge \gamma_1 \wedge \ldots \wedge \gamma_n)$ than π. As $\Diamond(\beta_k \wedge \gamma_1 \wedge \ldots \wedge \gamma_n) \in \Delta(T)$, the result holds.

If (3) holds, then $(\gamma_1 \wedge \ldots \wedge \gamma_n) \models (\xi_k \vee \phi_1 \vee \ldots \vee \phi_k)$. So $\Box(\gamma_1 \wedge \ldots \wedge \gamma_n) \models$ $\Box(\xi_k \vee \phi_1 \vee \ldots \vee \phi_k)$. As $\Box(p \to q) \models \Diamond p \to \Diamond q$, so $\Box(\neg p \vee q) \models \neg \Diamond p \vee \Diamond q$. As $\neg \Diamond p \vee \Diamond q \equiv \Box \neg p \vee \Diamond q$ so $\Box(\neg p \vee q) \models \Box \neg p \vee \Diamond q$. So taking $\neg p = p_1$ we get $\Box(p_1 \vee q) \models \Box p_1 \vee \Diamond q$. So, $\Box(\xi_k \vee \phi_1 \vee \ldots \vee \phi_k) \models \Box \xi_k \vee \Diamond(\phi_1 \vee \ldots \vee \phi_k)$. As $\models \Diamond(\phi_1 \vee \ldots \vee \phi_k) \leftrightarrow (\Diamond \phi_1 \vee \ldots \vee \Diamond \phi_k)$, so $\Box(\gamma_1 \wedge \ldots \wedge \gamma_n) \models (\Box \xi_k \vee \Diamond \phi_1 \vee \ldots \vee \Diamond \phi_k)$, so $\Box(\gamma_1 \wedge \ldots \wedge \gamma_n) \models \pi$. Now we have to prove $\Box(\gamma_1 \wedge \ldots \wedge \gamma_n)$ is an implicate of T, i.e, we have to prove $T \models \Box(\gamma_1 \wedge \ldots \wedge \gamma_n)$, i.e, we have to prove $\alpha_1 \wedge \ldots \wedge \alpha_l \wedge \Diamond \beta_1 \wedge \ldots \wedge \Diamond \beta_m \wedge \Box \gamma_1 \wedge \ldots \wedge \Box \gamma_n \models \Box(\gamma_1 \wedge \ldots \wedge \gamma_n)$. This is true iff $\alpha_1 \wedge \ldots \wedge \alpha_l \wedge \Diamond \beta_1 \wedge \ldots \wedge \Diamond \beta_m \wedge \Box \gamma_1 \wedge \ldots \wedge \Box \gamma_n \wedge \Diamond \neg(\gamma_1 \wedge \ldots \wedge \gamma_n)$ is unsatisfiable. Then by Theorem 1, one of the following must hold. So either $\alpha_1 \wedge \ldots \wedge \alpha_l \models \bot$, or $\beta_k \wedge \gamma_1 \wedge \ldots \wedge \gamma_n \models \bot$ or $\neg(\gamma_1 \wedge \ldots \wedge \gamma_n) \wedge \gamma_1 \wedge \ldots \wedge \gamma_n \models \bot$. As the third part is true, so $\Box(\gamma_1 \wedge \ldots \wedge \gamma_n)$ is an implicate of T. So π must be equivalent to $\Box(\gamma_1 \wedge \ldots \wedge \gamma_n)$ otherwise we get a stronger implicate than π. Since $\Box(\gamma_1 \wedge \ldots \wedge \gamma_n)$ is one of the elements of $\Delta(T)$, so we are done.

Theorem 7. *Every prime implicate of $\mathcal{T}_i$ is equivalent to some element in $CAND_i$.*

Proof. Let $\mathcal{T}_i = (\alpha_{11} \wedge \ldots \wedge \alpha_{1u} \wedge \Diamond \beta_{11} \wedge \ldots \wedge \Diamond \beta_{1v} \wedge \Box \gamma_{11} \wedge \ldots \wedge \Box \gamma_{1w}) \vee (\alpha_{21} \wedge \ldots \wedge \alpha_{2u} \wedge \Diamond \beta_{21} \wedge \ldots \wedge \Diamond \beta_{2v} \wedge \Box \gamma_{21} \wedge \ldots \wedge \Box \gamma_{2w}) \vee \ldots \vee (\alpha_{n1} \wedge \ldots \wedge \alpha_{nu} \wedge \Diamond \beta_{n1} \wedge \ldots \wedge \Diamond \beta_{nv} \wedge \Box \gamma_{n1} \wedge \ldots \wedge \Box \gamma_{nw})$. Let Let $\pi = \psi_1 \vee \ldots \vee \psi_p \vee \Diamond \phi_1 \vee \ldots \vee \Diamond \phi_q \vee \Box \xi_1 \vee \ldots \vee \Box \xi_r$ be a prime implicate of $\mathcal{T}_i$. As π ia an implicate of $\mathcal{T}_i$ so $\mathcal{T}_i \models \pi$. This implies $((\alpha_{11} \wedge \ldots \wedge \alpha_{1u} \wedge \Diamond \beta_{11} \wedge \ldots \wedge \Diamond \beta_{1v} \wedge \Box \gamma_{11} \wedge \ldots \wedge \Box \gamma_{1w}) \vee (\alpha_{21} \wedge \ldots \wedge \alpha_{2u} \wedge \Diamond \beta_{21} \wedge \ldots \wedge \Diamond \beta_{2v} \wedge \Box \gamma_{21} \wedge \ldots \wedge \Box \gamma_{2w}) \vee \ldots \vee (\alpha_{n1} \wedge \ldots \wedge \alpha_{nu} \wedge \Diamond \beta_{n1} \wedge \ldots \wedge \Diamond \beta_{nv} \wedge \Box \gamma_{n1} \wedge \ldots \wedge \Box \gamma_{nw})) \wedge \neg \psi_1 \wedge \ldots \wedge \neg \psi_p \wedge \Box \neg \phi_1 \wedge \ldots \wedge \Box \neg \phi_q \wedge \Diamond \neg \xi_1 \wedge \ldots \wedge \Diamond \neg \xi_r$ is unsatisfiable. By distributive law $(\wedge_{j_i=11 \leq i \leq n}^u (\vee_{k=1}^n \alpha_{kj_k})) \wedge (\wedge_{j_i=11 \leq i \leq n}^v (\vee_{k=1}^n \Diamond \beta_{kj_k})) \wedge (\wedge_{j_i=11 \leq i \leq n}^w (\vee_{k=1}^n \Box \gamma_{kj_k})) \wedge (\wedge_{1 \leq j \leq u 1 \leq k \leq v} ((\vee_{i=1i \neq s}^n \alpha_{ij}) \vee (\vee_{s=1i \neq s}^n \Diamond \beta_{sk}))) \wedge (\wedge_{1 \leq j \leq u 1 \leq k \leq w} ((\vee_{i=1i \neq s}^n \alpha_{ij}) \vee (\vee_{s=1i \neq s}^n \Box \gamma_{sk}))) \wedge (\wedge_{1 \leq j \leq v 1 \leq k \leq w} ((\vee_{i=1i \neq s}^n \Diamond \beta_{ij}) \vee (\vee_{s=1i \neq s}^n \Box \gamma_{sk}))) \wedge (\wedge_{1 \leq t \leq u 1 \leq j \leq v 1 \leq k \leq w} ((\vee_{m=1m \neq i \neq s}^n \alpha_{mt}) \vee (\vee_{i=1m \neq i \neq s}^n \Diamond \beta_{ij}) \vee (\vee_{\substack{s=1 \\ m \neq i \neq s}}^n \Box \gamma_{sk}))) \wedge \neg \psi_1 \wedge \ldots \wedge \neg \psi_p \wedge \Box \neg \phi_1 \wedge \ldots \wedge \Box \neg \phi_q \wedge \Diamond \neg \xi_1 \wedge \ldots \wedge \Diamond \neg \xi_r$ is unsatisfiable.

Then by Theorem 1, one of the following will hold.

1. $(\vee_{k=1}^n \alpha_{kj_k}) \wedge \neg \psi_1 \wedge \ldots \wedge \neg \psi_p \models \bot$ for some j_k such that $1 \leq j_k \leq u$
2. $(\vee_{k=1}^n \beta_{kj_k}) \wedge \neg \phi_1 \wedge \ldots \wedge \neg \phi_q \models \bot$ for some j_k such that $1 \leq j_k \leq v$.
3. $(\vee_{k=1}^n \gamma_{kj_k}) \wedge \neg \xi_m \wedge \neg \phi_1 \wedge \ldots \wedge \neg \phi_q \models \bot$ for some j_k such that $1 \leq j_k \leq w$ and for some m.
4. $((\vee_{\substack{i=1 \\ i \neq s}}^n \alpha_{ij}) \vee (\vee_{\substack{s=1 \\ i \neq s}}^n \beta_{sk})) \wedge \neg \phi_1 \wedge \ldots \wedge \neg \phi_q \models \bot$ for some j and k.
5. $((\vee_{\substack{i=1 \\ i \neq s}}^n \alpha_{ij}) \vee (\vee_{\substack{s=1 \\ i \neq s}}^n \gamma_{sk})) \wedge \neg \xi_m \wedge \neg \phi_1 \wedge \ldots \wedge \neg \phi_q \models \bot$ for some j, k and m.
6. $((\vee_{\substack{i=1 \\ i \neq s}}^n \beta_{ij}) \vee (\vee_{\substack{s=1 \\ i \neq s}}^n \gamma_{sk})) \wedge \neg \phi_1 \wedge \ldots \wedge \neg \phi_q \models \bot$ for some j and k.

7. $((\vee^n_{\substack{m=1 \\ m\neq i\neq s}} \alpha_{mt}) \vee (\vee^n_{\substack{i=1 \\ m\neq i\neq s}} \beta_{ij}) \vee (\vee^n_{\substack{s=1 \\ m\neq i\neq s}} \gamma_{sk})) \wedge \neg\phi_1 \wedge \ldots \wedge \neg\phi_q \models \bot$ for some t, j and k.

If (1) holds, then $(\vee^n_{k=1}\alpha_{kj_k}) \models \psi_1 \vee \ldots \vee \psi_p$. This implies $(\vee^n_{k=1}\alpha_{kj_k}) \models \pi$. So π must be equivalent to $(\vee^n_{k=1}\alpha_{kj_k})$, i.e, otherwise we are getting a stronger implicate $(\vee^n_{k=1}\alpha_{kj_k})$ than π contradicting our assumption that π is a prime implicate. Since $(\vee^n_{k=1}\alpha_{kj_k}) \in CAND_i$, an element of $CAND_i$ is equivalent to a prime implicate of $\mathcal{T}_i$.

If (2) holds, then $(\vee^n_{k=1}\beta_{kj_k}) \models \phi_1 \vee \ldots \vee \phi_q$. This implies $\Diamond(\vee^n_{k=1}\beta_{kj_k}) \models \Diamond(\phi_1 \vee \ldots \vee \phi_q)$. As $\models \Diamond(p \vee q) \leftrightarrow \Diamond p \vee \Diamond q$, so $(\vee^n_{k=1}\Diamond\beta_{kj_k}) \models \Diamond\phi_1 \vee \ldots \vee \Diamond\phi_q$. Hence, $(\vee^n_{k=1}\Diamond\beta_{kj_k}) \models \pi$. So π must be equivalent to $(\vee^n_{k=1}\Diamond\beta_{kj_k})$, otherwise we are getting a stronger implicate $(\vee^n_{k=1}\Diamond\beta_{kj_k})$ than π contradicting our assumption that π is a prime implicate. Since $(\vee^n_{k=1}\Diamond\beta_{kj_k}) \in CAND_i$, an element of $CAND_i$ is equivalent to a prime implicate of $\mathcal{T}_i$.

If (3) holds, then $(\vee^n_{k=1}\gamma_{kj_k}) \models \xi_m \vee \phi_1 \vee \ldots \vee \phi_q$. This implies $\Box(\vee^n_{k=1}\gamma_{kj_k}) \models \Box(\xi_m \vee \phi_1 \vee \ldots \vee \phi_q)$. As $\Box(p \to q) \models \Diamond p \to \Diamond q$, so $\Box(\neg p \vee q) \models \neg\Diamond p \vee \Diamond q$. As $\neg\Diamond p \vee \Diamond q \equiv \Box\neg p \vee \Diamond q$ so $\Box(\neg p \vee q) \models \Box\neg p \vee \Diamond q$. So taking $\neg p = p_1$ we get $\Box(p_1 \vee q) \models \Box p_1 \vee \Diamond q$. So $\Box(\xi_m \vee \phi_1 \vee \ldots \vee \phi_q) \models \Box\xi_m \vee \Diamond(\phi_1 \vee \ldots \vee \phi_q) \models \Box\xi_m \vee \Diamond\phi_1 \vee \ldots \vee \Diamond\phi_q$. Hence $\Box(\vee^n_{k=1}\gamma_{kj_k}) \models \Box\xi_m \vee \Diamond\phi_1 \vee \ldots \vee \Diamond\phi_q$. So $\Box(\vee^n_{k=1}\gamma_{kj_k}) \models \pi$. So π must be equivalent to $\Box(\vee^n_{k=1}\gamma_{kj_k})$ otherwise we are getting a stronger implicate $\Box(\vee^n_{k=1}\gamma_{kj_k})$ than π contradicting our assumption that π is a prime implicate. Since $\Box(\vee^n_{k=1}\gamma_{kj_k}) \in CAND_i$, an element of $CAND_i$ is equivalent to a prime implicate of $\mathcal{T}_i$.

If (4) holds, then $((\vee^n_{i=1i\neq s}\alpha_{ij}) \vee (\vee^n_{s=1i\neq s}\beta_{sk})) \models \phi_1 \vee \ldots \vee \phi_q$. This implies $\Diamond((\vee^n_{i=1i\neq s}\alpha_{ij}) \vee (\vee^n_{s=1i\neq s}\beta_{sk})) \models \Diamond(\phi_1 \vee \ldots \vee \phi_q)$. As $\models \Diamond(p \vee q) \leftrightarrow \Diamond p \vee \Diamond q$, so $((\vee^n_{i=1i\neq s}\Diamond\alpha_{ij}) \vee (\vee^n_{s=1i\neq s}\Diamond\beta_{sk})) \models \Diamond\phi_1 \vee \ldots \vee \Diamond\phi_q$. As $\models p \to \Diamond p$, so $(\vee^n_{i=1i\neq s}\alpha_{ij}) \models (\vee^n_{i=1i\neq s}\Diamond\alpha_{ij})$. This implies $((\vee^n_{i=1i\neq s}\Diamond\alpha_{ij}) \vee (\vee^n_{s=1i\neq s}\Diamond\beta_{sk})) \models ((\vee^n_{i=1i\neq s}\Diamond\alpha_{ij}) \vee (\vee^n_{s=1i\neq s}\Diamond\beta_{sk})) \models \Diamond\phi_1 \vee \ldots \vee \Diamond\phi_q$. So, we have $((\vee^n_{i=1i\neq s}\alpha_{ij}) \vee (\vee^n_{s=1i\neq s}\Diamond\beta_{sk})) \models \pi$. So π must be equivalent to $((\vee^n_{i=1i\neq s}\alpha_{ij}) \vee (\vee^n_{s=1i\neq s}\Diamond\beta_{sk}))$ otherwise we are getting a stronger implicate $((\vee^n_{i=1i\neq s}\alpha_{ij}) \vee (\vee^n_{s=1i\neq s}\Diamond\beta_{sk}))$ than π contradicting our assumption that π is a prime implicate. Since $((\vee^n_{i=1i\neq s}\alpha_{ij}) \vee (\vee^n_{s=1i\neq s}\Diamond\beta_{sk})) \in CAND_i$, an element of $CAND_i$ is equivalent to a prime implicate of $\mathcal{T}_i$.

If (5) holds, then $((\vee^n_{i=1i\neq s}\alpha_{ij}) \vee (\vee^n_{s=1i\neq s}\gamma_{sk})) \models \xi_m \vee \phi_1 \vee \ldots \vee \phi_q$. This implies $\Box((\vee^n_{i=1i\neq s}\alpha_{ij}) \vee (\vee^n_{s=1i\neq s}\gamma_{sk})) \models \Box(\xi_m \vee \phi_1 \vee \ldots \vee \phi_q)$. As $\models (\Box p \vee \Box q) \to \Box(p \vee q)$, so $((\vee^n_{i=1i\neq s}\Box\alpha_{ij}) \vee (\vee^n_{s=1i\neq s}\Box\gamma_{sk})) \models \Box((\vee^n_{i=1i\neq s}\alpha_{ij}) \vee (\vee^n_{s=1i\neq s}\gamma_{sk}))$. Hence, $((\vee^n_{i=1i\neq s}\Box\alpha_{ij}) \vee (\vee^n_{s=1i\neq s}\Box\gamma_{sk})) \models \Box(\xi_m \vee \phi_1 \vee \ldots \vee \phi_q)$. In the proof of case (3) we have shown that $\Box(p_1 \vee q) \models \Box p_1 \vee \Diamond q$. So $((\vee^n_{i=1i\neq s}\Box\alpha_{ij}) \vee (\vee^n_{s=1i\neq s}\Box\gamma_{sk})) \models \Box\xi_m \vee \Diamond(\phi_1 \vee \ldots \vee \phi_q) \models \Box\xi_m \vee \Diamond\phi_1 \vee \ldots \vee \Diamond\phi_q$. Hence $((\vee^n_{i=1i\neq s}\Box\alpha_{ij}) \vee (\vee^n_{s=1i\neq s}\Box\gamma_{sk})) \models \pi$. So π must be equivalent to $((\vee^n_{i=1i\neq s}\Box\alpha_{ij}) \vee (\vee^n_{s=1\,i\neq s}\Box\gamma_{sk}))$ otherwise we are getting a stronger implicate $((\vee^n_{i=1i\neq s}\Box\alpha_{ij}) \vee (\vee^n_{s=1\,i\neq s}\Box\gamma_{sk}))$ than π contradicting our assumption that π is a prime implicate. Since $((\vee^n_{i=1\,i\neq s}\Box\alpha_{ij}) \vee (\vee^n_{s=1\,i\neq s}\Box\gamma_{sk})) \in CAND_i$, an element of $CAND_i$ is equivalent to a prime implicate of $\mathcal{T}_i$.

If (6) holds, then $((\vee_{i=1 i\neq s}^{n}\beta_{ij}) \vee (\vee_{s=1 i\neq s}^{n}\gamma_{sk})) \models \phi_1 \vee \ldots \vee \phi_q$. This implies $\Diamond((\vee_{i=1 i\neq s}^{n}\beta_{ij}) \vee (\vee_{s=1 i\neq s}^{n}\gamma_{sk})) \models \Diamond(\phi_1 \vee \ldots \vee \phi_q)$. As $\models \Diamond(p \vee q) \leftrightarrow \Diamond p \vee \Diamond q$, So $((\vee_{i=1 i\neq s}^{n}\Diamond\beta_{ij}) \vee (\vee_{s=1 i\neq s}^{n}\Diamond\gamma_{sk})) \models \Diamond\phi_1 \vee \ldots \vee \Diamond\phi_q$. As $\models \Box p \rightarrow \Diamond p$. so $((\vee_{i=1 i\neq s}^{n}\Diamond\beta_{ij}) \vee (\vee_{s=1 i\neq s}^{n}\Box\gamma_{sk})) \models ((\vee_{i=1 i\neq s}^{n}\Diamond\beta_{ij}) \vee (\vee_{s=1 i\neq s}^{n}\Diamond\gamma_{sk})) \models \Diamond\phi_1 \vee \ldots \vee \Diamond\phi_q$. Hence $((\vee_{i=1 i\neq s}^{n}\Diamond\beta_{ij}) \vee (\vee_{s=1 i\neq s}^{n}\Box\gamma_{sk})) \models \pi$. So π must be equivalent to $((\vee_{i=1 i\neq s}^{n}\Diamond\beta_{ij}) \vee (\vee_{s=1 i\neq s}^{n}\Box\gamma_{sk}))$ otherwise we are getting a stronger implicate $((\vee_{i=1 i\neq s}^{n}\Diamond\beta_{ij}) \vee (\vee_{s=1 i\neq s}^{n}\Box\gamma_{sk}))$ than π contradicting our assumption that π is a prime implicate. Since $((\vee_{i=1 i\neq s}^{n}\Diamond\beta_{ij}) \vee (\vee_{\substack{s=1 \\ i\neq s}}^{n}\Box\gamma_{sk})) \in CAND_i$, an element of $CAND_i$ is equivalent to a prime implicate of $\mathcal{T}_i$.

If (7) holds, then $((\vee_{m=1 m\neq i\neq s}^{n}\alpha_{mt}) \vee (\vee_{i=1 m\neq i\neq s}^{n}\beta_{ij}) \vee (\vee_{s=1 m\neq i\neq s}^{n}\gamma_{sk})) \models \phi_1 \ldots \vee \phi_q$. So, $\Diamond((\vee_{m=1 m\neq i\neq s}^{n}\alpha_{mt}) \vee (\vee_{i=1 m\neq i\neq s}^{n}\beta_{ij}) \vee (\vee_{s=1 m\neq i\neq s}^{n}\gamma_{sk})) \models \Diamond(\phi_1 \vee \ldots \vee \phi_q)$. As $\models \Diamond(p \vee q) \leftrightarrow \Diamond p \vee \Diamond q$, so $((\vee_{m=1 m\neq i\neq s}^{n}\Diamond\alpha_{mt}) \vee (\vee_{i=1 m\neq i\neq s}^{n}\Diamond\beta_{ij}) \vee (\vee_{s=1 m\neq i\neq s}^{n}\Diamond\gamma_{sk})) \models \Diamond\phi_1 \vee \ldots \vee \Diamond\phi_q$. As $\models p \rightarrow \Diamond p$, $\models \Diamond p \rightarrow \Diamond p$ and $\models \Box p \rightarrow \Diamond p$, hence $((\vee_{m=1 m\neq i\neq s}^{n}\alpha_{mt}) \vee (\vee_{i=1 m\neq i\neq s}^{n}\Diamond\beta_{ij}) \vee (\vee_{s=1 m\neq i\neq s}^{n}\Box\gamma_{sk})) \models ((\vee_{m=1 m\neq i\neq s}^{n}\Diamond\alpha_{mt}) \vee (\vee_{i=1 m\neq i\neq s}^{n}\Diamond\beta_{ij}) \vee (\vee_{s=1 m\neq i\neq s}^{n}\Diamond\gamma_{sk})) \models \Diamond\phi_1 \vee \ldots \vee \Diamond\phi_q$. Hence $((\vee_{m=1 m\neq i\neq s}^{n}\alpha_{mt}) \vee (\vee_{i=1 m\neq i\neq s}^{n}\Diamond\beta_{ij}) \vee (\vee_{s=1 m\neq i\neq s}^{n}\Box\gamma_{sk})) \models \pi$. Hence π must be equivalent to $((\vee_{m=1 m\neq i\neq s}^{n}\alpha_{mt}) \vee (\vee_{i=1 m\neq i\neq s}^{n}\Diamond\beta_{ij}) \vee (\vee_{s=1 m\neq i\neq s}^{n}\Box\gamma_{sk}))$ else we are getting a stronger implicate $((\vee_{m=1 m\neq i\neq s}^{n}\alpha_{mt}) \vee (\vee_{i=1 m\neq i\neq s}^{n}\Diamond\beta_{ij}) \vee (\vee_{s=1 m\neq i\neq s}^{n}\Box\gamma_{sk}))$ than π contradicting our assumption that π is a prime implicate. Since $((\vee_{\substack{m=1 \\ m\neq i\neq s}}^{n}\alpha_{mt}) \vee (\vee_{\substack{i=1 \\ m\neq i\neq s}}^{n}\Diamond\beta_{ij}) \vee (\vee_{\substack{s=1 \\ m\neq i\neq s}}^{n}\Box\gamma_{sk})) \in CAND_i$, an element of $CAND_i$ is equivalent to a prime implicate of $\mathcal{T}_i$. Hence it is proved.

From the above theorem we conclude that every prime implicate of $\mathcal{T}_i$ is equivalent to some element of $CAND_i$, so every prime implicate of $\mathcal{T}$ is equivalent to some element of $\cup_i^n CAND_i$, i.e, equivalent to some element of $CANDID$. Now by the help of following theorem 8, we delete the clauses entailed by other clauses in $CANDID$ using residue of subsumption operation to remain with the set of prime implicates of $\mathcal{T}$. As $\mathcal{T} = MODPI(\Pi(X), F)$ and by lemma 1 we know $\Pi(X \wedge F) = \Pi(\Pi(X) \wedge F)$ so we find the prime implicates of $X \wedge F$.

Theorem 8. $\Pi(X \wedge F) = Res(CANDID) = CANDIDATES$, *i.e, the algorithm ITER_MODPI computes exactly the set of prime implicates.*

Proof. Let $C \in \Pi(X \wedge F)$. Since C is also an implicate of $X \wedge F$, there exists $D \in CANDID$ such that $D \models C$. If $C \notin CANDID$ then D an implicate of $X \wedge F$ entails C and $D \neq C$. This contradicts that C is prime. Hence $C \in CANDID$. This proves that $\pi(X \wedge F) \subseteq CANDID$. Let $\Psi(X \wedge F)$ be the set of implicates of $X \wedge F$. Since the Distribution property (refer [1]) is satisfied by definition 4 (refer [1]), so $CANDID \subseteq \Psi(X \wedge F)$. If $D \in \Psi(X \wedge F) - CANDID$ then D is entailed by some clause $C \in CANDID$. Therefore $Res(CANDID) = Res(\Psi(X \wedge F))$. As $Res(\Psi(X \wedge F)) = \Pi(X \wedge F)$, this implies $Res(CANDID) = \Pi(X \wedge F)$.

Theorem 9. *The algorithm ITER_MODPI terminates all the time.*

Proof. The algorithm $MODPI$ outputs only a finite set of terms. This implies the set $\Delta(T)$ contains only finite number of elements for each T. Thus $CANDID$

has also finite set of elements. In the final step, the algorithm takes residue of subsumption on $CANDID$ by comparing each pair of elements exactly once. As the comparision always comes to an end because of finite number of pairs in $CANDID$, so the algorithm $ITER\_MODPI$ terminates all the time.

Theorem 10. *The length of the smallest clausal representation of a prime implicate of a formula $\Pi \wedge F$ does not exceed $(|\Pi| + 2|F|) * n3^{|\Pi \wedge F|} + (n3^{|\Pi \wedge F|} - 1)$.*

Proof. The number of prime implicates of $\Pi \wedge F$ generated by $ITER\_MODPI$ does not exceed $n3^{|\Pi \wedge F|}$ disjuncts as $MODPI$ outputs $n3^{|\Pi \wedge F|}$ formulas on input $\Pi \wedge F$ by Theorem 4. The length of each disjunct does not exceed $\Pi| + 2|F|$ by Theorem 3. So the total number of symbols is $(\Pi| + 2|F|) \times n3^{|\Pi \wedge F|}$. But there exists $n3^{|\Pi \wedge F|} - 1$ disjunction symbols which connect $n3^{|\Pi \wedge F|}$ disjuncts. So the length of the smallest clausal representation of a prime implicate of a formula $\Pi \wedge F$ does not exceed $(\Pi| + 2|F|) * n3^{|\Pi \wedge F|} + (n3^{|\Pi \wedge F|} - 1)$.

Theorem 11. *The number of prime implicates, which are not equivalent, of a formula $\Pi \wedge F$ does not exceed $|\Pi \wedge F|^{n3^{|\Pi \wedge F|}}$.*

Proof. We know by Theorem 6 that every prime implicate of $\Pi \wedge F$ is equivalent to some clause output by $ITER\_MODPI$ and those clauses are arranged in the form $\vee_{T \in MODPI} \Gamma_T$ where $\Gamma_T \in \Delta(T)$. By Theorem 4, we know that the number of terms in $MODPI$ does not exceed $n3^{|\Pi \wedge F|}$. So these clauses cann't exceed $n3^{|\Pi \wedge F|}$ disjuncts. Since $|\Delta(T)|$ is bounded above by $|\Pi \wedge F|$, so Γ_T has $|\Pi \wedge F|$ alternatives. This implies, the number of clauses output by $ITER\_MODPI$ does not exceed $|\Pi \wedge F|^{n3^{|\Pi \wedge F|}}$. Hence the number of prime implicates, which are not equivalent, does not exceed $|\Pi \wedge F|^{n3^{|\Pi \wedge F|}}$.

Example 2. Consider a formula $X = a \wedge ((\Diamond(b \wedge c) \wedge \Diamond b) \vee (\Diamond b \wedge \Diamond(c \vee d) \wedge \Box e \wedge \Box f))$. The set of prime implicates of X as computed in [1] is $\Pi(X) = \{a \vee a, \Diamond(b \wedge c) \vee \Box(e \wedge f), \Diamond(b \wedge c) \vee \Diamond(b \wedge e \wedge f), \Diamond(b \wedge c) \vee \Diamond((c \vee d) \wedge e \wedge f)\}$. As Example 1 is very big we consider a new formula here. Let $F = \Box(a \wedge c)$ be a new formula. Now the prime implicates of $X \wedge F$ is computed from $\Pi(X) \wedge F$ as given below.

Compute $\mathcal{T}_1 = \{T_1, T_2\}$ where $T_1 = a \wedge \Box(a \wedge c)$ and also $T_2 = a \wedge \Box(a \wedge c)$. Then $A_{T_1} = \{a\}, B_{T_1} = \emptyset, C_{T_1} = a \wedge c.$ $\Delta(T_1) = \{a, \Box(a \wedge c), \Diamond(a \wedge c)\}$ and also $\Delta(T_2) = \{a, \Box(a \wedge c), \Diamond(a \wedge c)\}$. Then $CAND_1 = \Delta(T_1) \vee \Delta(T_2) = \{a \vee a, a \vee \Box(a \wedge c), a \vee \Diamond(a \wedge c), \Box(a \wedge c) \vee a, \Box(a \wedge c) \vee \Box(a \wedge c), \Box(a \wedge c) \vee \Diamond(a \wedge c), \Diamond(a \wedge c) \vee a, \Diamond(a \wedge c) \vee \Box(a \wedge c), \Diamond(a \wedge c) \vee \Diamond(a \wedge c)\} = \{a \vee a, a \vee \Box(a \wedge c), a \vee \Diamond(a \wedge c), \Box(a \wedge c), \Box(a \wedge c) \vee \Diamond(a \wedge c), \Diamond(a \wedge c)\}$.

Now compute $\mathcal{T}_2 = \{T_3, T_4\}$, where $T_3 = \Diamond(b \wedge c) \wedge \Box(a \wedge c)$ and $T_4 = \Box(e \wedge f) \wedge \Box(a \wedge c)$. Then $A_{T_3} = \emptyset, B_{T_3} = b \wedge c, C_{T_3} = a \wedge c.$ $\Delta(T_3) = \{\Box(a \wedge c), \Diamond(b \wedge c \wedge a \wedge c)\}$ and $A_{T_4} = \emptyset, B_{T_4} = \emptyset, C_{T_4} = e \wedge f \wedge a \wedge c.$ $\Delta(T_4) = \{\Box(e \wedge f \wedge a \wedge c), \Diamond(e \wedge f \wedge a \wedge c)\}$. Then $CAND_2 = \Delta(T_3) \vee \Delta(T_4) = \{\Box(a \wedge c) \vee \Box(e \wedge f \wedge a \wedge c), \Box(a \wedge c) \vee \Diamond(e \wedge f \wedge a \wedge c), \Diamond(b \wedge c \wedge a \wedge c) \vee \Box(e \wedge f \wedge a \wedge c), \Diamond(b \wedge c \wedge a \wedge c) \vee \Diamond(e \wedge f \wedge a \wedge c)\}$

Similarly, $\mathcal{T}_3 = \{T_5, T_6\}$, where $T_5 = \Diamond(b \wedge c) \wedge \Box(a \wedge c)$ and $T_6 = \Diamond(b \wedge e \wedge f) \wedge \Box(a \wedge c)$. So $A_{T_5} = \emptyset, B_{T_5} = b \wedge c, C_{T_5} = a \wedge c.$ $\Delta(T_5) = \{\Box(a \wedge c), \Diamond(b \wedge c \wedge a \wedge c)\}$

and $A_{T_6} = \emptyset, B_{T_6} = b \wedge e \wedge f, C_{T_6} = a \wedge c.$ $\Delta(T_6) = \{\Box(a \wedge c), \Diamond(b \wedge e \wedge f \wedge a \wedge c)\}.$
$CAND_3 = \Delta(T_5) \vee \Delta(T_6) = \{\Box(a \wedge c) \vee \Box(a \wedge c), \Box(a \wedge c) \vee \Diamond(b \wedge e \wedge f \wedge a \wedge c), \Diamond(b \wedge c \wedge a \wedge c) \vee \Box(a \wedge c), \Diamond(b \wedge c \wedge a \wedge c) \vee \Diamond(b \wedge e \wedge f \wedge a \wedge c)$

Similarly, $\mathcal{T}_4 = \{T_7, T_8\}$, where $T_7 = \Diamond(b \wedge c) \wedge \Box(a \wedge c)$ and $T_8 = \Diamond((c \vee d) \wedge e \wedge f) \wedge \Box(a \wedge c)$. Then $A_{T_7} = \emptyset, B_{T_7} = b \wedge c, C_{T_7} = a \wedge c.$ $\Delta(T_7) = \{\Box(a \wedge c), \Diamond(b \wedge c \wedge a \wedge c)\}$ and $A_{T_8} = \emptyset, B_{T_8} = (c \vee d) \wedge e \wedge f, C_{T_8} = a \wedge c.$ $\Delta(T_8) = \{\Box(a \wedge c), \Diamond((c \vee d) \wedge e \wedge f \wedge a \wedge c)\}.$ $CAND_4 = \Delta(T_7) \vee \Delta(T_8) = \{\Box(a \wedge c) \vee \Box(a \wedge c), \Box(a \wedge c) \vee \Diamond((c \vee d) \wedge e \wedge f \wedge a \wedge c), \Diamond(b \wedge c \wedge a \wedge c) \vee \Box(a \wedge c), \Diamond(b \wedge c \wedge a \wedge c) \vee \Diamond((c \vee d) \wedge e \wedge f \wedge a \wedge c).$

Now $CANDID = CAND_1 \cup CAND_2 \cup CAND_3 \cup CAND_4.$

So the set of prime implicates is $CANDIDATES = Res(CANDID) = \{a \vee a, \Box(a \wedge c), \Diamond(b \wedge c \wedge a \wedge c) \vee \Box(e \wedge f \wedge a \wedge c), \Diamond(b \wedge c \wedge a \wedge c) \vee \Diamond(b \wedge e \wedge f \wedge a \wedge c), \Diamond(b \wedge c \wedge a \wedge c) \vee \Diamond((c \vee d) \wedge e \wedge f \wedge a \wedge c)\}.$

4 Conclusion

In this paper, we have suggested an incremental algorithm to compute the set of prime implicates of a knowledge base KB and a new knowledge base (i.e, a formula) F. We have also proved the correctness of the algorithm. When new clauses or terms are added, computation of the prime implicates uses the primeness of the already computed clauses or terms. If we compute the prime implicates of $KB \wedge F$ directly by using the algorithm from [1], we obtain the same prime implicates, though it involves wasteful computations. Efficiency of the proposed algorithm $ITER\_MODPI$ results in exploiting the properties of $\Pi(X)$ instead of using X directly. By Implicant-Implicate duality (refer [1]) any algorithm for computing prime implicates can be used to compute prime implicants. So our algorithm can be used to compute prime implicants incrementally.

References

1. Bienvenu, M.: Prime implicates and prime implicants: From propositional to modal logic. J. Artif. Intell. Res. (JAIR) 36, 71–128 (2009)
2. Bienvenu, M.: Consequence Finding in Modal Logic. PhD Thesis, Universit Paul Sabatier (May 7, 2009)
3. Cook, S.A.: The complexity of theorem-proving procedures. In: Proc. 3rd ACM Symp. on the Theory of Computing, pp. 151–158. ACM Press (1971)
4. Cadoli, M., Donini, F.M.: A survey on knowledge compilation. AI Communications-The European Journal for Articial Intelligence 10, 137–150 (1998)
5. Coudert, O., Madre, J.: Implicit and incremental computation of primes and essential primes of boolean functions. In: Proceedings of the 29th ACM/IEEE Design Automation Conference, pp. 36–39. IEEE Computer Society Press (1991)
6. Darwiche, A., Marquis, P.: A knowledge compilation map. Journal of Artificial Intelligence Research 17, 229–264 (2002)
7. Jackson, P., Pais, J.: Computing prime implicants. In: Stickel, M.E. (ed.) CADE 1990. LNCS, vol. 449, pp. 543–557. Springer, Heidelberg (1990)

8. Kean, A., Tsiknis, G.: An incremental method for generating prime implicants/implicates. J. Symb. Comput. 9(2), 185–206 (1990)
9. de Kleer, J.: An assumption-based tms. In: Ginsberg, M.L. (ed.) Readings in Nonmonotonic Reasoning, pp. 280–297. Kaufmann, Los Altos (1987)
10. de Kleer, J.: An improved incremental algorithm for generating prime implicates. In: Proceedings of the Tenth National Conference on Artificial Intelligence, AAAI 1992, pp. 780–785. AAAI Press (1992)
11. Ngair, T.H.: A new algorithm for incremental prime implicate generation. In: Proc. of the 13th IJCAI, Chambery, France, pp. 46–51 (1993)
12. Raut, M.K., Singh, A.: Prime implicates of first order formulas. IJCSA 1(1), 1–11 (2004)
13. Reiter, R., de Kleer, J.: Foundations of assumption-based truth maintenance systems. In: Proceedings of the Sixth National Conference on Artificial Intelligence (AAAI 1987), pp. 183–188 (1987)
14. Shiny, A.K., Pujari, A.K.: Computation of prime implicants using matrix and paths. J. Log. Comput. 8(2), 135–145 (1998)
15. Slagle, J.R., Chang, C.L., Lee, R.C.T.: A new algorithm for generating prime implicants. IEEE Trans. on Comp. C-19(4), 304–310 (1970)
16. Strzemecki, T.: Polynomial-time algorithm for generation of prime implicants. Journal of Complexity 8, 37–63 (1992)
17. Tison, P.: Generalized consensus theory and application to the minimisation of boolean functions. IEEE Trans. on Elec. Comp. EC-16(4), 446–456 (1967)
18. Pagnucco, M.: Knowledge compilation for belief change. In: Sattar, A., Kang, B.-H. (eds.) AI 2006. LNCS (LNAI), vol. 4304, pp. 90–99. Springer, Heidelberg (2006)
19. Przymusinski, T.C.: An algorithm to compute circumscription. Artif. Intell. 38(1), 49–73 (1989)
20. Blackburn, P., de Rijke, M., Venema, Y.: Modal Logic. Cambridge University Press, Cambridge (2002)
21. Blackburn, P., van Benthem, J., Wolter, F.: Handbook of modal logic. Elsevier, Amsterdam (2007)
22. Jackson, P.: Computing prime implicants incrementally. In: Proceedings of the 11th International Conference on Automated Deduction, vol. 607, pp. 253–267 (1992)
23. Raut, M.K.: An incremental knowledge compilation in first order logic. CoRR, abs/1110.6738 (2011)

Approximation Algorithms
for the Weight-Reducible Knapsack Problem[*]

Marc Goerigk[1], Yogish Sabharwal[2], Anita Schöbel[3], and Sandeep Sen[4]

[1] University of Kaiserslautern, Germany
[2] IBM Research Delhi, India
[3] University of Göttingen, Germany
[4] IIT Delhi, India

Abstract. We consider the *weight-reducible knapsack problem*, where we are given a limited budget that can be used to decrease item weights, and we would like to optimize the knapsack objective value using such weight improvements.

We develop a pseudo-polynomial algorithm for the problem, as well as a polynomial-time 3-approximation algorithm based on solving the LP-relaxation.

Furthermore, we consider the special case of one degree of improvement with equal improvement costs for each item, and present a linear-time 3-approximation algorithm based on solving a cardinality-constrained and a classic knapsack problem, and show that the analysis of the polynomial-time 3-approximation algorithm can be improved to yield a 2-approximation.

Keywords: knapsack problem, approximation algorithms, improvable optimization problems.

1 Introduction

We consider the notion of *improvable* optimization problems, in which there are two kinds of problem data: Firstly, the "nominal" or un-improved data, and secondly, "improved" data of several stages that would allow solutions with better performance. We are allowed to use some of the improved data, but each of the improvements has a cost, and a budget constraint on the improvements have to be considered.

This kind of general problem extension can be considered for a wide range of applications: As examples, consider the shortest path problem, where we can improve some of the arcs in the graph; the maximum flow problem, where we may increase arc capacities; or the knapsack problem, where item weights can be reduced.

Typically, such a problem extension will increase the complexity of a given polynomially solvable problem to being NP-hard. It can easily be shown that

[*] Partially supported by grants SCHO 1140/6-3 within the Indo-German DST-DFG Programme and SCHO 1140/3-2 within the DFG programme Algorithm Engineering.

T V Gopal et al. (Eds.): TAMC 2014, LNCS 8402, pp. 203–215, 2014.

this is the case, e.g., for shortest path, spanning tree, or max flow. In the following, we will consider the improvable knapsack problem in more detail as a starting point of analysis; specifically, we will develop solution approaches for the weight-reducible knapsack problem. Practical motivations for this kind of problem appear whenever it is possible to reduce the weight of an item by spending money, e.g. a hiking group may buy lighter sleeping bags instead of taking the heavier ones which are already there. The situation is similar to the time-cost trade-off problem where the duration of an activity can be shortened by e.g. hiring additional workers. Applications may include loading of train wagons, where we have a limited amount of improved containers available with less weight than the standard model, or an additional sponsor for reducing the costs of a given number of projects competing for the same budget. Another line of application is robust planning for uncertain weights: Here, the maximal weight has to be taken into account. However, for a limited set of items, additional measurements can be taken which reveal their true weights.

The knapsack problem itself is one of the classic problems in discrete optimization and has been intensely studied in many variants; [1] gives an overview on the topic. In its basic form, it can be written as an integer (Boolean) linear program (IP) in the following way:

$$\text{(KP)}\quad \max\left\{\sum_{i=1}^{n} p_i x_i : \sum_{i=1}^{n} w_i x_i \leq B, x \in \mathbb{B}^n\right\},$$

where $p_i \in \mathbb{R}$ and $w_i \in \mathbb{R}$ denote the profit and the weight of item i, respectively, and $B \in \mathbb{R}$ the available budget.

In the variant we consider here, we are able to decrease the weight of items, but only for a limited number of items.

In particular, we assume that we are given multiple weights for every item i: A weight $w_{i,0}$ that corresponds to the "unimproved" case, and weights $w_{i,j(i)} \leq \ldots \leq w_{i,1} \leq w_{i,0}$ that specify "improved" weights of different levels. However, each improvement has its cost, and we are given a budget on the total value of improvement that we may make.

Contributions and Overview. We formally introduce weight-reducible knapsack problems in Section 2. In Sections 3 and 4, we develop a pseudo-polynomial solution algorithm, and a polynomial-time 3-approximation algorithm, respectively. We then consider the special case of a single possible improvement per item with equal improvement costs in Section 5, and derive a linear-time 3-approximation algorithm, based on solving a cardinality-constrained and a classic knapsack problem. Furthermore, we show that the analysis of the 3-approximation algorithm from Section 4 can be refined in this case, to yield a 2-approximation. This work is concluded in Section 6, and pointers to future research are given.

2 Problem Definition and Notation

In the following description we assume that a set of n items $[n] := \{1, \ldots, n\}$ with profits $p_i, i = 1, \ldots, n$ and a budget B is given. Let $\text{KP}(p, w)$ denote the classic knapsack problem with weight vector w and $KP^*(p, w)$ its optimal objective value. As we will mostly be interested in problems for varying weights w, we will write $KP(w)$ and $KP^*(w)$ for short, if the profits are clear from the context. Furthermore, we shall write $p(J)$ to denote the sum of profits of the items $J \subseteq [n]$.

An instance of the *weight-reducible knapsack problem* is given by the same set of items $[n]$ with profits $p_i, i \in [n]$, a budget B, and original weights $w_{i,0}, i \in [n]$. Additionally, we given given reduced weights $w_{i,1} \geq \ldots \geq w_{i,j(i)}$ for all $i \in [n]$, along with increasing improvement costs $q_{i,1} \leq \ldots \leq q_{i,j(i)}$ and an improvement budget k.

The goal is to determine the degree of improvement $\ell_i \in \{0, \ldots, j(i)\}$ for each item i, such that

$$\sum_{i=1}^{n} q_{i,\ell_i} \leq k,$$

where $q_{i,0} = 0$, and then to solve the resulting knapsack problem with the resulting weights w_{i,ℓ_i}, i.e., to determine a set of items with maximal profit. Note that any improvement on items that are not used in the knapsack problem solution would be redundant.

We denote this problem as WRKP, and will also write $WRKP^*$ for its optimal objective value.

Note that we may assume that $w_{i,j(i)} \leq B$ for all $i \in [n]$ (otherwise just delete item i in a preprocessing step, since it can never be packed). It is easily seen that WRKP is NP-complete, as it corresponds to a classic knapsack problem if $k = 0$.

We now present an integer programming (IP) formulation for the problem. As before, let a variable x_i denote if item $i \in [n]$ is packed or not. We additionally introduce variables $z_{i,\ell}, \ell \in [j(i)]$, to model the degree of weight-improvement for item i.

$$\max \sum_{i=1}^{n} p_i x_i \tag{1}$$

$$\text{s.t.} \sum_{i=1}^{n} w_{i,0} x_i + \sum_{\ell=1}^{j(i)} (w_{i,\ell} - w_{i,\ell-1}) z_{i,\ell} \leq B \tag{2}$$

$$\sum_{i=1}^{n} \sum_{\ell=1}^{j(i)} (q_{i,\ell} - q_{i,\ell-1}) z_{i,\ell} \leq k \tag{3}$$

$$z_{i,1} \leq x_i \ \forall i = 1, \ldots, n \tag{4}$$

$$z_{i,\ell+1} \leq z_{i,\ell} \ \forall i = 1, \ldots, n \text{ and } \forall \ell = 1, \ldots, j(i) - 1 \tag{5}$$

$$x_i, z_{i,\ell} \in \mathbb{B} \ \forall i = 1, \ldots, n \text{ and } \forall \ell = 1, \ldots, j(i) \tag{6}$$

Constraint (2) models the item weight budget; while the left term $\sum_{i=1}^{n} w_{i,0} x_i$ captures the item weight without improvements, the telescope sum $\sum_{\ell=1}^{j(i)} (w_{i,\ell} - w_{i,\ell-1}) z_{i,\ell}$ reduces this weight to the desired degree of improvement. To this end, we demand that for a certain variable $z_{i,\ell}$ to be one, also the preceding variables in $z_{i,\ell-1}, \ldots, z_{i,1}, x_i$ have to be one using Constraints (4) and (5). Finally, we model the improvement budget using Constraint (3).

Note that we may consider this problem as a special case of a multi-dimensional knapsack problem with conflict constraints (for the one-dimensional case, we refer to [2,3]). For a constant number of knapsack constraints, as is the case here, one can apply the randomized rounding methods proposed in [4] to obtain a randomized $e/(e-1)$-approximation algorithm[1]. However, as this rounding approach can be applied to a broad class of problems, it is also significantly more complex than the easily implementable and deterministic methods we propose here. It may be noted that, no explicit analysis is presented in [4] except for a claim that the method takes polynomial time.

3   A Pseudo-Polynomial Algorithm

We begin with an algorithm for integral profits, and subsequently using scaling techniques to obtain a more efficient variation at the expense of $1 + \epsilon$ approximation in the objective value. The basic idea of updating a table with relevant problem information can be found in, e.g., [5] for the knapsack problem.

3.1   Dynamic Programming for Integral Profits

Let $W(i, q, r)$ denote the minimum weight of objects among $\{x_1 \ldots x_i\}$ that can attain profit r using at most q weight-improvement budget. The following observations are immediate.

1. $W(i, 0, r)$ is the standard version of the knapsack problem where the optimal solution is $\arg \max_r \{W(n, 0, r) \leq B\}$ where the weights are $w_{i,0}$ and $r \leq P = \sum_i p_i$.
2. $W(i, q+1, r) \leq W(i, q, r)$, viz., more weight reductions cannot decrease the the value of the solution.

We can now write the following recurrence for $1 \leq n, 1 \leq q \leq k, 1 \leq r \leq P$: For an item i, there are weight reductions with increasing costs $q_{i,1} \leq \ldots \leq q_{i,j(i)}$ that yields (decreasing) weights $w_{i,1} \geq \ldots \geq w_{i,j(i)}$. We can now write the following dynamic programming recurrence

$$W(i, q, r) = \min \begin{cases} W(i-1, q, r), & \text{do not use } i \\ W(i-1, q, r-p_i) + w_{i,0}, & i \text{ is not reduced} \\ W(i-1, q-q_{i,1}, r-p_i) + w_{i,1}, & \text{it costs } q_{i,1} \text{ for } w_{i,1} \\ W(i-1, q-q_{i,2}, r-p_i) + w_{i,2}, & \text{it costs } q_{i,2} \text{ for } w_{i,2} \\ \ldots, & \text{it costs } q_{i,\ell} \text{ for } w_{i,\ell} \\ W(i-1, q-q_{i,j(i)}, r-p_i) + w_{i,j(i)} & \text{it costs } q_{i,j(i)} \text{ for } w_{i,j(i)} \end{cases}$$
$$(7)$$

---

[1] We thank the anonymous reviewer for pointing out this reference.

It may be noted that reducing the weight of the i-th item and not choosing it is worse than the first term, and hence need not be considered. Let $W(i, x, r) = -\infty$ for $x < 0$ so that we do not consider terms in the dynamic programming where the query cost exceeds the current query budget. Use the base case as $W(1, 0, r) = w_{1,0}$ for $r = p_1$ and 0 otherwise. We assume that $q_{i,j}$ for all i, j are integral and each entry of the table can be computed in $\max_i j(i) \leq Q$ steps.

Algorithm 1. Pseudo-polynomial Algorithm for WRKP

Require: A problem instance of WRKP with integer weights.
1: Initialize the table $W = n \times k \times P$ to $-\infty$. Set $W(1, 0, r) = w_{1,0}$ for $r = p_1$, and 0
      otherwise.
2: **for** $q = 0$ to C **do**
3:     **for** $i = 1$ to n **do**
4:         **for** $r = 1$ to P **do**
5:             $W(i, q, r) =$ according to Equation 7
6:         **end for**
7:     **end for**
8: **end for**
9: **return** $\arg\max_r \{W(n, k, r) \leq B\}$.

Lemma 1. *Algorithm 1 takes time $O(nkQP)$ where k is the budget for weight improvement queries.*

Proof. Each entry can be computed in Q steps where the order of computation proceeds from $q = 0$ to C and for a fixed q, we compute the entries in increasing order of i and r (for a fixed i, in increasing order of r).

3.2 Faster Approximation Algorithms Using Profit Scaling

Using profit scaling, we now convert the previous algorithm into a more efficient version by compromising with an approximation factor in the objective function. Suppose we want to compute a solution with an objective value of at least $(1-\epsilon)\text{WRKP}^*$. We use the scaling method, namely for any object x_i, we consider its new profit $p_i' = \lfloor \frac{p_i}{K} \rfloor$ where $K = \frac{\epsilon \cdot p_{\max}}{n}$ and[2] use this to run the dynamic programming equation.

    Using $p_{\max}' = O(n/\epsilon)$, the running time of the algorithm is $O(nkQ \cdot n \cdot \frac{n}{\epsilon})$. which is similar to the classic FPTAS for Knapsack [6].

Theorem 1. *The dynamic programming algorithm for multiple weight reduction returns a solution with objective value at least $(1 - \epsilon)\,WRKP^*$ in $O(\frac{n^3 \cdot Qk}{\epsilon})$ time.*

Remark. If the total improvement budget k is bounded by a polynomial in n this is an FPTAS.

---

[2] $K \leq \epsilon \cdot \text{WRKP}^*/n$ suffices.

Algorithm 2. FPTAS for WRKP

Require: A problem instance of WRKP, and $\epsilon > 0$.

1: Set $K = \frac{\epsilon p_{\max}}{n}$. Let $p'_i = \lfloor \frac{p_i}{K} \rfloor$
2: Solve the instance WRKP(p', k, W, Q) using Algorithm 1. Let (x, z) be the resulting solution.
3: **return** (x, z)

---

4   A Polynomial-Time 3-Approximation Algorithm

We now present a polynomial time approximation algorithm for the weight-reducible knapsack problem. This is achieved at a cost of relaxing the approximation to factor 3.

To see this, we consider the LP relaxation obtained by relaxing constraints (6) to

$$x_i \leq 1 \quad \forall i = 1, \ldots, n \tag{8}$$

$$z_{i,j(i)} \geq 0 \quad \forall i = 1, \ldots, n \tag{9}$$

Thus, there are $j(i) + 2$ constraints associated with every item – obtained from constraints (8), (9) above combined with constraints (4) and (5) recalled below:

$$z_{i,1} \leq x_i \quad \forall i = 1, \ldots, n \tag{4}$$

$$z_{i,\ell+1} \leq z_{i,\ell} \quad \forall i = 1, \ldots, n \;\&\; \forall \ell = 1, \ldots, j(i) - 1 \tag{5}$$

In addition, we have the knapsack and k-constraints. Therefore, the total number of constraints is

$$2 + \sum_{i=1}^{n} (j(i) + 2).$$

As there are $j(i) + 1$ variables associated with every item, the total number of variables is

$$\sum_{i=1}^{n} (j(i) + 1).$$

Moreover the LP is bounded. Therefore the number of tight constraints in an optimal basic feasible solution must be

$$\sum_{i=1}^{n} (j(i) + 1).$$

This implies that at most $n + 2$ constraints can be non-tight in a basic feasible solution. Let us see how the items contribute non-tight constraints. The important observation is that for any item i, all $j(i) + 2$ constraints cannot be simultaneously tight as this would imply that

$$0 = z_{i,j(i)} = \ldots z_{i,\ell+1} = z_{i,\ell} = \ldots = x_i = 1$$

which is not possible. Thus every item must contribute at least 1 non-tight constraint. Since the total number of non-tight constraints can be at most $n + 2$, at most 2 items can contribute more than 1 non-tight constraint; all the remaining items must contribute only 1 non-tight constraint.

Now consider an item that contributes exactly 1 non-tight constraints. Then one of the cases holds depending on which constraint is non-tight:

- If $z_{i,j(i)} > 0$, then
$$z_{i,j(i)} = \ldots = z_{i,1} = x_i = 1.$$

- If $z_{i,k+1} < z_{i,k}$ for some $1 \leq k \leq j(i) - 1$, then
$$0 = z_{i,j(i)} = \ldots = z_{i,k+1} \text{ and } z_{i,k} = \ldots = z_{i,1} = x_i = 1.$$

- If $z_{i,1} < x_i$, then
$$0 = z_{i,j(i)} = \ldots = z_{i,1} \text{ and } x_i = 1.$$

- If $x_i < 1$, then
$$0 = z_{i,j(i)} = \ldots = z_{i,1} = x_i.$$

Thus, if an item contributes exactly 1 non-tight constraint, then all the variables associated with this item must be integral. We call such items to be *integral*.

Now, since at most 2 items can contribute more than 1 non-tight constraint, it implies that there can be at most two items that are not integral. We create 3 integral solutions from the LP solution: One consisting of all the integral items in the LP solution and one each corresponding to the two items that are not integral. Clearly the one with the best profit is a 3-approximate solution. We summarize this approach as Algorithm 3.

Algorithm 3

Require: A problem instance of WRKP.
  1: Compute an optimal basic solution of the LP relaxation of WRKP. Let J be the indices of integral items that are packed, and let Z denote the accompanying vector of integral improvements. Let x_{f_1} and x_{f_2} denote the fractional items of the solution, if they exist.
  2: **return** $\arg\max\{p(J, Z), p(x_{f_1}), p(x_{f_2})\}$

---

5   The Special Case of One Improvement per Item

We now assume consider the weight-reducible knapsack problem with the special case that $j(i) = 1$, and $q_{i,1} = 1$ for all $i \in [n]$; i.e., each item can be improved at most once, and the number of such improvements is bounded by k. For the ease of presentation, we will write $\overline{w}_i := w_{i,0}$ for the unimproved weights, and $\hat{w}_i := w_{i,1}$ for the improved weights.

We discuss two ways to formulate this special case as an integer program. Both IP formulations are used later on. In the first formulation we use Boolean variables $\bar{x}_i$ and $\hat{x}_i$ which are equal to one if item i is chosen with its original, or its reduced weight, respectively.

$$\max \ \sum_{i=1}^{n} p_i(\bar{x}_i + \hat{x}_i) \tag{10}$$

$$\text{s.t.} \ \sum_{i=1}^{n} \overline{w}_i \bar{x}_i + \sum_{i=1}^{n} \hat{w}_i \hat{x}_i \leq B$$

$$\sum_{i=1}^{n} \hat{x}_i \leq k$$

$$\bar{x}_i + \hat{x}_i \leq 1 \quad \forall i = 1, \ldots, n$$

$$\bar{x}, \hat{x} \in \mathbb{B}^n$$

The second formulation uses Boolean variables x_i, indicating if item i has been chosen, and Boolean variables z_i to specify if the reduced weight of item i is taken.

$$\max \ \sum_{i=1}^{n} p_i x_i \tag{11}$$

$$\text{s.t.} \ \sum_{i=1}^{n} \overline{w}_i x_i \leq B + \sum_{i=1}^{n} (\overline{w}_i - \hat{w}_i) z_i$$

$$\sum_{i=1}^{n} z_i \leq k$$

$$z_i \leq x_i \quad \forall i = 1, \ldots, n$$

$$z, x \in \mathbb{B}^n$$

Note that this special case of weight-reducible knapsack problem is related to the cardinality-constrained knapsack problem (k-CKP), where we are allowed to pack at most k items. To model it as an integer program, we simply add a constraint limiting the number of items to (KP):

$$(\text{CKP}) \quad \max \left\{ \sum_{i=1}^{n} p_i x_i : \sum_{i=1}^{n} w_i x_i \leq B, \ \sum_{i=1}^{n} x_i \leq k, x \in \mathbb{B}^n \right\}.$$

[7] and [8] present algorithms for this kind of problem. However, note that a feasible solution for the weight-reducible knapsack problem may contain more than k items. The k-CKP is important for some of our results. We will denote the knapsack problem with cardinality constraint as $\text{CKP}(k, w)$ or simply k-CKP, and its optimal objective value as CKP^*.

5.1   A Linear-Time 3-Approximation Algorithm

We first note that a linear-time 4-approximation algorithm can be easiliy achieved: Following [7], there is a 2-approximation algorithm for CKP, with linear runtime due to [9]. Using that

$$\text{WRKP}^* \leq 2 \max \left\{ \text{CKP}^*(k, \hat{w}), \text{KP}^*(\overline{w}) \right\},$$

we can immediately construct a 4-approximative solution by solving the cardinality constrained knapsack problem with improved weights, and the classic knapsack problem with unimproved weights using 2-approximations. As the linear runtime result from [9] assumes a constant number of constraints, it does not apply to the algorithm presented in Section 4.

We now consider an approach that is based on creating a cardinality-constrained knapsack problem again. In particular, given a WRKP instance with weights $\overline{w}$ and $\hat{w}$, we create a CKP instance by doubling all items; i.e., we create an instance consisting of $2n$ items, where the first n items have weight $\overline{w}$, and the next n items have weight $\hat{w}$. As a slight modification of the original CKP definition, we assume that the cardinality constraint only applies to the items with weight $\hat{w}$. The problem we consider is denoted as CKP':

$$\max \sum_{i=1}^{n} p_i(\bar{x}_i + \hat{x}_i) \tag{12}$$

$$\text{s.t.} \sum_{i=1}^{n} \overline{w}_i \bar{x}_i + \sum_{i=1}^{n} \hat{w}_i \hat{x}_i \leq B$$

$$\sum_{i=1}^{n} \hat{x}_i \leq k$$

$$\bar{x}, \hat{x} \in \mathbb{B}^n.$$

CKP' is a relaxation of (10), hence CKP'$^* \geq$ WRKP*. Solving the LP-relaxation of CKP' results in a basic solution with a set of integer variables $J_I = \bar{J}_I \cup \hat{J}_I$ and a set of fractional variables J_F. Note that, as before, $|J_F| \leq 2$.

Lemma 2. *Let $(\bar{x}, \hat{x})$ be a basic solution of the LP relaxation of CKP'. If there are two fractional variables, then these are $\hat{x}_i$ and $\hat{x}_j$ with $\hat{x}_i + \hat{x}_j = 1$ for some i, j.*

Proof. Let there be two fractional variables. We consider the following cases:

1. If $\bar{x}_i$ and $\bar{x}_j$ are fractional, we can improve the solution by increasing the variable with better profit to weight ratio, and decreasing the other, until one of them is either 0 or 1.
2. If $\bar{x}_i$ and $\hat{x}_j$ are fractional, the cardinality constraint cannot be tight. We hence can improve the solution by increasing the variable with better profit to weight ratio as in 1. until either one of the variables reaches 0 or 1.

3. If $\hat{x}_i$ and $\hat{x}_j$ are fractional, and the cardinality constraint is not tight, we may proceed as in 2., until one of the variables reaches 0 or 1, or the cardinality constraint becomes tight.
4. If $\hat{x}_i$ and $\hat{x}_j$ are fractional, and the cardinality constraint is tight, we have $\hat{x}_i + \hat{x}_j = 1$.

We use these properties to construct the following feasible solutions for WRKP:

1. If $J_F = \emptyset$, we construct the two solutions $(\bar{J}_I, \emptyset)$ and $(\emptyset, \hat{J}_I)$.
2. If $J_F = \{i\}$, we use the three solutions $(\emptyset, \{i\})$, $(\bar{J}_I, \emptyset)$, and $(\emptyset, \hat{J}_I)$.
3. Finally, if $J_F = \{i, j\}$, where w.l.o.g. $\hat{w}_i \geq \hat{w}_j$, we use $(\emptyset, \{i\})$, $(\bar{J}_I, \emptyset)$, and $(\emptyset, \hat{J}_I \cup \{j\})$.

Note that these solutions are feasible for WRKP, and the sum of their objective values is larger than WRKP*. Thus, choosing the solution with the maximal objective value yields a 3-approximation. We recapitulate this approach in Algorithm 4.

Algorithm 4

Require: A problem instance of WRKP.
1: Solve the LP relaxation of CKP'. Let $J_I = \bar{J}_I \cup \hat{J}_I$ and J_F denote the item indices with integer values packed with original or reduced weights, and the item indices with fractional values in a basic solution.
2: **if** $J_F = \emptyset$ **then**
3:     **return** $\arg\max\{p(\bar{J}_I, \emptyset), p(\emptyset, \hat{J}_I)\}$.
4: **else if** $J_F = \{i\}$ for some $i \in [n]$ **then**
5:     **return** $\arg\max\{p(\emptyset, \{i\}), p(\bar{J}_I, \emptyset), p(\emptyset, \hat{J}_I)\}$
6: **else if** $J_F = \{i, j\}$ for some $i, j \in [n]$ with $\hat{w}_i \geq \hat{w}_j$ **then**
7:     **return** $\arg\max\{p(\emptyset, \{i\}), p(\bar{J}_I, \emptyset), p(\emptyset, \hat{J}_I \cup \{j\})\}$
8: **end if**

Note that the LP relaxation of CKP' can be solved in linear time [9]. Thus we can state the following theorem.

Theorem 2. *Algorithm 4 has an approximation ratio of at most 3 for WRKP, and runs in linear time.*

5.2   A Polynomial-Time 2-Approximation Algorithm

We now show that a factor 2 approximation can be achieved by running in polynomial time. Recall that for the generalized case, we are able to achieve a factor 3-approximation algorithm by considering the LP relaxation of the problem and characterizing the basic feasible solutions of the relaxed LP. We show that for the special case of one improvement per item with unit costs, we can better characterize the basic feasible solutions of the relaxed LP yielding an improved factor 2 approximation. For this, we consider the the LP relaxation of (11). Note that the linear-time result of [9] does not apply here due to the non-constant number of constraints. The LP relaxation can be written as:

$$\max \sum_{i=1}^{n} c_i x_i \tag{13}$$

$$\text{s.t.} \sum_{i=1}^{n} \overline{w}_i x_i \leq B + \sum_{i=1}^{n} (\overline{w}_i - \hat{w}_i) z_i \tag{14}$$

$$\sum_{i=1}^{n} z_i \leq k \tag{15}$$

$$z_i \leq x_i \ \forall i = 1, \ldots, n \tag{16}$$

$$x_i \leq 1 \ \forall i = 1, \ldots, n \tag{17}$$

$$z_i \geq 0 \ \forall i = 1, \ldots, n \tag{18}$$

The LP has $2n$ variables and $3n + 2$ constraints comprising of the knapsack-constraint (14), the k-constraint (15) and three constraints for each item, (16), (17) and (18). Observe that the item constraints imply that the feasible region is bounded. For any basic feasible solution there must be $2n$ linearly independent constraints that are tight. We categorize the items based on the number of tight constraints among (16),(17), and (18) it can contribute, see Table 1.

Table 1. Item categorization

Case	Type	Num of Tight Constraints	Tight Constraints	Num of non-integral variables
i	T1	0	None	2
ii	T2	1	$x_i = z_i$	2
iii	T3	1	$z_i = 1$	1
iv	T3	1	$x_i = 1$	1
v	T4	2	$x_i = 1, z_i = 0$	0
vi	T4	2	$z_i = 0, x_i = z_i$	0
vii	T4	2	$x_i = 1, x_i = z_i$	0
viii	T5	3	$z_i = 0, x_i = z_i, x_i = 1$	Not Possible

We observe that an item cannot contribute more than 2 tight constraints, i.e., constraints (16), (17) and (18) cannot simultaneously be all tight for the same item (case viii).

We consider two scenarios: either the k-constraint (15) is tight or not.

In case it is not tight, then discounting the knapsack constraint, we see that $2n-1$ of the tight constraints must be constraints of type (16), (17) and (18). This implies that at least $n - 1$ items must be of type T4. Therefore n items can contribute at least $2n - 1$ tight constraints only under one of the following scenarios:

A. n items of type T4
B. $(n - 1)$ items of type T4 and 1 item of type T1, T2 or T3.

In case, the k-constraint is tight, then discounting the k-constraint and the knapsack constraint, we see that $2n - 2$ of the tight constraints must come from constraints of type (16), (17) and (18). This implies that at least $n - 2$ items must be of type T4. Therefore, n items can contribute at least $2n - 2$ tight constraints only under one of the following scenarios:

C. n items of type T4
D. $(n - 1)$ items of type T4 and 1 item of type T1, T2 or T3.
E. $(n - 2)$ items of type T4 and 2 items of type T2 or T3

In Cases A and C, all the variables are integral and therefore the solution is integral yielding the exact optimal.

In Cases B and D, we form 2 solutions – one consisting of all the type T4 items (which are already integral) and the other consisting of the remaining item that is either of type T1, T2 or T3 in the weight-reduced form. The first solution is clearly integral feasible, as it is a subset of the fractional optimal. The second solution is integral as every item under consideration is feasible in its weight-reduced form. We simply pick the better of the two solutions yielding a 2-approximate solution.

In case E, let i and j be the two items of type T1/T2/T3. We note that the k-constraint must be tight. Thus, we have that $z_i + z_j = 1$. Without loss of generality, let $\hat{w}_i \le \hat{w}_j$. We therefore form two solutions – one consisting of all the type T4 items along with i in weight-reduced form and the other consisting of j in weight-reduced form. We again pick the best of the two solutions to yield a 2-approximation.

Thus we obtain a 2-approximation algorithm. Note that unlike the 3-approximation algorithm for the generalized case, the relaxation to unit costs allows us to utilize the tightness of the k-constraint in a meaningful way to obtain a better approximation.

Algorithm 5

Require: A problem instance of WRKP.
 1: Compute an optimal basic solution of the LP relaxation of WRKP. Let $(J_i^{\overline{w}}, J_i^{\hat{w}})$, $i = 1, 2, 3, 4$, denote the unimproved and improved item indices of type T_i, respectively.
 2: **if** $|T_4| = n$ **then**
 3:    **return** the (optimal) WRKP solution $(J_4^{\overline{w}}, J_4^{\hat{w}})$.
 4: **else if** $|T_4| = n - 1$ and $|T_1 \cup T_2 \cup T_3| = \{i\}$ **then**
 5:    **return** $\arg\max\{p(J_4^{\overline{w}}, J_4^{\hat{w}}),\ p(\emptyset, \{i\})\}$.
 6: **else if** $|T_4| = n - 2$ and $|T_2 \cup T_3| = \{i, j\}$ **then**
 7:    W.L.O.G., let $\hat{w}_i \le \hat{w}_j$.
 8:    **return** $\arg\max\{p(J_4^{\overline{w}}, J_4^{\hat{w}} \cup \{i\}),\ p(\emptyset, \{j\})\}$.
 9: **end if**

Theorem 3. *Algorithm 5 has an approximation ratio of at most 2 for WRKP, and runs in polynomial time; more specifically, in time required to solve an LP.*

6   Conclusion and Further Research

In this work, we introduced the concept of improvable problems, where one can use limited improvement resources to modify the original problem data. Using the knapsack problem as a starting point for the analysis, we developed a pseudo-polynomial solution algorithm, as well as a polynomial-time 3-approximation that is based on solving the corresponding LP relaxation.

Considering the special case of only one improvement per item and constant improvement costs, we further presented a linear-time 3-approximation based on the algorithm of [7] for the related k-CKP, and strengthened the analysis of the previous 3-approximation for the general problem, to be a 2-approximation for the special case.

Further research with regard to knapsack problems includes the application of these results on the *uncertain* weight-reducible knapsack problem, where the value $\hat{w}$ is not exactly known [10], and a bicriteria problem, in which we simultaneously minimize the costs for reducing item weights, and maximize the knapsack value.

Furthermore, the generality of the proposed concept of improvable problems allows the analysis of many other possible applications, including classic combinatorial problems as spanning tree or shortest paths, as well as real-world problems where improvements are possible. It remains to analyze which of the algorithms presented for the knapsack case in this paper can be generalized to other such problems.

References

1. Kellerer, H., Pferschy, U., Pisinger, D.: Knapsack problems. Springer (2004)
2. Yamada, T., Kataoka, S.: Heuristic and exact algorithms for the disjunctively constrained knapsack problem. Information Processing Society of Japan Journal 43(9), 2864–2870 (2002)
3. Pferschiy, U., Schauer, J.: The knapsack problem with conflict graphs. Journal of Graph Algorithms and Applications 13(2), 233–249 (2009)
4. Chekuri, C., Vondrák, J., Zenklusen, R.: Dependent randomized rounding via exchange properties of combinatorial structures. In: 51st Annual IEEE Symposium on Foundations of Computer Science (FOCS), pp. 575–584 (2010)
5. Ibarra, O.H., Kim, C.E.: Fast approximation algorithms for the knapsack and sum of subset problems. J. ACM 22(4), 463–468 (1975)
6. Vazirani, V.V.: Approximation Algorithms. Springer, Berlin (2001)
7. Caprara, A., Kellerer, H., Pferschy, U., Pisinger, D.: Approximation algorithms for knapsack problems with cardinality constraints. European Journal of Operational Research 123(2), 333–345 (2000)
8. Mastrolilli, M., Hutter, M.: Hybrid rounding techniques for knapsack problems. Discrete Applied Mathematics 154(4), 640–649 (2006)
9. Megiddo, N., Tamir, A.: Linear time algorithms for some separable quadratic programming problems. Operations Research Letters 13, 203–211 (1993)
10. Goerigk, M., Gupta, M., Ide, J., Schöbel, A., Sen, S.: The uncertain knapsack problem with queries. Technical Report 2013-02, Institute for Numerical and Applied Mathematics, University of Göttingen (2013)

Polynomial-Time Algorithms
for SUBGRAPH ISOMORPHISM
in Small Graph Classes of Perfect Graphs*

Matsuo Konagaya, Yota Otachi, and Ryuhei Uehara

School of Information Science, Japan Advanced Institute of Science and Technology,
Ishikawa, 923-1292, Japan
{matsu.cona,otachi,uehara}@jaist.ac.jp

Abstract. Given two graphs, SUBGRAPH ISOMORPHISM is the problem
of deciding whether the first graph (the *base graph*) contains a subgraph
isomorphic to the second graph (the *pattern graph*). This problem is
NP-complete for very restricted graph classes such as connected proper
interval graphs. Only a few cases are known to be polynomial-time solvable even if we restrict the graphs to be perfect. For example, if both
graphs are co-chain graphs, then the problem can be solved in linear
time.

   In this paper, we present a polynomial-time algorithm for the case
where the base graphs are chordal graphs and the pattern graphs are
co-chain graphs. We also present a linear-time algorithm for the case
where the base graphs are trivially perfect graphs and the pattern graphs
are threshold graphs. These results answer some of the open questions
of Kijima et al. [*Discrete Math.* 312, pp. 3164–3173, 2012]. To present
a complexity contrast, we then show that even if the base graphs are
somewhat restricted perfect graphs, the problem of finding a pattern
graph that is a chain graph, a co-chain graph, or a threshold graph is
NP-complete.

Keywords: Subgraph isomorphism, Graph class, Polynomial-time algorithm, NP-completeness.

1   Introduction

The problem SUBGRAPH ISOMORPHISM is a very general and extremely hard
problem which asks, given two graphs, whether one graph (the *base graph*)
contains a subgraph isomorphic to the other graph (the *pattern graph*). The problem generalizes many other problems such as GRAPH ISOMORPHISM, HAMILTO-
NIAN PATH, CLIQUE, and BANDWIDTH. Clearly, SUBGRAPH ISOMORPHISM is
NP-complete in general. Furthermore, by slightly modifying known proofs [8,5],
it can be shown that SUBGRAPH ISOMORPHISM is NP-complete when G and H

---

* Partially supported by JSPS KAKENHI Grant Numbers 23500013, 25730003, and
  by MEXT KAKENHI Grant Number 24106004.

T V Gopal et al. (Eds.): TAMC 2014, LNCS 8402, pp. 216–228, 2014.

Table 1. NP-complete cases of SPANNING SUBGRAPH ISOMORPHISM

Base	Pattern	Complexity	Reference
Bipartite Permutation	NP-complete	[16]
Proper Interval	NP-complete	[16]
Trivially Perfect	NP-complete	[16]
Chain	Convex	NP-complete	[16]
Co-chain	Co-bipartite	NP-complete	[16]
Threshold	Split	NP-complete	[16]
Bipartite	Chain	NP-complete	This paper
Co-convex	Co-chain	NP-complete	This paper
Split	Threshold	NP-complete	This paper

are disjoint unions of paths or of complete graphs. Therefore, it is NP-complete even for small graph classes of perfect graphs such as proper interval graphs, bipartite permutation graphs, and trivially perfect graphs, while GRAPH ISO-MORPHISM can be solved in polynomial time for them [4,18]. For these graph classes, Kijima et al. [16] showed that even if both input graphs are connected and have the same number of vertices, the problem remains NP-complete. They call the problem with such restrictions SPANNING SUBGRAPH ISOMORPHISM.

Kijima et al. [16] also found polynomial-time solvable cases of SUBGRAPH ISOMORPHISM in which both graphs are chain, co-chain, or threshold graphs. Since these classes are proper subclasses of the aforementioned hard classes, those results together give sharp contrasts of computational complexity of SUBGRAPH ISOMORPHISM. However, the complexity of more subtle cases, like the one where the base graphs are proper interval graphs and the pattern graphs are co-chain graphs, remained open.

1.1 Our Results

In this paper, we study the open cases of Kijima et al. [16], and present polynomial-time algorithms for the following cases:

- the base graphs are chordal graphs and the pattern graphs are co-chain graphs,
- the base graphs are trivially perfect graphs and the pattern graphs are threshold graphs.

We also show that even if the pattern graphs are chain, co-chain, or threshold graphs and the base graphs are somewhat restricted perfect graphs, the problem remains NP-complete. The problem of finding a chain subgraph in a bipartite permutation graph, which is an open case of Kijima et al. [16], remains unsettled. See Tables 1 and 2 for the summary of our results.

1.2 Related Results

SUBGRAPH ISOMORPHISM for trees can be solved in polynomial time [22], while it is NP-complete for connected outerplanar graphs [26]. Therefore, the problem

Table 2. Polynomial-time solvable cases of SUBGRAPH ISOMORPHISM

Base	Pattern	Complexity	Reference
Chain		$O(m+n)$	[16]
Co-chain		$O(m+n)$	[16]
Threshold		$O(m+n)$	[16]
Bipartite permutation	Chain	Open	
Chordal	Co-chain	$O(mn^2 + n^3)$	This paper
Trivially perfect	Threshold	$O(m+n)$	This paper

is NP-complete even for connected graphs of bounded treewidth. On the other hand, it can be solved in polynomial time for 2-connected outerplanar graphs [17]. More generally, it is known that SUBGRAPH ISOMORPHISM for k-connected partial k-trees can be solved in polynomial time [21,11]. Eppstein [7] gave a $k^{O(k)}n$-time algorithm for SUBGRAPH ISOMORPHISM on planar graphs, where k and n are the numbers of the vertices in the pattern graph and the base graph, respectively. Recently, Dorn [6] has improved the running time to $2^{O(k)}n$. For other general frameworks, especially for the parameterized ones, see the recent paper by Marx and Pilipczuk [19] and the references therein.

Another related problem is INDUCED SUBGRAPH ISOMORPHISM which asks whether the base graph has an induced subgraph isomorphic to the pattern graph. Damaschke [5] showed that INDUCED SUBGRAPH ISOMORPHISM on cographs is NP-complete. He also showed that INDUCED SUBGRAPH ISOMORPHISM is NP-complete for the disjoint unions of paths, and thus for proper interval graphs and bipartite permutation graphs. Marx and Schlotter [20] showed that INDUCED SUBGRAPH ISOMORPHISM on interval graphs is W[1]-hard when parameterized by the number of vertices in the pattern graph, but fixed-parameter tractable when parameterized by the numbers of vertices to be removed from the base graph. Heggernes et al. [14] showed that INDUCED SUBGRAPH ISOMORPHISM on proper interval graphs is NP-complete even if the base graph is connected. Heggernes et al. [15] have recently shown that INDUCED SUBGRAPH ISOMORPHISM on proper interval graphs and bipartite permutation graphs can be solved in polynomial time if the pattern graph is connected. Belmonte et al. [1] showed that INDUCED SUBGRAPH ISOMORPHISM on connected trivially perfect graphs is NP-complete. This result strengthens known results since every trivially perfect graph is an interval cograph. They also showed that the problem can be solved in polynomial time if the base graphs are trivially perfect graphs and the pattern graphs are threshold graphs.

2   Preliminaries

All graphs in this paper are finite, undirected, and simple. Let $G[U]$ denote the subgraph of $G = (V, E)$ induced by $U \subseteq V$. For a vertex $v \in V$, we denote by $G - v$ the graph obtained by removing v from G; that is, $G - v = G[V \setminus \{v\}]$. The *neighborhood* of a vertex v is the set $N(v) = \{u \in V \mid \{u, v\} \in E\}$. A vertex $v \in V$ is *universal* in G if $N(v) = V \setminus \{v\}$. A vertex $v \in V$ is *isolated* in G if $N(v) = \emptyset$. A set $I \subseteq V$ in $G = (V, E)$ is an *independent set* if for all $u, v \in I$, $(u, v) \notin E$. A set $S \subseteq V$ in $G = (V, E)$ is a *clique* if for all $u, v \in S$, $(u, v) \in E$. A pair (X, Y) of sets of vertices of a bipartite graph $H = (U, V; E)$ is a *biclique* if for all $x \in X$ and $y \in Y$, $(x, y) \in E$. A *component* of a graph G is an inclusion maximal connected subgraph of G. A component is *non-trivial* if it contains at least two vertices. The *complement* of a graph $G = (V, E)$ is the graph $\bar{G} = (V, \bar{E})$ such that $\{u, v\} \in \bar{E}$ if and only if $\{u, v\} \notin E$. The *disjoint union* of two graphs $G = (V_G, E_G)$ and $H = (V_H, E_H)$ is the graph $(V_G \cup V_H, E_G \cup E_H)$, where $V_G \cap V_H = \emptyset$. For a map $\eta \colon V \to V'$ and $S \subseteq V$, let $\eta(S)$ denote the set $\{\eta(s) \mid s \in S\}$.

2.1 Definitions of the Problems

A graph $H = (V_H, E_H)$ is *subgraph-isomorphic* to a graph $G = (V_G, E_G)$ if there exists an injective map η from V_H to V_G such that $\{\eta(u), \eta(v)\} \in E_G$ holds for each $\{u, v\} \in E_H$. We call such a map η a *subgraph-isomorphism* from H to G. Graphs G and H are called the *base graph* and the *pattern graph*, respectively. The problems SUBGRAPH ISOMORPHISM and SPANNING SUBGRAPH ISOMORPHISM are defined as follows:

Problem 2.1. SUBGRAPH ISOMORPHISM
Instance: A pair of graphs $G = (V_G, E_G)$ and $H = (V_H, E_H)$.
Question: Is H subgraph-isomorphic to G?

Problem 2.2. SPANNING SUBGRAPH ISOMORPHISM
Instance: A pair of connected graphs $G = (V_G, E_G)$ and $H = (V_H, E_H)$, where $|V_G| = |V_H|$.
Question: Is H subgraph-isomorphic to G?

2.2 Graph Classes

Here we introduce the graph classes we deal with in this paper. For their inclusion relations, see the standard textbooks in this field [3,9,25]. See Fig. 1 for the class hierarchy.

A bipartite graph $B = (X, Y; E)$ is a *chain* graph if the vertices of X can be ordered as $x_1, x_2, \ldots, x_{|X|}$ such that $N(x_1) \subseteq N(x_2) \subseteq \cdots \subseteq N(x_{|X|})$. A graph $G = (V, E)$ with $V = \{1, 2, \ldots, n\}$ is a *permutation graph* if there is a permutation π over V such that $\{i, j\} \in E$ if and only if $(i - j)(\pi(i) - \pi(j)) < 0$. A *bipartite permutation graph* is a permutation graph that is bipartite. A bipartite graph $H = (X, Y; E)$ is a *convex* graph if one of X and Y can be ordered such that the neighborhood of each vertex in the other side is consecutive in the ordering. It is known that a chain graph is a bipartite permutation graph, and that a bipartite permutation graph is a convex graph.

A graph is a *co-chain* graph if it is the complement of a chain graph. An *interval graph* is the intersection graph of a family of closed intervals of the real line. A *proper interval graph* is the intersection graph of a family of closed

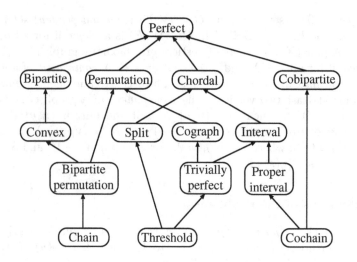

Fig. 1. Graph classes

intervals of the real line where no interval is properly contained in another. A graph is *co-bipartite* if its complement is bipartite. In other words, co-bipartite graphs are exactly the graphs whose vertex sets can be partitioned into two cliques. From the definition, every co-chain graph is co-bipartite. It is known that every co-chain graph is a proper interval graph.

A graph is a *threshold* graph if there is a positive integer T (the *threshold*) and for every vertex v there is a positive integer $w(v)$ such that $\{u, v\}$ is an edge if and only if $w(u) + w(v) \geq T$. A graph is *trivially perfect* if the size of the maximum independent set is equal to the number of maximal cliques for every induced subgraph. It is known that a threshold graph is a trivially perfect graph, and that a trivially perfect graph is an interval graph.

A *split graph* is a graph whose vertex set can be partitioned into a clique and an independent set. A graph is chordal if every induced cycle is of length 3. Clearly, every threshold graph is a split graph, and every split graph is a chordal graph. It is known that every interval graph is a chordal graph. It is easy to see that any split graph (and thus any threshold graph) has at most one non-trivial component.

A graph is *perfect* if for any induced subgraph the chromatic number is equal to the size of a maximum clique. Graphs in all classes introduced in this section are known to be perfect.

3   Polynomial-Time Algorithms

In this section, we denote the number of the vertices and the edges in a base graph by n and m, respectively. For the input graphs G and H, we assume that $|V_G| \geq |V_H|$ and $|E_G| \geq |E_H|$, which can be checked in time $O(m + n)$.

3.1   Finding Co-chain Subgraphs in Chordal Graphs

It is known that co-chain graphs are precisely $\{I_3, C_4, C_5\}$-free graphs [13]; that is, graphs having no vertex subset that induces I_3, C_4, or C_5, where I_3 is the empty graph with three vertices and C_k is the cycle of k vertices. Using this characterization, we can show the following simple lemma.

Lemma 3.1. *A graph is a co-chain graph if and only if it is a co-bipartite chordal graph.*

Proof. To prove the if-part, let G be a co-bipartite chordal graph. Since G is co-bipartite, it cannot have I_3 as its induced subgraph. Since G is chordal, it does not have C_4 or C_5 as its induced subgraph. Therefore, G is $\{I_3, C_4, C_5\}$-free.

To prove the only-if-part, let G be a co-chain graph, and thus it is a co-bipartite graph. Suppose that G has an induced cycle C of length $k \geq 4$. Then k cannot be 4 or 5 since it does not have C_4 or C_5. If $k \geq 6$, then the first, third, and fifth vertices in the cycle form I_3.

Now we can solve the problem as follows.

Theorem 3.2. SUBGRAPH ISOMORPHISM *is solvable in $O(mn^2 + n^3)$ time if the base graphs are chordal graphs and the pattern graphs are co-chain graphs.*

Proof. Let $G = (V_G, E_G)$ be the base chordal graph and $H = (V_H, E_H)$ be the pattern co-chain graph. We assume that G is not complete, since otherwise the problem is trivial.

Algorithm. We enumerate all the maximal cliques $C_1, \ldots, C_k$ of G. For each pair (C_i, C_j), we check whether H is subgraph-isomorphic to $G[C_i \cup C_j]$. If H is subgraph-isomorphic to $G[C_i \cup C_j]$ for some i and j, then we output "yes." Otherwise, we output "no."

Correctness. It suffices to show that H is subgraph-isomorphic to G if and only if there are two maximal cliques C_i and C_j of G such that H is subgraph-isomorphic to $G[C_i \cup C_j]$. The if-part is obviously true. To prove the only-if-part, assume that there is a subgraph-isomorphism η from H to G. Observe that for any clique C of H, there is a maximal clique C' of G such that $\eta(C) \subseteq C'$. Thus, since H is co-bipartite, there are two maximal cliques C_i and C_j such that $\eta(V_H) \subseteq C_i \cup C_j$. That is, H is subgraph-isomorphic to $G[C_i \cup C_j]$.

Running time. It is known that a chordal graph of n vertices with m edges has at most n maximal cliques, and all the maximal cliques can be found in $O(m+n)$ time [2,12]. Since $G[C_i \cup C_j]$ is a co-chain graph by Lemma 3.1, testing whether H is subgraph-isomorphic to $G[C_i \cup C_j]$ can be done in $O(m + n)$ time [16]. Since the number of pairs of maximal cliques is $O(n^2)$, the total running time is $O(mn^2 + n^3)$.

3.2   Finding Threshold Subgraphs in Trivially Perfect Graphs

Here we present a linear-time algorithm for finding a threshold subgraph in a trivially perfect graph. To this end, we need the following lemmas.

Lemma 3.3. *If a graph G has a universal vertex u_G, and a graph H has a universal vertex u_H, then H is subgraph-isomorphic to G if and only if $H - u_H$ is subgraph-isomorphic to $G - u_G$.*

Proof. To prove the if-part, let η' be a subgraph-isomorphism from $H - u_H$ to $G - u_G$. Now we define $\eta \colon V_H \to V_G$ as follows:

$$\eta(w) = \begin{cases} u_G & \text{if } w = u_H, \\ \eta'(w) & \text{otherwise.} \end{cases}$$

Let $\{x, y\} \in E_H$. If $u_H \notin \{x, y\}$, then $\{\eta(x), \eta(y)\} = \{\eta'(x), \eta'(y)\} \in E_G$. Otherwise, we may assume that $x = u_H$ without loss of generality. Since u_G is universal in G, it follows that $\{\eta(x), \eta(y)\} = \{\eta(u_H), \eta(y)\} = \{u_G, \eta'(y)\} \in E_G$.

  To prove the only-if-part, assume that η' is a subgraph-isomorphism from H to G. If there is no vertex $v \in V_H$ such that $\eta'(v) = u_G$, then we are done. Assume that $\eta'(v) = u_G$ for some vertex $v \in V_H$. Now we define $\eta \colon V_H \backslash \{u_H\} \to V_G \backslash \{u_G\}$ as follows:

$$\eta(w) = \begin{cases} \eta'(u_H) & \text{if } w = v, \\ \eta'(w) & \text{otherwise.} \end{cases}$$

Let $\{x, y\} \in E_H$. If $v \notin \{x, y\}$, then $\{\eta(x), \eta(y)\} = \{\eta'(x), \eta'(y)\} \in E_G$. Otherwise, we may assume without loss of generality that $v = x$. Since u_H is universal in H, it follows that $\{\eta(x), \eta(y)\} = \{\eta'(u_H), \eta'(y)\} \in E_G$.

  A component of a graph is *maximum* if it contains the maximum number of vertices among all the components of the graph. If a split graph has a nontrivial component, then the component is the unique maximum component of the graph.

Lemma 3.4. *A split graph H with a maximum component C_H is subgraph-isomorphic to a graph G if and only if $|V_H| \le |V_G|$ and there is a component C_G of G such that C_H is subgraph-isomorphic to C_G.*

Proof. First we prove the only-if-part. Let η be a subgraph-isomorphism from H to G. We need $|V_H| \le |V_G|$ to have an injective map from V_H to V_G. Since C_H is connected, $G[\eta(V(C_H))]$ must be connected. Thus there is a component C_G such that $\eta(V(C_H)) \subseteq V(C_G)$. Then $\eta|_{V(C_H)}$, the map η restricted to $V(C_H)$, is a subgraph isomorphism from C_H to C_G.

  To prove the if-part, let η' be a subgraph-isomorphism from C_H to C_G. Let $R_H = V_H \backslash V(C_H) = \{u_1, \ldots, u_r\}$, and let $R_G = V_G \backslash \eta'(V(C_H)) = \{w_1, \ldots, w_s\}$. Since $|V_H| \le |V_G|$ and $|V(C_H)| = |\eta'(V(C_H))|$, it holds that $r \le s$. Now we define $\eta \colon V_H \to V_G$ as follows:

$$\eta(v) = \begin{cases} w_i & \text{if } v = u_i \in R_H, \\ \eta'(v) & \text{otherwise.} \end{cases}$$

Since H is a split graph, any component of H other than C_H cannot have two or more vertices. Thus the vertices in R_H are isolated in H. Therefore, the map η is a subgraph-isomorphism from H to G.

The two lemmas above already allows us to have a polynomial-time algorithm. However, to achieve a linear running time, we need the following characterization of trivially perfect graphs.

A *rooted tree* is a directed tree with a unique in-degree 0 vertex, called the *root*. Intuitively, every edge is directed from the root to leaves in a directed tree. A *rooted forest* is the disjoint union of rooted trees. The *comparability graph* of a rooted forest is the graph that has the same vertex set as the rooted forest, and two vertices are adjacent in the graph if and only if one of the two is a descendant of the other in the forest. Yan et al. [28] showed that a graph is a trivially perfect graph if and only if it is the comparability graph of a rooted forest, and that such a rooted forest can be computed in linear time. We call such a rooted forest a *generating forest* of the trivially perfect graph. If a generating forest is actually a rooted tree, then we call it a *generating tree*.

Theorem 3.5. SUBGRAPH ISOMORPHISM *is solvable in $O(m + n)$ time if the base graphs are trivially perfect graphs and the pattern graphs are threshold graphs.*

Proof. Let $G = (V_G, E_G)$ be the base trivially perfect graph and $H = (V_H, E_H)$ be the pattern threshold graph.

Algorithm The pseudocode of our algorithm can be found in Algorithm 1. We use the procedure SGI which takes a trivially perfect graph as the base graph and a threshold graph as the pattern graph, and conditionally answers whether the pattern graph is subgraph-isomorphic to the base graph. The procedure SGI requires that

– both the graphs are connected, and
– the base graph has at least as many vertices as the pattern graph.

To use this procedure, we first attach a universal vertex to both G and H. This guarantees that both graphs are connected. We call the new graphs G' and H', respectively. By Lemma 3.3, (G', H') is a yes-instance if and only if so is (G, H). After checking that $|V_{G'}| \geq |V_{H'}|$, we use the procedure SGI.

In SGI(G, H), let u_G and u_H be universal vertices of G and H, respectively. There are such vertices since G and H are connected trivially perfect graphs [27]. Let C_H be a maximum component of $H - u_H$. For each connected component C_G of $G - u_G$, we check whether C_H is subgraph-isomorphic to C_G, by recursively calling the procedure SGI itself. If at least one of the recursive calls returns "yes," then we return "yes." Otherwise we return "no."

Correctness. It suffices to prove the correctness of the procedure SGI. If $|V_H| = 1$, then H is subgraph-isomorphic to G since $|V_G| \geq |V_H|$ in SGI. By Lemmas 3.3 and 3.4, H is subgraph-isomorphic to G if and only if there is a component C_G of

---

Algorithm 1. Finding a threshold subgraph H in a trivially perfect graph G

1: $G' := G$ with a universal vertex
2: $H' := H$ with a universal vertex
3: **if** $|V_{G'}| \geq |V_{H'}|$ **then**
4:      **return** SGI(G', H')
5: **else**
6:      **return** no

Require: G and H are connected, and $|V_G| \geq |V_H|$
7: **procedure** SGI(G, H)
8:      **if** $|V_H| = 1$ **then**
9:          **return** yes
10:      $u_G :=$ a universal vertex of G
11:      $u_H :=$ a universal vertex of H
12:      $C_H :=$ a maximum component of $H - u_H$
13:      **for all** components C_G of $G - u_G$ **do**
14:          **if** $|V(C_G)| \geq |V(H - u_H)|$ **then**
15:              **if** SGI(C_G, C_H) = yes **then**
16:                  **return** yes
17:      **return** no

$G - u_G$ such that C_H is subgraph-isomorphic to C_G. (Recall that any threshold graph is a split graph.) The procedure just checks these conditions. Also, when SGI recursively calls itself, the parameters C_G and C_H satisfy its requirements; that is, C_G and C_H are connected, and $|V(C_G)| \geq |V(C_G)|$.

Running time For each call of SGI(G, H), we need the following:

- universal vertices u_G and u_H of G and H, respectively,
- a maximum component C_H of $H - u_H$,
- the components C_G of $G - u_G$, and
- the numbers of the vertices of C_G and $H - u_H$.

We show that they can be computed efficiently by using generating forests. Basically we apply the algorithm to generating forests instead of graphs.

Before the very first call of SGI(G, H), we compute generating trees of G and H in linear time. Additionally, for each node in the generating trees, we store the number of its descendants. This can be done also in linear time in a bottom-up fashion.

At some call of SGI(G, H), assume that we have generating trees of G and H. It is easy to see that the root of the generating trees are universal vertices. Hence we can compute u_G and u_H in constant time. By removing these root nodes from the generating trees, we obtain generating forests of $G - u_G$ and $H - u_H$. Each component of the generating forests corresponds to a component of the corresponding graphs. Thus we can compute the components of $G - u_G$ and a maximum component of $H - u_H$, with their generating trees, in time proportional to the number of the children of u_G and u_H. The numbers of the vertices of C_G and $H - u_H$ can be computed easily in constant time, because we know the number of the descendants of each node in generating trees.

The recursive calls of SGI take only $O(n)$ time in total since it is proportional to the number of edges in the generating trees. Therefore, the total running time is $O(m + n)$.

4   NP-completeness

It is known that for perfect graphs, CLIQUE can be solved in polynomial time [10]. Since co-chain graphs and threshold graphs are very close to complete graphs, one may ask whether the problem of finding co-chain graphs or threshold graphs can be solved in polynomial time for perfect graphs. In this section, we show that this is not the case. More precisely, we show that even the specialized problem SPANNING SUBGRAPH ISOMORPHISM is NP-complete for the case where the base graphs are somewhat restricted perfect graphs and the pattern graphs are co-chain or threshold graphs.

It is known that MAXIMUM EDGE BICLIQUE, the problem of finding a biclique with the maximum number of edges, is NP-complete for bipartite graphs [24]. This implies that SUBGRAPH ISOMORPHISM is NP-complete if the base graphs are connected bipartite graphs and the pattern graphs are connected chain graphs, because complete bipartite graphs are chain graphs. We sharpen this hardness result by showing that the problem is still NP-complete if we further restrict the pattern chain graphs to have the same number of vertices as the base graph. That is, we show that SPANNING SUBGRAPH ISOMORPHISM is NP-complete when the base graphs are bipartite graphs and the pattern graphs are chain graphs.

Since the problem SPANNING SUBGRAPH ISOMORPHISM is clearly in NP for any graph class, we only show its NP-hardness here. All the results in this section are based on the following theorem and lemma taken from Kijima et al. [16].

Theorem 4.1 (Kijima et al. [16]). SPANNING SUBGRAPH ISOMORPHISM *is NP-complete if*

1. *the base graphs are chain graphs and the pattern graphs are convex graphs,*
2. *the base graphs are co-chain graphs and the pattern graphs are co-bipartite graphs, or*
3. *the base graphs are threshold graphs and the pattern graphs are split graphs.*

Lemma 4.2 (Kijima et al. [16]). *If* $|V_H| = |V_G|$, *then* H *is subgraph-isomorphic to* G *if and only if* $\bar{G}$ *is subgraph-isomorphic to* $\bar{H}$.

For a graph class $\mathcal{C}$, let co-$\mathcal{C}$ denote the graph class $\{\bar{G} \mid G \in \mathcal{C}\}$. The next lemma basically shows that if $\mathcal{C}$ satisfies some property, then the hardness of SPANNING SUBGRAPH ISOMORPHISM for $\mathcal{C}$ implies the hardness for co-$\mathcal{C}$.

Lemma 4.3. *Let* $\mathcal{C}$ *and* $\mathcal{D}$ *be graph classes such that* co-$\mathcal{C}$ *and* co-$\mathcal{D}$ *are closed under universal vertex additions. If* SPANNING SUBGRAPH ISOMORPHISM *is NP-complete when the base graphs belong to* $\mathcal{C}$ *and the pattern graphs belong to* $\mathcal{D}$, *then the problem is NP-complete also when the base graphs belong to* co-$\mathcal{D}$ *and the pattern graphs belong to* co-$\mathcal{C}$.

Proof. Given two connected graphs $G \in \mathcal{C}$ and $H \in \mathcal{D}$ with $|V_G| = |V_H|$, it is NP-complete to decide whether H is subgraph-isomorphic to G. By Lemma 4.2, H is subgraph-isomorphic to G if and only if $\bar{G}$ is subgraph-isomorphic to $\bar{H}$. By Lemma 3.3, $\bar{G}$ is subgraph-isomorphic to $\bar{H}$ if and only if $\bar{G}'$ is subgraph-isomorphic to $\bar{H}'$, where $\bar{G}'$ and $\bar{H}'$ are obtained from $\bar{G}$ and $\bar{H}$, respectively, by adding a universal vertex. Therefore, H is subgraph-isomorphic to G if and only if $\bar{G}'$ is subgraph-isomorphic to $\bar{H}'$. Clearly, $\bar{G}' \in$ co-$\mathcal{C}$ and $\bar{H}' \in$ co-$\mathcal{D}$, they are connected, and they have the same number of vertices. Thus the lemma holds.

A graph is a *co-convex* graph if its complement is a convex graph. Clearly co-convex graphs are closed under additions of universal vertices.

Corollary 4.4. SPANNING SUBGRAPH ISOMORPHISM *is NP-complete if*

1. *the base graphs are co-convex graphs and the pattern graphs are co-chain graphs,*
2. *the base graphs are bipartite graphs and the pattern graphs are chain graphs, or*
3. *the base graphs are split graphs and the pattern graphs are threshold graphs.*

Proof. The NP-completeness of the case (1) is a corollary to Theorem 4.1 (1) and Lemma 4.3. To prove (3), we need Theorem 4.1 (3), Lemma 4.3, and the well-known facts that threshold graphs and split graphs are self-complementary [9]. That is, the complement of a threshold graph is a threshold graph, and the complement of a split graph is a split graph.

For (2), we cannot directly apply the combination of Theorem 4.1 (2) and Lemma 4.3 since bipartite graphs and chain graphs are not closed under universal vertex additions. Fortunately, we can easily modify the proof of Theorem 4.1 (2) in Kijima et al. [16] so that the complements of the base graphs and the pattern graphs are also connected. Then, Lemma 4.2 implies the statement. Since it will be a repeat of a known proof with a tiny difference, we omit the detail.

5   Conclusion

We have studied (SPANNING) SUBGRAPH ISOMORPHISM for classes of perfect graphs, and have shown sharp contrasts of its computational complexity. An interesting problem left unsettled is the complexity of SUBGRAPH ISOMORPHISM where the base graphs are bipartite permutation graphs and the pattern graphs are chain graphs. It is known that although the maximum edge biclique problem is NP-complete for general bipartite graphs [24], it can be solved in polynomial time for some super classes of bipartite permutation graphs (see [23]). Therefore, it might be possible to have a polynomial-time algorithm for SUBGRAPH ISOMORPHISM when the pattern graphs are chain graphs and the base graphs belong to an even larger class like convex graphs.

References

1. Belmonte, R., Heggernes, P., van 't Hof, P.: Edge contractions in subclasses of chordal graphs. Discrete Appl. Math. 160, 999–1010 (2012)
2. Blair, J.R.S., Peyton, B.: An introduction to chordal graphs and clique trees. In: George, A., Gilbert, J.R., Liu, J.W.H. (eds.) Graph Theory and Sparse Matrix Computation. The IMA Volumes in Mathematics and its Applications, vol. 56, pp. 1–29. Springer (1993)
3. Brandstädt, A., Le, V.B., Spinrad, J.P.: Graph Classes: A Survey. SIAM (1999)
4. Colbourn, C.J.: On testing isomorphism of permutation graphs. Networks 11, 13–21 (1981)
5. Damaschke, P.: Induced subgraph isomorphism for cographs is NP-complete. In: Möhring, R.H. (ed.) WG 1990. LNCS, vol. 484, pp. 72–78. Springer, Heidelberg (1991)
6. Dorn, F.: Planar subgraph isomorphism revisited. In: STACS 2010. LIPIcs, vol. 5, pp. 263–274 (2010)
7. Eppstein, D.: Subgraph isomorphism in planar graphs and related problems. J. Graph Algorithms Appl. 3, 1–27 (1999)
8. Garey, M.R., Johnson, D.S.: Computers and Intractability: A Guide to the Theory of NP-Completeness. W.H. Freeman and Company (1979)
9. Golumbic, M.C.: Algorithmic Graph Theory and Perfect Graphs, 2nd edn. Annals of Discrete Mathematics, vol. 57. North Holland (2004)
10. Grötschel, M., Lovász, L., Schrijver, A.: The ellipsoid method and its consequences in combinatorial optimization. Combinatorica 1, 169–197 (1981)
11. Gupta, A., Nishimura, N.: The complexity of subgraph isomorphism for classes of partial k-trees. Theoret. Comput. Sci. 164, 287–298 (1996)
12. Heggernes, P.: Treewidth, partial k-trees, and chordal graphs. Partial curriculum in INF334 - Advanced algorithmical techniques, Department of Informatics, University of Bergen, Norway (2005)
13. Heggernes, P., Kratsch, D.: Linear-time certifying recognition algorithms and forbidden induced subgraphs. Nordic J. Comput. 14, 87–108 (2007)
14. Heggernes, P., Meister, D., Villanger, Y.: Induced subgraph isomorphism on interval and proper interval graphs. In: Cheong, O., Chwa, K.-Y., Park, K. (eds.) ISAAC 2010, Part II. LNCS, vol. 6507, pp. 399–409. Springer, Heidelberg (2010)
15. Heggernes, P., van 't Hof, P., Meister, D., Villanger, Y.: Induced subgraph isomorphism on proper interval and bipartite permutation graphs. Submitted manuscript
16. Kijima, S., Otachi, Y., Saitoh, T., Uno, T.: Subgraph isomorphism in graph classes. Discrete Math. 312, 3164–3173 (2012)
17. Lingas, A.: Subgraph isomorphism for biconnected outerplanar graphs in cubic time. Theoret. Comput. Sci. 63, 295–302 (1989)
18. Lueker, G.S., Booth, K.S.: A linear time algorithm for deciding interval graph isomorphism. J. ACM 26, 183–195 (1979)
19. Marx, D., Pilipczuk, M.: Everything you always wanted to know about the parameterized complexity of subgraph isomorphism (but were afraid to ask). CoRR, abs/1307.2187 (2013)
20. Marx, D., Schlotter, I.: Cleaning interval graphs. Algorithmica 65, 275–316 (2013)
21. Matoušek, J., Thomas, R.: On the complexity of finding iso- and other morphisms for partial k-trees. Discrete Math. 108, 343–364 (1992)
22. Matula, D.W.: Subtree isomorphism in $O(n^{5/2})$. In: Alspach, B., Hell, P., Miller, D. (eds.) Algorithmic Aspects of Combinatorics. Annals of Discrete Mathematics, vol. 2, pp. 91–106. Elsevier (1978)

23. Nussbaum, D., Pu, S., Sack, J.-R., Uno, T., Zarrabi-Zadeh, H.: Finding maximum edge bicliques in convex bipartite graphs. Algorithmica 64(2), 311–325 (2012)
24. Peeters, R.: The maximum edge biclique problem is NP-complete. Discrete Appl. Math. 131, 651–654 (2003)
25. Spinrad, J.P.: Efficient Graph Representations. Fields Institute monographs, vol. 19. American Mathematical Society (2003)
26. Sysło, M.M.: The subgraph isomorphism problem for outerplanar graphs. Theoret. Comput. Sci. 17, 91–97 (1982)
27. Wolk, E.S.: A note on "The comparability graph of a tree". Proc. Amer. Math. Soc. 16, 17–20 (1965)
28. Yan, J.-H., Chen, J.-J., Chang, G.J.: Quasi-threshold graphs. Discrete Appl. Math. 69(3), 247–255 (1996)

A Pseudo-Random Bit Generator
Based on Three Chaotic Logistic Maps
and IEEE 754-2008 Floating-Point Arithmetic

Michael François[1], David Defour[2], and Pascal Berthomé[1]

[1] INSA Centre Val de Loire, Univ. Orléans, LIFO EA 4022, Bourges, France
{michael.francois,pascal.berthome}@insa-cvl.fr
[2] Univ. Perpignan Via Domitia, DALI F-66860, LIRMM UMR 5506 F-34095,
Perpignan, France
david.defour@univ-perp.fr

Abstract. A novel pseudo-random bit generator (PRBG), combining three chaotic logistic maps is proposed. The IEEE 754-2008 standard for floating-point arithmetic is adopted and the binary64 double precision format is used. A more efficient processing is applied to better extract the bits, from outputs of the logistic maps. The algorithm enables to generate at each iteration, a block of 32 random bits by starting from three chosen seed values. The performance of the generator is evaluated through various statistical analyzes. The results show that the output sequences possess high randomness statistical properties for a good security level. The proposed generator lets appear significant cryptographic qualities.

Keywords: PRBG, Pseudo-random, Logistic map, Chaotic map, IEEE 754-2008.

1 Introduction

The generation of pseudo-random bits (or numbers) plays a crucial role in a large number of applications such as statistical mechanics, numerical simulation, gaming industry, communication or cryptography [1]. The term "pseudo-random" is used to indicate that, the bits (or numbers) appear to be random and are generated from an algorithmic process so-called generator. From a single initial parameter (or seed), the generator will always produce the same pseudo-random sequence. The main advantages of such generators are the rapidity and the repeatability of the sequences and require less memory for algorithm storage. Some fundamental methods are typically used to implement pseudo-random number generators, such as: non-linear congruences [2], linear feedback shift registers (LFSR) [3], discrete logarithm problem [4], quadratic residuosity problem [5], cellular automata [6], etc. In general, the security of a cryptographic pseudo-random number generator (PRNG), is based on the difficulty to solve the related mathematical problem. That usually makes the algorithm much slower, due to heavy computational instructions. For example, the Blum Blum Shub

T V Gopal et al. (Eds.): TAMC 2014, LNCS 8402, pp. 229–247, 2014.
© Springer International Publishing Switzerland 2014

algorithm [5] has a security proof, assuming the intractability of the quadratic residuosity problem. The algorithm is also proven to be secure, under the assumption that the integer factorization problem is difficult. However, the algorithm is very inefficient and impractical unless extreme security is needed. The Blum-Micali algorithm [4] has also an unconditional security proof based on the difficulty of the discrete logarithm problem, but is also inefficient.

Another interesting way to design such generators is connected to chaos theory [7]. That theory focuses primarily on the description of these systems that are often very simple to define, but whose dynamics appears to be very confused. Indeed, chaotic systems are characterized by their high sensitivity to initial conditions and some properties like ergodicity, pseudo-random behavior and high complexity [7]. The extreme sensitivity to the initial conditions (i.e. a small deviation in the input can cause a large variation in the output) makes chaotic system very attractive for implementing pseudo-random number generators. Obviously, chaos-based generators do not enjoy universal mathematical proofs compared with cryptographic ones, but represent a serious alternative that needs to be exploited. Moreover, during the last decade, several pseudo-random number generators have been proposed [8–14]. However, a rigorous analysis is necessary to evaluate the randomness level and the global security of the generator.

In this paper, a new PRBG using a standard chaotic logistic map is presented. It combines three logistic maps involving binary64 floating-point arithmetic and generates a block of 32 random bits at each iteration. The novelty of the paper is mainly based on the extraction mechanism of bits from the outputs of chaotic logistic maps. The produced pseudo-random sequences have successfully passed the various statistical tests. The assets of the generator are: high sensitivity to initial seed values, high level of randomness and good throughput. The paper is structured as follows, a brief introduction on floating-point arithmetic and the used chaotic logistic map is given in Sect. 2. Section 3 presents a detailed description of the algorithm. The statistical analysis applied on two groups of generated pseudo-random sequences is given in Sect. 4. The global security analysis of the PRBG is achieved in Sect. 5, before concluding.

2   Background

2.1   IEEE 754-2008 Standard

Digital computers represent numbers in sets of binary digits. For real numbers, two formats of representation can be distinguished: fixed-point format and floating-point format. The fixed-point format is designed to represent and manipulate integers or real numbers with a fixed precision. In the case of real numbers with variable precision, the representation is made through the floating-point format. There exists a standard that defines the arithmetic formats, the rounding rules, the operations and the exception handling for floating-point arithmetic.

The IEEE 754-2008 [15] is the current version of the technical standard, used by hardware manufacturer to implement floating-point arithmetic. Among them, binary32 (single precision) and binary64 (double precision) are the two most widely used and implemented formats. As the generator described herein relies exclusively on binary64, we will only consider this format in the rest of the article.

Binary64 comprises two infinities, two kinds of NaN (Not a Number) and the set of finite numbers. Each finite number is uniquely described by three integers: s a sign represented on 1 bit, e a biased exponent represented on 11 bits and m a mantissa represented on 52 bits, where the leading bit of the significand is implicitly encoded in the biased exponent (see Fig. 1). To make the encoding unique, the value of the significand m is maximized by decreasing e until either $e = emin$ or $m \geq 1$. After this process is done, if $e = emin$ and $0 < m < 1$, the floating-point number is subnormal. Subnormal numbers (and zero) are encoded with a reserved biased exponent value. Interested readers will find a good introduction to floating point arithmetic and issues that arise while using it in [16].

Fig. 1. Floating-point representation in double precision format (64 bits)

2.2 The Chaotic Logistic Map

The generator uses a chaotic logistic map given by:

$$F(X) = \lambda X(1 - X) , \tag{1}$$

with λ between 3.57 and 4.0 [17]. This function has been widely studied [18] and several pseudo-random number generators have already used such logistic map [12, 17, 19–22]. To avoid non-chaotic behavior (island of stability, oscillations, ...), the value of λ is fixed to 3.9999 that corresponds to a highly chaotic case [23]. The logistic map can be used under the iterative form:

$$X_{n+1} = 3.9999X_n(1 - X_n), \forall n \geq 0 , \tag{2}$$

where the initial seed X_0 is a real number belonging to the interval $]0, 1[$. All the output elements X_n are also real numbers in $]0, 1[$.

3 The Proposed Generator

The main idea of the PRBG is to combine several chaotic logistic maps and carefully arrange them in the same algorithm in order to increase the security level. A block of 32 random bits per iteration is produced using the following three logistic maps:

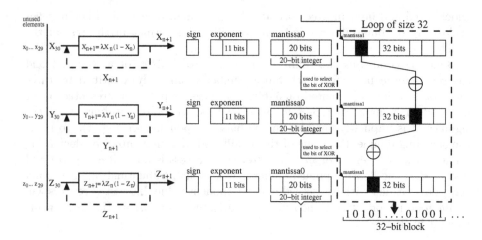

Fig. 2. Graphical description of the PRBG. In each mantissa1, the used bit (in xor) is moved at the end of chain and such process is not explicit on the scheme.

$$X_{n+1} = 3.9999X_n(1 - X_n), \forall n \geq 0 ,\tag{3}$$

$$Y_{n+1} = 3.9999Y_n(1 - Y_n), \forall n \geq 0 ,\tag{4}$$

$$Z_{n+1} = 3.9999Z_n(1 - Z_n), \forall n \geq 0 .\tag{5}$$

For the three chaotic maps, the same value of λ is chosen to maintain its surjectivity in the same interval. The graphical description of the generator is shown in Fig. 2. The technical details of the implementation in C, using definitions from the file *ieee754.h* are given in Algorithm 1. The algorithmic principle of the PRBG consists in three steps:

1. Line 2: three different seed values X_0, Y_0 and Z_0 are chosen to initiate the generation process (see Sect. 3.1).
2. Line 3–8: the results of the 30 first iterations are discarded to decorrelate the beginning of the output sequences (see Sect. 3.2).
3. Line 9–50: a loop of size N is started, with N being the length of the output sequence in block of 32 bits, then:
   (a) line 10–12: iterate the three logistic maps,
   (b) line 13–21: in each case, the bits of mantissa0 and mantissa1 are saved in two variables,
   (c) line 23–48: start another loop of size 32, and one bit is selected at a time from each mantissa1, according to the value of mantissa0. For more security, the value of the mantissa0 of Z_n is used to index the bits of the mantissa1 of X_n, the value of the mantissa0 of X_n to index the bits of mantissa1 of Y_n, and the value of mantissa0 of Y_n to index those of the mantissa1 in Z_n. Indeed, the bits of mantissa0 form a 20-bit integer, and by making a regressive modulo from 32, that allows to fix the position of the bit to be used. Thus, the three selected bits are combined by a

Algorithm 1. The PRBG algorithm

Require: $X_0; Y_0; Z_0; N;$
Ensure: A sequence of N blocks of 32 bits
1: **Declaration:** $union\ ieee754\_double *F_1, *F_2, *F_3;$
2: **Initialization:** $i = 1; j = 1; X = X_0; Y = Y_0; Z = Z_0;$
3: **while** $i \leq 30$ **do**
4:     $X \leftarrow 3.9999 \times X \times (1 - X)$
5:     $Y \leftarrow 3.9999 \times Y \times (1 - Y)$
6:     $Z \leftarrow 3.9999 \times Z \times (1 - Z)$
7:     $i \leftarrow i + 1$
8: **end while**
9: **while** $j \leq N$ **do**
10:     $X \leftarrow 3.9999 \times X \times (1 - X)$
11:     $Y \leftarrow 3.9999 \times Y \times (1 - Y)$
12:     $Z \leftarrow 3.9999 \times Z \times (1 - Z)$
13:     $F_1 \leftarrow (union\ ieee754\_double *)\ \&\ X$
14:     $F_2 \leftarrow (union\ ieee754\_double *)\ \&\ Y$
15:     $F_3 \leftarrow (union\ ieee754\_double *)\ \&\ Z$
16:     $M_{0X} \leftarrow F_1-> ieee.mantissa0$
17:     $M_{1X} \leftarrow F_1-> ieee.mantissa1$
18:     $M_{0Y} \leftarrow F_2-> ieee.mantissa0$
19:     $M_{1Y} \leftarrow F_2-> ieee.mantissa1$
20:     $M_{0Z} \leftarrow F_3-> ieee.mantissa0$
21:     $M_{1Z} \leftarrow F_3-> ieee.mantissa1$
22:     $k \leftarrow 32$
23:     **while** $k > 0$ **do**
24:         $l \leftarrow k - 1$
25:         $P_X \leftarrow M_{0Z} \bmod k$
26:         $P_Y \leftarrow M_{0X} \bmod k$
27:         $P_Z \leftarrow M_{0Y} \bmod k$
28:         $B_x \leftarrow (M_{1X} >> (P_X)\ \&\ 1)$
29:         $B_y \leftarrow (M_{1Y} >> (P_Y)\ \&\ 1)$
30:         $B_z \leftarrow (M_{1Z} >> (P_Z)\ \&\ 1)$
31:         $B \leftarrow (B_x + B_y + B_z) \bmod 2$ {output bit}
32:         $b_x \leftarrow (M_{1X} >> (l)\ \&\ 1)$
33:         $b_y \leftarrow (M_{1Y} >> (l)\ \&\ 1)$
34:         $b_z \leftarrow (M_{1Z} >> (l)\ \&\ 1)$
35:         **if** $bx \neq B_x$ **then**
36:             $M_{1X} \leftarrow M_{1X}\ \hat{}\ (1 << (l))$
37:             $M_{1X} \leftarrow M_{1X}\ \hat{}\ (1 << (P_X))$
38:         **end if**
39:         **if** $by \neq B_y$ **then**
40:             $M_{1Y} \leftarrow M_{1Y}\ \hat{}\ (1 << (l))$
41:             $M_{1Y} \leftarrow M_{1Y}\ \hat{}\ (1 << (P_Y))$
42:         **end if**
43:         **if** $bz \neq B_z$ **then**
44:             $M_{1Z} \leftarrow M_{1Z}\ \hat{}\ (1 << (l))$
45:             $M_{1Z} \leftarrow M_{1Z}\ \hat{}\ (1 << (P_Z))$
46:         **end if**
47:         $k \leftarrow k - 1$
48:     **end while**
49:     $j \leftarrow j + 1$
50: **end while**

xor to give the output bit (line 31). From each mantissa1, the selected bit is then permuted with the bit at the end of chain to not be used again (line 35–46). At the end of this loop, a block of 32 random bits is produced. Such mechanism is definitely costly for the algorithm, but it allows to better decorrelate the outputs of the PRBG, especially in case of a possible collision.

3.1   Seed Selection

The input and output values of the logistic map belong to $]0, 1[$. To increase the robustness of the generator, three identical logistic maps are then combined. To preserve such robustness, one must avoid constructing identical chaotic trajectories, that may occur when using inappropriate initial seeds. To understand such mechanism, it is important to know how a difference δ between two computed values X_n and Y_n at a given iteration n will propagate to the next iteration. Without loss of generality, we can assume that $Y_n = X_n(1 + \delta)$. From (2) we know that:

$$X_{n+1} = \lambda X_n(1 - X_n) \quad and \quad Y_{n+1} = \lambda Y_n(1 - Y_n) ,$$

which is equivalent to:

$$Y_{n+1} = X_{n+1}\left(1 + \frac{\delta - 2\delta X_n - \delta^2 X_n}{(1 - X_n)}\right) .$$

Therefore, the difference between Y_{n+1} and X_{n+1} is:

$$Y_{n+1} - X_{n+1} = \lambda \delta X_n(1 - 2X_n - \delta X_n) .$$

We can deduce that, the smallest difference between Y_{n+1} and X_{n+1} is reached when $\delta = (1 - 2X_n)/X_n$. Finally, as X_n approaches 2^{-1} we obtain:

$$\lim_{X_n \to 2^{-1}} (Y_{n+1} - X_{n+1}) = -\frac{\lambda \delta^2}{4} .$$

To avoid identical representations, this difference must be representable in binary64, that means $\lambda \delta^2/4$ must be greater than 2^{-53}. By hypothesis, we set $\lambda = 3.9999$, then $\delta > 2^{-26.5}$. In this case, such value of δ allows to start with different chaotic trajectories, but it does not prevent a possible collision of elements at a certain rank n, which is a rare phenomenon but not impossible. In binary64 floating-point arithmetic, the computed value $(1 - x)$ is equal to 1.0 for any $x \in]0, 2^{-53}[$. This means that, for an initial seed selected in the interval $]0, 2^{-53}[$ the computed value of (2) is equivalent to λX_n. To avoid such problem, initial seeds have to be chosen in the interval $]2^{-53}, 2^{-1}[$.

Overall, the first seed X_0 is a random floating-point number representable in binary64 in the interval $]2^{-53}, 2^{-1}[$. The two other seeds Y_0 and Z_0 are constructed by randomly choosing two binary64 floating-point numbers. However, the minimum gap between each pair of seeds must be greater than 2^{-53} to avoid identical representations.

3.2   Initial Chaotic Behavior

The trajectory of the logistic map, from a small starting value (here $10^{-15} \approx 2^{-49.82}$) is plotted in Fig. 3. This value can also be considered as the minimum gap between two initial seeds. The aim is to analyze the evolution of this gap, through the iterative process. One can remark that, for the first iterations the trajectory is not chaotic. Indeed, a small initial difference between two seeds, spreads slowly toward the leading bits of mantissa. This problem does not occur, when the initial seeds are very different. However, to decorrelate the beginning of the output sequences in both cases, it is necessary to discard the first iterations before starting the generation. Thus, to decorrelate the outputs and increase the security level of the PRBG, we choose that the generation will start from the 31st iteration.

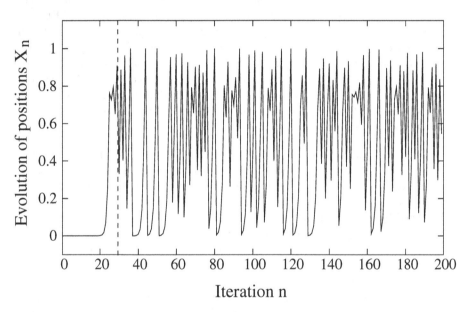

Fig. 3. Trajectory of the chaotic logistic map given in (2), for $X_0 = 10^{-15}$ and $n = 200$

4   Statistical Analysis

The output sequences of a PRBG must have a high level of randomness and be completely uncorrelated from each other. Therefore, a statistical analysis based on the randomness level and correlation should be carefully conducted to prove the quality of the sequences.

4.1 Randomness Evaluation

The analysis consists in evaluating the randomness quality of the sequences produced by the generator. Therefore, the sequences are evaluated through statistical tests suite NIST (National Institute of Standards and Technology of the U.S. Government). Such suite consists in a statistical package of fifteen tests developed to quantify and to evaluate the randomness of binary sequences produced by cryptographic random or pseudo-random number generators [24]. For each statistical test, a set of p_{value} is produced and compared to a fixed significance level $\alpha = 0.01$. A p_{value} of zero indicates that, the tested sequence appears to be not random. A p_{value} larger than α means that, the tested sequence is considered to be random with a confidence level of 99%. Therefore, a sequence passes a statistical test for $p_{value} \geq \alpha$ and fails otherwise. If at the same time more than one sequence is tested, each statistical test defines a proportion η as the ratio of sequences passing successfully the test relatively to the total number of tested sequences T (i.e. $\eta = n[p_{value} \geq 0.01]/T$). The proportion η is compared to an acceptable proportion η_{accept} which corresponds to the ratio of sequences that should pass the test. The range of acceptable proportions, excepted for the tests *Random Excursion-(Variant)* is determined by using the confidence interval defined as $(1 - 0.01) \pm 3\sqrt{0.01(1 - 0.01)/T}$ [24]. To analyze various aspects of the sequences, the NIST tests are applied on: individual sequences, the concatenated sequence and resulting sequences.

1. Individual sequences: all the produced sequences are individually tested and the results are given as ratio of success relatively to the threshold η_{accept}. Such test indicates the global randomness level of generated sequences.
2. Concatenated sequence: a new sequence of binary size $32 \times N \times T$ is constructed by concatenating all the individual sequences. The randomness level of the constructed sequence is also analyzed with the NIST tests. In the case of truly uncorrelated random sequences, the concatenated sequence should also be random.
3. Resulting sequences: are the sequences obtained from the columns, if the produced sequences are superimposed on each other. Thus, N resulting sequences of binary size $32 \times T$ are constructed, by collecting for each position $1 \leq j \leq N$, the 32-bit bloc of each sequence. The NIST tests are used to analyze such resulting sequences. This approach is interesting especially for sequences generated with successive seed values and can show whether there is some hidden linear structures between the original sequences.

4.2 Correlation Evaluation

The correlation evaluation is achieved in two different ways. Firstly, the correlation between the generated sequences is analyzed globally by computing the Pearson's correlation coefficient of each pair of sequences [25]. Consider a pair of sequences given by: $S_1 = [x_0, \ldots, x_{N-1}]$ and $S_2 = [y_0, \ldots, y_{N-1}]$. Therefore, the corresponding correlation coefficient is:

$$C_{S_1,S_2} = \frac{\sum\limits_{i=0}^{N-1}(x_i - \overline{x}) \cdot (y_i - \overline{y})}{\left[\sum\limits_{i=0}^{N-1}(x_i - \overline{x})^2\right]^{1/2} \cdot \left[\sum\limits_{i=0}^{N-1}(y_i - \overline{y})^2\right]^{1/2}} , \qquad (6)$$

where x_i and y_i are 32-bit integers, $\overline{x} = \sum\limits_{i=0}^{N-1} x_i/N$ and $\overline{y} = \sum\limits_{i=0}^{N-1} y_i/N$ the mean values of S_1 and S_2, respectively. For two uncorrelated sequences, $C_{S_1,S_2} = 0$. A strong correlation occurs for $C_{S_1,S_2} \simeq \pm 1$. The coefficients C_{S_1,S_2} are computed for each pair of produced sequences and the distribution of the values is presented by a histogram.

In the second approach, a correlation based directly on the bits of sequences is analyzed. The Hamming distance between two binary sequences (of the same length M) is the number of places where they differ, i.e., the number of positions where one has a 0 and the other a 1. Thus, for two binary sequences S_1^b and S_2^b, the corresponding Hamming distance is:

$$d(S_1^b, S_2^b) = \sum_{j=0}^{M-1}(x_j \oplus y_j) , \qquad (7)$$

where x_j (resp. y_j) are the elements of S_1^b (resp. S_2^b). In the case of truly random binary sequences, such distance is typically around $M/2$, which gives a proportion (i.e. $d(S_1^b, S_2^b)/M$) of about 0.50. For each pair of produced sequences, this proportion is determined and all values are represented through a histogram. The interest of both approaches is to check the correlation for generated sequences mainly from nearby or successive seed values.

4.3   Analysis of Pseudo-Random Sequences

In the case of very distant seed values, the chaotic trajectories are very different, which usually allows to obtain good pseudo-random sequences. That is why the analysis is achieved on sequences produced from nearby or successive seed values. Here, two groups of pseudo-random sequences are considered. The binary length of each sequence is $32 \times N$ with $N = 1024$ and the total number of sequences per group is $T = 15000$. The first group (GRP1) is generated from the seed values $X_0 = 1 \times 10^{-15}, Y_0 = 2 \times 10^{-15}$ and $Z_0 = 3 \times 10^{-15}$ where each new sequence is obtained with the same values of X_0, Y_0 and by incrementing of 10^{-15} the last seed value Z_0. For the second group (GRP2), the same strategy is applied to the starting seeds $X_0' = 0.325873724698325, Y_0' = 0.325873724698326$ and $Z_0' = 0.325873724698327$. A simple loop on the latest seed values Z_0 and Z_0' allows to generate the two groups of sequences GRP1 and GRP2. The aim is to show whatever the structure of the initial seeds, the PRBG produces sequences of high quality.

Results of Randomness Evaluation. The results of NIST tests obtained on the two groups of 15000 sequences are presented in Table 1 and Table 2, respectively. For individual sequences (resp. resulting sequences), the acceptable proportion should lie above $\eta_{accept} = 98.75\%$ (resp. $\eta'_{accept} = 98.04\%$). For the tests *Non-Overlapping* and *Random Excursions-(Variant)*, only the smallest percentage of all under tests is presented. In the case of individual sequences, the *Universal* test is not applicable due to the size of sequences. Table 1 and Table 2 show that, all the tested sequences pass successfully the NIST tests.

Table 1. Results of the NIST tests on the 15000 generated sequences of GRP1. The ratio η (resp. η') of p_{value} passing the tests are given for individual (resp. resulting) sequences and the p_{value} is given for the concatenated sequence.

Test Name	Indiv. Seq.		Concat. Seq.		Result. Seq.	
	η	Result	p_{value}	Result	η'	Result
Frequency	99.06	Success	0.338497	Success	99.31	Success
Block-Frequency	99.11	Success	0.673515	Success	98.92	Success
Cumulative Sums (1)	99.08	Success	0.589087	Success	99.21	Success
Cumulative Sums (2)	99.00	Success	0.408891	Success	99.21	Success
Runs	98.93	Success	0.343876	Success	99.02	Success
Longest Run	98.99	Success	0.417880	Success	99.02	Success
Rank	98.86	Success	0.788352	Success	98.63	Success
FFT	98.90	Success	0.609162	Success	98.24	Success
Non-Overlapping	99.26	Success	0.012083	Success	98.04	Success
Overlapping	99.00	Success	0.175000	Success	98.92	Success
Universal	-	-	0.366163	Success	98.43	Success
Approximate Entropy	98.93	Success	0.138980	Success	98.14	Success
Random Excursions	98.75	Success	0.100729	Success	98.12	Success
Random Ex-Variant	98.75	Success	0.043821	Success	97.71	Success
Serial (1)	98.91	Success	0.158943	Success	99.12	Success
Serial (2)	99.06	Success	0.367717	Success	98.92	Success
Linear Complexity	98.98	Success	0.975515	Success	98.73	Success

Results of Correlation Evaluation. Concerning the correlation analysis, the Pearson's correlation coefficient between each pair of the 15000 produced sequences is computed. For each group, the corresponding histogram is presented in Fig. 4. One can see that, the two histograms have the same shape and show that the computed coefficients are very close to 0. For the group GRP1 (resp. GRP2), around 99.02% (resp. 99.00%) of the coefficients have an absolute value smaller than 0.08. The histograms show that, the correlation between the produced sequences is very small. About the correlation analysis using the Hamming distance, the histograms are presented in Fig. 5. The distributions show that, all the proportions are around 50%. For the group GRP1 (resp. GRP2), around 98.45% (resp. 99.98%) of the coefficients belong to $]0.488, 0.512[$. The values for GRP2 are better, due to the entropy of seed values.

Table 2. Results of the NIST tests on the 15000 produced sequences of GRP2. The ratio η (resp. η') of p_{value} passing the tests are given for individual (resp. resulting) sequences and the p_{value} for the concatenated sequence is also given.

Test Name	Indiv. Seq.		Concat. Seq.		Result. Seq.	
	η	Result	p_{value}	Result	η'	Result
Frequency	99.09	Success	0.408718	Success	98.73	Success
Block-Frequency	99.14	Success	0.458897	Success	98.63	Success
Cumulative Sums (1)	99.16	Success	0.239274	Success	98.63	Success
Cumulative Sums (2)	99.00	Success	0.494016	Success	99.02	Success
Runs	98.99	Success	0.025894	Success	98.82	Success
Longest Run	98.92	Success	0.281249	Success	99.12	Success
Rank	99.01	Success	0.806842	Success	99.12	Success
FFT	98.75	Success	0.673608	Success	98.73	Success
Non-Overlapping	99.27	Success	0.012472	Success	98.04	Success
Overlapping	99.02	Success	0.711625	Success	99.12	Success
Universal	-	-	0.149652	Success	98.53	Success
Approximate Entropy	99.02	Success	0.532585	Success	98.33	Success
Random Excursions	96.29	Success	0.060350	Success	98.52	Success
Random Ex-Variant	97.53	Success	0.134550	Success	98.73	Success
Serial (1)	98.92	Success	0.291906	Success	99.60	Success
Serial (2)	99.04	Success	0.196383	Success	99.51	Success
Linear Complexity	98.99	Success	0.215418	Success	99.60	Success

5 Security Analysis

The global security analysis of the generator is carefully conducted. The analysis is based on all the critical points allowing to detect weaknesses in the generator. The investigated points are: the size of key space, key sensitivity, quality of outputs, weak or degenerate keys, speed performance and period length of the logistic map. Even if all the existing attacks can not be tested, the PRBG must resist to some basic-known attacks. In the present case, the resistance to three basic attacks (brute-force attack, differential and guess-and-determine attacks) is discussed.

5.1 Key Space

It is generally accepted that, today a key space of size smaller than 2^{128} is not secure enough. A good PRBG should have a large key space, to have a high diversity of choices for the generation. The proposed generator combines three chaotic logistic maps. A key is then a combination of three initial seeds, used to generate a pseudo-random bit sequence. We have set the conditions for seed selection in Sect. 3.1. The seed X_0 is a binary64 floating-point number selected from the interval $]2^{-53}, 2^{-1}[$. That corresponds to 2^{52} different combinations of mantissa times 51 different values for the exponent, which gives 51×2^{52} different seeds. The seeds Y_0 and Z_0 are selected such that there is a minimum

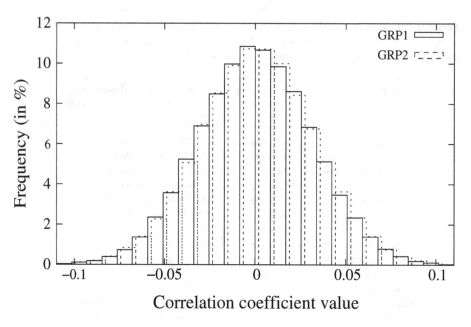

Fig. 4. Histogram of Pearson's correlation coefficient values on interval $[-0.1, 0.1]$ for the group GRP1 (resp. GRP2)

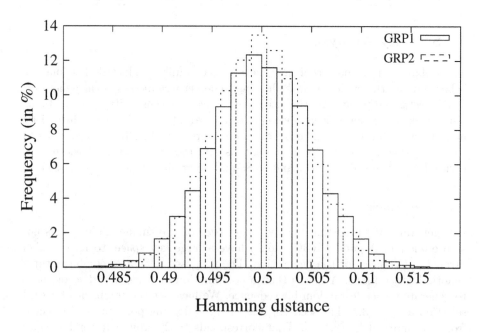

Fig. 5. Histogram of Hamming distance on interval $[0.485, 0.515]$ for the group GRP1 (resp. GRP2)

gap of $2^{-49.82}$ among each seeds. This means that Y_0 is selected in a space of $51 \times 2^{52} - 2^{49.82}$ possible seeds and Z_0 in a space of $51 \times 2^{52} - 2^{50.82}$ different numbers. The total space of seeds is approximately 2^{173}.

5.2   Key Sensitivity

The sensitivity related to the key (i.e. the seeds) is an essential aspect for chaos-based PRBG. Indeed, a small deviation in the starting seeds should cause a large change in the pseudo-random sequences. Actually in the test of correlation (Sect. 4.3), the seed sensitivity was already tested due to the successive seed values. To bring an additional analysis, large pseudo-random sequences of size $N = 5000000$ (i.e. 160000000 random bits) are considered. A sequence S_1 is produced by using the seed values $X_0 = 1 \times 10^{-15}, Y_0 = 2 \times 10^{-15}$ and $Z_0 = 4 \times 10^{-15}$. Two others sequences S_2 and S_3 are produced with $X_0' = X_0, Y_0' = Y_0, Z_0' = 3 \times 10^{-15}$ and $X_0'' = X_0, Y_0'' = Y_0, Z_0'' = 5 \times 10^{-15}$, respectively. The set of the three produced sequences is denoted KS1. The same approach is achieved from another set of sequences denoted KS2. The first sequence is generated with $X_0 = 0.328964524728163, Y_0 = 0.423936234268352$ and $Z_0 = 0.267367904037358$. The two supplementary sequences are obtained by decrementing and incrementing of 10^{-15} the last seed. In both cases, the analysis is done using the linear correlation coefficient of Pearson, the correlation coefficient of Kendall [26] and the Hamming distance. The same analysis is conducted on the sets KS1 and KS2 by using the algorithm proposed by Patidar et al. [2009], with the parameter $\lambda = 3.9999$. As the algorithm uses only two chaotic logistic maps, for each set of sequences only the last two seed values are considered. The results are given in Table 3 and show that, for the proposed algorithm the correlation coefficient values are close to 0 and the proportion of elements that differ in sequences are around 50%. The results show also that, the sequences are highly correlated for the Patidar's algorithm.

Another test of correlation using the randomness of the sequences is achieved. The test is to concatenate the three generated sequences and evaluate the obtained sequence through the NIST tests. The results are presented in Table 4. In each case, all the p_{value} are larger than 0.01 for the current PRBG. Therefore, the concatenated sequence can be viewed as a random sequence, which prove that the sequences S_1, S_2 and S_3 are completely uncorrelated. The results of the Patidar et al. algorithm are added to show that, it is not enough just to combine multiple chaotic logistic maps to build a secure generator.

5.3   Quality of Pseudo-Random Sequences

The strength of any generator is undeniably related to the quality of its outputs. Indeed, whichever way the algorithm is designed, the produced sequences must be strong (i.e. random, uncorrelated and sensitive). In the literature, various statistical tests are available to analyze the randomness of sequences. In fact, the NIST proposes a battery of tests, that must be applied on the binary sequences [24]. One can also find other batteries of tests, such as TestU01 [27] or

Table 3. Pearson's and Kendall's correlation coefficients and Hamming distance (in term of proportion) between large output sequences S_1, S_2, S_3 produced from slightly different initial seeds

PRBG	Set	Test	S_1/S_2	S_1/S_3	S_2/S_3
Proposed algorithm	KS1	Pearson Corr.	−0.000422	−0.000201	0.000127
		Kendall Corr.	−0.000150	−0.000437	−0.000141
		Ham. Dist.	0.499985	0.500064	0.500033
	KS2	Pearson Corr.	−0.000423	0.000235	0.000583
		Kendall Corr.	−0.000116	−0.000025	0.000199
		Ham. Dist.	0.500002	0.500055	0.499931
Patidar et al. algorithm	KS1	Pearson Corr.	0.329043	0.329214	0.329024
		Kendall Corr.	0.233170	0.231704	0.231653
		Ham. Dist.	0.333416	0.333362	0.333366
	KS2	Pearson Corr.	0.329542	0.329417	0.330055
		Kendall Corr.	0.231709	0.232413	0.231693
		Ham. Dist.	0.333284	0.333324	0.333354

the DieHARD suites [28]. Here, the NIST tests are adopted and all the produced sequences passed successfully the tests. The correlation between the outputs is evaluated and the results showed that, only a very small (or negligible) correlation exists between sequences. The proposed PRBG is also very sensitive to starting seeds, even when using slightly different seed values. That shows the quality of the pseudo-random sequences produced by the proposed generator.

5.4 Weak or Degenerate Keys

A crucial element for any PRBG is to ensure that, the output sequences are always produced from strong keys. Here, a careful study of the chaotic regions from the seed space, is necessary for avoiding weak keys. However, the first task is to choose a parameter λ of the logistic map, that contributes to have an excellent chaotic behavior. To avoid similar chaotic trajectories, the seed values must be chosen in $]2^{-53}, 2^{-1}[$, with a representable difference in binary64. The various statistical tests clearly showed the quality of tested sequences, from successive seed values. Thus, these regions are considered as homogeneously chaotic, allowing to choose independently the seed values in $]2^{-53}, 2^{-1}[$. Therefore, the proposed PRBG should not present weak or degenerate keys.

5.5 Speed Analysis

Beyond the randomness aspect, it is also necessary to have a fast generator. Indeed, in real-time applications, the temporal constraint in the execution of a

Table 4. Results of the NIST tests on the concatenated sequence obtained from the sequences of the set KS1 (resp. KS2) for the proposed and Patidar et al. algorithm

Test Name	Proposed algo.		Patidar et al. algo.	
	KS1	KS2	KS1	KS2
	p_{value}	p_{value}	p_{value}	p_{value}
Frequency	0.888272	0.189097	0.218594	0.933359
Block-Frequency	0.013966	0.409511	1.000000	1.000000
Cumulative Sums (1)	0.851269	0.250461	0.343496	0.679001
Cumulative Sums (2)	0.723802	0.375858	0.284913	0.757413
Runs	0.239428	0.196619	0.000000	0.000000
Longest Run	0.341867	0.124501	0.000000	0.000000
Rank	0.690933	0.468857	0.611764	0.788756
FFT	0.704824	0.837336	0.000000	0.000000
Non-Overlapping	0.014372	0.017263	0.000000	0.000000
Overlapping	0.544746	0.513071	0.000000	0.000000
Universal	0.693543	0.467674	0.000000	0.000000
Approximate Entropy	0.534042	0.087565	0.000000	0.000000
Random Excursions	0.016321	0.014831	0.013588	0.000622
Random Ex-Variant	0.014383	0.013285	0.125754	0.038526
Serial (1)	0.532881	0.383964	0.000000	0.000000
Serial (2)	0.508815	0.828262	0.000000	0.000000
Linear Complexity	0.956706	0.189871	0.817657	0.400909

process is as important as the result of the process. Thus, for a fast generator, the domain of its applications can be extended. The speed performance analysis is achieved on a work computer with processor: Intel(R) Xeon(R) CPU E5410 @ 2.33 GHz × 4. The source code is compiled using GCC 4.6.3 on Ubuntu (64 bits). The proposed generator enables to produce around 7 Gbits/s. This is an advantage for applications requiring a good security level and a fast execution time.

5.6 Period Length of the Logistic Map

Here, the period length of the logistic map is discussed. A PRBG should have a reasonably long period before its output sequence repeats itself. The idea is to build the trajectories formed by the different seed values and then compute the lengths of cycles. In a period-p cycle, $X_k = F^p(X_k)$ for some X_k, where F^p is the pth iterate of F. To analyze the evolution of cycles of the logistic map, the mantissa bits are modified by using the GNU MPFR library [29]. Figure 6 shows the curve representing the length of longest (resp. smallest) cycles, when the mantissa bits are varied between 10 and 25. One can see that, the logistic map has very small cycle lengths. For example under binary32 format, the computed longest (resp. smallest) cycle length is equal to 3055 (resp. 1). Such format is not appropriate for generating pseudo-random numbers. To obtain long periods, it is necessary to consider more bits in the mantissa, in other words increase the precision. Thus, by using the binary64 format, the cycle lengths should be much longer. Indeed, the length of the longest cycle is 40037583 ($\approx 2^{25.25}$), while for the smallest cycle is 2169558 ($\approx 2^{21.04}$). In this case, only a given set of randomly chosen seeds is tested due to the large size of the binary format. Approximately we found the same cycle lengths than those given in [30]. For the proposed PRBG, three logistic maps are used during the generation process. In this way, the length of the global resulting cycle is given by the LCM of the three cycle lengths. As one can see, the best way to use this PRBG and then avoid the problem of short period, is to produce sequences of small sizes. However, if needed, long sequences can be obtained by concatenating several ones. In the case of maximum security, it might be better to limit the length of output sequences to the smallest cycle length.

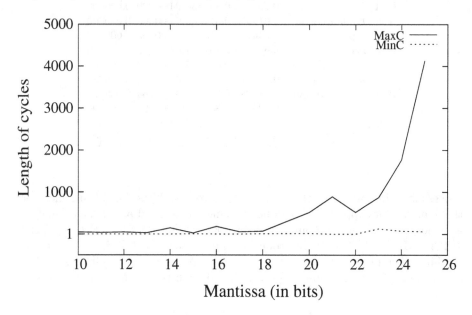

Fig. 6. The curve "MaxC" (resp. "MinC") representing the length of longest (resp. smallest) cycles, when the mantissa bits vary between 10 and 25

5.7   Basic Attacks

Here, the resistance of the generator against three basic attacks, such as brute-force attack, differential and guess-and-determine attacks is discussed.

Brute-force Attack. A brute-force attack [7] is a standard attack that can be used against any PRNG. The strategy consists in checking systematically all possible keys, until the correct key is found. In the worst case, all the combinations are tested, that necessitates to try all the key space. On average, just half of the key space needs to be tested to find the original key. Such an attack might be utilized when it is not possible to detect any weakness in the algorithm, that would make the task easier. To resist this kind of attack, the size of the key space must be large. It is generally accepted that, a key space of size larger than 2^{128} is computationally secure against such attack. In this case, the size of the key space is around 2^{173}, which clearly allows to resist the brute force-attack.

Differential Attack. Such technique of cryptanalysis was introduced by Biham and Shamir [31]. As a chosen-plaintext attack, its principle is to analyze and exploit the effect of a small difference in input pairs on the difference of corresponding output pairs. This strategy allows to find the most probable key that was used to produce the pseudo-random sequence. Given two inputs I and I' (e.g. X_0, Y_0, Z_0 and X_0', Y_0', Z_0') and the corresponding outputs O and O', the most commonly used differences are:

1. Subtraction modulus: the differences related to both inputs and outputs are defined by $\Delta_{in} = |I - I'|$ and $\Delta_{out} = |O - O'|$, respectively. Here, for inputs the difference can be computed between (X_0, X_0'), (Y_0, Y_0') and (Z_0, Z_0') and for outputs, between the bits of pseudo-random sequences.
2. Xor difference: defined by $\Delta_{in} = I \oplus I'$ and $\Delta_{out} = O \oplus O'$.

The diffusion aspect on the initial conditions is then measured by a differential probability. However, in the design of the algorithm, the decorrelation of outputs was taken into account by choosing the seed values in $]2^{-53}, 2^{-1}[$, and by making 30 iterations before starting the generation. Moreover, even with slightly different seeds, the produced sequences are almost independent from each other. Therefore, the proposed PRBG should resist to the differential attack.

Guess-and-determine Attack. Such kind of attack is proven to be effective against word-oriented stream ciphers [32]. As it comes from the name, in guess-and-determine attack, the strategy is to guess firstly the value of few unknown variables of the cipher and then, the remaining unknown variables are deduced by iterating the system a few times and by comparing the output sequence with the original sequence. If these two sequences are the same, then the guessed values are correct and the generator is broken, otherwise the attack should be repeated with new guessed values. It seems that the attack discussed in reference [32] can not be directly applied on the proposed algorithm, which is not of the same family of involved stream ciphers. Indeed, the internal structure of the cipher algorithm is completely different from a Linear Feedback Shift Register (LFSR). Here the algorithm starts with three seed values and generates a 32-bit bloc after each iteration. An alternative way to apply such attack would be to guess and fix the two seed values X_0 and Y_0, then iterate the algorithm to find the seed Z_0. Knowing that the algorithm is very sensitive to starting seeds, one should try in the worst case $2^{57.67}$ different values. Once all the comparisons made without success, the two initial seeds (X_0 and Y_0) are guessed again and the process is repeated in the same way until success. However, this approach has almost the same complexity than a classic brute-force attack.

6   Conclusions

A novel pseudo-random bit generator based on the combination of three chaotic logistic maps was presented. The generator uses the IEEE 754-2008 standard for floating-point arithmetic and especially the binary64 double precision format. For three given initial seeds, the algorithm produces a pseudo-random sequence formed of 32-bit blocks. The main strength of the generator is based on a special mechanism allowing to effectively extract the random bits. Such a generator has shown its ability to produce a very large number of pseudo-random sequences. The advantages of this PRBG are: a high sensitivity related to the initial seed values, a high randomness level of output sequences and the rapidity of the algorithm. The proposed scheme can be considered to be a serious alternative for generating pseudo-random bit sequences.

References

1. Sun, F., Liu, S.: Cryptographic pseudo-random sequence from the spatial chaotic map. Chaos Solit. Fract. 41(5), 2216–2219 (2009)
2. Eichenauer, J., Lehn, J.: A non-linear congruential pseudo random number generator. Statistische Hefte 27(1), 315–326 (1986)
3. Rose, G.: A stream cipher based on linear feedback over $GF(2^8)$. In: Boyd, C., Dawson, E. (eds.) ACISP 1998. LNCS, vol. 1438, pp. 135–146. Springer, Heidelberg (1998)
4. Blum, M., Micali, S.: How to generate cryptographically strong sequences of pseudorandom bits. SIAM J. Comput. 13(4), 850–864 (1984)
5. Blum, L., Blum, M., Shub, M.: A simple unpredictable pseudo-random number generator. SIAM J. Comput. 15(2), 364–383 (1986)
6. Tomassini, M., Sipper, M., Zolla, M., Perrenoud, M.: Generating high-quality random numbers in parallel by cellular automata. Future Gener. Comput. Syst. 16(2), 291–305 (1999)
7. Álvarez, G., Li, S.: Some basic cryptographic requirements for chaos-based cryptosystems. Int. J. Bifurcat. Chaos 16(8), 2129–2151 (2006)
8. Guyeux, C., Wang, Q., Bahi, J.M.: A pseudo random numbers generator based on chaotic iterations: Application to watermarking. In: Wang, F.L., Gong, Z., Luo, X., Lei, J. (eds.) Web Information Systems and Mining. LNCS, vol. 6318, pp. 202–211. Springer, Heidelberg (2010)
9. Zheng, F., Tian, X., Song, J., Li, X.: Pseudo-random sequence generator based on the generalized Henon map. J. China Univ. Posts Telecommun. 15(3), 64–68 (2008)
10. Pareschi, F., Setti, G., Rovatti, R.: A fast chaos-based true random number generator for cryptographic applications. In: Proceedings of the 32nd European Solid-State Circuits Conference, ESSCIRC 2006, pp. 130–133. IEEE (2006)
11. Pareek, N., Patidar, V., Sud, K.: A random bit generator using chaotic maps. Int. J. Netw. Secur. 10(1), 32–38 (2010)
12. Patidar, V., Sud, K., Pareek, N.: A pseudo random bit generator based on chaotic logistic map and its statistical testing. Informatica (Slovenia) 33(4), 441–452 (2009)

13. López, A.B.O., Marañon, G.Á., Estévez, A.G., Dégano, G.P., García, M.R., Vitini, F.M.: Trident, a new pseudo random number generator based on coupled chaotic maps. In: Herrero, Á., Corchado, E., Redondo, C., Alonso, Á. (eds.) Computational Intelligence in Security for Information Systems 2010. AISC, vol. 85, pp. 183–190. Springer, Heidelberg (2010)
14. François, M., Grosges, T., Barchiesi, D., Erra, R.: A new pseudo-random number generator based on two chaotic maps. Informatica 24(2), 181–197 (2013)
15. 754-2008 IEEE standard for floating-point arithmetic. IEEE Computer Society Std (August 2008)
16. Goldberg, D.: What every computer scientist should know about floating-point arithmetic. ACM Comput. Surv. 23(1), 5–48 (1991)
17. Bose, R., Banerjee, A.: Implementing symmetric cryptography using chaos functions. In: Proc. 7th Int. Conf. Advanced Computing and Communications, pp. 318–321. Citeseer (1999)
18. Weisstein, E.: Logistic map (2013),
http://mathworld.wolfram.com/LogisticMap.html
19. Baptista, M.: Cryptography with chaos. Phys. Lett. A 240(1), 50–54 (1998)
20. Cecen, S., Demirer, R., Bayrak, C.: A new hybrid nonlinear congruential number generator based on higher functional power of logistic maps. Chaos Solit. Fract. 42(2), 847–853 (2009)
21. Xuan, L., Zhang, G., Liao, Y.: Chaos-based true random number generator using image. In: 2011 International Conference on Computer Science and Service System (CSSS), pp. 2145–2147. IEEE (2011)
22. François, M., Grosges, T., Barchiesi, D., Erra, R.: Pseudo-random number generator based on mixing of three chaotic maps. Commun. Nonlinear Sci. Numer. Simul. 19(4), 887–895 (2014)
23. Pareek, N., Patidar, V., Sud, K.: Image encryption using chaotic logistic map. Image Vis. Comput. 24(9), 926–934 (2006)
24. Rukhin, A., Soto, J., Nechvatal, J., Smid, M., Barker, E., Leigh, S., Levenson, M., Vangel, M., Banks, D., Heckert, A., Dray, J., Vo, S.: A statistical test suite for random and pseudorandom number generators for cryptographic applications. Technical report, NIST Special Publication Revision 1a (2010)
25. Patidar, V., Pareek, N., Purohit, G., Sud, K.: A robust and secure chaotic standard map based pseudorandom permutation-substitution scheme for image encryption. Opt. Commun. 284(19), 4331–4339 (2011)
26. Kendall, M.: Rank correlation methods, 4th edn. Griffin, London (1970)
27. L'ecuyer, P., Simard, R.: Testu01: A C library for empirical testing of random number generators. ACM Trans. Math. Softw. 33(4), 22–es (2007)
28. Marsaglia, G.: Diehard: a battery of tests of randomness (1996),
http://stat.fsu.edu/geo/diehard.html
29. The GNU MPFR library, http://www.mpfr.org
30. Keller, J., Wiese, H.: Period lengths of chaotic pseudo-random number generators. In: Proceedings of the Fourth IASTED International Conference on Communication, Network and Information Security, CNIS 2007, pp. 7–11 (2007)
31. Biham, E., Shamir, A.: Differential Cryptanalysis of the Data Encryption Standard. Springer, New York (1993)
32. Ahmadi, H., Eghlidos, T.: Heuristic guess-and-determine attacks on stream ciphers. Inf. Secur. IET 3(2), 66–73 (2009)

Set Cover, Set Packing and Hitting Set
for Tree Convex and Tree-Like Set Systems[*]

Min Lu[1], Tian Liu[1,**], Weitian Tong[2], Guohui Lin[2,**], and Ke Xu[3,**]

[1] Key Laboratory of High Confidence Software Technologies, Ministry of Education,
Institute of Software, School of Electronic Engineering and Computer Science,
Peking University, Beijing 100871, China
{lummy,lt}@pku.edu.cn
[2] Department of Computing Science, University of Alberta Edmonton,
Alberta T6G 2E8, Canada
{weitian,guohui}@ualberta.ca
[3] National Lab of Software Development Environment,
Beihang University, Beijing 100191, China
kexu@nlsde.buaa.edu.cn

Abstract. A set system is a collection of subsets of a given finite universe. A *tree convex* set system has a tree defined on the universe, such that each subset in the system induces a subtree. A *circular convex* set system has a circular ordering defined on the universe, such that each subset in the system induces a circular arc. A *tree-like* set system has a tree defined on the system, such that for each element in the universe, all subsets in the system containing this element induce a subtree. A *circular-like* set system has a circular ordering defined on the system, such that for each element in the universe, all subsets in the system containing this element induce a circular arc. In this paper, we restrict the trees to be *stars*, *combs*, *triads*, respectively, and restrict the set system to be *unweighted*. We show tractability of Triad Convex Set Cover, Circular-like Set Packing, and Triad-like Hitting Set, intractability of Comb Convex Set Cover and Comb-like Hitting Set. Our results not only complement the known results in literatures, but also rise interesting questions such as which other kind of trees will lead to tractability or intractability results of Set Cover, Set Packing and Hitting Set for tree convex and tree-like set systems.

Keywords: Tree convex set systems, tree-like set systems, set cover, set packing, hitting set, polynomial time, $\mathcal{NP}$-complete.

1   Introduction

Set Cover (SC), Set Packing (SP) and Hitting Set (HS) are three closely related computational problems. They naturally arise in many areas such as combinatorics,

---

[*] Partially supported by National 973 Program of China (Grant No. 2010CB328103), Natural Science Foundation of China (Grant Nos. 61370052 and 61370156) and NSERC.
[**] Corresponding authors.

T V Gopal et al. (Eds.): TAMC 2014, LNCS 8402, pp. 248–258, 2014.

bioinformatics, and optimization. A set system is a collection of subsets of a given finite universe. Given a finite universe U and a set system S of subsets of U, SC asks for a minimum cardinality subsystem C of S such that each element in U is contained in at least one of the subsets in C. As a dual problem to SC, SP asks for a maximum cardinality subsystem C of S such that any two subsets in C are disjoint. As an equivalent problem to SC, HS asks for a minimum cardinality subset W of U such that each subset in S contains at least one element of W. The equivalence between SC and HS roots at the symmetry between the roles of U and S [11,5]. We consider the *unweighted* version of all these three problems.

On their computational complexity, the decision versions of SC, SP and HS are among Karp's twenty-one classical $\mathcal{NP}$-complete problems [11]. These three problems are also hard to approximate [4] and parameterized intractable [3]. In the literature, most works were done to find ways to solve them efficiently for certain *restricted* set systems.

Definition 1. *A set system S is called* tree convex, *if there is a tree T_U on U, such that each subset Y in S induces a subtree. When the tree T_U is restricted to be a path, a star, a comb, and a triad (see Figure 1) respectively, S is called* convex, star convex, comb convex, *and* triad convex *respectively. A set system S is called* circular convex, *if there is a circular ordering T_U on U, such that each subset Y in S induces a circular arc of T_U.*

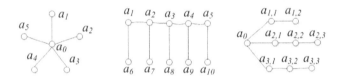

Fig. 1. An illustration of a star, a comb and a triad

Definition 2. *A set system S is called* tree-like, *if there is a tree T_S on S, such that for each elements x in U, all subsets in S containing x induce a subtree of T_S. When the tree T_S is restricted to be a path, a star, a comb, and a triad respectively, S is called* convex-like, star-like, comb-like, *and* triad-like *respectively. A set system S is called* circular-like, *if there is a circular ordering T_S on S, such that for each elements x in U, all subsets in S containing x induce a circular arc of T_S.*

A set system (U, S), where $U = \{x_1, x_2, \ldots, x_{|U|}\}$ and $S = \{Y_1, Y_2, \ldots, Y_{|S|}\}$, can be represented by a matrix $M_{|U| \times |S|}$ with entry $m_{i,j} = 1$ if and only if $x_i \in Y_j$. In this representation, circular convex or convex set systems are also said having *circular ones property for column* (*Circ1P for column*) or *consecutive ones property for column* (*C1P for column*), and circular-like or convex-like set systems are also said having *circular ones property for row* (*Circ1P for row*) or *consecutive ones property for row (C1P for row)*. One may refer to [2] for more details on C1P and Circ1P.

Due to the duality between SC and HS and the duality between tree convex and tree-like, SC on tree convex or tree-like set systems can be cast equivalently as HS on the corresponding tree-like or tree convex set systems; this applies the same to circular convex and circular-like, again due to their duality. On the above introduced restricted set systems, known tractability and intractability results of SC, SP and HS are summarized in Table 1.

Table 1. Complexity results for SC, SP and HS on various restricted set systems, here 'P' refers to polynomial time solvable and 'NPC' refers to $\mathcal{NP}$-complete. The '↓' and '↑' indicate the same complexity results since a path, a star, a comb, and a triad are all special trees. Entries marked with * are obtained in this paper.

Set Systems	SP	SC	HS
Tree convex	P [16]	↑	P [6]
Star convex	↓	NPC [16]	↓
Comb convex	↓	NPC*	↓
Triad convex	↓	P*	↓
Convex (Column-convex, Column C1P)	↓	P[1]	↓
Circular convex (Column-circular, Column Circ1P)	P [1]	P [1]	P [2]
Tree-like	P [7]		
Star-like	↓		
Comb-like	↓		
Triad-like	↓		
Convex-like (Row-convex, Row C1P)	↓		
Circular-like (Row-circular, Row Circ1P)	P*		

Chronically, Trick (1988) showed that SP is tractable on tree convex set systems and SC is intractable on star convex set systems [16], where tree convex set systems are called *induced subtrees of a tree*. Then, Boctor and Renaud (2000) showed that SP and SC are tractable on circular convex set systems under some condition [1], where circular convex set systems are called *column-circular*, and the condition is that the optimal solution is such that there is at least two consecutive elements which are not simultaneously covered by any of the selected subsets. Later, Guo and Niedermeier (2006) proved that SC is tractable on tree-like set systems [6] (equivalently, HS is tractable on tree convex set systems), and Dom (2009) proved that SC is tractable on circular-like set systems [2] (equivalently, HS is tractable on circular convex set systems), where circular-like is called *Circ1P for row*. Most recently, SP is shown tractable on tree-like set systems by Gulek and Toroslu (2010) [7].

In this paper, we complete the above summary table by filling in the last three entries with two tractability and one intractability results. We first prove that SC is tractable on triad convex set systems. Note that this result is incomparable with the result that SC is tractable on circular convex set systems in [1,2], since triad convex set systems and circular convex set systems are not comparable. Secondly, we show that SP is tractable on circular-like set systems. We again remark that this result is incomparable with the result that SP is shown tractable on tree-like set systems in [7], since circular-like set systems and tree-like set systems are incomparable. Lastly, we prove that SC is $\mathcal{NP}$-complete for comb convex set systems. This negative result claims no hope of SC tractability on tree convex set systems even when the tree has a maximum degree as small as three, though SC is tractable on triad convex set systems. A borderline separating tractable and intractable SC and HS, as well as inclusion relationship for various restricted set systems, is shown in Figure 2.

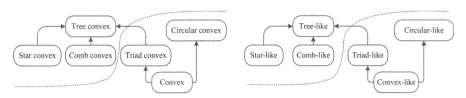

Fig. 2. Various restricted set systems and borderlines separating tractable and intractable SC, and tractable and intractable HS

The main contribution of this paper is to further restrict the trees to be stars, combs and triads respectively and show tractability results for triad and intractability results for stars and combs respectively. These results rise interesting questions such as which kind of trees will lead to tractability or intractability results. In the literature, there are several similar tractability and intractability results for other graph problems, including *feedback vertex Set*, variants of *domination, hamiltonian circuit, hamiltonian path* and *treewidth*, for *tree convex bipartite graphs* and *circular convex bipartite graphs* [8,10,15,18,9,13,12,14,17].

This paper is structured as follows. In Section 2, we recall some necessary definitions and notations mainly for graphs and sets. In Sections 3 and 4, we give an explicit polynomial algorithm for SC on triad convex set systems, and an explicit polynomial algorithm for SP on circular-like set systems, respectively, and prove its correctness. In Section 5, we first present a reduction from the general Set Cover problem to SC on star convex set systems, and then a slightly modified reduction from the general Set Cover problem to SC on comb convex set systems, which shows the $\mathcal{NP}$-completeness of SC on comb convex set systems. Finally are Conclusions.

2   Preliminaries

In this section, we will review some relevant concepts and notations for graphs and sets which will be used later.

A graph G is a tuple (V, E) where V and E are sets of vertices (or nodes) and edges respectively, and each edge is incident with two vertices which are called adjacent to each other. For each vertex v, its neighborhood is defined as

$$N(v) = \{u|u \text{ is adjacent to } v\},$$

and its closed neighborhood is $N[v] = N(v) \cup \{v\}$. For a subset V' of vertices, $N(V') = \bigcup_{v \in V'} N(v)$. A tree, a cycle and a (simple) path are defined as usual. For a subset V' of vertices, the induced subgraph $G[V'] = (V', E')$, where $E' = \{e \in E | e \subseteq V'\}$. A graph (V_1, E_1) is a subgraph of (V, E) if $V_1 \subseteq V$ and $E_1 \subseteq E$. Given a tree, a subtree is defined as a connected subgraph of the tree. A graph on a set S is a graph whose vertex set is exactly S.

The cardinality of a set S, i.e. the number of elements in S, is denoted by $|S|$. The difference of two sets X and Y is denoted by $X \backslash Y = \{x|x \in X \text{ and } x \notin Y\}$. The empty set is denoted by $\emptyset$. When there is an ordering on a set, it is denoted by $\prec$.

A set system $(U, \mathcal{S})$ with $U = \{x_1, x_2, \ldots, x_n\}$, $\mathcal{S} = \{Y_1, Y_2, \ldots, Y_m\}$ can also be represented by a bipartite graph $G = (U, \mathcal{S}, E)$ such that $(x_i, Y_j) \in E$ if and only if $x_i \in Y_j$. We mostly use this representation in the sequel.

3   Triad Convex Set Cover

In this section, we show that SC on triad convex set systems is polynomial time solvable, by reducing it to SC on convex set systems, which is linear time solvable using a greedy method. The greedy method always select the next subset which contains as many as possible uncovered elements and also contains the first uncovered elements under the linear ordering. The reduction is a Cook reduction, that is, a polynomial time Turing reduction.

We give some definitions first. Recall that a triad consists of three paths with a common end. Given an instance $I = (V, \mathcal{S})$ of SC on triad convex set system with a triad on the universe V, we divide V into four parts: V_1, V_2, V_3, and $\{v_0\}$, such that $V_i \cup \{v_0\}$ induces a path of the triad for each i. We assume without loss of generality that

$$V_i = \{v_{i,1}, v_{i,2}, \ldots, v_{i,n_i}\},$$

where $\sum_{i=1}^{3} n_i = |V| - 1$ and

$$v_0 v_{i,1} v_{i,2} \ldots v_{i,n_i}$$

is a path of the triad with the common end v_0. For ease of presentation, we also use $v_{i,0}$ to denote v_0.

We classify the elements in $\mathcal{S}$ into 4 disjoint sets $\mathcal{S}_0, \mathcal{S}_1, \mathcal{S}_2, \mathcal{S}_3$ (see Figure 3), where

$$\mathcal{S}_0 = \{s|s \in \mathcal{S} \text{ and } v_0 \in s\},$$
$$\mathcal{S}_i = \{s|s \in \mathcal{S} \text{ and } s \subseteq V_i\}, \text{ for } i = 1, 2, 3.$$

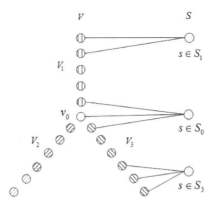

Fig. 3. An instance of SC on triad convex set systems

For each vertex $v_{i,j_i} \in V_i$ with $1 \le j_i \le n_i$, we define $I_{v_{i,j_i}}$ as follows (see Figure 4):

$$I_{v_{i,j_i}} = (V_{v_{i,j_i}}, S_{v_{i,j_i}}), \text{ where}$$
$$V_{v_{i,j_i}} = \{v_{i,j_i+1}, v_{i,j_i+2}, \ldots, v_{i,n_i}\} \text{ and}$$
$$S_{v_{i,j_i}} = \{s \cap V_{v_{i,j_i}} \mid s \in S_i, s \cap V_{v_{i,j_i}} \ne \emptyset\}.$$

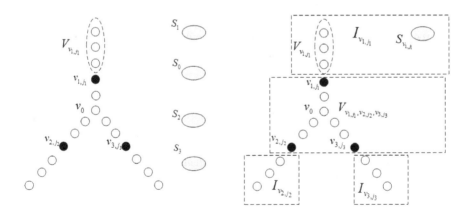

Fig. 4. Decomposition of instance I of SC on triad convex set systems

Lemma 1. $I_{v_{i,j_i}} = (V_{v_{i,j_i}}, S_{v_{i,j_i}})$ *is a convex set system.*

Proof. Since instance I is a triad convex set system, for each $s \in S_i$, the non-empty $s \cap V_{v_{i,j_i}}$ is a path on $V_{v_{i,j_i}}$. Thus, we can define a linear ordering denoted as $\prec_i$ on $V_{v_{i,j_i}}$ as $v_{i,j_i+1} \prec_i v_{i,j_i+2} \prec_i \ldots \prec_i v_{i,n_i}$. In this ordering, each element of $S_{v_{i,j_i}}$ maps to an interval. $\square$

For every triple (j_1, j_2, j_3), we define a set

$$V_{v_{1,j_1}, v_{2,j_2}, v_{3,j_3}} = V \setminus \left(\cup_{i=1}^3 V_{v_{i,j_i}} \right).$$

Lemma 2. *Let C be the minimum set cover of instance I. We have $|C \cap S_0| \leq 3$.*

Proof. Let $C_0 = C \cap S_0$ and $U' = \cup_{s \in C_0} s$. The induced subgraph of the triad on set V by set U' is also a triad. Assume the three paths of this sub-triad end at v_{1,j_1}, v_{2,j_2}, v_{3,j_3}, respectively, other than the common end v_0. Then there are at most three sets s_1, s_2, s_3 in C_0, such that $v_{1,j_1} \in s_1, v_{2,j_2} \in s_2, v_{3,j_3} \in s_3$, respectively, and thus $U' \subseteq \cup_{i=1}^3 s_i$. Since C be the minimum set cover of instance I, we conclude that $|C_0| \leq 3$.   □

Theorem 1. *SC on triad convex set systems is solvable in $O(|V||S|^3)$ time.*

Proof. The algorithm for finding the minimum set cover on a triad convex set system $I = (V, S)$ consists of the following four steps. The first step is to pre-process I to partition S into S_0, S_1, S_2 and S_3 as defined above. In the second step, a subset C_0 of S_0 is selected, which contains at most three elements of S_0; the ends of the (at most) three paths of the sub-triad associated with C_0 are determined, and assume they are v_{1,j_1}, v_{2,j_2}, v_{3,j_3}. In the third step, three convex set systems are generated: $I_{v_{1,j_1}}$, $I_{v_{2,j_2}}$, $I_{v_{3,j_3}}$; and on each the minimum set cover is computed by a greedy method, denoted as C_1, C_2, C_3, respectively. Lastly, $C = \cup_{i=0}^3 C_i$ is a candidate set cover on I; and the algorithm loops through all possible C_0's and return the set cover C of the minimum size.

Lemma 2 tells that it is sufficient to enumerate all subsets of S_0 of size at most three, and Lemma 1 guarantees the minimum set cover on each sub set systems generated. The overall minimum is due to the disjointness of the three subproblems on convex set systems.

The first step of preprocessing takes linear time $O(|V| + |S|)$ time. Upon C_0 is selected, generating the three subproblems on convex set systems and solving them take only $O(|V|)$ time. It follows that looping through to generate all candidate set covers can be done in $O(|V||S|^3)$ time.   □

Corollary 1. *HS on triad-like set systems is solvable in $O(|V|^3|S|)$ time.*

4   Circular-Like Set Packing

In this section, we show that SP on circular-like set systems is polynomial time solvable, by reducing it to SP on convex-like set systems. Given an instance $I = (V, S)$ of SP on circular-like set systems with a circle on $S = \{s_1, s_2, \ldots, s_m\}$ clockwise, we can define a linear ordering $\prec$ on S: $s_1 \prec s_2 \prec \ldots \prec s_m$. We next classify the elements of V into two disjoint sets V_c, V_{nc} (see Figure 5), where the elements in V_c are intervals under this linear ordering $\prec$, and V_{nc} contains all the other elements. Since I is a circular-like set system, all elements in V_{nc} have the form of $\{s_1, \ldots, s_i, s_j, \ldots, s_m\}$ for some $i < j - 1$.

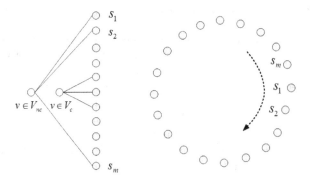

Fig. 5. An instance I of SP on circular-like set systems

For each pair of vertices s_i and s_j with $i < j-1$ in $\mathcal{S}$, we define some notations as follows (see Figure 5):

$$\mathcal{S}_{s_i} = \{s_1, s_2, \ldots, s_i\},$$
$$\mathcal{S}_{s_j} = \{s_j, s_{j+1}, \ldots, s_m\}, \text{ and}$$
$$I_{s_i,s_j} = (V_{s_i,s_j}, \mathcal{S}_{s_i,s_j}) \text{ where}$$
$$\mathcal{S}_{s_i,s_j} = \{s_k | s_k \cap (s_i \cup s_j) = \emptyset, i < k < j\},$$
$$V_{s_i,s_j} = \cup_{s_k \in \mathcal{S}_{s_i,s_j}} s_k.$$

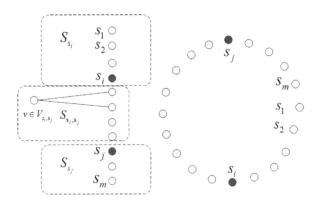

Fig. 6. Decomposition of instance I of SP on circular-like set systems

Lemma 3. *For each pair of vertices s_i and s_j with $i < j - 1$ and $s_i \cap s_j = \emptyset$, I_{s_i,s_j} is a convex-like set system.*

Proof. Since $s_1, s_2, \ldots, s_m$ is a circle on $\mathcal{S}$, for each pair of vertices s_i and s_j with $i < j-1$, every element v of V_{s_i,s_j} is contained in some subset in $\mathcal{S}_{s_i,s_j}$. We define a linear ordering $\prec$ on $\mathcal{S}_{s_i,s_j}$ as $s_{i+1} \prec \ldots \prec s_{j-1}$. This way, all subsets in $\mathcal{S}_{s_i,s_j}$ containing v induce a path under this linear ordering $\prec$. □

Theorem 2. *SP on circular-like set systems is solvable in $O(|\mathcal{S}|^3)$ time.*

Proof. Our algorithm for finding the maximum set packing on a circular-like set system I consists of the following steps. Firstly, we find a pair of vertices s_i and s_j such that $i < j - 1$, $s_i \cap s_j = \emptyset$, and indices i and j cannot be decreased and increased respectively. Such a pair (s_i, s_j) is called *nearest* to pair (s_1, s_m). Note that there are at most $O(|\mathcal{S}|)$ nearest pairs and all of them can be determined in $O(|\mathcal{S}|^2)$ time. Secondly, for each such pair, construct the sub-instance I_{s_i, s_j} as above. Then we compute the maximum set packing C^* for sub-instance I_{s_i, s_j} using the dynamic programming algorithm for SP on convex-like set systems in [7]. $C^* \cup \{s_i, s_j\}$ is a candidate maximum set packing for instance I, conditioning on (s_i, s_j) being a nearest pair. Lastly, let C denote the maximum cardinality set packing among all the candidates, and return it as the final solution.

Since the dynamic programming algorithm for finding the maximum set packing for a convex-like set system runs in linear time. The overall running time of our algorithm is $O(|\mathcal{S}|^3)$. □

5   Comb Convex Set Cover

The general Set Cover problem is $\mathcal{NP}$-complete. The following lemma says that it remains $\mathcal{NP}$-complete on star convex set systems [16].

Lemma 4. *SC on star convex set systems is $\mathcal{NP}$-complete.*

Proof. We reduce the general SC to SC on star convex set systems. Given an instance $I = (V, \mathcal{S})$ of the general SC, where $V = \{v_1, v_2, \ldots, v_n\}$ and $\mathcal{S} = \{s_1, s_2, \ldots, s_m\}$, we construct an instance I' by adding a dummy element v_0 to every set $s_j \in \mathcal{S}$ and regarding v_0 as the center of the star. Clearly, the resultant set system is star convex and I has a set cover of cardinality k if and only if I' has a set cover of cardinality k. Our reduction is done in $O(|\mathcal{S}|)$-time. □

Corollary 2. *HS on star-like set systems is $\mathcal{NP}$-complete.*

With a slight modification, we can transform the star constructed above into a new tree whose maximum degree is at most three. We first split the single center vertex v_0 into n vertices labeled as $w_1, w_2, \ldots, w_n$, respectively. Then we link $w_1, w_2, \ldots, w_n$ to form a path and add an edge between v_i and w_i for all $i = 1, 2, \ldots, n$ (see Figure 7). For each subset $s_j \in \mathcal{S}$, add all $w_i(i = 1, \ldots, n)$ to s_j.

Theorem 3. *SC on comb convex set systems is $\mathcal{NP}$-complete.*

Proof. Given an instance $I = (V, \mathcal{S})$ of the general SC, where $V = \{v_1, v_2, \ldots, v_n\}$ and $\mathcal{S} = \{s_1, s_2, \ldots, s_m\}$, we continue on the above modification that constructs a comb convex set system I'. This reduction is done in $O(|V||\mathcal{S}|)$ time. Again, one clearly see that I has a set cover of cardinality k if and only if I' has a set cover of cardinality k. □

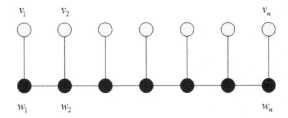

Fig. 7. A tree where the maximum degree is at most three

Corollary 3. *SC on tree convex set systems is $\mathcal{NP}$-complete, even when the tree has a maximum degree of three.*

Corollary 4. *HS on comb-like set systems is $\mathcal{NP}$-complete; HS on tree-like set systems is $\mathcal{NP}$-complete, even when the tree has a maximum degree of three.*

6   Conclusions

In this paper, we have restricted the trees to be *stars*, *combs*, *triads*, respectively, and restricted the set system to be *unweighted*, and shown tractability of Triad Convex Set Cover, Circular-like Set Packing, and Triad-like Hitting Set, intractability of Comb Convex Set Cover and Comb-like Hitting Set. These results not only complement the known results in literatures, but also rise interesting questions such as which other kind of trees will lead to tractability or intractability results of Set Cover, Set Packing and Hitting Set for tree convex and tree-like set systems.

Also recently in the literature, there are several similar tractability and intractability results for other graph problems, including *feedback vertex Set*, variants of *domination, hamiltonian circuit, hamiltonian path* and *treewidth*, for *tree convex bipartite graphs* and *circular convex bipartite graphs* [8,10,15,18,9,13,12,14,17]. The same questions as above on complexity classification of these graph problems for tree convex bipartite graphs based on different kind of restrictions on the trees are also largely open.

Acknowledgments. Min Lu and Tian Liu thank Prof Kaile Su for encouragement and supports. We thank Prof Francis Y.L. Chin for bringing our attention to the notion of circular convex during FAW-AAIM 2011. This work was partially done during a workshop co-organized by Prof Binhai Zhu following COCOON 2013, and we thank the participants and the discussions.

References

1. Boctor, F.F., Renaud, J.: The column-circular, subsets-selection problem: complexity and solutions. Computers & OR 27, 383–398 (2000)
2. Dom, M.: Algorithmic aspects of the consecutive-ones property. Bulletin of the EATCS 98, 27–59 (2009)
3. Downey, R.G., Fellows, M.R.: Parameterized Complexity. Monographs in Computer Science. Springer (1999)
4. Du, D., Ko, K., Hu, X.: Design and Analysis of Approximation Algorithms. Springer (2012)
5. Garey, M.R., Johnson, D.S.: Computers and Intractability, A Guide to the Theory of NP-Completeness. W.H. Freeman and Company (1979)
6. Guo, J., Niedermeier, R.: Exact algorithms and applications for tree-like weighted set cover. J. Discrete Algorithms 4, 608–622 (2006)
7. Gulek, M., Toroslu, I.H.: A dynamic programming algorithm for tree-like weighted set packing problem. Information Sciences 180, 3974–3979 (2010)
8. Jiang, W., Liu, T., Ren, T., Xu, K.: Two hardness results on feedback vertex sets. In: Atallah, M., Li, X.-Y., Zhu, B. (eds.) FAW-AAIM 2011. LNCS, vol. 6681, pp. 233–243. Springer, Heidelberg (2011)
9. Jiang, W., Liu, T., Wang, C., Xu, K.: Feedback vertex sets on restricted bipartite graphs. Theor. Comput. Sci. 507, 41–51 (2013)
10. Jiang, W., Liu, T., Xu, K.: Tractable feedback vertex sets in restricted bipartite graphs. In: Wang, W., Zhu, X., Du, D.-Z. (eds.) COCOA 2011. LNCS, vol. 6831, pp. 424–434. Springer, Heidelberg (2011)
11. Karp, R.: Reducibility among combinatorial problems. In: Complexity of Computer Computations, pp. 85–103. Plenum Press, New York (1972)
12. Lu, M., Liu, T., Xu, K.: Independent Domination: Reductions from Circular- and Triad-Convex Bipartite Graphs to Convex Bipartite Graphs. In: Fellows, M., Tan, X., Zhu, B. (eds.) FAW-AAIM 2013. LNCS, vol. 7924, pp. 142–152. Springer, Heidelberg (2013)
13. Lu, Z., Liu, T., Xu, K.: Tractable Connected Domination for Restricted Bipartite Graphs (Extended Abstract). In: Du, D.-Z., Zhang, G. (eds.) COCOON 2013. LNCS, vol. 7936, pp. 721–728. Springer, Heidelberg (2013)
14. Lu, Z., Lu, M., Liu, T., Xu, K.: Circular convex bipartite graphs: Feedback vertex set. In: Widmayer, P., Xu, Y., Zhu, B. (eds.) COCOA 2013. LNCS, vol. 8287, pp. 272–283. Springer, Heidelberg (2013)
15. Song, Y., Liu, T., Xu, K.: Independent domination on tree convex bipartite graphs. In: Snoeyink, J., Lu, P., Su, K., Wang, L. (eds.) FAW-AAIM 2012. LNCS, vol. 7285, pp. 129–138. Springer, Heidelberg (2012)
16. Trick, M.A.: Induced subtrees of a tree and the set packing problem. IMA Preprint Series, 377 (1988)
17. Wang, C., Chen, H., Lei, Z., Tang, Z., Liu, T., Xu, K.: NP-Completeness of Domination, Hamiltonicity and Treewidth for Restricted Bipartite Graphs (submitted, 2014)
18. Wang, C., Liu, T., Jiang, W., Xu, K.: Feedback vertex sets on tree convex bipartite graphs. In: Lin, G. (ed.) COCOA 2012. LNCS, vol. 7402, pp. 95–102. Springer, Heidelberg (2012)

Efficient Algorithms for the Label Cut Problems

Peng Zhang*

School of Computer Science and Technology,
Shandong University, Jinan 250101, China
algzhang@sdu.edu.cn

Abstract. Given a graph with labels defined on edges and a source-sink pair (s, t), the Label s-t Cut problem asks a minimum number of labels such that the removal of edges with these labels disconnects s and t. Similarly, the Global Label Cut problem asks a minimum number of labels such that its removal disconnects G itself. For these two problems we give some efficient algorithms that are useful in practice. In particular, we give a combinatorial $l_{\max}$-approximation algorithm for the Label s-t Cut problem, where $l_{\max}$ is the maximum s-t length. We show the Global Label Cut problem is polynomial-time solvable in several special cases, including graphs with bounded treewidth, planar graphs, and instances with bounded label frequency.

1 Introduction

The Label s-t Cut problem is an edge-classification minimum s-t cut problem and attracts a good deal of attention from researchers recently (c.f. [1–7]). As introduced in [7], this problem comes from system security, in particular from intrusion detection and from the generation and analysis of attack graphs [4, 5]. In this application, an attack graph describes the attack of an intruder on a system, in which vertices representing various states of the intruder, with a pair of special vertices s and t representing the initial state and the success state of the intruder. A directed edge (u, v) with label ℓ means the intruder's state changes from u to v by carrying out an "atomic attack" named ℓ. Once the intruder arrives at state t, it means the intruder has successfully intruded into the system. To disable an atomic attack incurs some cost. The computational task here is to find a subset of atomic attacks of minimum cardinality (or minimum total cost), such that the removal of all edges labeled by these atomic attacks disconnects s and t. This gives the Label s-t Cut problem.

Definition 1 (The Label s-t Cut problem). *In the problem we are given a (directed or undirected) graph $G = (V, E)$, a source-sink pair (s, t), and a label set $L = \{\ell_1, \ell_2, \cdots, \ell_q\}$. Each edge $e \in E$ has a label $\ell(e) \in L$; many edges may*

* Supported by the State Scholarship Fund of China, Natural Science Foundation of Shandong Province (ZR2012Z002 and ZR2011FM021), and the Independent Innovation Foundation of Shandong University (2012TS072). The work was done when the author was visiting University of California at Riverside, USA.

T V Gopal et al. (Eds.): TAMC 2014, LNCS 8402, pp. 259–270, 2014.

have the same label. A label s-t cut $L' \subseteq L$ is a subset of labels such that the removal of all edges with these labels from G disconnects s and t. The goal of the problem is to find a label s-t cut of the minimum size.

The Label s-t Cut problem is sufficiently natural that it can appear in many contexts. It is easy to see that the Label s-t Cut problem is a generalization of the well-known Min s-t Cut problem, since each edge in the latter problem can be viewed as having a distinct label. By a simple reduction from the Set Cover problem, it is easy to prove that the Label s-t Cut problem is NP-hard even in very restricted graphs [4]. Hence people usually seek approximation algorithms [1, 6, 7] and parameterized algorithms for this problem.

Just like that the Global Min Cut problem is the global version of the Min s-t Cut problem, it is natural to define the Global Label Cut problem.

Definition 2 (The Global Label Cut problem). *In the problem we are given a (directed or undirected) graph $G = (V, E)$ and a label set $L = \{\ell_1, \ell_2, \cdots, \ell_q\}$. Each edge $e \in E$ has a label $\ell(e) \in L$. A global label cut $L' \subseteq L$ is a subset of labels such that G is disconnected after removing all edges with labels in L'. The goal of the problem is to find a global label cut of the minimum size.*

Notations. Given an instance $\mathcal{I}$ of an optimization problem such as Label s-t cut and Global Label Cut, we use $OPT(\mathcal{I})$ (or simply OPT when $\mathcal{I}$ is clearly known from the context) to denote the optimal value of the instance. For the Label s-t Cut problem, its instance is denoted by (G, L, ℓ, s, t); for the Global Label Cut problem, its instance is denoted by (G, L, ℓ). Given an edge subset $E' \subseteq E(G)$, we use $L(E')$ to denote the set $\{\ell(e) \mid e \in E'\}$ of labels of all edges in E'. We note here that we abuse slightly the letter ℓ: When we say a label $\ell \in L$, ℓ is a variable; when we say an edge e has label $\ell(e)$, ℓ is a function mapping edges to labels. We do not introduce more symbols to distinguish these two cases.

1.1 Our Results

In this paper, we identify some parameters of the Label s-t Cut problem and the Global Label Cut problem, and show how they affect the complexity of the problems. In several cases, we get combinatorial approximation algorithms and polynomial time exact algorithms.

First consider the s-t length as a parameter. Let $l_{\max}$ be the maximum s-t length in the Label s-t Cut instance. We give an $l_{\max}$-approximation algorithm for the Label s-t Cut problem (Theorem 2). The algorithm is purely combinatorial, and has a primal-dual explanation. This algorithm is simpler and faster than the $l_{\max}$-approximation algorithm for Label s-t Cut in [6], which first solves a linear program for the problem and then rounds the fractional solution to an integer solution.

We show that when $l_{\max}$ is bounded and the Label s-t Cut problem is parameterized by k, the number of labels in a solution, the problem is fixed parameter tractable in time $O^*(l_{\max}{}^k)$ (Theorem 5). Note that the Label s-t Cut problem is already NP-hard even when $l_{max} = 2$ (Theorem 1).

Next consider as a parameter the label frequency, that is, the number of appearances of a label in the input graph. Let $f_{\max}$ be the maximum label frequency. We show Label s-t Cut can be approximated within a factor of $f_{\max}$ (Theorem 3). When $f_{\max}$ is bounded, Global Label Cut is polynomial time solvable (Theorem 4). In contrast, Label s-t Cut is already NP-hard when $f_{\max} = 2$ (Theorem 1).

The last considered parameter is treewidth of the input graph. We show Global Label Cut is polynomial time solvable with bounded treewidth (Theorem 4). This is contrasted with the NP-hardness of Label s-t Cut even when treewidth is two (Theorem 1).

1.2 Related Work

It is well known that Minimum Set Cover has a greedy polynomial time $(1+\ln n)$-approximation algorithm, where n is the size of the underlying set. The same algorithm can be translated to Hitting Set by duality. In [4], Jha et al. express the Label s-t Cut problem as an *implicit* Minimum Hitting Set problem as follows: Let $U = \{S \mid S$ is the set of labels appearing on an s-t path$\}$; these are the sets to hit. A set of labels is a label cut if and only if it intersects (hits) every $S \in U$. The authors [4] then translate the greedy algorithm for Set Cover to get an approximation algorithm for Label s-t Cut with approximation guarantee of $1 + \ln |U|$. The issue here is that in general U may be of exponential size.

In [7], Zhang et al. present the first (polynomial-time) approximation algorithm for the Label s-t Cut problem. The approximation ratio is $O(m^{\frac{1}{2}})$, where m is the number of edges in the input graph. The authors also show that it is NP-hard to approximate Label s-t Cut within a factor of $2^{(\log |I|)^{1-(\log\log |I|)^{-c}}}$ for any constant $c < 1/2$, where $|I|$ is the input length of the problem. The essentially same approximation hardness result appears independently in [1].

In [6], Tang et al. give the first approximation algorithm for the Label s-t Cut problem whose approximation factor is in terms of n, the number of vertices in the input graph. The approximation factor of the algorithm [6] is $O(n^{\frac{2}{3}}/OPT^{\frac{1}{3}})$. The authors also present an $O((m/OPT)^{\frac{1}{2}})$-approximation algorithm for the Label s-t Cut problem, improving the $O(m^{\frac{1}{2}})$ result in [7]. Moreover, they show that the linear program relaxation for Label s-t Cut, on which they obtain the approximation results, has integrality gap $\Omega((m/OPT)^{\frac{1}{2}-\epsilon})$ for any small constant $\epsilon > 0$.

Fellows et al. [2] consider the Label s-t Cut problem in the view of parameterized complexity. They prove that when parameterized by the number of used labels (that is, $|L'|$ for a label s-t cut $L' \subseteq L$), the Label s-t Cut problem is W[2]-hard in graphs with pathwidth at most 3, and when parameterized by the number of used edges (that is, $|E'|$ for a label s-t cut $L(E')$), the Label s-t Cut problem is W[1]-hard in graphs with pathwidth at most 4. The above results mean that Label s-t Cut in the corresponding graphs is unlikely fixed parameter tractable.

Given a subset $E' \subseteq E$ of edges, define $g(E')$ to be the number of labels appeared in E'. It is easy to see that g is a *submodular* function, that is, g satisfies $\forall X, Y \subseteq E$, $g(X) + g(Y) \geq g(X \cup Y) + g(X \cap Y)$. Jegelka et al. [3] study a more general cut problem called Cooperative s-t Cut, in which the objective function defined on 2^E can be an arbitrary submodular function. The Cooperative s-t Cut problem finds an s-t cut such that the objective function is minimized. Similarly, the (global) Cooperative Cut problem finds a global cut such that the objective function is minimized. Obviously, Label s-t Cut (Global Label Cut, resp.) is a special case of Cooperative s-t Cut (Cooperative Cut, resp.). Jegelka et al. [3] prove the NP-hardness of the Cooperative Cut problem, and an approximation hardness factor of $\Omega(n^{1/3-\epsilon})$ for the Cooperative s-t Cut problem, where n is the number of vertices in the input graph. Note that these two results do not extend to the corresponding label cut problems.

2   Preliminaries

In the following we give the definitions of three parameters we will consider in the paper, i.e., the maximum label frequency $f_{\max}$, the maximum s-t length $l_{\max}$, and the treewidth of an undirected graph.

Definition 3 (Label frequency and $f_{\max}$). *Given a Label s-t Cut instance (G, L, ℓ, s, t) or a Global Label Cut instance (G, L, ℓ), the label frequency $f(\ell)$ of label $\ell \in L$ is the number of edges in G whose label is ℓ. The maximum label frequency $f_{\max}$ of the instance is defined as $\max\{f(\ell) \mid \ell \in L\}$.*

Definition 4 ($l_{\max}$). *Given a graph $G = (V, E)$ with source $s \in V$ and sink $t \in V$, the maximum s-t length $l_{\max}$ is defined as the length of a longest simple s-t path in terms of the number of edges.*

Treewidth is a useful measure of graph and many intractable problems become tractable in bounded treewidth graphs. Some examples of bounded treewidth graphs includes cactus graphs, series-parallel graphs, and outer-planar graphs. Before giving the definition of treewidth, first we give the definition of tree decomposition.

Definition 5 (Tree decomposition). *A tree decomposition of an undirected graph G is a pair $(T, \{V_x : x \in V(T)\})$ in which each vertex x of T corresponds to a vertex subset V_x of G called piece, satisfying the following three properties.*

1. *Every vertex of G belongs to at least one piece.*
2. *Every edge of G has at least one piece containing both its two ends.*
3. *Let x_1, x_2, and x_3 be three vertices of T such that x_2 lies on the path from x_1 to x_3. Then, any vertex belonging to both V_{x_1} and V_{x_3} must also belong to V_{x_2}.*

Definition 6 (Treewidth). *The width of a tree decomposition $(T, \{V_x : x \in V(T)\})$ is defined as $\max_x\{|V_x|\} - 1$. The treewidth $tw(G)$ of a graph G is defined as the minimum width of its any tree decomposition.*

3 Hardness Results

In this section, we show some hardness results of the Label s-t Cut problem in terms of the parameters $f_{\max}$, $l_{\max}$ and treewidth.

Given a graph $G = (V, E)$, the Vertex Cover problem asks for a vertex subset $V' \subseteq V$ of minimum cardinality such that V' touches at least one endpoint of every edge. Denote by $\mathrm{VC}(d)$ the Vertex Cover problem in graphs whose maximum vertex degree is d. It is known that $\mathrm{VC}(3)$ is already NP-hard [8].

Theorem 1. *The Label s-t Cut problem is NP-hard if either*

(i) $l_{\max} = 2$ in both undirected and directed graphs, or
(ii) $f_{\max} = 2$ in both undirected and directed graphs, or
(iii) the treewidth of the input (undirected) graph is 2.

Proof. (i) First we reduce Vertex Cover on undirected graph G to Label s-t Cut on graph G'.

Initially G' contains two vertices s and t. Then, for each edge $e = (v_i, v_j) \in E(G)$, add an s-t path P_{ij} of length 2 to G'. The two edges on path P_{ij} are labeled with v_i and v_j, respectively. See Figure 1 for an illustration. The label set L is just $V(G)$. Then a minimum vertex cover for G corresponds to a minimum label s-t cut for G', and vice versa. Obviously for graph G' we have $l_{\max} = 2$.

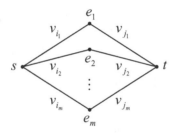

Fig. 1. Reduce Vertex Cover to Label s-t Cut

(ii) Next we reduce MAX SAT to Label s-t Cut to prove the latter problem is NP-hard even if $f_{\max} = 2$. While the reduction is modified from the one in [1] from MAX E3SAT (MAX SAT in which each clause has exactly three literals) to the Label Path problem, the notion of $f_{\max}$ is not mentioned therein.

Let ϕ be a conjunctive normal formula containing n variables $x_1, x_2, \cdots, x_n$ and m clauses $C_1, C_2, \cdots, C_m$, with each clause having exactly three literals. For each variable x_i we construct a variable gadget as shown in Figure 2(a). Suppose x_i appears in n_i clauses $C_{j_1}, C_{j_2}, \cdots, C_{j_{n_i}}$ (including both positive and negative appearances). In the variable gadget we have four groups of labels $\{T_{ij_k}\}$, $\{T'_{ij_k}\}$, $\{F_{ij_k}\}$, and $\{F'_{ij_k}\}$, where T_{ij} (F_{ij}, resp.) means x_i is assigned true (false, resp.) in C_j. For each clause C_j, we construct a clause gadget as shown in Figure 2(b). The gadget contains k_j labels that reflect the appearances

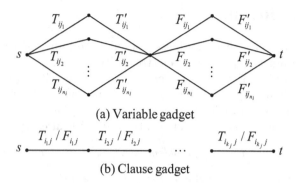

(a) Variable gadget

(b) Clause gadget

Fig. 2. Reduce MAX SAT to Label s-t Cut

of literals in the clause; for example, if $C_j = (x_{i_1} \vee x_{i_2} \vee \neg x_{i_3})$, then the three labels should be $T_{i_1 j}$, $T_{i_2 j}$ and $F_{i_3 j}$, Finally, we merge all vertices s in the gadgets into a single source s, and all vertices t a single sink t, finishing the construction of the graph G' in Label s-t Cut. It is easy to verify that in the resulting instance we have $f_{\max} = 2$.

Let m^* be the optimum of the MAX SAT instance, and q^* the optimum of the Label s-t Cut instance. We claim that for any integer $m' \geq 0$, $m^* \geq m' \Longleftrightarrow q^* \leq \sum_{i=1}^{n} n_i + m - m'$.

($\Longrightarrow$) Suppose ϕ has a truth assignment τ satisfying $\geq m'$ clauses. For each x_i, if x_i is true, then pick all labels of type $T_{i\cdot}$; otherwise pick all labels of type $F_{i\cdot}$. In this way we pick in total $\sum n_i$ labels and disconnect all the variable gadgets. Since τ satisfies $\geq m'$ clauses, by the above picked labels $\geq m'$ clause gadgets have already been disconnected. For each of the remaining connected clause gadgets, pick any label in it to disconnect the gadget. Thus we get a label s-t cut with $\leq \sum n_i + m - m'$ labels.

($\Longleftarrow$) Suppose G' has a label s-t cut L' of size $\leq \sum n_i + m - m'$. Note that all labels in L' appear in the variable gadgets. By replacing all labels of type T'_{ij} with the corresponding labels T_{ij}, and all labels of type F'_{ij} with the corresponding labels F_{ij}, we can assume that L' does not contain labels of type T'_{ij} or F'_{ij}.

By the construction of variable gadget, a variable gadget is disconnected if and only if all its T_{ij} labels, or all its F_{ij} labels, are picked in L'. Let L° be the set of such labels. If for some variable gadget, both of its all T_{ij} labels and its all F_{ij} labels are picked in L', then we arbitrarily pick a type, to say, all its T_{ij} labels into L°.

Since all variables appear $\sum n_i$ times in ϕ, we know $|L^\circ| = \sum n_i$ and hence $|L' - L^\circ| \leq m - m'$. So, $L' - L^\circ$ can only disconnect at most $m - m'$ clause gadgets, and therefore, at least m' clause gadgets are disconnected by L°. This implies that a truth assignment indicated by L° satisfies $\geq m'$ clauses in ϕ.

(iii) In the reductions of (i) and (ii), the resulting graphs are both series-parallel graphs, having treewidth exactly 2. □

Remarks. (a) In the proof of Theorem 1(i), if we use a reduction from VC(3), then we know that Label s-t Cut is NP-hard even in graphs G with $l_{max} = 2$, $f_{max} = 3$, $tw(G) = 2$. In the proof of Theorem 1(ii), if we use a reduction from MAX E3SAT [9, Problem 9.5.5], then we know that Label s-t Cut is NP-hard even in graphs G with $l_{max} = 4$, $f_{max} = 2$, $tw(G) = 2$. In addition, since Label s-t Cut is NP-hard with bounded l_{max}, f_{max} or treewidth, there is no hope to seek parameterized algorithms by these parameters (assuming P $\neq$ NP).

(b) When $f_{max} = 1$, Label s-t Cut degenerates to the well-known Min s-t Cut problem, which can be solved in polynomial time by several methods. When $f_{max} \geq 2$, Label s-t Cut is NP-hard by Theorem 1(ii). So $f_{max} = 2$ just acts as a threshold of the computational complexity of Label s-t Cut.

Corollary 1. *Label s-t Cut is APX-hard even restricted in graphs G with $l_{max} = 2$, $f_{max} = 4$, and $tw(G) = 2$.*

Proof. A series of L-reductions from the MAX 3SAT problem to the VC(4) problem is given in [9]. Since MAX 3SAT is APX-hard [10], this means VC(4) is APX-hard. It is easy to verify the reduction from Vertex Cover to Label s-t Cut in the proof of Theorem 1 is an L-reduction. Using this reduction, VC(4) L-reduces to Label s-t Cut in graphs G with $l_{max} = 2$, $f_{max} = 4$, and $tw(G) = 2$. □

The APX-hardness means Label s-t Cut does not admit PTAS even in graphs G with $l_{max} = 2$, $f_{max} = 4$, and $tw(G) = 2$, if P $\neq$ NP. On the other hand, starting from the UG-hardness of Vertex Cover [11], Theorem 1 already means Label s-t Cut with $l_{max} = 2$ cannot be approximated within $2 - \epsilon$ for any constant $\epsilon > 0$, if the Unique Games Conjecture is true.

4   Approximation Algorithms

In this section, we first develop a polynomial time l_{max}-approximation algorithm for the Label s-t Cut problem, where l_{max} is the length of the longest s-t path. In [6], it is shown that Label s-t Cut can be approximated within a factor of l_{max} via LP-rounding. In contrast, our algorithm is purely combinatorial (does not need to solve linear program) and is faster and simpler than the l_{max}-approximation algorithm in [6].

The approximation algorithm for Label s-t Cut is shown as Algorithm $\mathcal{A}$.

Algorithm $\mathcal{A}$. Find any s-t path P from the current graph. Pick all the labels in $L(P)$ and remove all edges whose labels are in $L(P)$. Repeat the above procedure until s and t are disconnected.

We give two analyses for Algorithm $\mathcal{A}$, with one being purely combinatorial and the other being based primal-dual. Here is the first one.

Theorem 2. *Algorithm $\mathcal{A}$ is an $l_{\max}$-approximation algorithm for the Label s-t Cut problem in both directed and undirected graphs.*

Proof. An s-t path can be easily found in a directed or undirected graph by breadth-first search. So Algorithm $\mathcal{A}$ runs in polynomial time.

Let $P_1, P_2, \cdots, P_h$ be the paths found in the algorithm. The number of labels picked by the algorithm is thus $\leq l_{\max}h$. A crucial observation is that all the $L(P_i)$'s $(1 \leq i \leq h)$ are disjoint. So the optimal solution must pick at least one label from each $L(P_i)$, implying $OPT \geq h$. Therefore, the approximation ratio is $l_{\max}$. The theorem follows. □

The second analysis is a linear programming explanation for Algorithm $\mathcal{A}$. While the algorithm has nothing to do with linear programming in its form, its analysis can be primal-dual. The linear program relaxation for Label s-t Cut and its dual are given below.

$$\min \quad \sum_{\ell \in L} x_\ell \tag{LP}$$

$$\text{s.t.} \quad \sum_{\ell \in L(P)} x_\ell \geq 1, \quad \forall P \in \mathcal{P}_{st} \tag{1}$$

$$x_\ell \geq 0, \qquad \forall \ell \in L$$

$$\max \quad \sum_{P \in \mathcal{P}_{st}} y_P \tag{DP}$$

$$\text{s.t.} \quad \sum_{P \in \mathcal{P}_{st} : \ell \in L(P)} y_P \leq 1, \quad \forall \ell \in L$$

$$y_P \geq 0, \qquad \forall P \in \mathcal{P}_{st}$$

In (LP), for each label ℓ we have a decision variable x_ℓ to indicate whether ℓ is included in the solution. The notation $\mathcal{P}_{st}$ denotes the set of all (simple) s-t paths. Constraint (1) says that for every s-t path P, there is at least one label in $L(P)$ picked in the solution. This implies a feasible solution x to the integer version of (LP) is really a label s-t cut of the input graph. (DP) is just the dual program of (LP).

Proof (the second proof of Theorem 2). Let x, y be the solutions to (LP) and (DP), respectively. Initially, $\forall \ell, x_\ell = 0$ and $\forall P, y_P = 0$. So x is infeasible and y is feasible. When we pick a path P and all its labels in the algorithm, we set $y_P = 1$ and $x_\ell = 1$ for all $\ell \in L(P)$. When the algorithm terminates, there is no s-t path in the current graph. So, x becomes a feasible solution to (LP). Since we pick all labels in $L(P)$ when we find an s-t path, for each label $\ell \in L$ there is only at most one s-t path P whose y_P is one, implying that y remains feasible to (DP) when the algorithm terminates.

Let L' be the set of labels we pick in the algorithm. Recalling that l_{max} is the length of a longest s-t path, it is easy to get

$$|L'| = \sum_{\ell \in L'} x_\ell \le \sum_{P:\, y_P=1} |P| = \sum_P |P| y_P \le l_{max} \sum_P y_P \le l_{max} OPT.$$

$\square$

We end this section with a simple approximation algorithm for Label s-t Cut.

Theorem 3. *Label s-t Cut can be approximated in polynomial time within a factor of f_{max} in both undirected and directed graphs.*

Proof. We just compute a minimum s-t cut E' and use $L(E')$ as the label s-t cut. It is well-known that the undirected minimum s-t cut and directed minimum s-t cut can be computed in polynomial time. Note that a directed s-t cut (A, B) only contains the edges from the s-side (say A) to the t-side (that is, B).

Let L^* be an optimal label s-t cut and E^* be the corresponding edge cut. E^* is obviously a feasible s-t cut and contains at most $f_{max}|L^*|$ edges. So we have

$$|L(E')| \le |E'| \le |E^*| \le f_{max} OPT.$$

$\square$

5   Algorithms for Tractable Cases

5.1   Polynomial Algorithms

We identify some tractable cases for the Global Label Cut problem. Let n be the number of vertices and m the number of edges in a given graph.

Theorem 4. *Global Label Cut is polynomial-time solvable when*

  (i) *the input graph has bounded treewidth, or*
 (ii) *the input graph is planar, or*
(iii) *the problem has bounded f_{max}.*

Proof. (i) Suppose the input graph G has treewidth w. We will show below that there is a global label cut with labels at most w^2. If so, an optimal global label cut can be found by enumerating and testing all $\binom{q}{w^2}$ possible optimal solutions. The time complexity of the above procedure is $O(q^{w^2}(m+n))$, where $O(m+n)$ is the time complexity to test graph connectivity by a graph search algorithm, to say, breadth-first search.

Let T (rooted at any vertex) be a decomposition tree of G whose maximum piece size is $w + 1$. Let x be any leaf vertex of T, whose parent is y. Then removing edge (x, y) breaks T into two connected components $X = \{x\}$ and $Y = V(T) - X$. By [12, Theorem (10.14)], deleting $V_x \cap V_y$ disconnects $V(G)$ into two parts $V_X - (V_x \cap V_y)$ and $V_Y - (V_x \cap V_y)$, so that there is neither common vertex belonging to these two parts, nor edge running between them. That is, $V_x \cap V_y$ is a separator of G. Note that by the property (3) of decomposition tree, $V_x \cap V_y$ cannot be empty.

Thus, by defining $A = V_X - (V_x \cap V_y) = V_x - V_y$ and $B = V(G) - A$, we get a cut (A, B) of G whose cut edges are just the ones connecting $V_x - V_y$ and $V_x \cap V_y$. Since $|V_x| \leq w + 1$, we get that $|(A, B)| \leq w^2$. This means that an optimal global label cut of G contains at most w^2 labels.

(ii) In general, planar graphs do not have bounded treewidth. However, planar graphs possess a good property which says that in each planar graph there is a vertex of degree at most 5. This immediately suggests that the optimum is at most 5 and hence Global Label Cut is polynomial-time solvable in planar graphs.

(iii) Let E^* and L^* be respectively the edge set and the label set in an optimal solution. Since each label associates with at most f_{max} edges, we have $|E^*| \leq f_{max}|L^*|$. Let c^* be the capacity (number of edges) of a minimum cut C^* of the input graph G. Since C^* is a feasible global label cut of G, we have $|L^*| \leq |L(C^*)| \leq c^*$. Together by these two inequalities, we have $|E^*| \leq f_{max}c^*$. This means a minimum global label cut is an f_{max}-approximate cut, where by α-approximate cut we mean a cut whose capacity is at most α times the capacity of a minimum cut.

By [13], there are only $O(n^{2\alpha})$ α-approximate cuts in an undirected graph. By [14], all these α-approximate cuts can be found in $O(m^2n + mn^{2\alpha})$ time. In our setting, $\alpha = f_{max}$ is upper bounded, implying that all f_{max}-approximate cuts can be found in polynomial time. Among these cuts, the one with the minimum labels must be an optimal global label cut. The time complexity of the above algorithm is $O(mn^{2f_{max}})$.   □

Remarks. Theorem 4 says that Global Label Cut is tractable either in graphs with bounded treewidth, or in planar graphs, or in instance with bounded f_{max}. In contrast, by Theorem 1, Label s-t Cut is NP-hard in each of these cases.

5.2   FPT Algorithms

When parameterized on the number of labels in the solution, Label s-t Cut is W[2]-hard [2]. In this section, we show that the problem is FPT if it is with bounded l_{max}.

Theorem 5. *When parameterized on the number of labels in the solution, Label s-t Cut with bounded l_{max} is fixed-parameter tractable in both undirected and directed graphs.*

Proof. Let k be the parameter, that is, the number of labels that can be used in a solution. Then Label s-t Cut with bounded l_{max} can be solved using the standard bounded search tree strategy. We pick any s-t path P from the graph; it has at most l_{max} edges and hence $|L(P)| \leq l_{max}$. We branch at each label $\ell \in L(P)$ by removing all edges whose label is ℓ. The depth of the search tree is at most k, and the number of children of each internal node in the tree is at most l_{max}. This shows that Label s-t Cut parameterized on k can be solved in $O^*(l_{max}^{\ k})$ time.   □

6 Discussions

While some polynomial time tractable cases for Global Label Cut are presented in the paper, the exact computational complexity of the problem is still unknown. Unlike the situation of the Global Min Cut problem and the Min s-t Cut problem—they are both polynomial time solvable, the Global Label Cut problem seems very different from the Label s-t Cut problem.

Recalling from Section 1.2, the objective function g in Global Label Cut is a submodular function on the power set 2^E. On the other hand, if we define a function h on 2^V that counts the number of labels in a cut resulted from a vertex subset A, that is, $h(A) = g(\delta(A))$ where $\delta(A) = (A, V - A)$, then we can easily verify that h is *not* submodular. So it is still unclear whether and how to make use of submodularity in solving Global Label Cut, leaving this as an interesting open problem.

Acknowledgements. We thank an anonymous reviewer for showing us the combinatorial proof of Theorem 2.

References

1. Coudert, D., Datta, P., Perennes, S., Rivano, H., Voge, M.E.: Shared risk resource group: complexity and approximability issues. Parallel Processing Letters 17, 169–184 (2007)
2. Fellows, M., Guo, J., Kanj, I.: The parameterized complexity of some minimum label problems. Journal of Computer and System Sciences 76(8), 727–740 (2010)
3. Jegelka, S., Bilmes, J.: Cooperative cuts: graph cuts with submodular edge weights. Technical Report TR-189 (2010)
4. Jha, S., Sheyner, O., Wing, J.M.: Two formal analyses of attack graphs. In: Proceedings of the 15th IEEE Computer Security Foundations Workshop (CSFW), pp. 49–63. IEEE Computer Society (2002)
5. Sheyner, O., Haines, J., Jha, S., Lippmann, R., Wing, J.: Automated generation and analysis of attack graphs. In: Proceedings of the IEEE Symposium on Security and Privacy, Oakland, CA, pp. 273–284 (2002)
6. Tang, L., Zhang, P.: Approximating minimum label s-t cut via linear programming. In: Fernández-Baca, D. (ed.) LATIN 2012. LNCS, vol. 7256, pp. 655–666. Springer, Heidelberg (2012)
7. Zhang, P., Cai, J.Y., Tang, L., Zhao, W.: Approximation and hardness results for label cut and related problems. Journal of Combinatorial Optimization 21(2), 192–208 (2011)
8. Karp, R.: Reducibility among combinatorial problems. In: Miller, R.E., Thatcher, J.W. (eds.) Complexity of Computer Computations, pp. 85–103. Plenum Press, New York (1972)
9. Papadimitriou, C.: Computational Complexity. Addison-Wesley Publishing Company, Inc. (1994)

10. Papadimitriou, C., Yannakakis, M.: Optimization, approximation, and complexity classes. Journal of Computer and System Sciences 43, 425–440 (1991)
11. Khot, S., Regev, O.: Vertex cover might be hard to approximate to within $2 - \epsilon$. Journal of Computer and System Sciences 74(3), 335–349 (2008)
12. Kleinberg, J., Tardos, E.: Algorithm design. Addison-Wesley (2006)
13. Karger, D., Stein, C.: A new approach to the minimum cut problem. Journal of the ACM 43(4), 601–640 (1996)
14. Nagamochi, H., Nishimura, K., Ibaraki, T.: Computing all small cuts in an undirected network. SIAM Journal on Discrete Mathematics 10(3), 469–481 (1997)

A Dynamic Approach
to Frequent Flyer Program

Rajiv Veeraraghavan, Rakesh Kashyap, Archita Chopde,
and Swapan Bhattacharya

Department of Computer Science
National Institute of Technology Karnataka, Surathkal
{r1rajiv92,rakeshkashyap123,architachopde}@gmail.com,
bswapan2000@yahoo.co.in

Abstract. The frequent flyer algorithms adopted are static in nature, that is the points awarded to a frequent flyer is proportional only to the miles traveled. In static approach, there is neither an incentive for the frequent flyer to travel more (increase profitability) nor does it ensure customer satisfaction. In this paper, we propose a dynamic approach that considers time varying factors such as competition from rival airliners, number of travels made so far, load factor etc and prove that it can not only improve profitability but at the same time ensure customer satisfaction.

Keywords: frequent flyer, dynamic approach, customer satisfaction, profitability, static algorithm, peak seasons, load factor, rival airlines.

1 Introduction

Innovation and strategies play a crucial role in sustaining a profitable company. There is a huge difference between actually lowering the prices and seeming to lower the price. Most retailers play on the psychology of the customer. A perfect example is pricing a commodity at $9.99 instead of $10. According to a 1997 study published in the *Marketing Bulletin*, approximately 60% of prices in advertising material ended in the digit 9![1] These days, concept of giving a bonanza is not only used to attract customers to their company and but also increase their loyalty to that particular brand.

The Airlines industry is one of the pioneers in adopting such strategies[2]. Almost all airlines these days have their own frequent flyer programs. Passengers have to register themselves as a frequent flyer(FF) with a particular airliner. With every flight they take, the airliner awards them some points (also called as miles), which they can redeem for tickets on any of their subsequent travels. The idea is to not only increase the loyalty of the customer to that particular airliner, but also make them travel more. The airliners again play on the psychology of the passenger by appearing to give them more points but on the other hand, they have the luxury of slightly increasing the actual ticket price without the fear of customers changing their loyalties.

T V Gopal et al. (Eds.): TAMC 2014, LNCS 8402, pp. 271–279, 2014.

The algorithm adopted by these airliners to award points is usually static and the concept of dynamism has rarely been explored. At this stage, we would like to cite an example from the Miles and More Program of the Lufthansa airlines[3]. It has about 20 million passengers registered for this program as of Feb 2011. They have normally two types of members, the first are normal members, whose points expire after every 36 months. And the second are higher class members, whose points never expire but the higher class members need to pay a certain amount to register.

Lufthansa and its partner airlines offer points using a static frequent flyer algorithm , ie- 50% * miles flown on economy and 300% * miles flown on business class for intercontinental flights and a fixed 125 and 1500 points for inter-Europe flights. They offer double miles for customers traveling on new routes as promotional offer. To redeem the miles, they have a chart which shows the number of miles required to travel between any 2 airports.

So, as we see it from the outset, they have adopted a program which is neither dependent on the demand-supply nor the time component. In order to enhance profit and at the same time satisfy the customer, it is important to change prices at real time[6]. The price that you pay now also depends on the demand-supply ratio, competition from other brands, time of purchase and a lot of other factors specific to the commodity. The price that X pays for a certain object can be manifold times the price paid by Y for an identical object under different circumstances. The concept of dynamic pricing is typical to the airliner industry. Travellers have come to terms with the fact that the person seated next to him may have paid much lesser for the ticket but still enjoys the same facilities that the airlines has to offer [4]. But most airliners have not embraced dynamism while awarding Frequent Flyer points.

In the airline industry, loyalty or customer satisfaction is defined as the likelihood of a customer becoming a repeat customer and that customer's willingness to behave as a partner to the airline[5]. It is with this intention of building loyalty and attracting new customers, Frequent Flyer Programs were introduced in the 1980s. Since then, FFPs have evolved through time. Morrison and Winston's [7] model of joint airline and route choice using a sample of origin and destination data of individual trips showed that FFPs had a significant effect upon airline and route choice.

Customer satisfaction without compromising on profitability is the most important aspect to be considered while designing a FFP algorithm . Customer satisfaction in the case of FFP can be achieved if the customer perceives to be receiving higher points with each travel. Profitability can be defined in terms of total number of frequent flyer points awarded to the customer. If the customer is awarded less FF points, profitability is more. So in order to improve the profitability, the total frequent flyer points awarded should be less than the static FF algorithm and at the same the time catering to customer satisfaction (giving more points with each travel)

There is another important distinction we need to look at before we propose our approach to the frequent flyer program. This is with regard to the Business travelers and Leisure travelers. According to Basso et al [8], the number of leisure travelers signed up for FFPs is negligible compared to business travelers as leisure travelers pay of their pockets and will certainly opt for the cheapest ticket available whereas for a businessman, it is a third party which is sponsoring the ticker and hence they dont mind paying a little more if they can accumulate more frequent flyer points. So, in our approach we consider only business travelers.

So, in this paper, we try and develop our own FFP which is dynamic in nature. We compare this with the static approach and prove that despite the passengers perceiving to receive more points, the airliners profit more in this case. In section 2, we formally define our problem statement. In subsequent sections 3 and 4, we elucidate our approach and discuss the results we obtain respectively. In section 5, we conclude the paper.

2   Problem Statement

A Dynamic Frequent flyer algorithm is developed which depends on demand-supply and the time factor. Our aim, as stated above, is to maximize the profit of the airliners but at the same time, the passenger should perceive to be receiving more points with each travel. Most frequent flyer algorithms fix a stipulated period of time within which the passenger has to apply for redemption. For the Lufthansas Miles and More Program, the period is 36 months. This is necessary as it will prevent a frequent flyer from accumulating points or miles infinitely and prompts the frequent flyer to redeem the points he has earned as early as possible. This facilitates more travel by the frequent flyer which is always beneficial to the airliner.

Depending on the trips made by the customer, the algorithm awards points to the customer. A database maintaining history of the frequent flyer, information pertaining to every source destination pair and the flight characteristics is used as input to the algorithm. For every source destination pair, parameters such as distance, number of flights in a day and number of flights operating by rival airlines in a day are defined. For every frequent flyer, information about his previous travels date on which he traveled, class of travel (economy, business or first), number of passengers on that flight, flight characteristics (total strength of flight, flight id, class of travels available).

The unique feature of our approach is that we consider the seasonal congestion between a pair of cities. Preceding years data is analyzed to identify the peak seasons where the congestion between a pair of cities is expected to be high i.e most of flights during that period run on full capacity. In such cases, a slightly modified algorithm is adopted to ensure that the airliners profit is maximized. But our algorithm does not take into account unexpected outburst of traffic due to reasons such as natural calamities, sporting events, national conventions etc.

The primary aim of the algorithm is to ensure customer satisfaction. We define a customer to be satisfied solely based on the points he is awarded as the satisfiability due to other benefits such as a separate queue for frequent flyers, allow more baggage weight, give them first preference for choice of seats, discount on other products of the airliner, etc.[3] is not quantifiable. A customer is satisfied if he receives more points with each travel he makes. We also do not emphasis on the other services offered by the airliner like hotels, restaurants, shopping malls, etc. We leave out this part as we want to generalize the algorithm to suit any airliner.

The output of the algorithm will just be the number of points awarded to a frequent flyer after travel. We deliberately mention after travel because we calculate the points after the completion of the journey and not after booking the tickets.At the time of booking, the passengers will be provided with an approximate estimation of the FF points that they will receive. In the course of this paper, we prove that our algorithm is more profitable to the airliners than the static frequent flyer algorithm without compromising customer satisfiability.

3   Dynamic Approach

The static frequent flyer algorithm as in the Lufthansa Mile and More Program depends only on the class of travel and miles traveled by the frequent flyer. The number of frequent flyer points awarded is defined a follows

$$FFP = class(Cl) * distance(D) \tag{1}$$

where distance is the miles traveled and

$$class = \begin{cases} 1 \text{ for economy class} \\ 2 \text{ for business class} \\ 3 \text{ for first class} \end{cases}$$

A good frequent flyer program should not only attract passengers but also refrain the passengers from travelling by rival airliners. Keeping in mind the above Miles and More program of the lufthansa Airlines (Most of the other airlines also follow a similar program), we develop an algorithm incorporating the following dynamic factors to improve profitability. The relationship between FFP and the 2 parameters :{class,distance} has already been established [3] and we incorporate the same.

3.1   Competition from Rival Airlines (C)

We define competition to be the total number of flights from rival airliners operating between a source and destination pair on an average day. We use the term average day to eliminate discrepancies in the number of flights operating between a pair of cities due to inclement weather conditions or occurrence of a special event at the destination. The FFP points awarded is going to be directly proportional to the competition. This is because most of the frequent flyers are registered in more than one frequent flyer program [9] and they always have the option of opting for rival airlines. So, if the number of rival airlines increase, it is wise for the airliner to award more frequent flyer points to attract the customers. An assumption we need to make is that this number is constant for a given source-destination pair and for the stipulated period under consideration.

We define C_{max}, as the maximum number of rival flights that operate between any pair of source and destination. Hence,

$$C = \frac{\text{(Number of rival flights operating between source and dest)}}{C_{max}}.$$

So, C will always be value between 0 and 1.

$$FFP \propto C * Cl * D, \tag{2}$$

where Cl and D are same parameters as the static algorithm

3.2   Frequency (F)

Frequency which is defined as the number of travels made by the frequent flyer so far is going to play a vital role in deciding the FF points. We say that a customer is satisfied if he receives more points compared to his previous travel between a particular source and destination provided he is travelling in the same class. So, the FFP function should be monotonically increasing with frequency and we want the total FF points awarded within the 36 months period to be less than the points awarded by the static algorithm to ensure profitability. In contrast, the FF points awarded after some point in time (preferably after certain number of travels-avg/2) should be more than the points given by the static algorithm in order to satisfy the customer. A monotonically increasing function whose area under the curve is minimum guaranteeing the above condition is a sigmoidal function.

$$FFP \propto \frac{1}{(1 + e^{-f + \frac{avg}{2}})} * C * Cl * D, \tag{3}$$

where avg is the average number of travels made by a frequent flyer. If the number of travels made by the frequent flyer is more than $avg/2$, he will be awarded more points compared to the static algorithm.

$$for f > avg/2, \frac{1}{(1 + e^{-f + \frac{avg}{2}})} > 0.5 \tag{4}$$

3.3   Load Factor (LF)

It is the ratio of the number of seats that are filled up to the total capacity of the flight. We assume that a flight usually operates only when the number of travelers is least 40% of its capacity. Otherwise, the passengers are combined with another flight. We suggest the frequent flyer points awarded to be inversely proportional to the Load factor at the time of travel. To describe in simple words, there is no need for the airliner to give out sops for a flight for which there is already so much of demand. On the other hand, the airliner will not

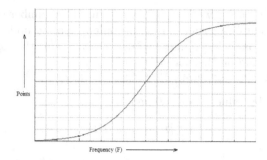

Fig. 1. Relationship between frequent flyer points and frequency

mind giving away more points if it can fill up a flight which usually has lesser demand (Presumably under loss). We determine the value of n later on.

$$FFP \propto C * Cl * D * \left(\frac{1}{(1 + e^{-f + \frac{avg}{2}})} + \frac{1}{LF^n}\right), \tag{5}$$

3.4 Determining the Value of n

The FFP function has to satisfy the condition that if frequency increases for a particular source destination pair, the FFP awarded should increase irrespective of the load factor. Here, we consider a boundary situation where the frequency increase from 1 to 2 (the increase in sigmiodal function is minimum when frequency changes from 1 to 2) and load factor changes from 0.4 to 1. Even for the smallest increase in sigmoidal function and for the largest decrease in load factor, the FF points awarded should increase. This is done by solving the following equation.

$$C * Cl * D * \left[\left(\frac{1}{(1 + e^{-(f+1) + \frac{avg}{2}}} + 1/(1)^n\right) - \left(\frac{1}{(1 + e^{-f + avg/2})} + \frac{1}{(0.4)^n}\right)\right] > 0 \tag{6}$$

Solving the above equation we get n to be 0.1

3.5 Peak Seasons

In case of peak seasons, for a particular source-destination route where the congestion is maximum, a slightly modified approach is followed to enhance the profitability. These peak seasons are identified for every source-destination pair using history data where we plot a graph of load factor versus time . The plateaus (elevated high part of the graph) obtained in this graph are termed as peak seasons. For every source-destination pair, there must ideally exist a peak of season of about 2 weeks in a year.

$$FFP \begin{cases} \propto C * Cl * D * \left(\frac{1}{1 + e^{-f + \frac{avg}{2}}} + \frac{1}{LF^n}\right), iff < avg/2 \\ \propto Cl * D, iff > avg/2 \end{cases}$$

If frequency is greater than avg/2, we are awarding more points compared to static algorithm. In case of peak season, there is no necessity to award more points to attract customers as there is a high demand for ticket and instead we could award the same points as the static algorithm.

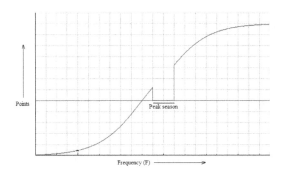

Fig. 2. Modified relationship between frequent flyer points and frequency

3.6   Estimating FFP at Time of Booking

According to our algorithm, the points awarded can be notified to the customer only after the flight takes off. This is because the actual load factor of the flight can be determined only at the time of take-off. Hence, at the time of booking, we can estimate the approximate number of points that the customer can expect to obtain. This should be done in order to ensure that the customer is not left in the dark about the number of points that he shall receive. A parameter called average load factor defined for every source destination pair can be used to determine the number of points that the customer can expect to receive. An approximate estimation of frequent flyer points to be awarded is given as:

$$FFP \propto C * Cl * D * \left(\frac{1}{(1 + e^{-f+avg/2})} + \frac{1}{LF_{avg}^n} \right), \tag{7}$$

We have incorporated the same static redemption algorithms as Lufthansa Miles and more where redeeming points is based on a chart i.e for every source destination pair, fixed number of points is required to fully redeem the ticket. These points as given in the chart are predetermined using the parameters namely class and distance. Since the Dynamic approach awards lesser points in total when compared to the static algorithm, adopting this approach will enhance the profitability to the airliner as the frequent flyer can only redeem fewer travels.

3.7   Algorithm

Input

- For every source destination pair:
  1. Source_destination id
  2. Distance(D)

3. Average Load factor($\dot{LF}_{avg}$)
4. Maximum number of flights operating by rival airlines on a day (C_{max})
- For every flight:
  1. Flight id
  2. Maximum strength (max_strn)
  3. Classes available
- For a frequent flyer, details pertaining to travels are stored in a file
  1. Source_Destination id
  2. Flight id
  3. Number of passengers travelling (num_pass)
  4. Class of travel (Cl)
  5. Number of flights operating by rival airlines on that day (f_comp)

Points Awarded

$$\text{Load Factor(L.F)} = \frac{\text{num\_pass}}{\text{max\_strn}} \quad FFP = C * Cl * D * \left(\frac{1}{1+e^{-f+\frac{avg}{2}}} + \frac{1}{LF^n}\right)$$

At the time of booking, *approximate* frequent flyer points to be awarded

$$FFP = C * Cl * D * \left(\frac{1}{1+e^{-f+\frac{avg}{2}}} + \frac{1}{LF^n_{avg}}\right)$$

4   Experiment

We generated a database of cities, flights and travels made by a frequent flyer. For our experiment, we have considered 25 cities that results in 300 source-destination pairs, 50 flights operating under the airliner and a normal distribution function to determine the number of travels made by each frequent flyer. For each travel made by a frequent flyer, source, destination, num of passengers in the flight and class of travel are generated using a randomize function.

The static algorithm as well as our approach has been applied on the data for 50 different frequent flyers. It is evident from the results that the points awarded to the frequent flyer increases monotonically thus guaranteeing customer satisfaction. Compared to the static algorithm, the total points awarded to the frequent flyer using our approach is less. This ensures an enhancement

Distance	Class	Load factor	Competition
107.415085	1	0.8833333333	0.72
78.006409	1	0.5833333333	0.53
171.210403	2	0.4666666667	0.67
90.077744	2	0.6555555556	0.5
59.615433	3	0.5944444444	0.44
129.757462	2	0.7722222222	0.38
80.324341	1	0.8333333333	0.62
146.239532	3	0.8722222222	0.27
65.863495	3	0.6333333333	0.84
130.015381	3	0.5333333333	0.73
185.564011	2	0.6444444444	0.18

Fig. 3. Part of data pertaining to a frequent flyer detailing his travel history

in the profit as the points available to the frequent flyer to redeem is less and therefore the number of tickets that can be redeemed using the points is going to be less.

For the above generated data the total frequent flyer points awarded by the static approach was 8732, whereas our algorithm awards a total of 7402 showing an improvement in profitability without compromising customer satisfaction.

5 Conclusion

In our study, we have implemented a dynamic algorithm to award frequent flyer points. We have proved using several test cases that our algorithm awards lesser points than the static algorithm, despite ensuring customer satisfiability. Further research would be to consider nonquantifiable benefits offered by the airliners and to check the feasibility of a dynamic redemption algorithm (would it enhance the profitability even more).

References

1. Harris, C., Bray, J.: Price endings and consumer segmentation. Journal of Product and Brand Management 16(3), 200–205 (2007)
2. Luo, L., Peng, J.H.: Dynamic pricing model for airline revenue management under competition. Systems Engineering-Theory and Practice 27(11), 15–25 (2007)
3. Lufthansa Frequent flyer Program: Miles and more (2011)
4. Escobari, D.: Asymmetric Price Adjustments in Airlines. Managerial and Decision Economics (2012)
5. Kivetz, R., Simonson, I.: The idiosyncratic fit heuristic: Effort advantage as a determinant of consumer response to loyalty programs. Journal of Marketing Research, 454–467 (2003)
6. Elmaghraby, W., Keskinocak, P.: Dynamic pricing in the presence of inventory considerations: Research overview, current practices, and future directions. Management Science 49(10), 1287–1309 (2003)
7. Morrison, S., Winston, C.M.: The evolution of the airline industry. Brookings Institution Press (1995)
8. Basso, L.J., Clements, M.T., Ross, T.W.: Moral hazard and customer loyalty programs. American Economic Journal: Microeconomics 1(1), 101–123 (2009)
9. Burgos, A.: Perceived Value: How changes to frequent flyer rules and benefits can influence customer preferences (2011)

A Categorical Treatment of Malicious Behavioral Obfuscation

Romain Péchoux[1] and Thanh Dinh Ta[1,2]

[1] Université de Lorraine - Inria Project Team CARTE, Loria
[2] INRIA Nancy - Grand Est

Abstract. This paper studies malicious behavioral obfuscation through the use of a new abstract model for process and kernel interactions based on monoidal categories. In this model, program observations are considered to be finite lists of system call invocations. In a first step, we show how malicious behaviors can be obfuscated by simulating the observations of benign programs. In a second step, we show how to generate such malicious behaviors through a technique called path replaying and we extend the class of captured malwares by using some algorithmic transformations on morphisms graphical representation. In a last step, we show that all the obfuscated versions we obtained can be used to detect well-known malwares in practice.

Keywords: Behavioral obfuscation, malware detection, monoidal category.

1 Introduction

A traditional technique used by malware writers to bypass malware detectors is program transformation. Basically, the attacker applies some transformations (e.g. useless code injection, change of function call order, code encryption, ...) on a given malware in order to build a new version having the same malicious behavior, i.e. semantically equivalent relatively to a particular formal semantics. This version may bypass a malware detector succeeding in detecting the original malware if the transformation is cleverly chosen. This risk is emphasized in [20] *"an important requirement of a robust malware detection is to handle obfuscation transformation"*.

Currently, the works on *Code obfuscation* have been one of the leading research topic in the field of software protection [8]. By using code obfuscation, malwares can bypass code pattern-based detectors so that the detector's database has to be regularly updated in order to recognize obfuscated variants. As a consequence of Rice's theorem, verifying whether two programs are semantically equivalent is undecidable in general, which annihilates all hopes to write an ideal detector. Consequently, efficient detectors will have to handle code obfuscation in a convenient way while ensuring a good tractability. Most of recent researches have focused on *semantics-based* detection [6,20], where programs are described by abstractions independent from code transformations. Since the semantics of the

T V Gopal et al. (Eds.): TAMC 2014, LNCS 8402, pp. 280–299, 2014.
© Springer International Publishing Switzerland 2014

abstracted variants remains unchanged, the detection becomes more resilient to obfuscation.

In this paper, we will focus on *behavior-based* techniques [11,15] where programs are abstracted in terms of *observable behaviors*, that is interactions with the environment. Beside the works on detection (see [19] for an up-to-date overview), detection bypassing is also discussed academically in [16,23,21,10] and actively (but hardly accessible) in the underground. To our knowledge, there are only a few theoretical works on *behavioral obfuscation*. The lack of formalism and of general methods leads to some risks from the protection point of view: first, malwares deploying new attacks, that is attacks that were not practically handled before, might be omitted by current detectors. Second, the strength of behavior-based techniques might be overestimated, in particular if they have not a good resilience to code obfuscation.

As a illustrating example, consider the following sample, a variant of the trojan `Dropper.Win32.Dorgam` [3] whose malicious behavior consists in three consecutive stages:

- First, as illustrated in the listing of Figure 1, it unpacks two PE files whose paths are added into the registry value `AppInit_DLLs` so that they will be automatically loaded by the malicious codes downloaded later.
- Second, it creates the key `SOFTWARE\AD` and adds some entries as initialized values as illustrated by Figure 2.
- Third, it calls the function `URLDownloadToFile` of Internet Explorer (MSIE) to downloads other malicious codes from some addresses in the stored values.

```
NtCreateFile       (FileHdl=>0x00000734,RootHdl<=0x00000000,File<=\??\C:\WIND
OWS\system32\sys.sys)
NtWriteFile        (FileHdl<=0x00000734,BuffAddr<=0x0043DA2C,ByteNum<=11264)
NtFlushBuffersFile (FileHdl<=0x00000734)
NtClose            (Hdl<=0x00000734)
NtCreateFile       (FileHdl=>0x00000734,RootHdl<=0x00000000,File<=\??\C:\WINDO
WS\system32\intel.dll)
NtWriteFile        (FileHdl<=0x00000734,BuffAddr<=0x0041A22C,ByteNum<=145408)
NtFlushBuffersFile (FileHdl<=0x00000734)
NtClose            (Hdl<=0x00000734)
```

Fig. 1. File unpacking

Since the file unpacking at the first stage is general and the behaviors at the third stage are the same as those of the benign program MSIE, the only way for a behavior-based detector to detect the trojan is by examining its behaviors during the second stage in term of the syscall list[1]:

NtOpenKey, NtSetValueKey, NtClose, NtOpenKey, ...

corresponding to the consecutive syscalls of Figure 2 in order to detect this trojan. However, the `NtOpenKey` syscall associated to each `NtSetValueKey` syscall

---

[1] For readability, we omit the arguments in the syscall lists.

```
NtOpenKey      (KeyHdl=>0x00000730,RootHdl<=0x00000784,Key<=SOFTWARE\AD\)
NtSetValueKey  (KeyHdl<=0x00000730,ValName<=ID,ValType<=REG_SZ,ValEntry<=2062)
NtClose        (Hdl<=0x00000730)
NtOpenKey      (KeyHdl=>0x00000730,RootHdl<=0x00000784,Key<=SOFTWARE\AD\)
NtSetValueKey  (KeyHdl<=0x00000730,ValName<=URL,ValType<=REG_SZ,ValEntry<=
http://ad.***.com:82)
NtClose        (Hdl<=0x00000730)
NtOpenKey      (KeyHdl=>0x00000730,RootHdl<=0x00000784,Key<=SOFTWARE\AD\)
NtSetValueKey  (KeyHdl<=0x00000730,ValName<=UPDATA,ValType<=REG_SZ,ValEntry<=
http://t.***.com:82/***)
NtClose        (Hdl<=0x00000730)
NtOpenKey      (KeyHdl=>0x00000730,RootHdl<=0x00000784,Key<=SOFTWARE\AD\)
NtSetValueKey  (KeyHdl<=0x00000730,ValName<=LOCK,ValType<=REG_SZ,ValEntry<=
http://t.***.com?2062)
NtClose        (Hdl<=0x00000730)
......
```

Fig. 2. Registry initializing

is verbose and can be replaced by a single syscall. Moreover, the key handler can be obtained by duplicating a key handler located in another process, so the call NtOpenKey is not mandatory. Consequently, the following syscall lists are equivalent behaviors that the trojan could arbitrarily select in order to perform its malicious task:

NtOpenKey, NtSetValueKey, NtSetValueKey,...

NtDuplicateObject, NtSetValueKey, NtSetValueKey,...

The remainder of the paper will be devoted to modestly explain how such lists can be both generated and detected for restricted but challenging behaviors. For that purpose, our main contribution is to construct a formal framework explaining how semantics-preserving *behavioral transformations* may evolve. The underlying mathematical abstraction is the notion of monoidal category that we use to model syscall interactions and internal computations of a given process with respect to the kernel. In a first step, it allows us to formally define the behaviors in term of syscall observations. In a second step, it allows us to define a non-trivial subclass of behaviorally obfuscated programs on which detection becomes decidable. In a last step, we show that, apart from purely theoretical results, our model also leads to some encouraging experimental results since the aforementioned decidability result allows us to recognize distinct versions of malwares from the real-world quite efficiently.

This work was inspired by ideas of R. Milner in [18] where the importance of *effects* that influence each participant in interactions is emphasized. In fact, the kernel and process interaction semantics may be thought of as effects that an execution path has on the process and the kernel, respectively. The current work is an application of such ideas to a more specific context.

Outline In Section 2, we introduce a new abstract and simple model based on monoidal categories, that only requires some basic behavioral properties, and introduce the corresponding notions of observable behaviors. We provide several practical examples to illustrate that, though theoretically oriented, this model is very close to practical considerations. In Section 3, we present the main principles of behavioral obfuscation and some semantics-preserving transformations with respect to the introduced model. In Section 4, we introduce a practical implementation of our model and conclude by discussing related works and further developments.

2 Behavior Modeling

We assume the reader to be familiar with category theory (see [4], for an introduction) and, in particular, with the concept of monoidal categories [17] introducing a *tensor product* operator $\otimes$ to represent concurrent computations. As usual, $m, n, \dots$ will denote *objects* and $s, r, \dots$ will denote *morphisms* mapping a *source* object, noted $source(s)$, to a *target* object, noted $target(s) = n$, and will be represented by either $m \xrightarrow{s} n$ or $s : m \longrightarrow n$. Let also 1_m denote the identity morphism, for each object m, and let $\circ$ be the associative *composition*.

Morphism (resp. object) *terms* are terms built from basic morphisms (resp. objects) as variables, composition and tensor product. E.g. $(s_1 \circ s_2) \otimes s_3$ is a morphism term and $m_1 \otimes (m_2 \otimes m_3)$ is an object term.

2.1 Syscall Interaction Modeling

From a practical viewpoint, the computations and interactions between processes and the kernel can be divided in two main sorts, the *system calls* interactions and the process or kernel *internal computations*.

System calls are implemented by the trap mechanism where there is a mandatory control passing from the process (caller) to the kernel (callee). A syscall affects to and is affected by both process and kernel data in their respective memory spaces. Throughout the paper, we will distinguish *syscall names* (e.g. NtCreateFile) from *syscall invocations* (e.g. NtCreateFile(h, ...)). The former are just names while the later compute functions and will be the main concern of our study.

Internal computations are operations inside the process or kernel memory spaces. There is no control passing and they only affect to and are affected by data of the caller memory.

We will abstract this practical viewpoint by a categorical model where computations and interactions (i.e. both syscalls and internal computations) will be represented by morphisms on the appropriate objects. For that purpose, objects will consist in formal representations of physical memories.

Definition 1 (Memory space). *Let Addr be a fixed set of memory adresses. A* memory state *(or value) s is a mapping from a subset of memory addresses $B \subseteq Addr$ to memory bits in $\{0, 1\}$. The domain of s is defined by $dom(s) = B$. A* memory space *m is the set of all memory states corresponding to some fixed domain $B \subseteq Addr$, i.e. $m = \{s \mid dom(s) = B\}$. The domain of m is defined by $dom(m) = B$ and the codomain $codom(m)$ is the set of all binary words of length*

#B. *Given two memory spaces m and n, we write $m \subseteq n$ if $dom(m) \subseteq dom(n)$. In what follows, we will use the notation m^i, $i \in \{k, p\}$, to denote that m is either a kernel or a process memory space. Given two memory spaces m and n of disjoint domains, $m \sqcup n$ denotes their disjoint union.*

We now introduce the notion of *interaction category* in order to abstract syscall invocations and internal computations.

Definition 2 (Interaction category). *Let m^p, m^k be memory spaces satisfying $m^p \cap m^k = \emptyset$. The interaction category $\mathcal{C}\langle m^p, m^k \rangle$ is a category defined by:*

- *The set of objects is freely generated from process and kernel memory spaces n^i such that $n^i \subseteq m^i$, $i \in \{k, p\}$, and cartesian products $n^p \times n^k, \ldots$. The terminal unit object e consists in the empty set.*
- *The set of morphisms is freely generated from cartesian projections: π_i, $i \in \{k, p\}$, process and kernel internal computations: $s^i \colon n^i \to o^i$, $i \in \{k, p\}$, and syscall interactions: $s^{p\text{-}k} \colon n^p \times n^k \to o^p \times o^k$.*
- *The tensor product is partially defined on objects and morphisms by[2]:*
  - *if $n^i \cap o^i = \emptyset$ then $n^i \otimes o^i = n^i \sqcup o^i$,*
  - *if $n^p \otimes o^p$ and $n^k \otimes o^k$ are defined then:*

$$(n^p \times n^k) \otimes (o^p \times o^k) = (n^p \otimes o^p) \times (n^k \otimes o^k),$$

  - *if $n^p \otimes o^p$ or $n^k \otimes o^k$ are defined then:*

$$(n^p \times n^k) \otimes o^p = (n^p \otimes o^p) \times n^k \text{ or}$$
$$(n^p \times n^k) \otimes o^k = n^p \times (n^k \otimes o^k).$$

  - *given $s_1 \colon m_1 \to n_1$ and $s_2 \colon m_2 \to n_2$, then $s_1 \otimes s_2 \colon m_1 \otimes m_2 \to n_1 \otimes n_2$ is defined by $s_1 \otimes s_2(v_1 \otimes v_2) = s_1(v_1) \otimes s_2(v_2)$ whenever the following diagram commutes: (i.e. the tensor is defined on objects):*

$$
\begin{array}{ccc}
m_1 \otimes m_2 & \xrightarrow{\;s_1 \otimes 1_{m_2}\;} & n_1 \otimes m_2 \\
{\scriptstyle 1_{m_1} \otimes s_2}\Big\downarrow & & \Big\downarrow{\scriptstyle 1_{n_1} \otimes s_2} \\
m_1 \otimes n_2 & \xrightarrow{\;s_1 \otimes 1_{n_2}\;} & n_1 \otimes n_2
\end{array}
$$

Remark 1. The set notation $v \in m$ and the categorical notation $v \colon e \to m$ will be interchangeably used depending on the context in order to denote a value (memory state) v of a memory space m. So do the composition notation $s \circ v$ and the application notation $s(v)$ that denote the result of applying morphism s to value v.

---

[2] The tensor represents the concurrent accesses and modifications performed by both internal computations and syscall interactions on memory spaces. A necessary condition for these operations to be well-defined is that they do not interfere, that is they have to operate on disjoint domains.

We show that interaction categories enjoy the mathematical abstractions and properties of monoidal categories:

Proposition 1. *Each interaction category is a monoidal category (with a partially defined tensor product operator).*

A consequence is that all the abstract properties and graphical representations of monoidal categories can be used in the proofs and remainder of this paper.

Graphical representation Morphism and object terms can be given a standard graphical representation using *string diagrams* [22,14] defined as follows:

– nodes are morphisms (except for identity morphisms) and edges are objects:

– composition $s_j \circ s_i$:

– tensor product $s_i \otimes s_j$:

In (planar) monoidal categories, diagrams are *progressive* [22], namely edges are always oriented from left to right. Let $\preceq$ be (by abuse of notation) the reflexive and transitive closure of the relation defined by $s_i \preceq s_j$ holds if there is an edge from s_i to s_j.

Listing 1.1. Internal computation	**Listing 1.2.** Syscall invocation
```char *src = 0x00150500; char *dst = 0x00150770; strncpy(dst,src,10);```	```char *buf = 0x0015C898; HANDLE hdl = 0x00000730; NtWriteFile(hdl,...,buf,1024);```

Example 1. Listing 1.1 is an example of (process) internal computation. The function `strncpy` can be represented by the (process internal computation) morphism:

$$strncpy^p : [src] \otimes [dst] \longrightarrow [src] \otimes [dst],$$

where $[src]$ and $[dst]$ are 10 bytes memory spaces beginning at the addresses 0x150500 and 0x150770, respectively.

Listing 1.2 is an example of syscall invocation. The invocation of the syscall name `NtWriteFile` is represented by a (syscall interaction) morphism:

$$NtWriteFile^{p-k} : [buf] \times [hdl] \longrightarrow [buf] \times [hdl],$$

where $[buf]$ is a 1024 bytes memory space beginning at the address 0x15C898, and $[hdl]$ is a memory state identified by the handler 0x730.

In the interaction category $\mathcal{C}\langle m^p, m^k\rangle$, each internal computation s^p can be considered as a morphism $s^p\colon m^p \longrightarrow m^p$ since internal computations are memory modifiers operating on some previously allocated memory space. In the same spirit, each syscall interaction $s^{p\text{-}k}$ can be seen as a morphism $s^{p\text{-}k}\colon m^p \times n^k \longrightarrow m^p \times o^k$ such that we have either $dom(n^k) = dom(o^k)$ (memory modifier), or $dom(n^k) \subsetneq dom(o^k)$ (memory constructor) or $dom(o^k) \subsetneq dom(n^k)$ (memory destructor).

Example 2. The syscall `NtWriteFile(hdl,...)` of Example 1 is a memory modifier while `NtOpenKey(ph,...)` is a memory constructor, allocating a new memory space identified by `*ph`, and `NtClose(h)` is a memory destructor freeing the memory space identified by `h`.

2.2   Process Behaviors as Path Semantics

We now provide definitions of process behaviors in term of *paths*, namely lists of consecutive morphisms that processes realize during an execution. By assuming that processes can only be examined in finite time, the studied paths are finite.

Definition 3 (Execution path). *An* execution path $X \in \mathcal{X}$ *is a finite list of morphisms of the shape* $X = [s_1^{j_1}, s_2^{j_2}, \ldots]$, *with* $j_i \in \{p, p\text{-}k\}$, $\forall i$, *satisfying the following condition: for each* $s_i^{p\text{-}k} \in X$ *of the shape* $s_i^{p\text{-}k}\colon m^p \times n_i^k \longrightarrow m^p \times o_i^k$:

- *there is no memory duplication: if* $s_i^{p\text{-}k}$ *is a memory constructor then its constructed memory* $T_i^k = o_i^k \setminus n_i^k$ *is not duplicated, that is* $\forall s_j^{p\text{-}k} \in X$, *if* $j > i$ *then* $T_i^k \cap T_j^k = \emptyset$ *else* $T_i^k \cap n_j^k = T_i^k \cap o_j^k = \emptyset$.
- *there is no memory reuse: if* $s_i^{p\text{-}k}$ *is a memory destructor then its destructed memory* $U_i^k = n_i^k \setminus o_i^k$ *is not reused, that is* $\forall s_j^{p\text{-}k} \in X$, *if* $j > i$ *then* $U_i^k \cap n_j^k = U_i^k \cap o_j^k = \emptyset$.

Note that execution paths correspond to paths that are semantically meaningful: The first condition prevents the system from reallocating a memory address and from accessing to an unallocated one, while the second condition prevents it from accessing to a previously freed memory space.

Definition 4 (Observable path). *Given an execution path* X, *its* observable path $O \in \mathcal{O}$ *consists in the list of all syscall interactions in* X. *The function* $obs\colon \mathcal{X} \to \mathcal{O}$ *returns the observable path of an execution path given as input, e.g.* $obs([s_1^{p\text{-}k}, s_2^p, s_3^p, s_4^{p\text{-}k}]) = [s_1^{p\text{-}k}, s_4^{p\text{-}k}]$.

Execution paths will be used to study all the possible computations (internal computations and syscall interactions) at the process level while observable paths only consist in behaviors that can be grasped by an external observer, that is some sequence of syscall invocations, and will be the main concern of our study.

Example 3. Consider the following listings:

Listing 1.3. X_1	Listing 1.4. X_2

```
strncpy(dst,src1,10);          memcpy(src2,src1,1024);
strncpy(dst+10,src1+10,30);    strncpy(dst+2,src2,15);
NtOpenKey(h,...{...dst...});    strncpy(dst+17,src1+15,25);
memcpy(src2,src1,1024);        NtOpenKey(h,...{...dst+2...});
```

The corresponding execution paths X_1, X_2 are defined by:

$$X_1 = [strncpy_1^p, strncpy_2^p, NtOpenKey_3^{p-k}, memcpy_4^p]$$

$$X_2 = [memcpy_1^p, strncpy_2^p, strncpy_3^p, NtOpenKey_4^{p-k}]$$

and their respective observable paths are defined by:

$$obs(X_1) = [NtOpenKey_3^{p-k}]$$

$$obs(X_2) = [NtOpenKey_4^{p-k}]$$

The data modifications caused by an execution path X of a given interaction category can be represented by a morphism term built from the morphisms of X together with identity morphisms. In what follows, the morphism corresponding to these data modifications will be called the **path semantics** of X, denoted $s(X)$.

Proposition 2. *Given an execution path $X \in \mathcal{X}$ of the interaction category $\mathcal{C}\langle m^p, m^k \rangle$, the data modifications caused by X on data at memory adresses $dom(m^p)$ and $dom(m^k)$ is a morphism $s(X)$ of the shape:*

$$s(X): m^p \times n^k \longrightarrow m^p \times o^k, \text{ with } n^i, o^i \subseteq m^i.$$

that can be represented by a morphism term obtained by using the morphisms of X together with identity morphisms.

3   Behavioral Obfuscation

In this section, we show a theorem stating that if a benign path has the same effects on the kernel data as another (malicious) path, then there exist (malicious) paths having the same path semantics as the initial malicious one, and the same observations as the benign one. Though not surprising, this result has two main advantages. First, it gives a first formal treatment of camouflage techniques. Second, the proof of this theorem is constructive. It means that it does not only show the existence of such malicious paths but also allows us to build them in an automated way. This is a very first step towards an automated way of detecting such malicious paths.

3.1   Obfuscation

First, we need to give a clear definition of obfuscation: One of the main obfuscation techniques consists in camouflaging behaviors of malwares with those of a benign programs. Such a technique was partly illustrated by the trojan of our motivating example that was hiding some of its behaviors through the use of Internet Explorer functionalities.

Formally, given two execution paths X_1 and X_2 starting at some value $v_0^p \times v_0^k$ and ending at the same value $v_1^p \times v_1^k = s(X_2)\left(v_0^p \times v_0^k\right) = s(X_1)\left(v_0^p \times v_0^k\right)$, X_2 **obfuscates** X_1 (or $obs(X_2)$ **behaviorally obfuscates** $obs(X_1)$), denoted by $obs(X_2) \approx obs(X_1)$, if $obs(X_1) \neq obs(X_2)$.

Example 4. Consider the paths X_1', X_2' respectively consisting of the 3 last morphisms of X_1, X_2 in Listings 1.3 and 1.4, namely:

$$X_1' = [strncpy_2^p, NtOpenKey_3^{p-k}, memcpy_4^p]$$
$$X_2' = [strncpy_2^p, strncpy_3^p, NtOpenKey_4^{p-k}]$$

In general, $s(X_1') \neq s(X_2')$ but the equality holds if both execution paths start at values so that the data on $[src1] \cup [src2]^3$ are the same. For these particular values, we have:

$$obs(X_1') = \lfloor NtOpenKey_4^{p-k} \rfloor \approx \lfloor NtOpenKey_3^{p-k} \rfloor = obs(X_2')$$

Notice that the two syscall invocations have the same name but actually consist in two different morphisms.

3.2   Camouflage Theorem

In order to state the theorem, we need to introduce the notions of process and kernel (partial) semantics to distinguish the effects caused by a path on a kernel memory space from the ones caused on a process memory space.

Definition 5. *Given an execution path $X \in \mathcal{X}$ and its path semantics $s(X)$: $m^p \times n^k \longrightarrow m^p \times o^k$ wrt the interaction category $\mathcal{C}\langle m^p, m^k \rangle$, the:*

- *kernel semantics, noted $k(X)$, and kernel partial semantics at value v^p, noted $k(X)[v^p]$,*
- *process semantics, noted $p(X)$, and process partial semantics at value v^k, noted $p(X)[v^k]$,*

are defined to be the morphisms making the following diagram commute:

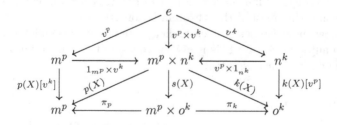

---

3 $[src1]$ and $[src2]$ denote memory spaces as explained in Example 1.

Example 5. The semantics of the path X_1 in Listing 1.3 is a morphism[4]:

$$s(X_1) : ([src1] \cup [src2] \cup [dst]) \times e \rightarrow ([src1] \cup [src2] \cup [dst]) \times [h]$$

and its process and kernel semantics are morphisms:

$$p(X_1) : ([src1] \cup [src2] \cup [dst]) \times e \longrightarrow [src1] \cup [src2] \cup [dst]$$
$$k(X_1) : ([src1] \cup [src2] \cup [dst]) \times e \longrightarrow [h]$$

The following theorem shows that if we first find an intermediate path $X_{1\text{-}2}$ having just the same kernel semantics as X_1 (i.e. the same effects on the kernel memory space), then we can later modify $X_{1\text{-}2}$ (while keeping its observable behaviors) to obtain X_2 having the same paths semantics as X_1.

Theorem 1 (Camouflage). *Let* $X_1 \in \mathcal{X}$ *and* $v^p \times v^k \in source\,(s\,(X_1))$, *for each* $X_{1\text{-}2} \in \mathcal{X}$ *such that* $p(X_{1\text{-}2})[v^k]$ *is monic (i.e. injective) and:*

$$k(X_{1\text{-}2})\left(v^p \times v^k\right) = k(X_1)\left(v^p \times v^k\right),$$

there exists $X_2 \in \mathcal{X}$ *satisfying* $obs\,(X_2) = obs\,(X_{1\text{-}2})$ *and:*

$$s(X_2)\left(v^p \times v^k\right) = s(X_1)\left(v^p \times v^k\right).$$

In other words, if X_1 is a malicious path and $X_{1\text{-}2}$ (possibly benign) has the same kernel semantics, then we can build X_2 so that $s(X_2)\left(v^p \times v^k\right) = s(X_1)\left(v^p \times v^k\right)$, namely X_2 is also malicious; but $obs\,(X_2) = obs\,(X_{1\text{-}2})$, namely it looks like a benign path.

Example 6. The paths X_1 and $X_{1\text{-}2}$ in Listings 1.3 and 1.5 have the same partial kernel semantics at the particular values `"\SYSTEM\CurrentControlSet\..."` $\cup$ $\ldots \cup \ldots$ of $[src1] \cup [src2] \cup [dst]$ but the process partial semantics are not (values on $[src1] \cup [src2] \cup [dst]$ are set to 0 in $X_{1\text{-}2}$). Consequently, it satisfies the hypothesis of Theorem 1 and we can generate a path having the same semantics as the one in Listing 1.3 and the same observations as the one in Listing 1.5.

Listing 1.5. $X_{1\text{-}2}$

```
NtOpenKey(h,..."\SYSTEM\CurrentControlSet\...");
memset(dst,0,1024);
memset(src1,0,1024);
memset(src2,0,1024);
```

3.3 Obfuscated Path Generation

As previously mentioned, the proof of Theorem 1 will allow us to generate paths with camouflaged behaviors through a procedure called *path replaying*. The intuition behind such a procedure is to transform a path X_1 by specializing some invoked values inside the process memory space m^p. For that purpose, the projection morphisms allowing us to extract the partial values of a given total value are introduced:

---

[4] $[src1], [src2], [dst]$ and $[h]$ still denote memory spaces.

Definition 6 (Projection morphisms). *Let* $v \in m_1 \otimes m_2$, *the partial values* $v_1 \in m_1$ *and* $v_2 \in m_2$ *are respectively defined by the morphisms* π_{m_1} *and* π_{m_2} *making the following diagram commute:*

$$
m_1 \xleftarrow{\quad \pi_{m_1} \quad} m_1 \otimes m_2 \xrightarrow{\quad \pi_{m_2} \quad} m_2
$$
$$
m_1 \xleftarrow{v_1} \quad \downarrow{v_1 \otimes v_2} \quad \xrightarrow{v_2} m_2
$$
$$
e
$$

Given an execution path $X = [s_1^{j_1}, s_2^{j_2}, \ldots, s_n^{j_n}]$ and a value $v^p \times v^k \in source$ $(s(X))$; for $1 \leq i \leq n$, define X_l to be the path containing the first l morphisms of X, i.e. $X_l = [s_1^{j_1}, s_2^{j_2}, \ldots, s_l^{j_l}]$. Consider a morphism $s_l^{p\text{-}k} \in obs(X)$, the source value $v_l^p \in m^p$ invoked by $s_l^{p\text{-}k}$ during the execution corresponding to path X can be computed by:

$$
v_l^p = \begin{cases} p(X_{l-1}) \left(\pi_{source(s(X_{l-1}))} \right) \left(v^p \times v^k \right) & \text{if } l > 1 \\ v^p & \text{otherwise} \end{cases}
$$

Definition 7 (Replay path). *Given an execution path* $X = [s_1^{j_1}, s_2^{j_2}, \ldots, s_n^{j_n}]$ *and a value* $v^p \times v^k \in source\,(s\,(X))$, *the* replay path $rep(X) = [r_1, r_2, \ldots, r_n]$ *of* X *at* $v^p \times v^k$ *is defined by:*

$$
r_i = \begin{cases} 1_{m^p} \times k(s_i^{j_i})[v_i^p] & \text{if } s_i^{j_i} \in obs(X) \\ s_i^{j_i} & \text{otherwise} \end{cases}
$$

Example 7. The path in Listing 1.6 is the replay of the path in Listing 1.4 at value "\SYSTEM\CurrentControlSet\..." of $[dst]$.

Listing 1.6. Replay $rep(X_2)$ of X_2

```
memcpy(src2,src1,1024);
strncpy(dst+2,src2,15);
strncpy(dst+17,src1+15,25);
NtOpenKey(h,..."\SYSTEM\CurrentControlSet\..."...);
```

Now we can state the following result:

Proposition 3. *Given an execution path* $X_1 \in \mathcal{X}$, *let* $rep\,(X_1)$ *be the replay of* X_1 *at values* $v^p \times v^k \in source\,(s\,(X_1))$. *For each* $X_{1\text{-}2} \in \mathcal{X}$ *satisfying* $s\,(X_{1\text{-}2}) = s\,(obs\,(rep\,(X_1)))$, $p\,(X_{1\text{-}2})\,[v^k]$ *is monic and the following properties hold:*

- $k(X_{1\text{-}2})\left(v^p \times v^k\right) = k(X_1)\left(v^p \times v^k\right)$,
- *if* $X_2 = X_{1\text{-}2}@[p(X_1)[v^k]]$, *where* @ *is the usual concatenation operator on lists, then:*

$$
s\,(X_2)\left(v^p \times v^k\right) = s\,(X_1)\left(v^p \times v^k\right).
$$

The former shows that $X_{1\text{-}2}$ satisfies the assumptions of Theorem 1 while the later explicitly builds an execution path X_2 such that $obs\,(X_2)$ *behaviorally obfuscates* $obs\,(X_1)$. Indeed, since $rep\,(X_1)$ is constructed out of X_1 by replacing each $s_i^{p\text{-}k} \in obs\,(X_1)$ by $1_{m^p} \times k(s_i^{p\text{-}k})[v_i^p]$, the observable paths $obs\,(rep\,(X_1))$ and $obs\,(X_1)$ have the shapes:

$$obs\,(X_1) = [s_i^{p\text{-}k}, \ldots, s_l^{p\text{-}k}]$$
$$obs\,(rep\,(X_1)) = [1_{m^p} \times k(s_i^{p\text{-}k})[v_i^p], \ldots, 1_{m^p} \times k(s_l^{p\text{-}k})[v_l^p]]$$

Proposition 3 provides a straightforward way of generating an obfuscated path X_2 of X_1 by setting:

$$X_2 = X_{1\text{-}2}@[p(X_1)[v^k]]$$

for some $X_{1\text{-}2}$ such that $X_{1\text{-}2} = obs\,(rep\,(X_1))$. The obtained path X_2 complies with $obs\,(X_2) = obs\,(rep\,(X_1))$ and $obs\,(X_2) \neq obs\,(X_1)$.

3.4 Graph-Based Path Transformation

Though having distinct syscall invocations, the replay paths obtained in previous section are "not that different" in the sense that involved syscall names are still identical (see Example 7). The general objective of this subsection is to show how to generate paths that are semantically equivalent to $obs\,(rep\,(X_1))$ but with distinct observations. For that purpose, the string diagram formalism induced by the considered monoidal categories and introduced at the end of Subsection 2.1 will be used throughout the remainder of this section in order to consider *semantics-preserving transformations* on the syscall invocations in $obs\,(rep\,(X_1))$.

By Proposition 2, the path semantics $s\,(obs\,(rep\,(X_1)))$ is represented by a morphism term constructed from the morphisms $1_{m^p} \times k(s_i^{p\text{-}k})[v_i^p]$. Hence, by Proposition 1, it has a graphical representation as a string diagram, moreover we can safely omit the identity morphism 1_{m^p} in considering these morphisms.

Among string diagrams, we will only consider **path diagrams**, namely the diagrams such that the projection of nodes on an horizontal axis is an injective function, so the projection allows us to define a total order on nodes. The reason for restricting our graphical representation to path diagrams is that they represent their corresponding paths in an unambiguous way.

Example 8. Consider the three following string diagrams: The string diagrams (b) and (c) are path diagrams representing the paths $[s_1, s_2, s_3]$ and $[s_1, s_3, s_2]$, respectively, but the string diagram (a) is not a path diagram.

The following theorem on coherence of progressive plane diagrams, when applied to the corresponding string (or path) diagrams, gives us a sound property on semantics-preserving transformations from one path to another.

Theorem 2 ([14,22]). *In monoidal categories, morphism terms equivalence can be deduced from axioms iff their corresponding string diagrams are planar isotopic.*

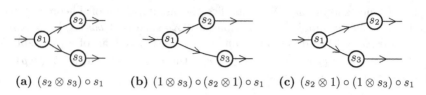

(a) $(s_2 \otimes s_3) \circ s_1$ (b) $(1 \otimes s_3) \circ (s_2 \otimes 1) \circ s_1$ (c) $(s_2 \otimes 1) \circ (1 \otimes s_3) \circ s_1$

Fig. 3. String diagrams

Following [22], we accept an informal definition of *planar isotopy* between string diagrams as "...one can be transformed to the other by continuously moving around nodes..." (but keep the diagram always progressive), the formal treatment can be referenced in [14], e.g. The three string diagrams of Example 8 are planar isotopic.

Between path diagrams, planar isotopy can be though of as moving the nodes but keeping the total oder compatible with the partial order $\preccurlyeq$ (see Section 2.1). Hence, a *linear extension* Y of $obs\,(rep\,(X_1))$, namely a permutation where the total order remains to be compatible with $\preccurlyeq$, will preserves the semantics of $obs\,(rep\,(X_1))$. This leads to the following Algorithm:

Input: an observable path $obs\,(rep\,(X_1))$
Output: a permutation Y satisfying $s\,(Y) = s\,(obs\,(rep\,(X_1)))$

begin
    $M_1 \leftarrow$ a morphism term of $s\,(obs\,(rep\,(X_1)))$;
    $G_1 \leftarrow$ a string diagram of M_1;
    $(obs\,(rep\,(X_1))\,,\preccurlyeq) \leftarrow$ a poset with order induced from G_1;
    $(Y, \leq) \leftarrow$ a linear extension of $(obs\,(rep\,(X_1))\,,\preccurlyeq)$;
end

Algorithm 1. Obfuscation by diagram deformation

Example 9. Consider the below listings corresponding to execution paths X_3 and X_4. They can be respectively represented by path diagrams (b) and (c). Consequently, given X_3, Algorithm 1 can generate X_4 (or the converse).

Listing 1.7. X_3

```
NtCreateKey(h,...{..."\SOFTWARE\AD\"...}...); /*s1*/
NtSetValueKey(h,...{..."DOWNLOAD"...}...,"abc"); /*s2*/
NtSetValueKey(h,...{..."URL"...}...,"xyz"); /*s3*/
```

Listing 1.8. X_4

```
NtCreateKey(h,...{..."\SOFTWARE\AD\"...}...); /*s1*/
NtSetValueKey(h,...{..."URL"...}...,"xyz"); /*s3*/
NtSetValueKey(h,...{..."DOWNLOAD"...}...,"abc"); /*s2*/
```

A variable in a morphism term (resp. a node in the string diagram) is also a *placeholder* [22] that can be substituted by another term (resp. another diagram) having the same semantics, so the below Algorithm can be derived from Algorithm 1:

Input: an observable path $obs\,(rep\,(X_1))$
Output: a new path Y satisfying $s\,(Y) = s\,(obs\,(rep\,(X_1)))$

begin
    $M_1 \leftarrow$ a morphism term of $obs\,(rep\,(X_1))$;
    $s \leftarrow$ a morphism of M_1;
    $X \leftarrow$ an execution path satisfying $s(X) = s$;
    $M \leftarrow$ a morphism term of X;
    $M_2 \leftarrow$ the morphism term $M_1\{M/s\}$;
    $G_2 \leftarrow$ a string diagram of M_2;
    $((obs\,(rep\,(X_1))) \setminus s) \cup X, \preccurlyeq) \leftarrow$ poset with order induced from G_2;
    $(Y, \leq) \leftarrow$ a linear extension of $((obs\,(rep\,(X_1))) \setminus s) \cup X, \preccurlyeq)$
end

Algorithm 2. Obfuscation by node replacement

Example 10. Consider the replacement of s_2 in Listing 1.7 by $X = [s_2', s_2]$ where:

$$s_2' = \texttt{NtSetValueKey}(h, \ldots \{\ldots \text{"DOWNLOAD"} \ldots\} \ldots, \text{"}a'b'c'\text{"})$$

Using this replacement, given the execution path X_3 in Listing 1.7, Algorithm 2 can generate the execution path X_5 corresponding to the following listing:

Listing 1.9. X_5

```
NtCreateKey(h,...{..."\SOFTWARE\AD\"...}...);  /*s1*/
NtSetValueKey(h,...{..."DOWNLOAD"}...,"a'b'c'");/*s'2*/
NtSetValueKey(h,...{..."URL"...}...,"xyz");  /*s3*/
NtSetValueKey(h,...{..."DOWNLOAD"}...,"abc");  /*s2*/
```

Proposition 4 (Soundness). *Given an execution path X_1, Algorithms 1 and 2 generate an observable path Y as output that behaviorally obfuscates $obs(rep(X_1))$ (provided that considered the linear extension is not the identity).*

4   Experiments and Detection

4.1   Experimental Implementation

Algorithms 1 and 2 have been applied to several sub-paths extracted from the malwares `Dropper.Win32.Dorgam` [3] and `Gen:Variant.Barys.159` [1]. The programs (written in C++ and Haskell) use Pin [13] for path tracing and FGL [9] for path transforming. The implemented pieces of code are available at the repository [2].

In the following experiments, the string diagrams of paths are illustrated as follows: The numbers appearing as node labels represent the total order in the original path. In each diagram, the fictitious nodes Input and Output are added as the *minimum* and the *maximum* in such a way that the path can be considered as a lattice. On a fixed line, the number appearing as edge labels represent the handlers which identify the corresponding memory space inside kernel. On different lines, the same numbers may identify different memory spaces. The obfuscated paths generated by Algorithm 1 are linear extensions which are compatible with the order defined in the lattice. Note that their corresponding diagrams are always path diagrams (but they are not illustrated here).

Experiment 1. The trojan Dropper.Win32.Dorgam has an execution path X corresponding to the trace in Figure 2 that consists in 24 morphims. Let $[h_i], i \in \{1, 4, 7, 10, 13, 16, 19, 22\}$ denote the memory spaces identified by the handlers of the accessed registry keys (i being the listing line number), the replay path is formulated by morphisms:

$$NtOpenKey_i^{p\text{-}k} : e \to [h_i]$$
$$NtSetValueKey_{i+1}^{p\text{-}k} : [h_i] \to [h_i]$$
$$NtClose_{i+2}^{p\text{-}k} : [h_i] \to e$$

Its string diagram is represented in Figure 4.

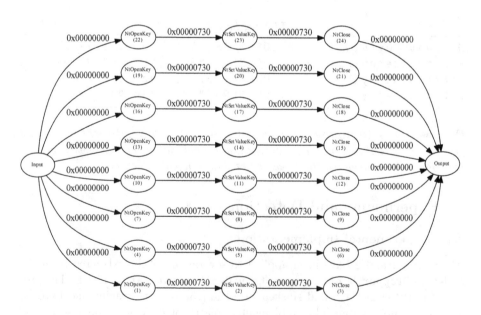

Fig. 4. Registry initializing string diagram

The number $e(X)$ of possible linear extensions of X is computed by:

$$e(X) = \binom{24}{3}\binom{21}{3}\binom{18}{3}\binom{15}{3}\binom{12}{3}\binom{9}{3}\binom{6}{3}\binom{3}{3} = \frac{24!}{6^8}$$

namely more than 369 quadrillion extensions (and paths) can be generated by Algorithm 1.

Experiment 2. The trojan also uses the trace in Listing 1.10 to create a copy of `iexplore.exe`, its replay path has the string diagram provided in Figure 5(a).

Listing 1.10. File copying

```
NtCreateFile (FileHdl=>0x00000730,File<=\??\C:\Program Files\
Internet Explorer\IEXPLORE.EXE)
NtCreateFile (FileHdl=>0x0000072C,File<=\??\C:\Program
Files\iexplore.exe)
NtReadFile   (FileHdl<=0x00000730,BuffAddr<=0x0015C898, ByteNum<=65536)
NtWriteFile  (FileHdl<=0x0000072C, BuffAddr<=0x0015C898, ByteNum<=65536)
......
NtReadFile   (FileHdl<=0x00000730,BuffAddr<=0x0015C898, ByteNum<=65536)
NtWriteFile  (FileHdl<=0x0000072C, BuffAddr<=0x0015C898, ByteNum<=48992)
NtReadFile   (FileHdl<=0x00000730, BuffAddr<=0x0015C898, ByteNum<=65536)
NtClose      (Hdl<=0x00000730)
NtClose      (Hdl<=0x0000072C)
```

This path can be considered as an obfuscated version (generated by Algorithm 2) of the path whose string diagram is in Figure 5(b), by considering the equivalences:

$$NtReadFile^{p\text{-}k}_{3(orig)} = [NtReadFile^{p\text{-}k}_3, NtReadFile^{p\text{-}k}_5, \dots]$$

$$NtWriteFile^{p\text{-}k}_{4(orig)} = [NtWriteFile^{p\text{-}k}_4, NtWriteFile^{p\text{-}k}_6, \dots]$$

It also means that a *behavior matching* detector can detect an obfuscated path, assuming the prior knowledge of both the original path and the semantics equivalences described above.

Experiment 3 We consider the ransomware `Gen:Variant.Barys.159` [1]. The extracted path in Listing 1.11 explains how the malware conceals itself by injecting code into file explorer process `explorer.exe`.

Listing 1.11. Code injecting

```
NtOpenProcess (ProcHdl=>0x00000780,DesiredAccess<=1080,ProcId<=0x00000240)
NtCreateSection (SecHdl=>0x00000778 ,AllocAttrs <=SEC_COMMIT ,FileHdl<=0x00000000)
NtMapViewOfSection (SecHdl <=0x00000778 ,ProcHdl <=0xFFFFFFFF ,BaseAddr <=0x02660000)
NtReadVirtualMemory (ProcHdl <=0x00000780 ,BaseAddr <=0x7C900000 ,BuffAddr <=0x026
60000 ,ByteNum <=729088)
NtMapViewOfSection (SecHdl <=0x00000778 ,ProcHdl <=0x00000780 ,BaseAddr <=0x7C900000)
```

The malware first obtains the handler `0x780` from the running instance (whose process id is `0x240`) of `explorer.exe` and then creates a section object identified by the handler `0x778`. It maps this section to the malware memory, it copies some data of the instance into the mapped memory, it performs data modification on this memory and, finally, it maps the section (now contains modified data) back to the instance.

Let $[h_1], [h_2]$ denote the memories identified by handlers of the opened process and of the created section, the replay path is formulated by morphisms:

$$NtOpenProcess_1^{p\text{-}k} : e \rightarrow [h_1]$$

$$NtCreateSection_2^{p\text{-}k} : e \rightarrow [h_2]$$

$$NtMapViewOfSection_3^{p\text{-}k} : [h_2] \rightarrow [h_2]$$

$$NtReadVirtualMemory_4^{p\text{-}k} : [h_1] \rightarrow [h_1]$$

$$NtMapViewOfSection_5^{p\text{-}k} : [h_1] \cup [h_2] \rightarrow [h_1] \cup [h_2]$$

and the corresponding string diagram is provided in Figure 5(c).

If the morphism $NtReadVirtualMemory_4^{p\text{-}k}$ is replaced by a path $[s_{4\text{-}1}^{p\text{-}k}, s_{4\text{-}2}^{p\text{-}k}]$ corresponding to the syscall invocations in Listing 1.12 then this replacement leads to the string diagram in Figure 5(d).

Listing 1.12. NtReadVirtualMemory

```
NtReadVirtualMemory(ProcHdl<=0x00000780,BaseAddr<=0x7C900000,BuffAddr<=0x026
60000,ByteNum<=9088)
NtReadVirtualMemory(ProcHdl<=0x00000780,BaseAddr<=0x7C909088,BuffAddr<=0x026
69088,ByteNum<=72000)
```

The numbers of linear extensions for the original string diagram and the obfuscated one are respectively:

$$e_1(X) = \binom{4}{2}\binom{2}{2} = 6 \qquad e_2(X) = \binom{5}{3}\binom{2}{2} = 10$$

4.2   Obfuscated Path Detection

We will now discuss the detection by using practical detectors introduced in previous existing works on *behavior matching* [7,15,12].

Basically, a behavior matching detector first represents an observable path by a directed acyclic graph (DAG) using the causal dependency between morphisms, a morphism s_j (directly or indirectly) depends on s_i if the sources values of s_j are (directly or indirectly) deduced from the target values of s_i. Then the detector decides a path is malicious or not by verifying whether there exists a malicious pattern occurring as a subgraph of the original DAG. Here the malicious pattern is a (sub-)DAG and it can be obfuscated to another semantics equivalent DAG.

Whereas Algorithm 1 can generate a large amount of paths, the verification of whether an obfuscated path is semantically equivalent to the original path is simple: it is an instance of the *DAG automorphism* problem where every vertex is mapped to itself. The instance can be decided in *P-time* by repeatedly verifying whether two paths have the same set of minimal elements, if they do then remove the set of minimal elements from both paths and repeat; if they do not then decide **No**; if the sets are both empty then decide **Yes** and stop.

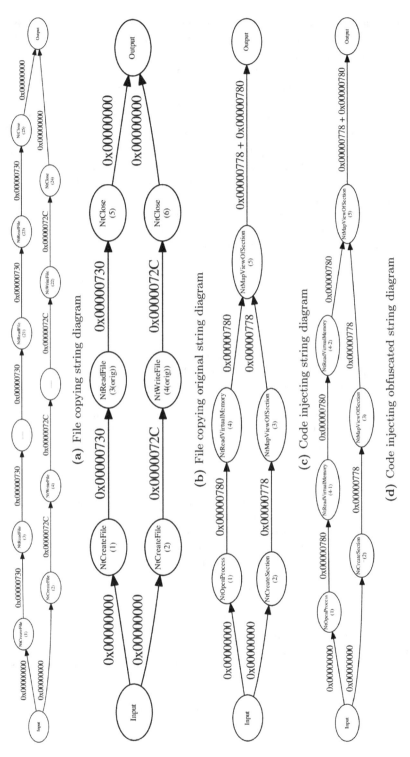

(a) File copying string diagram

(b) File copying original string diagram

(c) Code injecting string diagram

(d) Code injecting obfuscated string diagram

Fig. 5. Experiments string diagrams

The detection of obfuscated paths generated by Algorithm 2 is more challenging. When applied naively, the behavior matching does not work since the algorithm can generate paths of morphisms corresponding to syscall names and invocations distinct from those of the original path. More generally, it may be nonsense to compare an obfuscator and a detector which use different sets of behavioral transformations. In other words, as discussed in [19], a detector which abstracts behaviors by using the transformation set T, will be bypassed by an obfuscator which generates behaviors by using a set T' so that $T \cap T' \neq \emptyset$.

The original *behavior matching* techniques can be strengthened by generating (e.g. by using Algorithm 2) in prior a set of patterns that are semantically equivalent to the original one (see also the discussion in Experiment 2). Conversely, that means simplifying obfuscated paths to their original unique form, several simplifying techniques has been studied in some existing works on *semantics rewriting* (e.g. [5]). So we might suggest that a combination of *behavior matching* and *semantics rewriting* will improve the presented analysis. We reserve such an improvement as a future work.

References

1. Gen: Variant.Barys.159 (2013), `http://goo.gl/YDC1o`
2. Trace Transformation Tool (2013), `http://goo.gl/rqCSQ`
3. Trojan-Dropper.Win32.Dorgam.un (2013), `http://goo.gl/3e1AR`
4. Awodey, S.: Category Theory (Oxford Logic Guides), vol. 49. Oxford University Press, USA (2006)
5. Beaucamps, P., Gnaedig, I., Marion, J.-Y.: Behavior Abstraction in Malware Analysis. In: Barringer, H., et al. (eds.) RV 2010. LNCS, vol. 6418, pp. 168–182. Springer, Heidelberg (2010)
6. Christodorescu, M., Jha, S., Seshia, S.A., Song, D.X., Bryant, R.E.: Semantics-aware malware detection. In: Symposium on Security and Privacy, pp. 32–46. IEEE Computer Society (2005)
7. Christodorescu, M., Kruegel, C., Jha, S.: Mining Specifications of Malicious Behavior. In: ISEC, pp. 5–14. ACM (2008)
8. Collberg, C., Nagra, J.: Surreptitious Software: Obfuscation, Watermarking, and Tamperproofing for Software Protection. Addison-Wesley Professional (2009)
9. Erwig, M.: FGL/Haskell - A Functional Graph Library for Haskell (2008), `http://web.engr.oregonstate.edu/~erwig/fgl/haskell/`
10. Filiol, E.: Formalisation and implementation aspects of k-ary (malicious) codes. Journal in Computer Virology 3(2), 75–86 (2007)
11. Forrest, S., Hofmeyr, S.A., Somayaji, A., Longstaff, T.A.: A Sense of Self for Unix Processes. In: Symposium on Security and Privacy, pp. 120–128. IEEE (1996)
12. Fredrikson, M., Jha, S., Christodorescu, M., Sailer, R., Yan, X.: Synthesizing Near-Optimal Malware Specifications from Suspicious Behaviors. In: Symposium on Security and Privacy, pp. 45–60. IEEE (2010)
13. Intel. Pin - A Dynamic Binary Instrumentation Tool (2013), `http://software.intel.com/en-us/articles/pintool`
14. Joyal, A., Street, R.: The Geometry of Tensor Calculus, I. Advances in Mathematics 88, 55–112 (1991)

15. Kolbitsch, C., Milani Comparetti, P., Kruegel, C., Kirda, E., Zhou, X., Wang, X.: Effective and Efficient Malware Detection at the End Host. In: USENIX Security, pp. 351–366 (2009)
16. Ma, W., Duan, P., Liu, S., Gu, G., Liu, J.-C.: Shadow attacks: automatically evading system-call-behavior based malware detection. Journal in Computer Virology 8(1-2), 1–13 (2012)
17. Meseguer, J., Montanari, U.: Petri Nets are Monoids. Information and Computation 88(2), 105–155 (1990)
18. Milner, R.: Action Structures. Technical report (1992), http://www.lfcs.inf.ed.ac.uk/reports/92/ECS-LFCS-92-249/
19. Dalla Preda, M.: The grand challenge in metamorphic analysis. In: Dua, S., Gangopadhyay, A., Thulasiraman, P., Straccia, U., Shepherd, M., Stein, B. (eds.) ICISTM 2012. CCIS, vol. 285, pp. 439–444. Springer, Heidelberg (2012)
20. Dalla Preda, M., Christodorescu, M., Jha, S., Debray, S.K.: A semantics-based approach to malware detection. In: POPL, pp. 377–388. ACM (2007)
21. Ramilli, M., Bishop, M.: Multi-Stage Delivery of Malware. In: MALWARE, pp. 91–97 (2010)
22. Selinger, P.: A survey of graphical languages for monoidal categories. In: New Structures for Physics, pp. 289–355. Springer (2011)
23. Wagner, D., Soto, P.: Mimicry attacks on host-based intrusion detection systems. In: Conference on Computer and Communications Security, pp. 255–264. ACM (2002)

Space Complexity of Optimization Problems in Planar Graphs

Samir Datta[1] and Raghav Kulkarni[2]

[1] Chennai Mathematical Institute, India
[2] Center for Quantum Technologies, Singapore

Abstract. We initiate the study of space complexity of certain optimization problems restricted to planar graphs. We provide upper bounds and hardness results in space complexity for some of these well-studied problems in the context of planar graphs. In particular we show the following:

1. Max-Cut in planar graphs has a (UL∩co − UL)-approximation scheme;
2. Sparsest-Cut in planar graphs is in NL;
3. Max-Cut in planar graphs is NL-hard;
4. ⊕Directed-Spanning-Trees in planar graphs is ⊕L-complete.

For (1) we analyze the space complexity of the well known Baker's algorithm [1] using a recent result of Elberfeld, Jakoby, and Tantau [13] that gives a Log-space analogue of Courcelle's Theorem for MSO definable properties of bounded tree-width graphs.

For (2) we analyze the space complexity of the algorithm of Patel [21] that builds on a useful weighting scheme for planar graphs. Interestingly, the same weighting scheme has been crucially used in the totally different context of isolation in planar graphs [4,7].

For (3) we use a recent result of Kulkarni [17], which shows that Min-wt-PM in planar graphs is NL-hard.

For (4) we use the result by Datta, Kulkarni, Limaye, and Mahajan [8] that gives a reduction from the permanent in general graphs to planar graphs.

1 Introduction

1.1 Space Complexity

Perhaps the two most natural (and important) measures of complexity of computational problems are *running time* and *amount of space* used. The notion of efficiency for *time complexity* is captured by the complexity class P whereas the class L or Log-space (problems that can be solved on deterministic Turing machine using logarithmic amount of space) captures the notion of efficiency in the world of space complexity. In the algorithmic setting, especially for the optimization problems, time complexity has been the main focus so far; although there are examples of non-trivial space efficient algorithms such as the Log-space algorithm for undirected graph connectivity [22]. Recently Elberfeld, Jakoby, and

T V Gopal et al. (Eds.): TAMC 2014, LNCS 8402, pp. 300–311, 2014.

Tantau [13] obtained an analogue of the famous Courcelle's Theorem giving Log-space algorithms for several problems on bounded tree width graphs. Another interesting example is a recent result of [11] that shows Planar Graph Isomorphism is in Log-space. The focus of this paper is the space complexity of certain optimization problems whose time complexity has been well studied. This investigation is partly inspired by the paper [23] where Log-space optimization and Log-space approximation schemes were studied and by the paper [13] where a Log-space analogue of the Courcelle's Theorem was presented. We provide an application of the result of [13] in this paper. Besides Log-space, we consider another natural class NL (Non-deterministic Log-space) and its subclass UL (Unambiguous Log-space). NL is believed to be strictly larger than Log-space. Hence an NL algorithm is not as efficient as Log-space algorithm but the famous Savitch's Theorem tells us that NL is contained in $O(\log^2(n))$ space. The class UL is contained in NL and (as of today) may be considered *more space efficient* than NL; although whether or not UL $\overset{?}{=}$ NL remains an outstanding open question and it might well be the case that the two classes are in fact equal! Nevertheless we will use the class UL in order to point out more precise space bounds rather than NL whenever possible.

1.2 Planar Restrictions of Some Optimization Problems

Often imposing the natural restriction of planarity helps us to obtain more space efficient algorithms. For example: before Reingold's breakthrough result [22] that undirected graph reachability is in Log-space, a simple and elegant Log-space algorithm was known for planar graphs [3]. Recently Bourke, Tewari, and Vinodchandran [4] proved that Directed Planar Reachability is in UL as opposed to being NL-complete in general graphs. After this, space efficient algorithms were obtained for Matching in bipartite planar graphs and higher genus graphs [9,10,7]. Overall the structure of planarity seems to be ideal to exploit in the study of space complexity of many natural problems, see for instance [11,5,18]. In this paper we initiate the study of space complexity of some well studied optimization problems such as Max-Cut and Sparsest-Cut when restricted to planar graphs. Both problems are known to be in P (cf. [15,21]) as opposed to being NP-complete in general graphs. The goal is to identify such problems admitting highly space efficient (say Log-space) approximation algorithms or even approximation schemes when restricted to planar graphs and to exhibit hardness results towards obtaining better space efficient approximation. Although our progress towards this goal in the current paper is modest, we do provide some interesting examples of space efficient approximation as well as hardness. Our hope is that many more such algorithms and hardness results will be discovered later.

1.3 Our Main Results and Techniques

First we observe that certain optimization problems restricted to planar graphs admit space efficient approximation schemes. In particular:

Theorem 1. Max-Cut *in planar graphs has a* (UL ∩ co − UL)*-approximation scheme.*

We analyze the space complexity of the well known algorithm of Baker [1]. Baker's approach is broadly to decompose the planar graphs into bounded tree width graphs where the problem becomes easier to solve and then combine the solutions to obtain a good approximation. There are mainly two computational bottlenecks in Baker's approach: (1) distance computation in planar graphs is used for obtaining the decomposition into bounded tree width graphs, and (2) the optimization problem on bounded tree width graphs must have efficient *exact* solution. We observe that (1) can be performed in UL for planar graphs. For (2) we use a recent result of [13] that proves Log-space analogue of Courcelle's Theorem for *MSO definable* optimization problems: Max-Cut turns out to be one such. Our result in fact holds for a larger class of problems mentioned in [1] for which Baker's approach works. Moreover our result generalizes to the class of *bounded local tree width* graphs [14] (which include graphs embeddable on a fixed surface) where one can obtain a slightly weaker NL approximation schemes.

Next we exhibit a hardness result for Max-Cut in planar graphs.

In the spirit of identifying other optimization problems in planar graphs with space efficient algorithms we observe the following:

Theorem 2. Sparsest-Cut *in planar graphs is in* NL.

We analyze the algorithm of Patel [21] for this. Interestingly the algorithm makes use of a weighting scheme ([4], [7]) that has been recently used in an entirely different context namely isolation in planar graphs. The authors of [7] observe that the weighting scheme can be obtained in Log-space. Other than this the only non-trivial computational part in the algorithm of [21] is distance computation in graphs which can be performed in NL. Combining this gives an NL algorithm. It would be interesting to see if Sparsest-Cut in planar graphs is NL-hard, this would yield a natural NL-complete problem in planar graphs.

Theorem 3. Max-Cut *in planar graphs is* NL*-hard.*

We use a recent result of Kulkarni [17] that proves NL-hardness for minimum weight perfect matching in planar graphs. In the process we prove equivalence of Max-Cut and Min-wt-PM in planar graphs. This equivalence is interesting to note because constructing a perfect matching in planar graphs in NC is a long-standing open question [19,20] and our reduction indicates that it may be tied with constructing a maximum cut in planar graphs. Since Savitch's Theorem: NL ⊆ $O(\log^2(n))$-deterministic space, is believed to be optimal, our NL-hardness for Max-Cut in planar graphs means one would not expect a more efficient solution for Max-Cut in planar graphs than $O(\log^2(n))$ space. The possibility of a Log-space *approximation scheme* for the same remains an intriguing open question.

While for arbitrary graphs, Max-Cut has constant factor approximation and Sparsest-Cut is known to have only $\sqrt{\log n}$ factor approximation, the situation seems to reverse curiously for planar graphs in the context of space complexity.

We have NL hardness for Max-Cut whereas we are only able to show an NL upper bound for Sparsest-Cut in planar graphs.

Finally, in the spirit of noting hardness results for planar graphs, we conclude with another such result. Let ⊕Spanning-Trees denote the problem of deciding whether or not a graph has an even number of spanning trees and let ⊕Directed-Spanning-Trees denote the same problem in directed graphs: deciding whether or not there are even number of directed spanning trees rooted at (say) vertex 1. The edges of tree are directed away from the root vertex. Braverman, Kulkarni, and Roy [5] show the following somewhat surprising theorem.

Theorem. *(Braverman-Kulkarni-Roy [5])* ⊕Spanning-Trees *in planar graphs is computable in Log-space.*

It may be worthwhile to remind the reader that the authors of [9] use the above to obtain a Log-space algorithm for Unique-Perfect-Matching problem in outer-planar graphs.

In contrast to the above theorem, we show the following:

Theorem 4. ⊕Directed-Spanning-Trees *in planar graphs is* ⊕L-*complete.*

Although the above hardness result is not directly related to optimization problems in planar graphs, we found it worth noting in the context of space complexity of counting problems in planar graphs.

The organization of the paper is as follows: Section 2 contains preliminaries. Section 3 contains the proof of Theorem 1. Section 4 contains the proof of Theorem 2. Section 5 contains the proof of Theorem 3. Section 6 proves the equivalence between Max-Cut and Min-wt-PM in planar graphs. Section 7 contains the proof of Theorem 4. Section 8 contains open ends.

2   Preliminaries

2.1   Space Complexity Classes

Definition 1 (Log-space). *Log-space is the class of problems that can be solved by a deterministic Turing machine using $O(\log n)$ space, where n is the number of input bits.*

Definition 2 (NL). *Non-deterministic Log-space (NL) is the class of problems that can be solved by a non-deterministic Turing machine using $O(\log n)$ space, where n is the number of input bits.*

Definition 3 (UL). *Unambiguous Log-space (UL) is the class of problems that can be solved by an unambiguous non-deterministic Turing machine (non-deterministic Turing machine having at most one accepting path on every input) in $O(\log n)$ space where n is the number of input bits.*

Definition 4 (⊕L). *The parity L is the class of problems that can be reduced in Log-space to deciding whether or not the determinant of an integer matrix is even or odd.*

Technically the complexity classes above consists of *decision* problems, i.e., where the output is only one bit. We abuse the notion of space complexity classes for the problems that are not decision problems, i.e., when the number of output bits is more than one. For example, when we say Sparsest-Cut is in NL; it means that one can compute each output bit in NL. The optimization problems considered in this paper are restricted to have only polynomially bounded ($\leq n^{O(1)}$) weights.

2.2   Optimization Problems in Planar Graphs

In this paper we consider the following optimization problems restricted to planar graphs. We distinguish weighted and unweighted versions. For example: Max-Cut denotes unweighted version of the *maximum cut* problem whereas Max-wt-Cut denotes the weighted version of the same. The weights are always polynomially bounded. The problems below are *optimization versions*, i.e., find an optimum value. The construction version, i.e., outputting an optimum solution, will be indicated when necessary. For example: the construction version of *max cut* problem is denoted by Max-Cut (Construction).

Definition 5 (PM). Perfect-Matching *is the problem of deciding if a graph has a perfect matching or not.*

Min-wt-PM *is the problem of computing the weight of the minimum weight perfect matching in a graph.*

Definition 6 (Max-Cut). Max-Cut *is the problem of computing a partition of vertices* $(S, V - S)$ *such that the number of edges between S and $V - S$, i.e., $|E(S, V - S)|$ is maximum.*

Max-wt-Cut *is the problem of finding the weight of a maximum weight cut.*

Min-Bisection *is the problem of computing the minimum number of edges to remove from the graph so that the graph becomes bipartite.*

Definition 7 (Min-Odd-Vertex-Pairing). Min-OVP *is the problem of finding a collection of paths of minimum total length that pair the odd degree vertices in the graph.*

Definition 8 (Sparsest-Cut). Sparsest-Cut *is the problem of finding a cut* $(S, V - S)$ *such that* $|E(S, V - S)| / \min\{|S|, |V - S|\}$ *is minimum* $(S, V - S \neq \emptyset)$.

We will also be using the notions of tree-width and local tree-width. Please consult [12,14] for details.

3   Max-Cut in Planar Graphs has (UL ∩ co − UL) Approximation Scheme

In this section, we will prove the following theorem:

Theorem 5. *Let $\epsilon \in (0, 1]$ be any fraction, and let G be a planar graph (graph of bounded local tree-width), then we can compute a $(1 - \epsilon)$ approximate solution to* Max-Cut *in* UL ∩ co − UL ⊆ NL *(respectively,* NL*) (The space used is $O(1/\epsilon \times \log n)$).*

3.1   Proof Idea

We use Baker's algorithm [1] to give the desired upper bound. The novelty is in observing its space complexity.

Baker's algorithm works as follows: first we omit some edges from the graph so that the remaining edges can be partitioned into disjoint subgraphs of bounded diameter. This requires performing BFS on the graph and then omitting the edges in every t^{th} BFS layer (t is roughly $1/\epsilon$) starting from an offset. This step works for general graphs within NL. For planar graphs, in fact, BFS can in fact be performed in UL $\cap$ co $-$ UL [24]. By choosing the right offset, we can make sure that the number of edges removed affects the optimum solution at most by a fraction ϵ.

Next: Baker's algorithm requires obtaining an *exact solution* to the restriction of the problem on each of the pieces of bounded diameter. Planar graphs are *bounded local tree width* graphs, which means that the pieces of bounded diameter obtained from BFS layers are in fact graphs of bounded tree width. At this point we invoke the MSO-expressibility of Max-Cut. We can solve Max-Cut exactly (Section 3.2) on each piece obtained from the previous step, using the Log-space analogue [13] of Courcelle's Theorem [6].

Thus: modulo solving the problem on bounded tree width graphs, Baker's method amounts to performing multiple distance computations in an undirected planar graph and so can be done in UL $\cap$ co $-$ UL by [4,24]. In addition, using [13], MSO-computation on bounded tree-width graphs can be done in L. Hence the overall space complexity is $UL \cap$ co $-$ UL.

The same proof works for graphs of bounded local tree-width albeit with a slightly weaker bound of NL because the distance computation in such graphs requires NL. Because of bounded local tree-width property, the tree-width of the pieces is also bounded, which means one can appeal to the Log-space algorithm of [13].

3.2   Max-Cut in Bounded Tree-width Graphs is in L

Consider the following versions of Max-Cut:

Definition 9

> Count-Cut *Given a graph G and a positive integer c (in unary) output the number (in binary) of cuts of cardinality exactly c.*
> Cut(C) *Given a graph G does there exist a set of vertices S such that the cut $G[S, V \setminus S]$ consists of exactly the edges in C?*

The *histogram* of Cut is exactly the Count-Cut problem. Since histograms of MSO-problems on structures of bounded tree-width can be computed in L [13] we need the following to be able to compute Count-Cut on bounded tree-width graphs efficiently in bounded space:

Lemma 1. *The problem* Cut(C) *is expressible in* MSO-*logic.*

Proof. We just need to guess a set of vertices S and verify that that the set of edges with exactly one endpoint in S constitute exactly the set C. i.e.

$$\exists S \subseteq V \forall e \in E(e \in C \leftrightarrow (\exists u \in V \exists v \in V(u \neq v \wedge inc(u,e) \wedge inc(v,e) \wedge u \in S \wedge v \notin S))).$$

Here, $inc(u,e)$ is a built-in predicate that indicates that vertex u is incident on edge e.

The following is immediate, by finding the maximum cardinality cut with a non-zero count:

Proposition 1. Max-Cut *reduces to* Count-Cut *via a Log-space reduction.*

We have:

Theorem 6. *[13] (Log-space Cardinality Version of Courcelle's Theorem). For every $k \geq 1$ and every* MSO-*formula $\phi(X_1, \ldots, X_d)$, there is a logspace DTM that on input of any logical structure A of tree width at most k outputs histogram(A, ϕ)*

Notice that the histogram of [13] is nothing but a table of counts parameterized on the sizes of the free monadic second order variables. Thus, using Definition 9, Lemma 1, Proposition 1 and Theorem 6 we get that:

Lemma 2. Max-Cut *in bounded tree-width graphs is in* L.

4   Sparsest-Cut in Planar Graphs Is in NL

In this section we note an interesting consequence of the weighting scheme in [4] (more specifically [7]).

Theorem 7. Sparsest-Cut *in planar graphs is in* NL.

The following lemma is crucial.

Lemma 3 (cf. [21,4,10]). *One can assign skew-symmetric polynomially bounded weights $(w(i,j) = -w(j,i))$ to the edges of a planar graph in Log-space such that sum of the weights along any anti-clockwise simple cycle is equal to the number of faces in the interior.*

Interestingly the above lemma has found applications in a totally different context: isolation in planar graphs. Bourke et. al. used the above lemma to prove that Reachability in directed planar graphs is in UL (Unambiguous Log-space) as opposed to being NL-complete in general graphs.

Recently [21] used the above to show that Sparsest-Cut in planar graphs can be found in polynomial time. We analyze their algorithm and observe that it is in fact in NL. Patel constructs an auxiliary graph based on Lemma 3 such that the sparsest cut in the original graph can be computed by computing shortest distance between two points in the auxiliary graph. The auxiliary graph is constructed as follows: the vertices of the graph are denoted by (u, k) where u is a vertex of the original planar graph on n vertices and $k \in [-n^2, n^2]$. There is an edge between (u, k) and (v, k') if and only if $k' = k + w(u, v)$ where w is the weight from Lemma 3.

In [7] it is observed that weighting scheme in Lemma 3 can be computed in Log-space. Shortest distance in a graph with polynomially bounded weights can be computed in NL. Hence the algorithm in [21] works in NL.

5   Max-Cut in Planar Graphs is NL-Hard

In this section, we show that the Max-Cut problem restricted to planar graphs is NL-hard under Log-space reductions. Our proof has three crucial ingredients. The first ingredient is a reduction from Max-Cut in primal graph to Min-OVP in dual graph. This reduction was first observed by Hadlock [15] and since then has been used several times for obtaining efficient algorithms for Max-Cut in planar graphs. We use the same reduction for obtaining a hardness result instead. The second ingredient, the source of hardness, is the result by Kulkarni [17], which states that Min-wt-PM in planar graphs is NL-hard - even when weights are 0 or 1. The final ingredient to make our approach work is another reduction (due to Bampis et al. [2]) from Min-wt-PM in planar graphs to the same in 3-regular planar graphs. Below we put these ingredients together to obtain NL-hardness for Max-Cut.

Theorem 8. Max-Cut *in planar graphs is* NL-*hard.*

Proof: The theorem below [17] allows us to start with the following NL-hard instance: (0-1) Min-wt-PM in planar graphs.

Theorem 9 (Kulkarni [17]). Min-wt-PM *in planar graphs with edge weights* 0 *or* 1 *is* NL-*hard.*

We transform this NL-hard instance into an instance of Max-Cut in planar graphs. The transformation takes places in three steps.

Step 1: (Min-wt-PM) planar to 3-regular planar

Theorem 10 (Bampis et. al. [2]). *Given a weighted planar graph G one can obtain in Log-space another weighted planar graph G' such that: (a) G' is 3-regular; (b) the perfect matchings in G' are in bijection with the perfect matchings in G; (c) the bijection is weight preserving - in particular: minimum weight perfect matchings in G are mapped to those in G'.*

Part (a) and (b) of the above theorem are exactly as claimed in [2]. A glance at their proof in fact verifies Part (c) as well.

Step 2: (3-regular planar) Min-wt-PM to Min-wt-OVP

After Step 1 we have a 3-regular planar graph G' with 0-1 weights on edges. We obtain G'' from G' as follows: the vertex and edge sets of G'' are the same as that of G'. For an edge $e' \in E(G')$ let e'' denote the corresponding edge in $E(G'')$. We set $w(e'') := w(e') + N$, where $N = |V(G')|^2$. The weight of every perfect matching increases exactly by $N \cdot |V(G')|/2$. Thus: minimum weight perfect matchings in G' are mapped to those in G'' and vice versa.

In G'' note that every vertex has degree 3, which is odd. Moreoever: any odd-vertex-pairing in G'' must have weight at least $N \cdot |V|/2$. Hence, minimum weight odd-vertex-pairing in G'' corresponds to minimum weight perfect matching in G'', which in turn corresponds to minimum weight perfect matching in G'.

Step 3: (planar graphs) Min-wt-OVP to Max-wt-Cut

Lemma 4 (Hadlock [15]). *Minimum weight odd-vertex-pairing in a planar graph corresponds to maximum weight cut in the dual.*

This gives the desired hardness result for Max-wt-Cut. To reduce Max-wt-Cut (with polynomially bounded weights) to unweighted Max-Cut, we use the following simple gadget: replace every edge (u, v) of weight w by w paths, each of length three: $ux_1x_1'v, \dots, ux_wx_{w'}v$.

To see that the above gadget works, consider the problem Min-Bisection: minimum number of edges to remove so that the graph becomes bipartite. It is easy to check that the above gadget reduces Min-wt-Bisection to Min-Bisection in planar graphs. Since Min-Bisection is complimentary to Max-Cut (Min-Bisection + Max-Cut = Total Weight), we also get a reduction from Max-wt-Cut to Max-Cut in planar graphs. This completes the proof of Theorem 8.

6   Max-wt-Cut to Min-wt-PM in Planar Graphs

In this section we note that the reduction in previous section works in reverse direction as well. Thus the problems Max-wt-Cut and Min-wt-PM in planar graphs are equivalent up to Log-space reductions.

We transform an instance of Max-wt-Cut in planar graphs to an instance of Min-wt-PM in planar graphs. Our transformation works in three steps.

Step 1: (planar graphs) Triangulation

Add spurious edges to the planar graph so that every face is of length three. Set the weights of newly added edges to zero. This does not change the weight of maximum cut. Let G be the resulting graph.

Step 2: (planar graphs) Max-wt-Cut to Min-wt-OVP

Let G' be the dual of G.

Lemma 5 (Hadlock [15]). Max-wt-Cut *in planar graph reduces to* Min-wt-OVP *in its dual.*

Since G was triangulated, G' is 3-regular.

Step 3: (3-regular) Min-wt-OVP to Min-wt-PM

Replace each edge vertex of G' by a triangle (cycle of length 3) to obtain G''. Set high (say $N = |V|^2$) weight on the edges of these triangles. It is easy to see that minimum weight perfect matching in G'' corresponds to minimum weight odd-vertex-pairing in G'.

7   Hardness for $\oplus$Directed-Spanning-Trees

Proof of Theorem 4: The Kirchoff Matrix-Tree Theorem [16] for directed spanning trees (rooted at a vertex) immediately implies the $\oplus$L upper bound. It is known that $\oplus$Directed-Spanning-Trees in arbitrary directed graphs is $\oplus$L-hard [5]. Let G be a directed graph. Consider a straight line layout of G on the plane such that no three lines intersect at a point. A Log-space procedure for constructing such a layout is described in the Appendix. Now replace each crossing in G with the following gadget to obtain a directed planar graph H : Suppose the crossing consists of directed edges (a, b) and (c, d). We delete these two directed edges, add two new vertices x and y, add the edges $(a, x), (x, y), (y, b)$ and $(c, y), (y, x), (x, d)$. We claim that the number of directed spanning trees (rooted at vertex 1) is preserved modulo 2 under this transformation.

Let A_G denote the adjacency matrix of G.

Lemma 6. *[Datta, Kulkarni, Limaye, and Mahajan [8], Section 4.5]*

$$perm(A_G) = perm(A_H).$$

Let $L_G = D_G - A_G$ denote the Laplacian matrix of G where D_G is the diagonal matrix of out-degrees of the vertices of G. Let $L_G^{(1)}$ denote the matrix obtained from L_G by deleting the first row and the first column.

From the Matrix-Tree Theorem, we have:

$$\#\text{Directed-Spanning-Trees (rooted at 1)} = det(L_G^{(1)}).$$

Note that $det(L_G^{(1)}) = perm(L_G^{(1)})$ (mod 2). We view $L_G^{(1)}$ (mod 2) as the adjacency matrix $A_{G'}$ of a graph G'. Since each internal vertex in the gadget described above has out-degree 2 (which is an even number), we have:

$$perm(L_G^{(1)}) = det(A_{G'}) \stackrel{\text{Lemma 6}}{=} det(A_{H'}) = perm(L_H^{(1)}) \quad (\text{mod } 2).$$

Therefore,

$$det(L_G^{(1)}) = det(L_H^{(1)}) \quad (\text{mod } 2).$$

$\square$

8   Conclusion and Open Ends

We believe that our main contribution is to point out some natural examples of optimization problems in planar graphs which have space efficient algorithms. Our results suggest that studying the space complexity (both upper bound and hardness) for other natural problems in planar graphs might be fruitful. Below we list some potential questions to investigate.

Can our upper bound results be extended to handle bidimensional-problems?

Is Max-Cut (Construction) in planar graphs in NC? A positive answer would imply Perfect Matching Construction in planar graphs is in NC - a longstanding open question.

Does Max-Cut in planar graphs have Log-space approximation scheme?

Is Sparsest-Cut in planar graphs NL-hard?

References

1. Baker, B.S.: Approximation algorithms for NP-complete problems on planar graphs. J. ACM 41(1), 153–180 (1994)
2. Bampis, E., Giannakos, A., Karzonov, A., Manoussakis, Y., Milis, I.: Perfect matching in general vs. cubic graphs: A note on the planar and bipartite cases. Theoretical Computer Science and Applications 34(2), 87–97 (2000)
3. Blum, M., Kozen, D.: On the power of the compass (or why mazes are easier to search than graphs). In: FOCS, pp. 132–142 (1978)
4. Bourke, C., Tewari, R., Vinodchandran, N.V.: Directed planar reachability is in unambiguous log-space. TOCT 1(1) (2009)
5. Braverman, M., Kulkarni, R., Roy, S.: Space-efficient counting in graphs on surfaces. Computational Complexity 18(4), 601–649 (2009)
6. Courcelle, B.: The monadic second-order logic of graphs. I Recognizable Sets of Finite Graphs. Inf. Comput. 85(1), 12–75 (1990)
7. Datta, S., Gopalan, A., Kulkarni, R., Tewari, R.: Improved bounds for bipartite matching on surfaces. In: STACS, pp. 254–265 (2012)
8. Datta, S., Kulkarni, R., Limaye, N., Mahajan, M.: Planarity, determinants, permanents, and (unique) matchings. TOCT 1(3) (2010)
9. Datta, S., Kulkarni, R., Roy, S.: Deterministically isolating a perfect matching in bipartite planar graphs. Theory Comput. Syst. 47(3), 737–757 (2010)
10. Datta, S., Kulkarni, R., Tewari, R., Vinodchandran, N.V.: Space complexity of perfect matching in bounded genus bipartite graphs. J. Comput. Syst. Sci. 78(3), 765–779 (2012)
11. Datta, S., Limaye, N., Nimbhorkar, P., Thierauf, T., Wagner, F.: Planar graph isomorphism is in log-space. In: IEEE Conference on Computational Complexity, pp. 203–214 (2009)
12. Diestel, R.: Graph Theory (Graduate Texts in Mathematics). Springer (August 2005)
13. Elberfeld, M., Jakoby, A., Tantau, T.: Logspace versions of the theorems of bodlaender and courcelle. In: FOCS, pp. 143–152 (2010)
14. Eppstein, D.: Diameter and treewidth in minor-closed graph families. Algorithmica 27(3), 275–291 (2000)
15. Hadlock, F.: Finding a maximum cut of a planar graph in polynomial time. SIAM J. Comput. 4(3), 221–225 (1975)
16. Harris, J.M., Hirst, J.L., Mossinghoff, M.J.: Graph Theory (Undergraduate Texts in Mathematics). Springer (September 19, 2008)
17. Kulkarni, R.: On the power of isolation in planar graphs. TOCT 3(1), 2 (2011)
18. Limaye, N., Mahajan, M., Sarma, J.M.N.: Upper bounds for monotone planar circuit value and variants. Computational Complexity 18(3), 377–412 (2009)
19. Mahajan, M., Varadarajan, K.R.: A new NC-algorithm for finding a perfect matching in bipartite planar and small genus graphs. In: STOC, pp. 351–357 (2000)

20. Miller, G.L., Naor, J.: Flow in planar graphs with multiple sources and sinks. SIAM J. Comput. 24(5), 1002–1017 (1995)
21. Patel, V.: Determining edge expansion and other connectivity measures of graphs of bounded genus. In: de Berg, M., Meyer, U. (eds.) ESA 2010, Part I. LNCS, vol. 6346, pp. 561–572. Springer, Heidelberg (2010)
22. Reingold, O.: Undirected connectivity in log-space. J. ACM 55(4), 1–24 (2008)
23. Tantau, T.: Logspace optimization problems and their approximability properties. Theory Comput. Syst. 41(2), 327–350 (2007)
24. Thierauf, T., Wagner, F.: The isomorphism problem for planar 3-connected graphs is in unambiguous logspace. Theory Comput. Syst. 47(3), 655–673 (2010)

9 Appendix

9.1 Constructing a Non-degenerate Straight-Line Layout of K_n

We identify the vertices of K_n with n points with co-ordinates $v_i = (i, t \cdot i^2)(1 \leq i \leq n)$ that lie on a parabola $y = t \cdot x^2$ where t will be chosen later. Given any three lines $\ell_1 = (v_{i_1}, v_{j_1}), \ell_2 = (v_{i_2}, v_{j_2})$, and $\ell_3 = (v_{i_3}, v_{j_3})$, one can check whether they are collinear or not by computing a 3×3 determinant, which will be a polynomial of degree 3 in t, and testing if this determinant is zero or not. Let $p(t)$ denote the polynomial obtained by taking the product of all such polynomials over all possible triplets of the lines. Note that $p(t)$ is a polynomial of degree $O(n^6)$, having at most $O(n^6)$ roots. Hence there exists a value of $t = t_0$ in the range (say) $\{1, 2, ..., n\}$ such that $p(t_0) \neq 0$. Choosing this value of t gives us a straight-line layout of K_n such that no three lines intersect at a point. Since we have to try only polynomially many values of t, it is easy to check that the entire procedure works in Log-space, in fact the procedure works in a subclass of Log-space, namely TC_0.

Fine-Tuning Decomposition Theorem for Maximum Weight Bipartite Matching

Shibsankar Das and Kalpesh Kapoor

Department of Mathematics
Indian Institute of Technology Guwahati
Guwahati - 781039, India
{shibsankar,kalpesh}@iitg.ernet.in

Abstract. Let G be an undirected bipartite graph with non-negative integer weights on the edges. We refine the existing decomposition theorem originally proposed by Kao *et al.* in the context of maximum weight bipartite matching. We apply it to design an efficient version of the decomposition algorithm to compute the weight of a maximum weight bipartite matching of G in $O(\sqrt{n}W'/k(n, W'/m'))$-time by employing an algorithm designed by Feder and Motwani as a subroutine, where $n, m, m'(\leq m)$ denote number of nodes, number of edges and number of distinct edge weights of G, respectively. The parameter W' is smaller than the total edge weight W, essentially when the largest edge weight differs by more than one from the second largest edge weight in the current working graph in decomposition step of the algorithm. In best case $W' = O(m)$ and in worst case $W' = W$, i.e., $m \leq W' \leq W$.

Keywords: Graph algorithm, maximum weight bipartite matching, graph decomposition, minimum weight vertex cover.

1 Introduction

Let $G = (V, E, w)$ be an undirected and weighted graph with V and E as the set of vertices and edges, respectively, and has non-negative integer weights on the edges which are given by the weight function $w : E \to \mathbb{N}^0$, where $\mathbb{N}^0$ is the set of non-negative integers. We also assume that graph does not have any isolated vertex. For uniformity, we treat an unweighted graph as a weighted graph having unit weight for all edges.

A subset $M \subseteq E$ of edges is a *matching* if no two edges of M share a common vertex. A vertex $v \in V$ is said to be *covered* or *matched* by the matching M if it is incident with an edge of M; otherwise v is *unmatched* [2,3].

A matching M is called a *maximum (cardinality) matching* if there does not exist any matching with greater cardinality. We denote such a matching by $mm(G)$. The weight of a matching M is defined as $w(M) = \sum_{e \in M} w(e)$. A matching M is a *maximum weight matching*, denoted as $mwm(G)$, if $w(M) \geq w(M')$ for every other matching M' of graph G.

Observe that, if G is an unweighted graph then $mwm(G) = mm(G)$ and its weight is given by $w(mwm(G)) = |mm(G)|$. Similarly, if G is an undirected and

T V Gopal et al. (Eds.): TAMC 2014, LNCS 8402, pp. 312–322, 2014.
© Springer International Publishing Switzerland 2014

weighted graph with $w(e) = c$ for all edges e in G and c is a constant then also $mwm(G) = mm(G)$ with weight of the matching as $w(mwm(G)) = c * |mm(G)|$.

In this paper, we focus on computation of maximum weight matching in a bipartite graph using an exact algorithm. Let $G = (V = V_1 \cup V_2, E, w)$ be an undirected, weighted bipartite graph and without isolated nodes and having V_1 and V_2 as partition of vertex set V. Throughout the paper, we use the symbols N, W, n and m ($= \Omega(n)$) to denote the largest weight of any edge, total edge weight i.e. $\sum_{e \in E} w(e)$, number of nodes and number of edges of G, respectively.

Our contribution in this paper is a revised version of the existing decomposition theorem, as described in [13,14] and use it efficiently to design an improved version of the decomposition algorithm to estimate the weight of Maximum Weight Bipartite Matching (MWBM) of G in time $O(\sqrt{n}W'/k(n, W'/m'))$ by taking algorithm designed by Feder and Motwani [7] as base algorithm, where $k(x, y) = \log x / \log(x^2/y)$, m' ($\leq m$) is the number of distinct edge weights of G. The parameter W' is smaller than the total edge weight W, when the largest edge weight differs by more than one from the second largest edge weight in the current working graph during decomposition in each iteration of the algorithm. In best case, computation of maximum weight matching takes $O(\sqrt{n}m/k(n, m))$-time whereas in worst case it takes $O(\sqrt{n}W/k(n, W/m'))$, i.e., $m \leq W' \leq W$. However, it is very difficult and challenging to get rid of W or N from the complexity.

The modified algorithm works well for general W, but is the best known for $W' = o(m \log(nN))$. We also design a revised algorithm to construct minimum weight cover of a bipartite graph in time $O(\sqrt{n}W'/k(n, W'/m'))$ to identify the edges involved in maximum weight bipartite matching. It is also possible to use other algorithms as a subroutine, for example, algorithms given by Hopcroft and Karp [11] and Alt et al. [1] in which case the running times of our algorithm will be $O(\sqrt{n}W')$ and $O((n/\log n)^{1/2}W')$, respectively.

The rest of the paper is organized as follows. In Section 2, we give a detailed summary of existing maximum matching algorithms and their complexities for unweighted and weighted bipartite graphs. Section 3 describes modified decomposition theorem and an algorithm to compute the weight of MWBM. The complexity analysis of the algorithm is discussed in Section 4. The algorithm to compute minimum weight cover of a bipartite graph is given in Section 5 which is used to find the edges of a MWBM. We summarize the results in Section 6.

2 Related Work

The problem of computing maximum matching in a given graph is one of the fundamental algorithmic problem that has played an important role in the development of combinatorial optimization and algorithmics. A survey of some of the well known existing maximum (cardinality) matching and maximum weight matching algorithms for bipartite graph are summarized in Table 1 and Table 2, respectively. The algorithms with best asymptotic bound are indicated by "$*$" in the tables. A more detailed and technical discussion of the algorithms can be found in textbooks [15,19].

Table 1. Complexity survey of maximum (unweighted) bipartite matching algorithms

Year	Author(s)	Complexity
1973 *	Hopcroft and Karp [11]	$O(m\sqrt{n})$
1991	Alt, Blum, Mehlhorn and Paul [1]	$O(n^{1.5}\sqrt{m/\log n})$
1995 *	Feder and Motwani [7]	$O(m\sqrt{n}/k(n,m))$

For unweighted bipartite graphs, Hopcroft-Karp [11] algorithm, which is based on augmenting path technique, offers the best known performance for finding maximum matching in time $O(m\sqrt{n})$. In case of dense unweighted bipartite graphs, that is with $m = \Theta(n^2)$, slightly better algorithms exist. An algorithm by Alt et $al.$ [1] obtains a maximum matching in $O(n^{1.5}\sqrt{m/\log n})$-time. In case of $m = \Theta(n^2)$, this becomes $O(m\sqrt{n/\log n})$ and is also $\sqrt{\log n}$-factor faster than Hopcroft-Karp algorithm. This speed-up is obtained by an application of the fast adjacency matrix scanning technique of Cheriyan, Hagerup and Mehlhorn [4]. The algorithm given by Feder-Motwani [7] has the time complexity $O(m\sqrt{n}/k(n,m))$, where $k(x,y) = \log x/\log(x^2/y)$.

Table 2. Complexity survey of maximum weight bipartite matching algorithms

Year(s)	Author(s)	Complexity
1955, 1957	Kuhn [16], Munkres [17]	$O(n^4)$ (Hungarian method)
1960	Iri [12,19]	$O(n^2 m)$
1969	Dinic and Kronrod [5,19]	$O(n^3)$
1984, 1987 *	Fredman and Tarjan [8]	$O(n(m + n\log n))$
1985	Gabow [10]	$O(n^{3/4}m\log N)$
1989 *	Gabow and Tarjan [9]	$O(\sqrt{n}m\log(nN))$
1999	Kao, Lam, Sung and Ting [13]	$O(\sqrt{n}W)$
2001 *	Kao, Lam, Sung and Ting [14]	$O(\sqrt{n}W/k(n,W/N))$
This work *		$O(\sqrt{n}W')$
		$O((n/\log n)^{1/2}W')$
		$O(\sqrt{n}W'/k(n,W'/m'))$

Several algorithms have also been proposed for computing maximum weight bipartite matching, improving both theoretical and practical running times. The well known Hungarian method, the first polynomial time algorithm, was introduced by Kuhn [16] and Munkres [17]. Fredman and Tarjan [8] improved this with running time $O(n(m + n\log n))$ for sparse graph by using Fibonacci heaps. An $O(n^{3/4}m\log N)$-time scaling algorithm was proposed by Gabow [10] under the assumption that edge weights are integers. A different and faster scaling algorithm was given by Gabow and Tarjan [9] with running time $O(\sqrt{n}m\log(nN))$. Kao et $al.$ [14] proposed a $O(\sqrt{n}W/k(n,W/N))$-time decomposition technique under the assumptions that weights on the edges are positive and $W = o(m\log(nN))$.

In addition to the above exact algorithms, several randomized and approximate algorithms are also proposed, see for example [6,18].

3 Refined Decomposition Technique to Compute Weight of MWBM

We now propose a modified decomposition theorem which is generalized than the existing decomposition theorem originally proposed by Kao *et al.* [14] and use it to develop a revised version of the decomposition algorithm to decrease the number of iterations and speed-up the computation of the weight of MWBM. Let $G = (V = V_1 \cup V_2, E, w)$ be an undirected, weighted bipartite graph and without isolated nodes and having V_1 and V_2 as partition of vertex set V. Further, let $E = \{e_1, e_2, \ldots, e_m\}$ be set of edges with weights $w(e_i) = w_i$ for $1 \leq i \leq m$, where w_is are not necessarily distinct. As defined earlier, let N be the maximum edge weight, i.e., for all $i \in \{1, 2, \ldots, m\}$, $0 \leq w_i \leq N$, and $W = \sum_{1 \leq i \leq m} w_i$ be the sum of weights of all edges.

Our algorithm considers several intermediate graphs with lighter weight on edges. During this process it is possible that weights of some of the edges may be zero. An edge $e \in E$ is said to be *active* if its weight $w(e) > 0$, otherwise it is said to be *inactive* i.e. when $w(e) = 0$. Let there be m' ($\leq m$) distinct edge weights in current working graph where $w_1 < w_2 < \cdots < w_{m'-1} < w_{m'}$. We denote the first two distinct maximum edge weights in current working graph by H_1 and H_2 ($< H_1$), respectively. Assign $H_2 = 0$ in case $m' = 1$.

We first build two new graphs referred to as G_h and G_h^Δ from a given weighted bipartite graph G. For any integer $h \in [1, N]$, we decompose the weighted bipartite graph G into two lighter weighted bipartite graph G_h and G_h^Δ as proposed by Kao *et al.* [13,14]. Before we describe their construction, we describe minimum weight cover which is a dual of maximum weight matching [14]. A *cover* of G is a function $C : V_1 \cup V_2 \to \mathbb{N}^0$ such that $C(v_1) + C(v_2) \geq w(v_1, v_2) \; \forall \; v_1 \in V_1$ and $v_2 \in V_2$. Let $w(C) = \sum_{x \in V_1 \cup V_2} C(x)$. C is *minimum weight cover* if $w(C)$ is minimum.

Formation of G_h from G: The graph G_h is formed by including those edges $\{u, v\}$ of G whose weights $w(u, v)$ lie in the range $[N - h + 1, N]$. Each edge $\{u, v\}$ in graph G_h is assigned weight $w(u, v) - (N - h)$. For illustration, G_1 is constructed by the maximum weight edges of G and assigned unit weight to each edge.

Formation of G_h^Δ from G: Let C_h be the minimum weight cover of G_h. The graph G_h^Δ is formed by including every edge $\{u, v\}$ of G whose weight satisfies the condition $w(u, v) - C_h(u) - C_h(v) > 0$. The weight assigned to such an edge is $w(u, v) - C_h(u) - C_h(v)$.

Theorem 1 (The Decomposition Theorem [14]). *Let G be an undirected, weighted bipartite graph and without isolated nodes. Then*

a. for any integer $h \in [1, N]$, $w(mwm(G)) = w(mwm(G_h)) + w(mwm(G_h^\Delta))$,
b. in particular (trivial), for $h = 1$, $w(mwm(G)) = w(mm(G_1)) + w(mwm(G_1^\Delta))$.

Note that the Theorem 1(b) is derived from Theorem 1(a), since for $h = 1$, we have $mwm(G_1) = mm(G_1)$ and $w(mwm(G_1)) = w(mm(G_1)) = |mm(G_1)|$. The Theorem 1(b) is used recursively in the algorithm, originally proposed by Kao *et al.* in [14], to compute weight of $mwm(G)$.

Remark 1. G may not have all distinct edge weights. Consider the set of distinct edge weights of G. The algorithm works efficiently only when the largest edge weight differs by exactly one from the second largest edge weight of the current graph during an invocation of Theorem 1(b) in each iteration.

Remark 2. Observe that for arbitrary $h \in [1, N]$, $mwm(G_h)$ need not be equal to $mm(G_h)$, that is, we cannot always conclude that $mwm(G_h) = mm(G_h)$.

One of our objective is to investigate those values of h for which $mwm(G_h)$ is equal to $mm(G_h)$ apart from the trivial value of h as 1 in each iteration of algorithm to generate G_h having all its edge weights as 1.

In order to get the speed-up by decreasing the number of iterations, we revise the Theorem 1(b) and propose Theorem 2 which gives a domain of $h \in [1, N]$ where $mwm(G_h) = mm(G_h)$ and as a consequence of that $w(mwm(G_h)) = w(mm(G_h)) = h * |mm(G_h)|$. It works for $h = 1$ and performs well especially when the the largest edge weight differs by more than one from the second largest edge weight in the current graph in decomposition step during each iteration.

Theorem 2 (The Modified Decomposition Theorem). *The following equalities hold for any integer $h \in [1, H_1 - H_2]$ where H_1 and H_2 ($< H_1$) are the first two distinct maximum edge weights of graph G, respectively. We assign $H_2 = 0$ in case all edge weights are equal.*

a. $mwm(G_h) = mm(G_h)$,
b. $w(mwm(G)) = h * w(mm(G_h)) + w(mwm(G_h^\Delta))$.

Proof. The proof of the above statements are based on the construction of new graphs G_h and G_h^Δ from G and Theorem 1(a).

a. To prove that for any integer h, $1 \le h \le H_1 - H_2$, $mwm(G_h) = mm(G_h)$ holds true, it is enough to prove the same for the maximum value[1] of h, i.e., for $h = H_1 - H_2$. As specified earlier, the construction of G_h is done by

---

[1] For illustration, consider $h = c$ where $1 \le c \le H_1 - H_2$. Then as per the formation of G_h from G, G_c is built by choosing those edges of G that have weight $w(u, v) \in [N - (c - 1), N]$. Since, $c - 1 \ge 0$ and $N \in [N - (c - 1), N]$ for any $c \in [1, H_1 - H_2]$, G_c has only the heaviest edges of G. For optimization, choose $h = H_1 - H_2$, the maximum possible value of h.

choosing those edges $\{u, v\}$ of G that have weight $w(u, v) \in [N - h + 1, N] = [H_1 - (H_1 - H_2) + 1, H_1] = [H_2 + 1, H_1]$. Since $H_1 \in [H_2 + 1, H_1]$, G_h has only the heaviest edges of G and each such edge is assigned the same weight. Thus, $mwm(G_h) = mm(G_h)$ for $h = H_1 - H_2$.

b. Observe that $h \in [1, H_1 - H_2]$ and $[1, H_1 - H_2] \subseteq [1, N]$. So, by using Theorem 1(a) we have, $\forall h \in [1, H_1 - H_2]$,

$$w(mwm(G)) = w(mwm(G_h)) + w(mwm(G_h^\Delta)).$$

Also by using the Theorem 2 (a), $mwm(G_h) = mm(G_h)$ for all $h \in [1, H_1 - H_2]$. Weight of each such edges[2] of G in G_h is exactly $w(u, v) - (N - h) = H_1 - (H_1 - h) = h$. Therefore,

$$w(mwm(G_h)) = h * w(mm(G_h)) = h * |mm(G_h)|.$$

Hence for any integer $h \in [1, H_1 - H_2]$,
$$\begin{aligned} w(mwm(G)) &= w(mwm(G_h)) + w(mwm(G_h^\Delta)) \\ &= h * w(mm(G_h)) + w(mwm(G_h^\Delta)). \end{aligned}$$

□

Remark 3. The equality $mwm(G_h) = mm(G_h)$ in Theorem 2(a) is not true for $h > H_1 - H_2$ and $h \leq N$.

To show that for any h in interval $[H_1 - H_2 + 1, N]$ the equation $mwm(G_h) = mm(G_h)$ is not true, it is enough to show the same essentially for $h = H_1 - H_2 + 1$. Observe that $h = H_1 - H_2 + 1 \geq 2$, since $H_1 > H_2$. According to the construction of G_h, it is formed by edges $\{u, v\}$ of G whose weights $w(u, v) \in [N - h + 1, N] = [H_1 - (H_1 - H_2 + 1) + 1, H_1] = [H_2, H_1]$, i.e., G_h is built with the maximum weight edges and second maximum weight edges of G, because $\{H_1, H_2\} \in [H_2, H_1]$. Weight of each of the heaviest edge of G in G_h is exactly $w(u, v) - (N - h) = H_1 - (H_1 - h) = h \geq 2$ and that of the second heaviest edge of G in G_h is exactly $w(u, v) - (N - h) = H_2 - (H_1 - h) = (H_2 - H_1) + h = (1 - h) + h = 1$. Hence $mwm(G_h) \neq mm(G_h)$ for such a value of h.

Example 1. Consider the graph shown in the Figure 1(a). Let $h = H_1 - H_2 + 1$. So, $h = H_1 - H_2 + 1 = 9 - 4 + 1 = 6$. As shown in the Figure 1(b), G_h is formed by the edges $\{u, v\}$ whose weight $w(u, v) \in [N - h + 1, N] = [9 - 6 + 1, 9] = [4, 9]$ and their respective calculated weights are 6 and 1. Hence $mwm(G_h) \neq mm(G_h)$.

We use the modified decomposition Theorem 2 to design a recursive Algorithm 3.1 to compute the weight of $mwm(G)$.

---

[2] Only maximum weight edges of G are participating in G_h.

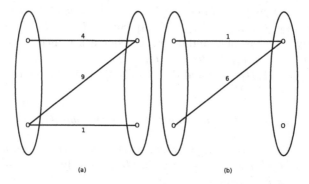

Fig. 1. (a) An undirected bipartite graph G with non-negative integer weights on the edges. **(b)** Considering $h = H_1 - H_2 + 1 = 6$, G_h is extracted, but $mwm(G_h) \neq mm(G_h)$.

Algorithm 3.1. Compute weight of the maximum weight matching of G

Input: A weighted, undirected, complete bipartite graph G with non-negative integer weights on the edges and without isolated nodes.
Output: Weight of the maximum weight matching of G, i.e. $w(mwm(G))$.

WT-MWBM(G)
    1. Assume that initially $w(mwm(G)) = 0$.
    2. Find $h = H_1 - H_2$ using current working graph G.
    3. Construct G_h from G.
    4. Compute $mm(G_h)$.
    5. Find minimum-weight-cover C_h of G_h.
    6. Construct G_h^{Δ} from G and C_h.
    7. **if** G_h^{Δ} is empty (i.e. G_h^{Δ} has no active edge)
        then return $h * |mm(G_h)|$
        else return $h * |mm(G_h)| + \text{WT-MWBM}(G_h^{\Delta})$.

Example 2. Consider the graph shown in Figure 2(a). The Algorithm 3.1 finds the weight of the MWBM in just two iterations, as the algorithm is designed for best h in every invocation of WT-MWBM(), whereas algorithm by Kao *et al.* [14] requires 500 iterations because it considers $h = 1$ in every invocation.

Correctness of the algorithm follows from the construction of G_h and G_h^{Δ} and the modified decomposition Theorem 2.

4   Complexity of the Algorithm

Let $G = (V = V_1 \cup V_2, E, w)$ be the initial input graph. Let N be the maximum edge weight, that is, for all $i \in \{1, 2, \ldots, m\}$, $0 \leq w_i \leq N$ and $W = \sum_{1 \leq i \leq m} w_i$

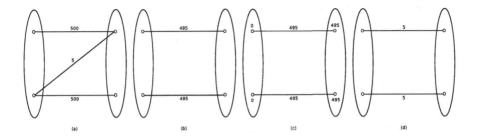

Fig. 2. (a) An undirected, weighted bipartite graph G with non-negative integer weights on the edges. Current $h = 495$. (b) G_h is extracted. (c) C_h is the weighted cover of G_h. (d) G_h^Δ is formed from G_h and C_h.

be the sum of weights of all edges. Further, let $w_1, \ldots, w_{m'}$ be the set of distinct edge weights, where $m' \leq m$.

Based on the construction of G_h, G_h^Δ, the modified decomposition Theorem 2 and the Algorithm 3.1, we can easily observe that maximum number of iterations of WT-MWBM() is the same as the number of distinct edge weights of the initial graph G which is nothing but $m'(\leq m)$ in our case. In worst case when all the edge weights of the initial graph G are distinct then that leads to m iterations in the above algorithm. In the best case, all the edge weights are the same, so the algorithm will terminate in the first iteration itself.

As the complexity analysis of the Algorithm 3.1 is almost similar to that presented elsewhere [14], we skip it. The algorithm takes $O(\sqrt{n}W'/k(n, W'/m'))$ time to compute the weight of $mwm(G)$ by using the algorithm by Feder and Motwani [7], as a subroutine.

Let L_i consist of edges of remaining G whose weights reduce in G_h^Δ in i-th iteration. Also let there be p iterations, $l_i = |L_i|$ for $i = 1, 2, \ldots, p \leq m' \leq N$ and $h_i = H_1 - H_2$ in the i-th iteration. From the detailed complexity analysis we have, $l_1 h_1 + l_2 h_2 + \cdots + l_p h_p = O(W)$. Let $l_1 + l_2 + \cdots + l_p = W'$. Observe that if $h_i = 1$ for all $i \in [1, p]$, then $W' = \sum_{i=1}^{p} l_i = W$. However, the parameter W' is smaller than the total edge weight W, essentially when the largest edge weight differs by more than one from the second largest edge weight in the current working graph in decomposition step during an iteration of the algorithm. In best case[3], it requires $O(\sqrt{n}m/k(n, m))$-time to compute maximum weight matching and in worst case $O(\sqrt{n}W/k(n, W/m'))$, i.e., $m \leq W' \leq W$. However, it is very difficult and challenging to get rid of W or N from the complexity. This modified algorithm works well for general W, but is the best known for $W' = o(m \log(nN))$.

Remark 4. In special case, the Algorithm 3.1 is independent of the total weight W of G. Even we increase all the weight of edges of G by some additional factor, say α, it is not going to affect the complexity of the Algorithm 3.1; whereas for

---

[3] In best case, all the edge weights of G are the same. So, the algorithm terminates in just one iteration and hence $W' = O(m)$.

complexity of algorithm of Kao *et al.* [14] will increase by an additional factor of $O(n\alpha m/k(n,\alpha m/N))$. Similar argument holds in case of increment in weight of each edge of G by some multiplicative factor. The Algorithm 3.1 depends on difference h between the largest edge weight and the second largest edge weight of the current working graph during an invocation of WT-MWBM() in each iteration. During the increment of all edge weights, the difference h in each invocation is going to remain constant.

Below we also analyze the complexity of the Algorithm 3.1 by considering the Hopcroft-Karp algorithm [11] and Alt-Blum-Mehlhorn-Paul algorithm [1] as base algorithms.

With respect to Hopcroft-Karp algorithm: Hopcroft-Karp algorithm [11] presents the best known worst-case performance for getting a maximum matching in a bipartite graph with runtime of $O(\sqrt{n}m)$. Hence the recurrence relation for running time of the algorithm with respect to Hopcroft-Karp algorithm is

$$T(n,W',m') = O(\sqrt{n}l_1) + T(n,W'',m'')$$
$$\text{and } T(n,0,0) = 0$$

$$\therefore T(n,W',m') = O(\sqrt{n}l_1) + O(\sqrt{n}l_2) + \cdots + O(\sqrt{n}l_p)$$
$$= O\left(\sqrt{n}\sum_{i=1}^{p} l_i\right) = O(\sqrt{n}W').$$

With respect to Alt-Blum-Mehlhorn-Paul algorithm: A bit better algorithm for dense bipartite graph is Alt-Blum-Mehlhorn-Paul Algorithm [1] which is $(\log n)^{1/2}$-factor faster[4] than Hopcroft-Karp algorithm for maximum bipartite matching. Hence the time complexity, with respect to Alt-Blum-Mehlhorn-Paul algorithm as a base algorithm, is $O((n/\log n)^{1/2}W')$. Hence it is $(\log n)^{1/2}$-factor faster than the above case.

5   Finding a Maximum Weight Matching

The Algorithm 3.1 computes only the weight of the $mwm(G)$ of a given graph G. To find the edges of $mwm(G)$, we first give a revised algorithm for constructing a Minimum Weight Cover (MWC) of G which is a dual of maximum weight matching. As mentioned before, a *cover* of G is a function $C : V_1 \cup V_2 \to \mathbb{N}^0$ such that $C(v_1) + C(v_2) \geq w(v_1, v_2) \; \forall \; v_1 \in V_1$ and $v_2 \in V_2$. Let $w(C) = \sum_{x \in V_1 \cup V_2} C(x)$. We say C is *minimum weight cover* if $w(C)$ is minimum. Let C be a MWC of a graph G.

---

[4] For dense graphs ($m = O(n^2)$) this improves on the $O(m\sqrt{n}) = O(n^{2.5})$ time algorithm of Hopcroft and Karp. The speed-up of this algorithm is obtained by an application of the fast adjacency matrix scanning technique of Cheriyan, Hagerup and Mehlhorn. It has time-complexity $O(n^{1.5}\sqrt{m/\log n}) = O(n(n/\log n)^{1/2}m^{1/2})$ $= O((n/\log n)^{1/2}m)$ and hence is $(\log n)^{1/2}$ factor faster than Hopcroft-Karp algorithm.

Lemma 1 ([14]). *Let C_h^Δ be any minimum weight cover of G_h^Δ. If C is a function on $V(G)$ such that for every $u \in V(G)$, $C(u) = C_h(u) + C_h^\Delta(u)$, then C is minimum weight cover of G.*

Using this lemma we design an $O(\sqrt{n}W'/k(n, W'/m'))$-time revised algorithm to compute a MWC of G. The correctness of this algorithm is clear form the Lemma 1 and the time complexity analysis is similar to that given in previous section.

Algorithm 5.1. Calculate a MWC C of G

Input: A weighted, undirected, complete bipartite graph G with non-negative integer weights on the edges and without isolated nodes.
Output: A minimum weight cover C of G.

MWC(G)
  1. Assume that initially $w(mwm(G)) = 0$.
  2. $h \leftarrow H_1 - H_2$ using current working graph G.
  3. Construct G_h from G.
  4. Compute $mm(G_h)$.
  5. Find minimum-weight-cover C_h of G_h.
  6. Construct G_h^Δ from G and C_h.
  7. **if** G_h^Δ is empty (i.e. G_h^Δ has no active edge)
    then return C_h
    else
        $C_h^\Delta \leftarrow$ MWC(G_h^Δ)
        return C, where $C(u) = C_h(u) + C_h^\Delta(u)$ for all nodes u in G.

Now as deduced by Kao *et al.* in [14], finding a maximum weight matching by using the given vertex cover takes $O(\sqrt{n}m/k(n, m))$-time. Since $m \le W' \le W$, so altogether $O(\sqrt{n}W'/k(n, W'/m'))$ time requires to find a MWBM of G.

6 Conclusions

We have fine-tuned the existing decomposition theorem originally proposed by Kao *et al.* in [14], in the context of maximum weight bipartite matching and applied it to design a revised version of the decomposition algorithm to compute the weight of a maximum weight bipartite matching in $O(\sqrt{n}W'/k(n, W'/m'))$-time by employing an algorithm designed by Feder and Motwani [7], as base algorithm. We have also analyzed the algorithm by using Hopcroft-Karp algorithm [11] and Alt-Blum-Mehlhorn-Paul algorithm [1] as base algorithms, respectively.

The algorithm performs well especially when the the largest edge weight differs by more than one from the second largest edge weight in the current working graph during an invocation of WT-MWBM() in each iteration. In best case $W' = O(m)$ and in worst case $W' = W$, i.e., $m \le W' \le W$. The algorithm works well for general W, but is the best known for $W' = o(m \log(nN))$.

References

1. Alt, H., Blum, N., Mehlhorn, K., Paul, M.: Computing a maximum cardinality matching in a bipartite graph in time $O(n^{1.5}\sqrt{m/\log n})$. Information Processing Letters 37(4), 237–240 (1991)
2. Bondy, J.A., Murty, U.S.R.: Graph theory with applications, Matchings, ch. 5, 5th edn., p. 70. North-Holland, NY (1982)
3. Bondy, J.A., Murty, U.S.R.: Graph Theory. Matchings, 1st edn., vol. 244, ch. 16 , p. 419. Springer (2008)
4. Cheriyan, J., Hagerup, T., Mehlhorn, K.: Can a maximum flow be computed in $o(nm)$ time? In: Paterson, M. (ed.) ICALP 1990. LNCS, vol. 443, pp. 235–248. Springer, Heidelberg (1990)
5. Dinic, E., Kronrod, M.: An algorithm for the solution of the assignment problem. Soviet Mathematics Doklady 10, 1324–1326 (1969)
6. Duan, R., Pettie, S.: Approximating maximum weight matching in near-linear time. In: Annual Symposium on Foundations of Computer Science, pp. 673–682. IEEE Computer Society, Washington, DC (2010)
7. Feder, T., Motwani, R.: Clique partitions, graph compression and speeding-up algorithms. Journal of Computer and System Sciences 51(2), 261–272 (1995)
8. Fredman, M.L., Tarjan, R.E.: Fibonacci heaps and their uses in improved network optimization algorithms. Journal of the ACM 34(3), 596–615 (1987)
9. Gabow, H.N., Tarjan, R.E.: Faster scaling algorithms for network problems. SIAM Journal on Computing 18(5), 1013–1036 (1989)
10. Gabow, H.N.: Scaling algorithms for network problems. Journal of Computer and System Sciences 31(2), 148–168 (1985)
11. Hopcroft, J.E., Karp, R.M.: An $n^{5/2}$ algorithm for maximum matchings in bipartite graphs. SIAM Journal on Computing 2(4), 225–231 (1973)
12. Iri, M.: A new method for solving transportation-network problems. Journal of the Operations Research Society of Japan 3, 27–87 (1960)
13. Kao, M.-Y., Lam, T.-W., Sung, W.-K., Ting, H.-F.: A decomposition theorem for maximum weight bipartite matchings with applications to evolutionary trees. In: Nešetřil, J. (ed.) ESA 1999. LNCS, vol. 1643, pp. 438–449. Springer, Heidelberg (1999)
14. Kao, M.Y., Lam, T.W., Sung, W.K., Ting, H.F.: A decomposition theorem for maximum weight bipartite matchings. SIAM Journal on Computing 31(1), 18–26 (2001)
15. Korte, B., Vygen, J.: Combinatorial Optimization: Theory and Algorithms, 4th edn. Springer (2007)
16. Kuhn, H.W.: The hungarian method for the assignment problem. Naval Research Logistics Quarterly 2, 83–97 (1955)
17. Munkres, J.: Algorithms for the assignment and transportation problems. Journal of the Society for Industrial and Applied Mathematics 5(1), 32–38 (1957)
18. Sankowski, P.: Maximum weight bipartite matching in matrix multiplication time. Theoretical Computer Science 410(44), 4480–4488 (2009)
19. Schrijver, A.: Combinatorial Optimization - Polyhedra and Efficiency, vol. 24 A. Springer (2003)

Intersection Dimension of Bipartite Graphs

Steven Chaplick[1,*], Pavol Hell[2,**], Yota Otachi[3,***],
Toshiki Saitoh[4,†], and Ryuhei Uehara[3,‡]

[1] Department of Applied Mathematics, Faculty of Mathematics and Physics, Charles
University, Malostranské náměstí 25, 118 00 Prague, Czech Republic
chaplick@kam.mff.cuni.cz
[2] School of Computing Science, Simon Fraser University,
Burnaby, B.C., Canada V5A 1S6
pavol@sfu.ca
[3] School of Information Science, Japan Advanced Institute of Science
and Technology, Asahidai 1-1, Nomi, Ishikawa 923-1292, Japan
{otachi,uehara}@jaist.ac.jp
[4] Graduate School of Engineering, Kobe University,
Rokkodai 1-1, Nada, Kobe, 657-8501, Japan
saitoh@eedept.kobe-u.ac.jp

Abstract. We introduce a concept of intersection dimension of a graph
with respect to a graph class. This generalizes Ferrers dimension, boxicity, and poset dimension, and leads to interesting new problems. We focus
in particular on bipartite graph classes defined as intersection graphs of
two kinds of geometric objects. We relate well-known graph classes such
as interval bigraphs, two-directional orthogonal ray graphs, chain graphs,
and (unit) grid intersection graphs with respect to these dimensions. As
an application of these graph-theoretic results, we show that the recognition problems for certain graph classes are NP-complete.

Keywords: Ferrers dimension, Boxicity, Unit grid intersection graph,
Segment-ray graphs, Orthogonal ray graph, NP-hardness.

1 Introduction

Given a family $\mathcal{F}$ of sets, the *intersection graph* of $\mathcal{F}$ is the graph in which each set
in $\mathcal{F}$ is a vertex, and two vertices are adjacent if and only if the corresponding sets
intersect. A typical example, when $\mathcal{F}$ is a family of intervals on a line, yields the

* Supported by the ESF GraDR EUROGIGA grant as project GACR
GIG/11/E023.
** Partially supported by NSERC (Canada) and ERCCZ LL 1201 Cores (Czech
Republic).
*** Partially supported by JSPS KAKENHI Grant Number 25730003 and MEXT
KAKENHI Grant Number 24106004.
† Partially supported by JSPS KAKENHI Grant Number 24700130.
‡ Partially supported by JSPS KAKENHI Grant Number 23500013 and MEXT
KAKENHI Grant Number 24106004.

T V Gopal et al. (Eds.): TAMC 2014, LNCS 8402, pp. 323–340, 2014.
© Springer International Publishing Switzerland 2014

well-known class of *interval graphs*. Interval graphs have linear time recognition algorithms [3,11], and nice forbidden structure characterizations. (For instance, the theorem of Lekkerkerker and Boland [24] characterizes interval graphs by the absence of induced cycles of length four and five, and the absence of asteroidal triples.)

It is natural to study a bipartite version of intersection graphs: given two families $\mathcal{F}$ and $\mathcal{F}'$ of sets, the *intersection bigraph* of $\mathcal{F}, \mathcal{F}'$ is the bipartite graph in which each set in $\mathcal{F}$ is a red vertex, each set in $\mathcal{F}'$ is a blue vertex, and a red vertex is adjacent to a blue vertex if and only if the corresponding sets intersect. When both $\mathcal{F}$ and $\mathcal{F}'$ are families of intervals on a line, we obtain *interval bigraphs* studied in [25,30]. We denote the class of interval bigraphs by IBG. While the recognition of interval bigraphs is polynomial (in time $O(n^{16})$ [25]), there is no efficient algorithm known, and no characterization in terms of forbidden substructures. It turns out that there are better bipartite analogues of interval graphs. A *two-directional orthogonal ray graph*, or 2DOR graph, is an intersection bigraph of a family $\mathcal{F}$ of upward rays, and a family $\mathcal{F}'$ of rightward rays, in the plane [32]. These graphs were introduced in connection with defect tolerance schemes for nano-programmable logic arrays [28,36]. There are several reasons these 2DOR graphs might be considered better bipartite analogues of interval graphs, including an ordering characterization [32,21], and a Lekkerkerker-Boland type characterization [12], both analogous to the characterizations for interval graphs. Moreover, it follows from [12] that the class 2DOR plays the same role for bigraphs as the class of interval graphs play for graphs, as far as polynomial solvability of certain constraint satisfaction problems is concerned. Other equivalent definitions, and forbidden structure characterizations of the class 2DOR can be found in [19,20,12].

Several other graph classes can be defined as intersection bigraphs of two families $\mathcal{F}, \mathcal{F}'$. When both $\mathcal{F}$ and $\mathcal{F}'$ are inclusion-free families of intervals on a line, we obtain the class of *proper interval bigraphs* which turns out to be the same as the better known class BPG of *bipartite permutation graphs* [20], see below. When $\mathcal{F}$ is a family of points, and $\mathcal{F}'$ a family of rightward rays, in a line, we obtain the class CHAIN of *chain graphs* (cf. below). When $\mathcal{F}$ is a family of vertical segments, and $\mathcal{F}'$ a family of horizontal segments, in the plane, we obtain the class GIG of *grid intersection graphs*. Several other examples are included in the paper.

We note that the following inclusions are well known or easy to derive

$$\mathsf{CHAIN} \subseteq \mathsf{BPG} \subseteq \mathsf{IBG} \subseteq \mathsf{2DOR} \subseteq \mathsf{GIG}.$$

We now introduce our concept of intersection dimension. Let $G = (V, E)$ and $G' = (V', E')$ be two graphs. The *intersection* $G \cap G'$ of G and G' is the graph $(V \cap V', E \cap E')$. For two graph classes $\mathcal{C}$ and $\mathcal{C}'$, we define the *pairwise intersection* of $\mathcal{C}$ and $\mathcal{C}'$ as $\mathcal{C} \boxtimes \mathcal{C}' = \{G \cap G' : G \in \mathcal{C}, G' \in \mathcal{C}'\}$. We also write $\mathcal{C}^k = \{G_1 \cap G_2 \cap \cdots \cap G_k : G_i \in \mathcal{C}$ for $1 \leq i \leq k\}$. If both $\mathcal{C}$ and $\mathcal{C}'$ are closed under taking induced subgraphs, it is easy to check that $\mathcal{C} \boxtimes \mathcal{C}' = \{G \cap G' : G \in \mathcal{C}, G' \in \mathcal{C}', V(G) = V(G')\}$. Since every graph class in this paper is closed under taking induced subgraphs, we shall from now on use the latter equality, and assume that the vertex sets of the two graphs are the same, when defining the pairwise intersection of graph classes.

The *dimension* of a graph G with respect to the graph class $\mathcal{C}$ is the minimum k such that $G \in \mathcal{C}^k$. In the discussion below we shall point out how this definition generalizes Ferrers dimension, boxicity, cubicity, and poset dimension. We are particularly interested in expressing one graph class as a (subset of a) power of another graph class.

It turns out that there are several natural statements of this kind. Among other results we will show that $2\mathsf{DOR} = \mathsf{CHAIN}^2$, $\mathsf{GIG} \subseteq \mathsf{CHAIN}^4$, and $\mathsf{UGIG} = \mathsf{BPG}^2$. We will also show that several of these inclusions are proper. See Fig. 1 for the summary of our results.

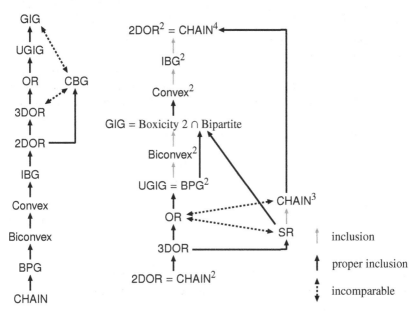

Fig. 1. (Left) Known hierarchy. (Right) New hierarchy based on intersection dimensions.

2 Preliminaries

A graph $G = (V, E)$ is a *bipartite graph* (or a *bigraph* for short) with bipartition (X, Y) if V is partitioned into X and Y in such a way that each edge of G has one endpoint in X and the other in Y. We denote such a bigraph by $(X, Y; E)$. A *biadjacency matrix* M_B of a bigraph $B = (X, Y; E)$ is a 0-1 matrix with the rows indexed by the vertices of X and the columns indexed by the vertices of Y such that $\{x, y\} \in E$ if and only if the corresponding entry of M_B is 1. For $m \times n$ 0-1 matrices M' and M'', their intersection $M = M' \cap M''$ is the 0-1 matrix such that $M_{i,j} = 1$ if and only if $M'_{i,j} = 1$ and $M''_{i,j} = 1$. The *neighborhood* of a vertex v in a graph G, denoted $N_G(v)$, is the vertices adjacent to v in G.

2.1   Graph Classes

Here we define the graph classes we deal with in this paper. We also introduce some important properties of them. For their inclusion relations and other known results for them, the readers can refer to the standard textbooks in this field [4,15,35].

For a graph class $\mathcal{C}$, the *recognition problem* of $\mathcal{C}$ is the problem deciding whether a given graph belongs to $\mathcal{C}$.

Chain Graphs and Ferrers Diagrams. A bipartite graph $B = (X, Y; E)$ is a *chain graph* if there is an ordering $(x_1, x_2, \ldots, x_p)$ on X such that $N_B(x_1) \supseteq N_B(x_2) \supseteq \cdots \supseteq N_B(x_p)$. It is easy to see that if there exists such an ordering on X, then there exists an ordering $(y_1, y_2, \ldots, y_q)$ on Y such that $N_B(y_1) \supseteq N_B(y_2) \supseteq \cdots \supseteq N_B(y_q)$. Chain graphs are also known as *difference graphs* and *Ferrers bigraphs*. It is known that chain graphs are exactly $2K_2$-free bigraphs [16]. The class of chain graphs is denoted by CHAIN.

A 0-1 matrix has the *Ferrers property* if its rows and columns can be reordered so that 1's in each row and column appear consecutively with the rows left-justified and the columns top-justified. The reorderd matrix is called a *Ferrers diagram*. It is easy to see that a matrix has the Ferrers property if and only if it has none of the following 2×2 matrices as a submatrix:

$$\begin{pmatrix} 0 & 1 \\ 1 & 0 \end{pmatrix}, \quad \begin{pmatrix} 1 & 0 \\ 0 & 1 \end{pmatrix}. \tag{1}$$

Since chain graphs are exactly the $2K_2$-free bigraphs, it is easy to see that chain graphs are exactly the bigraphs whose biadjacency matrices have the Ferrers property.

Bipartite Permutation Graphs, Convex Graphs, Biconvex Graphs, Interval Bigraphs, and Chordal Bipartite Graphs. A graph $G = (V, E)$ with $V = \{1, 2, \ldots, n\}$ is a *permutation graph* if there is a permutation π over V such that $\{i, j\} \in E(G)$ if and only if $(i - j)(\pi(i) - \pi(j)) < 0$. A graph is a *bipartite permutation graph* if it is bipartite and a permutation graph. The class of bipartite permutation graphs is denoted by BPG. Several equivalent definitions of the class BPG are collected in [20].

An ordering $<$ of X in a bipartite graph $B = (X, Y; E)$ has the *adjacency property* if for every vertex y in Y, $N(y)$ consists of vertices that are consecutive in the ordering $<$ of X. A bipartite graph $(X, Y; E)$ is *convex* if there is an ordering of X or Y that fulfills the adjacency property. A bipartite graph $(X, Y; E)$ is *biconvex* if there are orderings of X and Y that fulfill the adjacency property. We denote the classes of convex bipartite graphs and biconvex bipartite graphs by Convex and Biconvex, respectively.

A *bi-interval representation* of a bigraph $B = (U, V; E)$ is a pair $(\mathcal{I}_U, \mathcal{I}_V)$ of sets of closed intervals such that $\mathcal{I}_U = \{I_u = [\ell_u, r_u] : u \in U\}$ and $\mathcal{I}_V = \{I_v = [\ell_v, r_v] : v \in V\}$, and $\{u, v\} \in E$ for $u \in U$ and $v \in V$ if and only if $I_u \cap I_v \neq \emptyset$.

A bi-interval representation $(\mathcal{I}_U, \mathcal{I}_V)$ is *unit* if for each interval $[\ell, r] \in \mathcal{I}_U \cup \mathcal{I}_V$, $r - \ell = 1$.

A bigraph is a *chordal bipartite graph* if every induced cycle is of length four. The class of chordal bipartite graphs is denoted by CBG.

Orthogonal Ray Graphs. A bipartite graph $B = (X, Y; E)$ is an *orthogonal ray graph* if there is a pair $(\mathcal{R}_X, \mathcal{R}_Y)$ of families of rays (or half-lines) such that $\mathcal{R}_X = \{R_x : x \in X\}$ is a family of pairwise non-intersecting horizontal rays, $\mathcal{R}_Y = \{R_y : y \in Y\}$ is a family of pairwise non-intersecting vertical rays, and $\{x, y\} \in E$ if and only if R_x and R_y intersect. We call such a pair $(\mathcal{R}_X, \mathcal{R}_Y)$ an *orthogonal ray representation* of B. We denote the class of orthogonal ray graphs by OR.

Note that in a representation of an orthogonal ray graph horizontal rays can go rightward and leftward and vertical rays can go upward and downward. If we restrict horizontal rays to be only rightwards, then we have *3-directional orthogonal ray graphs*. Furthermore, if we restrict horizontal rays to be only rightwards and vertical rays to be only upwards, then we have *2-directional orthogonal ray graphs*. We denote the classes of 3-directional orthogonal ray graphs and 2-directional orthogonal ray graphs by 3DOR and 2DOR, respectively.

For the class 2DOR, several nice characterizations are known (see, for example, [21,12,29,30,31,19,32]). Among those characterizations, the followings are useful for our purpose. In this language they appear in [31,32], in an equivalent graph theoretic form they are given in [21,19].

Theorem 2.1. *For a bigraph B, the following conditions are equivalent:*

1. *B is a 2-directional orthogonal ray graph;*
2. *B is γ-freeable; that is, the rows and columns of a biadjacency matrix of B can be independently permuted so that no 0 has a 1 both below it and to its right;*
3. *B is of Ferrers dimension at most 2. (The Ferrers dimension of a bigraph is defined in Section 2.1.)*

There are other equivalent characterizations of the class 2DOR, as suggested in the introduction. In particular, 2DOR is precisely the class of bigraphs whose complements are circular arc graphs [32]; because of the characterizations of the latter class in [12,19,20], one obtains several other forbidden structure characterizations of 2DOR, in terms of the absence of induced cycles and bipartite versions of asteroids, in terms of the so-called invertible pairs, and in other terms.

It is known that the recognition of 2DOR can be done in polynomial time [12,32], while it is open for 3DOR and OR. Recently, Felsner, Mertzios, and Mustață [14] have shown that if the direction (right, left, up, or down) for each vertex is given, then it can be decided in polynomial time whether a given graph has an orthogonal ray representation in which each vertex has the given direction.

Grid Intersection Graphs. A bipartite graph $B = (X, Y; E)$ is a *grid intersection graph* if there is a pair $(\mathcal{S}_X, \mathcal{S}_Y)$ of families of segments such that

$\mathcal{S}_X = \{S_x : x \in X\}$ is a family of pairwise non-intersecting horizontal segments, $\mathcal{S}_Y = \{S_y : y \in Y\}$ is a family of pairwise non-intersecting vertical segments, and $\{x, y\} \in E$ if and only if S_x and S_y intersect. We call such a pair $(\mathcal{S}_X, \mathcal{S}_Y)$ a *grid intersection representation* of B. A bipartite graph is a *unit grid intersection graph* if it has a grid intersection representation in which each segment if of length 1. We denote the classes of grid intersection graphs and unit grid intersection graphs by GIG and UGIG, respectively.

Segment-Ray Graphs. A bipartite graph $B = (X, Y; E)$ is a *segment-ray graph* if there is a pair $(\mathcal{S}_X, \mathcal{R}_Y)$ of families of segments and rays such that $\mathcal{S}_X = \{S_x : x \in X\}$ is a family of pairwise non-intersecting horizontal segments, $\mathcal{R}_Y = \{R_y : y \in Y\}$ is a family of pairwise non-intersecting vertical upward rays, and $\{x, y\} \in E$ if and only if S_x and R_y intersect. We call such a pair $(\mathcal{S}_X, \mathcal{R}_Y)$ a *segment-ray representation* of B. We denote the class of segment-ray graphs by SR.

Recognition Problems and Inclusion Relations. For the graph classes introduced above, the following relations are known [4,27,32]: CHAIN $\subsetneq$ BPG $\subsetneq$ Biconvex $\subsetneq$ Convex $\subsetneq$ IBG $\subsetneq$ 2DOR $\subsetneq$ 3DOR $\subsetneq$ OR $\subsetneq$ UGIG $\subsetneq$ GIG. Also it is known that 2DOR $\subsetneq$ CBG [32], and that CBG is incomparable to 3DOR and GIG [27].

It is known that the recognition problems of CHAIN [18], BPG [33], Biconvex [35], Convex [35], IBG [25], 2DOR [32], and CBG [34] can be solved in polynomial time. On the other hand, it is known that the recognition problems of GIG [23] and UGIG [26,37] are NP-complete. The complexity of the recognition problems of 3DOR, OR, and SR is not known.

Note that even if three graph classes $\mathcal{A}$, $\mathcal{B}$, and $\mathcal{C}$ satisfy $\mathcal{A} \subseteq \mathcal{B} \subseteq \mathcal{C}$ and the recognition problems of $\mathcal{A}$ and $\mathcal{C}$ are both polynomial-time solvable (NP-hard), it does not mean the recognition problem of $\mathcal{B}$ is polynomial-time solvable (NP-hard, resp.).

Other Graphs. The *d-dimensional hypercube* H_d is the graph with 2^d vertices in which the vertices corresponds to the subsets of $\{1, \ldots, d\}$ and two vertices are adjacent if and only if the symmetric difference of the corresponding sets is of size 1.

Let $K_{a,b}$ denote the complete bipartite graph having a vertices in one side and b vertices in the other side. We denote by $K_{n,n} - nK_2$ the graph obtained by removing a perfect matching from the complete bipartite graph $K_{n,n}$.

Boxicity and Cubicity. An *interval graph* is the intersection graph of closed intervals on the real line. A *unit interval graph* is the intersection graph of closed unit intervals on the real line. We denote the classes of interval graphs and unit interval graphs by INT and UINT, respectively.

The *boxicity* of a graph G is the minimum integer k such that $G \in \mathsf{INT}^k$, and the *cubicity* of G is the minimum integer k such that $G \in \mathsf{UINT}^k$. It is known that given a graph, deciding whether its boxicity (or cubicity) is at most 2 is NP-complete [23,5].

Bigraph Intersection Dimension. For bipartite graph classes, if one of them is additionally closed under disjoint union, we may assume that the bipartitions of G and G' are the same when taking their intersection. More precisely, we have the following lemma.

Lemma 2.2. *Let $\mathcal{B}$ and $\mathcal{B}'$ be bipartite graph classes. If at least one of them is closed under disjoint union and taking induced subgraphs, then $\mathcal{B} \boxtimes \mathcal{B}' = \{(X, Y; E) \cap (X, Y; E') : (X, Y; E) \in \mathcal{B}, (X, Y; E') \in \mathcal{B}'\}$.*

Proof. Let $\mathcal{C} = \{(X, Y; E) \cap (X, Y; E') : (X, Y; E) \in \mathcal{B}, (X, Y; E') \in \mathcal{B}'\}$. Clearly, $\mathcal{C} \subseteq \mathcal{B} \boxtimes \mathcal{B}'$. In the following, we show that $\mathcal{B} \boxtimes \mathcal{B}' \subseteq \mathcal{C}$. By symmetry, we may assume that $\mathcal{B}'$ is closed under disjoint union and taking induced subgraphs.

Let $H = (X, Y; E) \in \mathcal{B}$ and $H' = (X', Y'; E') \in \mathcal{B}'$. Now let $H'' = (X, Y; E' \cap \{\{x, y\} : x \in X, y \in Y\})$. It is easy to see that $H \cap H' = H \cap H''$. Observe that H'' is the disjoint union of two induced subgraphs of H', where one is induced by $(X \cap X', Y \cap Y')$ and the other by $(X \cap Y', X \cap Y')$. Since $\mathcal{B}'$ is closed under disjoint union and taking induced subgraphs, it follows that $H'' \in \mathcal{B}'$. Since $H \cap H' = H \cap H''$, we have $H \cap H' \in \mathcal{C}$. $\square$

Unfortunately, CHAIN is not closed under disjoint union. For example, K_2 is a chain graph but $2K_2$ is not. It is the only exception in this paper. Fortunately, we have the following lemma for chain graphs.

Lemma 2.3. CHAIN$^2 = \{(X, Y; E) \cap (X, Y; E') : (X, Y; E), (X, Y; E') \in \text{CHAIN}\}$.

Proof. Let $\mathcal{C} = \{(X, Y; E) \cap (X, Y; E') : (X, Y; E), (X, Y; E') \in \text{CHAIN}\}$. Clearly, $\mathcal{C} \subseteq \text{CHAIN}^2$. In the following, we show that CHAIN$^2 \subseteq \mathcal{C}$.

Let $H_1 = (X_1, Y_1; E_1) \in \text{CHAIN}$ and $H_2 = (X_2, Y_2; E_2) \in \text{CHAIN}$. Now let $H_1' = (X_1, Y_1; E_1')$ and $H_2' = (X_1, Y_1; E_2')$, where

$$E_1' = E_1 \cup \{\{x, y\} : x \in X_1 \cap X_2, y \in Y_1 \cap X_2\} \setminus \{\{x, y\} : x \in X_1 \cap Y_2, y \in Y_1 \cap Y_2\},$$

$$E_2' = E_2 \cup \{\{x, y\} : x \in X_1 \cap Y_2, y \in Y_1 \cap Y_2\}, \setminus \{\{x, y\} : x \in X_1 \cap X_2, y \in Y_1 \cap X_2\}.$$

See Fig. 2. It is not difficult to see that $H_1 \cap H_2 = H_1' \cap H_2'$. Observe that both H_1' and H_2' are chain graphs. Therefore, $H_1 \cap H_2 = H_1' \cap H_2' \in \mathcal{C}$. $\square$

By Lemmas 2.2 and 2.3, we can assume that the bipartitions of two graphs are the same when we are defining the pairwise intersection of two graph classes, since, in this paper, either one of them is closed under disjoint union or both of them are the class of chain graphs.

Ferrers Dimension. The *Ferrers dimension* fd(B) of a bigraph B is the smallest number of Ferrers bigraphs whose intersection is B. That is, fd(B) is the minimum integer k such that $B \in \text{CHAIN}^k$. If $B = (X, Y; E)$ and fd$(B) = k$, then there are Ferrers bigraphs $B_i = (X, Y; E_i)$ for $1 \leq i \leq k$ such that

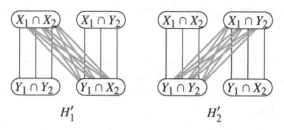

Fig. 2. Intersection of two chain graphs

$B = \bigcap_{1 \leq i \leq k} B_i$. That is, we can assume all the graphs B and B_i, $1 \leq i \leq k$ have the same bipartition.

A *Ferrers digraph* $D = (V, A)$ is a digraph whose adjacency matrix has the Ferrers property. The *Ferrers dimension* $\mathsf{fd}(D)$ *of a digraph* D is the smallest number of Ferrers digraphs whose intersection is D.

Poset Dimension. The *poset dimension* $\mathsf{pd}(P)$ of a poset P is the minimum integer k such that there exist k linear extensions of P such that for any two elements x, y of P, $x < y$ in P if and only if $x < y$ in all the linear extensions. The *Ferrers dimension* $\mathsf{fd}(P)$ *of a poset* P is the Ferrers dimension of the digraph defined in such way that the vertices are the elements of P and there is an arc (u, v) if and only if $u < v$. Cogis [10] showed that for any poset P, $\mathsf{fd}(P) = \mathsf{pd}(P)$.

A poset is of *height 2* if every element is either a minimal element or a maximal element. The *underlying graph* of a height-2 poset is the bigraph $B = (X, Y; E)$ such that X is the set of minimal elements, Y is the set of maximal elements, and $\{x, y\} \in E$ if and only if $x < y$. It is easy to see that any bigraph is the underlying graph of some poset of height 2.

3 $(P, Q; D)$-Bigraphs

We introduce the notion of $(P, Q; D)$-bigraphs, where a bigraph $B = (U, V, E)$ is said to be an $(P, Q; D)$-bigraph if and only if for some domain D (e.g., the real number line $\mathbb{R}$) each vertex in $u \in U$ can be represented as a type P subset P_u of D and each vertex $v \in V$ can be represented as a type Q subset Q_v of D such that for every $u \in U, v \in V$, $uv \in E$ if and only if $P_u \cap Q_v \neq \emptyset$. For example, in this setting, interval bigraphs are (interval, interval, $\mathbb{R}$)-bigraphs. We will use $(P, Q; D)$ to denote the class of $(P, Q; D)$-bigraphs.

Our discussion will focus on the cases when P, Q are the following subsets of $\mathbb{R}$: points, rays, unit-intervals, and intervals; and the following axis-aligned subsets of $\mathbb{R}^2$: points, rays, unit-segments, segments, squares, and rectangles. Note: for rays, we will use $\rightarrow, \downarrow, \leftarrow$, and $\uparrow$ to denote the *rightward, downward, leftward,* and *upward* rays respectively. Moreover, when we refer to a ray r (rather than using a specific arrow), r can be any axis-aligned ray from the domain.

3.1 $(P, Q; \mathbb{R})$-Bigraphs

We begin with some easy observations characterizing CHAIN, Convex, and Biconvex bigraphs as $(P, Q; D)$-bigraphs (see Proposition 3.1). This is followed by a couple essential lemmas that we will use to relate $(P, Q, \mathbb{R})$-bigraphs to $(P', Q', \mathbb{R}^2)$-bigraphs.

Proposition 3.1. *For a bigraph $B = (X, Y, E)$:*

1. *B is CHAIN if and only if B is (point, $\to$; $\mathbb{R}$).*
2. *B is Convex if and only if B is (point, interval; $\mathbb{R}$).*
3. *B is Biconvex if and only if B is both (point, interval; $\mathbb{R}$) and (interval, point; $\mathbb{R}$).*

Proof. These follow easily by definition. □

It is also known that a bigraph is a bipartite permutation graph (BPG) if and only if it is a unit-interval bigraph [20]; i.e., BPG = (unit-interval, unit-interval; $\mathbb{R}$). Interestingly, we observe that (unit-interval, unit-interval; $\mathbb{R}$)-bigraphs actually have a simpler representation. Specifically, (unit-interval, unit-interval; $\mathbb{R}$) = (point, unit-interval; $\mathbb{R}$) and we prove this via the following more general lemma.

Lemma 3.2. *For a bigraph $B = (U, V; E)$ and any $Q \in \{\to$, ray, unit-interval, interval$\}$, $B \in$ (unit-interval, Q; $\mathbb{R}$) if and only if $B \in$ (point, Q; $\mathbb{R}$).*

Proof. Notice that for any choice of Q each element of V is represented as an interval. Let $(\mathcal{I}_U, \mathcal{I}_V)$ be a (unit-interval, Q; $\mathbb{R}$) representation of B. Let $I_u = [\ell_u, \ell_u + 1] \in \mathcal{I}_U$ and $I_v = [\ell_v, r_v] \in \mathcal{I}_V$ be intervals corresponding to $u \in U$ and $v \in V$, respectively. It is easy to see that I_u and I_v intersect if and only if either $\ell_u \in I_v$ or $\ell_v - \ell_u \in [0, 1]$.

We define the following (point, Q; $\mathbb{R}$) representation $(\mathcal{I}'_U, \mathcal{I}'_V)$ as:

$$\mathcal{I}'_U = \{\{\ell_u\} : [\ell_u, \ell_u + 1] \in \mathcal{I}_U\},$$
$$\mathcal{I}'_V = \{[\ell_v - 1, r_v] : [\ell_v, r_v] \in \mathcal{I}_V\}.$$

Obviously $(\mathcal{I}'_U, \mathcal{I}'_V)$ represents B, since $\ell_u \in [\ell_v - 1, r_v]$ if and only if either $\ell_u \in I_v$ or $\ell_v - \ell_u \in [0, 1]$. It is easy to see that we now have a (point,Q;$\mathbb{R}$) representation of B. □

Lemma 3.2 allows us to equate several $(P, Q; \mathbb{R})$ classes. These are given in the following two corollaries.

Corollary 3.3. *For each $Q \in \{\to$, ray, unit-interval, interval$\}$, the following classes of bigraphs are the same: (point, Q; $\mathbb{R}$), ($\to$, Q; $\mathbb{R}$), (ray, Q; $\mathbb{R}$), (unit-interval, Q; $\mathbb{R}$).*

Corollary 3.4. *For each $P, Q \in \{$point, $\to$, $\leftarrow$, unit-interval$\}$, a bigraph B is $(P, Q; \mathbb{R})$ if and only if B is $(Q, P; \mathbb{R})$.*

Notice that the statement of Corollary 3.4 does not allow either of P or Q to be ray-type sets. This is because Lemma 3.2 cannot be used to give us the desired biconvexity-like when rays are allowed for a given set. However, by Lemma 3.2, we can transform any (ray, ray;$\mathbb{R}$) representation into a (point, ray;$\mathbb{R}$) representation. Thus, (ray,ray;$\mathbb{R}$) is a subset of the bigraphs which are both (point,ray;$\mathbb{R}$) and (ray,point;$\mathbb{R}$). One open question would be whether these are the same.

Moreover, the graph (P_7) given in Figure 3 is (point, ray; $\mathbb{R}$) but not both (point, ray; $\mathbb{R}$) and (ray, point; $\mathbb{R}$). This is easy to see since no three vertices in the same partition (say, X) can have pairwise incomparable neighborhoods; i.e., two of the three must be represented by rays in the same direction and thus must have nested neighborhoods. Moreover, the graph in Figure 3 has $a, b, c \in X$ such that their neighborhoods are pairwise incomparable. This is formalized in the following proposition.

Proposition 3.5. *If a bigraph $B = (X, Y; E)$ is (ray,point;$\mathbb{R}$) where each $x \in X$ is a ray then for every $\{x, x', x''\} \subseteq X$ and every $y \in Y$, there exists $x^* \in \{x, x', x''\}$ and $x^{**} \in \{x, x', x''\} \setminus \{x^*\}$ such that $N(x^*) \subseteq N(x^{**})$ or $N(x) \subseteq N(x'')$.*

Fig. 3. The path on seven vertices (P_7) and a (point, ray;$\mathbb{R}$) representation of it. Note: P_7 is not both (point, ray;$\mathbb{R}$) and (ray, point;$\mathbb{R}$) since the neighborhoods of a, b, and c are pairwise incomparable.

3.2 $(P, Q; \mathbb{R}^2)$-Bigraphs

In this subsection we consider the domain $\mathbb{R}^2$ and describe several classes of bigraphs as the intersection of one dimensional bigraph classes (i.e., as $(P, Q; \mathbb{R}) \cap (P', Q'; \mathbb{R})$). Notice that, for $P, Q \in \{$point, unit-interval, interval$\}$ $(P, Q; \mathbb{R})$ is hereditary and closed under disjoint union. Thus, by Lemma 2.2, for $P, Q \in \{$point, unit-interval, interval$\}$ and any choices of P' and Q', $B = (X, Y; E)$ is $(P, Q; \mathbb{R}) \cap (P', Q'; \mathbb{R})$ if and only if $B = (X, Y; E \cap E')$ for $(X, Y; E) \in (P, Q; \mathbb{R})$ and $(X, Y; E'') \in (P', Q'; \mathbb{R})$.

Theorem 3.6. $\mathsf{UGIG} = \mathsf{BPG}^2 = $*(point, unit-interval;$\mathbb{R}$)^2.*

Proof. First we show that $\mathsf{UGIG} \subseteq \mathsf{BPG}^2$. Let $G = (U, V; E) \in \mathsf{UGIG}$ and $\mathcal{R} = (\mathcal{U}, \mathcal{V})$ be a unit grid representation of G, where the horizontal segments $\mathcal{U}$ represent the vertices in U and the vertical segments $\mathcal{V}$ represent the vertices in V. That is, $\mathcal{U} = \{\{y_u\} \times [x_u, x_u + 1] : u \in U\}$, $\mathcal{V} = \{[y_v, y_v + 1] \times \{x_v\} : v \in V\}$, and $E = \{\{u, v\} : u \in U, v \in V, y_u \in [y_v, y_v + 1], x_v \in [x_u, x_u + 1]\}$. From $\mathcal{U}$, we construct two point-unit bi-interval representations $\mathcal{R}'$ and $\mathcal{R}''$ as follows:

$$\mathcal{R}' = (\{y_u : u \in U\}, \{[y_v, y_v + 1] : v \in V\}),$$
$$\mathcal{R}'' = (\{x_v : v \in V\}, \{[x_u, x_u + 1] : u \in U\}).$$

By Lemma 3.2, $\mathcal{R}'$ and $\mathcal{R}''$ represent the bipartite permutation graphs $G' = (U, V; E')$ and $G'' = (U, V; E'')$, respectively, where

$$E' = \{\{u, v\} : u \in U, v \in V, y_u \in [y_v, y_v + 1]\}, \text{ and}$$
$$E'' = \{\{u, v\} : u \in U, v \in V, x_v \in [x_u, x_u + 1]\}.$$

Since $\{u, v\} \in E' \cap E''$ for $u \in U$ and $v \in V$ if and only if $y_u \in [y_v, y_v + 1]$ and $x_v \in [x_u, x_u + 1]$, we have $E = E' \cap E''$. Therefore, $G = G' \cap G''$.

Next we show that $\mathsf{BPG}^2 \supseteq \mathsf{UGIG}$. Let $G' = (U, V; E')$ and $G'' = (U, V; E'')$ be bipartite permutation graphs. Let $\mathcal{R}'$ and $\mathcal{R}''$ be point-unit bi-interval representations of G' and G'', respectively, such that U is the point set of $\mathcal{R}'$ and the unit interval set of $\mathcal{R}''$. Such representations exist by Corollary 3.3. Let $u \in U$, and let p_u and $[\ell_u, \ell_u + 1]$ be the point in $\mathcal{R}'$ and the unit interval in $\mathcal{R}''$ representing the vertex u. We assign the unit horizontal segment $\{p_u\} \times [\ell_u, \ell_u + 1]$ to u. Similarly, for a vertex $v \in V$ with the unit interval $[\ell_v, \ell_v + 1]$ in $\mathcal{R}'$ and the point p_v in $\mathcal{R}''$, we assign the unit vertical segment $[\ell_v, \ell_v + 1] \times \{p_v\}$. The obtained unit grid representation represents $G = G' \cap G''$, since $\{p_u\} \times [\ell_u, \ell_u+1]$ and $[\ell_v, \ell_v+1] \times \{p_v\}$ intersect if and only if $p_u \in [\ell_v, \ell_v+1]$ and $p_v \in [\ell_u, \ell_u+1]$. $\square$

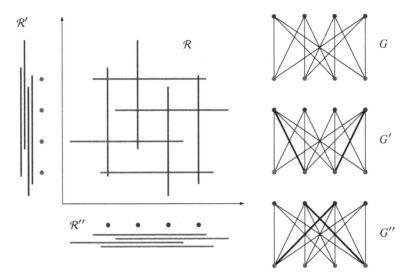

Fig. 4. UGIG = BPG^2

Using Theorem 3.6 and Corollary 3.4 the following is immediate.

Corollary 3.7. *(unit-square, unit-square;$\mathbb{R}^2$) = (point, unit-interval;$\mathbb{R}$)^2* = UGIG.

The corollary above implies that a bipartite graph of cubicity-2 is UGIG. It is easy to see that the star $K_{1,5}$ is UGIG, but its cubicity is more than 2. Therefore, we have the following corollary, which is a nice complement to the fact Boxicity-2 $\cap$ Bipartite = GIG [2].

Corollary 3.8. *Cubicity-2 $\cap$ Bipartite $\subsetneq$ UGIG.*

The proof of the following theorem is an easy modification of the proof of Theorem 3.6. The relation GIG $\neq$ Convex^2 is shown by Fig. 5.

Theorem 3.9. Biconvex^2 $\subseteq$ (Biconvex $\bowtie$ Convex) $\subseteq$ GIG $\subsetneq$ Convex^2.

Since Convex $\subset$ 2DOR, it holds that GIG $\subseteq$ 2DOR^2 = CHAIN^4. Therefore, every grid intersection graph has Ferrers dimension at most 4.

Corollary 3.10. *The recognition problems of* BPG^2, *Biconvex^2, and* Biconvex $\bowtie$ Convex *are NP-complete.*

Proof. The problems are in NP since the recognition problems of BPG and Biconvex are polynomial-time solvable and the intersection of two graphs can be computed in polynomial time.

Mustaţă and Pergel [26] showed that the recognition problem is NP-hard for any graph class $\mathcal{C}$ satisfying UGIG $\subseteq$ $\mathcal{C}$ $\subseteq$ GIG. By Theorems 3.6 and 3.9 and the fact that BPG $\subset$ Biconvex, it follows that UGIG = BPG^2 $\subseteq$ Biconvex^2 $\subseteq$ GIG. Therefore, the recognition problems are NP-hard for BPG^2 and Biconvex^2.   □

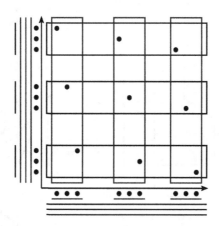

Fig. 5. A (point, interval)^2 representation of the full subdivision H of $K_{3,3}$; i.e., $H \in$ Convex^2. On the other hand, $H \notin$ GIG, since it is the full subdivision of a non-planar graph, and thus not a string graph.

4   Segment-Ray Graphs

Let F be a matrix with entries $0, 1, *$, where $*$ means "don't care." A matrix M is F-*free* if M does not have F as a submatrix ignoring $*$-entries. A bipartite graph is F-*freeable* if it has a F-free biadjacency matrix.

It is known that a bipartite graph is a chordal bipartite graph if and only if it is Γ-freeable (see [22]), a 2-directional orthogonal ray graph if and only if it is

γ-freeable [32], and a grid intersection graph if and only if it is cross-freeable [17], where the forbidden matrices are defined as follows:

$$\Gamma = \begin{pmatrix} 1 & 0 \\ 1 & 1 \end{pmatrix}, \qquad \gamma = \begin{pmatrix} 1 & 0 \\ * & 1 \end{pmatrix}, \qquad \text{cross} = \begin{pmatrix} * & 1 & * \\ 1 & 0 & 1 \\ * & 1 & * \end{pmatrix}.$$

In this section, using the following matrix V, we characterize segment-ray graphs:

$$\text{V} = \begin{pmatrix} 1 & 0 & 1 \\ * & 1 & * \end{pmatrix}.$$

Obviously, a matrix is cross-free if it is V-free, and V-free if it is γ-free.

The proof of the following proof is similar to the proofs of the cross-free characterization of GIG [17] and the γ-free characterization of 2DOR [32].

Theorem 4.1. *A bipartite graph is a segment-ray graph if and only if it is V-freeable.*

Proof. For the only-if part, let $B = (U, V; E)$ be a segment-ray graph and $\mathcal{R}$ be its segment-ray representation such that each vertex in U corresponds to a horizontal segment in $\mathcal{R}$, and each vertex in V corresponds to a vertical upward ray in $\mathcal{R}$. Let M be the bipartite adjacency matrix of B with the rows indexed by U and the columns indexed by V. Let S_u be the segment corresponding to $u \in U$ with y-coordinate b, and R_v be the ray corresponding to $v \in V$ with x-coordinate a. If S_u intersects with rays on both sides of $x = a$ and R_v intersects with a segment below $y = b$, then S_u and R_v must intersect at (a, b). Thus we can make M V-free by permuting the columns in nondecreasing order of the x-coordinates of the corresponding rays and the rows in nonincreasing order of the y-coordinates of the corresponding segments.

For the if part, let $B = (U, V; E)$ be a bipartite graph and M be its V-free bipartite adjacency matrix with the rows indexed by U and the columns indexed by V. For each $u \in U$, we put the horizontal segment with end points (i, j_1) and (i, j_2), where i is the row index of u and j_1, j_2 are the smallest and largest indices such that $M_{i,j} = 1$. For each $v \in V$, we put the vertical upward ray from the starting point (i, j), where j is the column index of v and i is the largest index such that $M_{i,j} = 1$. For any two vertices $u \in U$ and $v \in V$, it is clear that the corresponding segment and ray intersect if the vertices are adjacent. Conversely, if u and v are not adjacent, then the corresponding segment and ray cannot intersect since M is V-free. □

Now we show that every segment-ray graph has Ferrers dimension at most 3. To this end, we need the following simple fact.

Lemma 4.2. *An $m \times n$ 0-1 matrix M is V-free if and only if for each entry (i, j) with $M_{i,j} = 0$ at least one of the following holds:*

1. $M_{i,k} = 0$ *for all* $1 \le k \le j$;
2. $M_{i,k} = 0$ *for all* $j \le k \le n$;

3. $M_{k,j} = 0$ for all $i \leq k \leq m$.

Theorem 4.3. *Every segment-ray graph has Ferrers dimension at most* 3.

Proof. Let B be a segment-ray graph and M be its V-free bipartite adjacency matrix. Let $M^{(1)}$, $M^{(2)}$, $M^{(3)}$ be the following 0-1 matrices of the same size with M:

- $M_{i,j}^{(1)} = 0$ if and only if $M_{i,k} = 0$ for all $1 \leq k \leq j$;
- $M_{i,j}^{(2)} = 0$ if and only if $M_{i,k} = 0$ for all $j \leq k \leq n$;
- $M_{i,j}^{(3)} = 0$ if and only if $M_{k,j} = 0$ for all $i \leq k \leq m$.

It is easy to see that $M^{(1)}$, $M^{(2)}$, $M^{(3)}$ have the Ferrers property. By Lemma 4.2, it holds that $M^{(1)} \cap M^{(2)} \cap M^{(3)} = M$. This completes the proof. $\square$

Note that the upper bounds of the Ferrers dimension for GIG (≤ 4) and 2DOR (≤ 2) can be shown in similar ways by using the forbidden submatrix characterizations.

Corollary 4.4. OR *is incomparable to both* CHAIN^3 *and* SR.

Proof. By Theorem 4.3, it holds that SR $\subseteq$ CHAIN^3. Hence it suffices to show that OR $\not\subseteq$ CHAIN^3 and SR $\not\subseteq$ OR. Fig. 6a shows that $H_3 \in$ OR. From the definitions, it holds that $H_3 = K_{4,4} - 4K_2$. It is known that $\mathsf{fd}(K_{n,n} - nK_2) = n$ [38,39], and thus $\mathsf{fd}(H_3) = 4$. Thus OR $\not\subseteq$ CHAIN^3. It is known that $C_{2n} \notin$ OR if $n > 6$ [32]. On the other hand, it is easy to see that $C_{2n} \in$ SR for any n (see Fig. 6b). Thus SR $\not\subseteq$ OR. $\square$

Corollary 4.5. SR *is a proper subset of* GIG.

Proof. From the definition, SR is a subset of GIG. Since $H_3 \in$ OR $\subset$ GIG and $H_3 \notin$ CHAIN$^3 \supseteq$ SR, it holds that SR $\neq$ GIG. $\square$

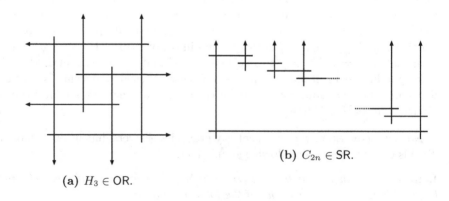

(a) $H_3 \in$ OR.

(b) $C_{2n} \in$ SR.

Fig. 6. Examples showing incomparabilities

5   Boxicity and Ferrers Dimension

Chatterjee and Ghosh [9] presented some relations between the boxicity of undirected graphs and the Ferrers dimension of the directed graphs obtained somehow from the undirected graphs. Here we present a similar but more direct relation between the boxicity and the Ferrers dimension of bigraphs.

If $\mathsf{fd}(B) = 1$, then $\mathsf{box}(B) \leq 2$. This is because, $\mathsf{fd}(B) = 1$ implies that B is a chain graph, and thus B is a grid intersection graph [27]. This bound is tight since $\mathsf{fd}(K_{n,n}) = 1$ and $\mathsf{box}(K_{n,n}) = 2$ for every $n \geq 2$.

Theorem 5.1. *Let B be a bigraph with $\mathsf{fd}(B) \geq 2$. It holds that*

$$\mathsf{box}(B) \leq \mathsf{fd}(B) \leq 2\mathsf{box}(B).$$

Proof. Adiga, Bhowmick, and Chandran [1] showed that for a poset Q of height 2 and its underlying graph H it holds that $\mathsf{box}(H) \leq \mathsf{pd}(Q) \leq 2\mathsf{box}(H)$ if $\mathsf{pd}(Q) \geq 2$. (Recently Felsner [13] has shown a more general result.) Since $\mathsf{fd}(Q) = \mathsf{pd}(Q)$ [10], it holds that $\mathsf{box}(H) \leq \mathsf{fd}(Q) \leq 2\mathsf{box}(H)$ if $\mathsf{fd}(Q) \geq 2$.

Let P be a poset that has B as the underlying graph. From the argument above, it follows that $\mathsf{box}(B) \leq \mathsf{fd}(P) \leq 2\mathsf{box}(B)$ if $\mathsf{fd}(P) \geq 2$. Hence it suffices to show that $\mathsf{fd}(P) = \mathsf{fd}(B)$.

Let M_B is a bipartite adjacency matrix of B. Then, an adjacency matrix M_P of the digraph corresponding to P can be represented by the following form:

$$M_P = \begin{pmatrix} M_B & \mathbf{0} \\ \mathbf{0} & \mathbf{0} \end{pmatrix}.$$

Thus it is easy to see that $\mathsf{fd}(P) \geq \mathsf{fd}(B)$ as M_B is a submatrix of M_P. On the other hand, let $B_1, \ldots, B_{\mathsf{fd}(B)}$ be Ferrers bigraphs that satisfy $B = \bigcap_{1 \leq i \leq \mathsf{fd}(B)} B_i$. Let M_{B_i} is the bipartite adjacency matrix of B_i in which the rows and columns are ordered as in M_B. Now we define M_{P_i} as follows:

$$M_{P_i} = \begin{pmatrix} M_{B_i} & \mathbf{0} \\ \mathbf{0} & \mathbf{0} \end{pmatrix}.$$

Clearly $M_P = \bigcap_{1 \leq i \leq \mathsf{fd}(B)} M_{P_i}$, and each M_{P_i} has the Ferrers property.. This implies that $\mathsf{fd}(P) \leq \mathsf{fd}(B)$. $\square$

The upper bound in Theorem 5.1 is tight. It is known that $\mathsf{box}(K_{n,n} - nK_2) = \lceil n/2 \rceil$ [6] and $\mathsf{fd}(K_{n,n} - nK_2) = n$ [38,39].

Bellatoni, Hartman, Przytycka, and Whitesides [2] showed that the grid intersection graphs are exactly the bigraphs of boxicity at most 2. This implies that the Ferrers dimension of a grid intersection graph is at most 4. We show that the converse is not true.

Theorem 5.2. $\mathsf{GIG} \subsetneq \mathsf{CHAIN}^4$.

Proof. We show that $H_4 \in \mathsf{CHAIN}^4 \setminus \mathsf{GIG}$. Chang and West [8] showed that H_4 cannot be represented as the intersection graph of axis-parallel rectangles in the plane. This implies that $H_4 \notin \mathsf{GIG}$. Let M and M' be the following matrices:

$$
M = \begin{pmatrix}
0 & 1 & 1 & 1 & 1 & 0 & 0 & 0 \\
1 & 0 & 1 & 1 & 0 & 1 & 0 & 0 \\
1 & 1 & 0 & 1 & 0 & 0 & 1 & 0 \\
1 & 1 & 1 & 0 & 0 & 0 & 0 & 1 \\
1 & 0 & 0 & 0 & 0 & 1 & 1 & 1 \\
0 & 1 & 0 & 0 & 1 & 0 & 1 & 1 \\
0 & 0 & 1 & 0 & 1 & 1 & 0 & 1 \\
0 & 0 & 0 & 1 & 1 & 1 & 1 & 0
\end{pmatrix},
\quad
M' = \begin{pmatrix}
a & 1 & 1 & 1 & 1 & a & a & a \\
1 & b & 1 & 1 & b & 1 & b & b \\
1 & 1 & c & 1 & c & c & 1 & c \\
1 & 1 & 1 & d & d & d & d & 1 \\
1 & b & c & d & d & 1 & 1 & 1 \\
a & 1 & c & d & 1 & c & 1 & 1 \\
a & b & 1 & d & 1 & 1 & b & 1 \\
a & b & c & 1 & 1 & 1 & 1 & a
\end{pmatrix}.
$$

The matrix M is a biadjacency matrix of H_4, and M' has the same 1-entries as M but has one of a, b, c, and d for each 0-entry of M. For $x \in \{a, b, c, d\}$, let M_x be the 0-1 matrix obtained from M' by replacing all x with 0 and replacing all other non-numeric entries with 1. It is easy to see that M_x, for all $x \in \{a, b, c, d\}$, has none of the forbidden 2×2 matrices in (1) as a submatrix, and thus has the Ferrers property. Since $M = M_a \cap M_b \cap M_c \cap M_d$, it holds that $H_4 \in \mathsf{CHAIN}^4$. □

Chandran, Francis, and Mathew [7] showed that boxicity is unbounded for chordal bipartite graphs. Thus we have the following.

Corollary 5.3. *Ferrers dimension is unbounded for chordal bipartite graphs.*

References

1. Adiga, A., Bhowmick, D., Chandran, L.S.: Boxicity and poset dimension. SIAM J. Discrete Math. 25, 1687–1698 (2011)
2. Bellatoni, S., Hartman, I.B.-A., Przytycka, T., Whitesides, S.: Grid intersection graphs and boxicity. Discrete Math. 114(1-3), 41–49 (1993)
3. Booth, K.S., Lueker, G.S.: Testing for the consecutive ones property, interval graphs and graph planarity using PQ-tree algorithms. Journal of Computer System Sciences 13, 335–379 (1976)
4. Brandstädt, A., Le, V.B., Spinrad, J.P.: Graph Classes: A Survey. SIAM (1999)
5. Breu, H.: Algorithmic aspects of constrained unit disk graphs. PhD thesis, The University of British Columbia, AAINN09049 (1996)
6. Chandran, L.S., Das, A., Shah, C.D.: Cubicity, boxicity, and vertex cover. Discrete Math. 309, 2488–2496 (2009)
7. Chandran, L.S., Francis, M., Mathew, R.: Chordal bipartite graphs with high boxicity. Graphs Combin. 27, 353–362 (2011)
8. Chang, Y.-W., West, D.B.: Rectangle number for hypercubes and complete multipartite graphs. In: 29th SE Conf. Comb., Graph Th. and Comp. Congr. Numer., vol. 132, pp. 19–28 (1998)
9. Chatterjee, S., Ghosh, S.: Ferrers dimension and boxicity. Discrete Math. 310, 2443–2447 (2010)

10. Cogis, O.: On the Ferrers dimension of a digraph. Discrete Math. 38, 47–52 (1982)
11. Corneil, D.G., Olariu, S., Stewart, L.: The LBFS structure and recognition of interval graphs. SIAM Journal on Discrete Mathematics 23, 1905–1953 (2009)
12. Feder, T., Hell, P., Huang, J.: List homomorphisms and circular arc graphs. Combinatorica 19, 487–505 (1999)
13. Felsner, S.: The order dimension of planar maps revisited. In: JCDCGG 2013, pp. 18–19 (2013)
14. Felsner, S., Mertzios, G.B., Mustaţă, I.: On the recognition of four-directional orthogonal ray graphs. In: Chatterjee, K., Sgall, J. (eds.) MFCS 2013. LNCS, vol. 8087, pp. 373–384. Springer, Heidelberg (2013)
15. Golumbic, M.C.: Algorithmic Graph Theory and Perfect Graphs, 2nd edn. Annals of Discrete Mathematics, vol. 57. North Holland (2004)
16. Hammer, P.L., Peled, U.N., Sun, X.: Difference graphs. Discrete Appl. Math. 28, 35–44 (1990)
17. Hartman, I.B.-A., Newman, I., Ziv, R.: On grid intersection graphs. Discrete Math. 87(1), 41–52 (1991)
18. Heggernes, P., Kratsch, D.: Linear-time certifying recognition algorithms and forbidden induced subgraphs. Nordic J. Comput. 14, 87–108 (2007)
19. Hell, P., Huang, J.: Two remarks on circular arc graphs. Graphs Combin. 13, 65–72 (1997)
20. Hell, P., Huang, J.: Interval bigraphs and circular arc graphs. J. Graph Theory 46, 313–327 (2004)
21. Hell, P., Mastrolilli, M., Nevisi, M.M., Rafiey, A.: Approximation of minimum cost homomorphisms. In: Epstein, L., Ferragina, P. (eds.) ESA 2012. LNCS, vol. 7501, pp. 587–598. Springer, Heidelberg (2012)
22. Klinz, B., Rudolf, R., Woeginger, G.J.: Permuting matrices to avoid forbidden submatrices. Discrete Appl. Math. 60, 223–248 (1995)
23. Kratochvíl, J.: A special planar satisfiability problem and a consequence of its NP-completeness. Discrete Appl. Math. 52(3), 233–252 (1994)
24. Lekkerkerker, C.G., Boland, J.C.: Representation of a finite graph by a set of intervals on the real line. Fund. Math. 51, 45–64 (1962)
25. Müller, H.: Recognizing interval digraphs and interval bigraphs in polynomial time. Discrete Appl. Math. 78(1-3), 189–205 (1997), Erratum is available at http:http://www.comp.leeds.ac.uk/hm/pub/node1.html
26. Mustaţă, I., Pergel, M.: Unit grid intersection graphs: Recognition and properties. CoRR, abs/1306.1855 (2013)
27. Otachi, Y., Okamoto, Y., Yamazaki, K.: Relationships between the class of unit grid intersection graphs and other classes of bipartite graphs. Discrete Appl. Math. 155, 2383–2390 (2007)
28. Rao, W., Orailoglu, A., Karri, R.: Logic mapping in crossbar-based nanoarchitectures. IEEE Des. Test 26, 68–77 (2009)
29. Saha, P.K., Basu, A., Sen, M.K., West, D.B.: Permutation bigraphs: An analogue of permutation graphs, http://www.math.uiuc.edu/~west/pubs/permbig.pdf
30. Sen, M., Das, S., Roy, A.B., West, D.B.: Interval digraphs: An analogue of interval graphs. J. Graph Theory 13, 189–202 (1989)
31. Sen, M.K., Sanyal, B.K., West, D.B.: Representing digraphs using intervals or circular arcs. Discrete Math. 147, 235–245 (1995)
32. Shrestha, A.M.S., Tayu, S., Ueno, S.: On orthogonal ray graphs. Discrete Appl. Math. 158, 1650–1659 (2010)
33. Spinrad, J.P., Brandstädt, A., Stewart, L.: Bipartite permutation graphs. Discrete Appl. Math. 18(3), 279–292 (1987)

34. Spinrad, J.P.: Doubly lexical ordering of dense 0-1 matrices. Inform. Process. Lett. 45, 229–235 (1993)
35. Spinrad, J.P.: Efficient Graph Representations. Fields Institute monographs, vol. 19. American Mathematical Society (2003)
36. Tahoori, M.B.: A mapping algorithm for defect-tolerance of reconfigurable nano-architectures. In: IEEE/ACM International Conference on Computer-Aided Design, pp. 668–672 (2005)
37. Takaoka, A., Tayu, S., Ueno, S.: On unit grid intersection graphs. In: JCDCGG 2013, pp. 120–121 (2013)
38. Trotter, W.T.: Dimension of the crown S_n^k. Discrete Math. 8, 85–103 (1974)
39. Trotter, W.T.: Partially ordered sets. In: Graham, R., Grötschel, M., Lovász, L. (eds.) Handbook of Combinatorics, pp. 433–480. Elsevier Science B. V. (1995)

On the Parameterized Complexity
for Token Jumping on Graphs*

Takehiro Ito[1], Marcin Kamiński[2], Hirotaka Ono[3],
Akira Suzuki[1], Ryuhei Uehara[4], and Katsuhisa Yamanaka[5]

[1] Graduate School of Information Sciences, Tohoku University,
Aoba-yama 6-6-05, Sendai, 980-8579, Japan
{takehiro,a.suzuki}@ecei.tohoku.ac.jp
[2] Dept. of Mathematics, Computer Science and Mechanics, University of Warsaw,
Banacha 2, 02-097, Warsaw, Poland
mjk@mimuw.edu.pl
[3] Faculty of Economics, Kyushu University,
Hakozaki 6-19-1, Higashi-ku, Fukuoka, 812-8581, Japan
hirotaka@en.kyushu-u.ac.jp
[4] School of Information Science, JAIST,
Asahidai 1-1, Nomi, Ishikawa, 923-1292, Japan
uehara@jaist.ac.jp
[5] Dept. of Electrical Engineering and Computer Science, Iwate University,
Ueda 4-3-5, Morioka, Iwate 020-8551, Japan
yamanaka@cis.iwate-u.ac.jp

Abstract. Suppose that we are given two independent sets I_0 and I_r of a graph such that $|I_0| = |I_r|$, and imagine that a token is placed on each vertex in I_0. Then, the TOKEN JUMPING problem is to determine whether there exists a sequence of independent sets which transforms I_0 into I_r so that each independent set in the sequence results from the previous one by moving exactly one token to another vertex. Therefore, all independent sets in the sequence must be of the same cardinality. This problem is PSPACE-complete even for planar graphs with maximum degree three. In this paper, we first show that the problem is W[1]-hard when parameterized only by the number of tokens. We then give an FPT algorithm for general graphs when parameterized by both the number of tokens and the maximum degree. Our FPT algorithm can be modified so that it finds an actual sequence of independent sets between I_0 and I_r with the minimum number of token movements.

1 Introduction

The TOKEN JUMPING problem was introduced by Kamiński et al. [13], which can be seen as a "dynamic" version of independent sets in a graph. Recall that an

* This work is partially supported by JSPS KAKENHI Grant Numbers 25106504 (Ito), 25104521 (Ono), 24106004 (Ono and Uehara), 24.3660 (Suzuki) and 25106502 (Yamanaka).

T V Gopal et al. (Eds.): TAMC 2014, LNCS 8402, pp. 341–351, 2014.

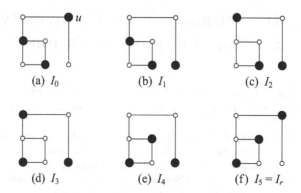

Fig. 1. A sequence $\langle I_0, I_1, \ldots, I_5 \rangle$ of independent sets of the same graph, where the vertices in independent sets are depicted by large black circles (tokens)

independent set of a graph G is a vertex-subset of G in which no two vertices are adjacent. (See Fig. 1 which depicts six different independent sets of the same graph.) Suppose that we are given two independent sets I_0 and I_r of a graph $G = (V, E)$ such that $|I_0| = |I_r|$, and imagine that a token (coin) is placed on each vertex in I_0. Then, the TOKEN JUMPING problem is to determine whether there exists a sequence $\langle I_0, I_1, \ldots, I_\ell \rangle$ of independent sets of G such that

   (a) $I_\ell = I_r$, and $|I_i| = |I_0| = |I_r|$ for all i, $1 \leq i \leq \ell$; and

   (b) for each index i, $1 \leq i \leq \ell$, I_i can be obtained from I_{i-1} by moving exactly one token on a vertex $u \in I_{i-1}$ to another vertex $v \in V \setminus I_{i-1}$, and hence $I_{i-1} \setminus I_i = \{u\}$ and $I_i \setminus I_{i-1} = \{v\}$.

Figure 1 illustrates a sequence $\langle I_0, I_1, \ldots, I_5 \rangle$ of independent sets which transforms I_0 into $I_r = I_5$.

Recently, this type of problems have been studied extensively in the framework of *reconfiguration problems* [8], which arise when we wish to find a step-by-step transformation between two feasible solutions of a problem such that all intermediate solutions are also feasible and each step abides by a prescribed reconfiguration rule (*i.e.*, an adjacency relation defined on feasible solutions of the original problem). For example, the TOKEN JUMPING problem can be seen as a reconfiguration problem for the (ordinary) INDEPENDENT SET problem: feasible solutions are defined to be all independent sets of the same cardinality in a graph; and the reconfiguration rule is defined to be the condition (b) above. This reconfiguration framework has been applied to several well-known problems, including INDEPENDENT SET [5,6,8,13,15], SATISFIABILITY [4,14], SET COVER, CLIQUE, MATCHING [8], VERTEX-COLORING [1,2,3], LIST EDGE-COLORING [9,11], LIST $L(2, 1)$-LABELING [10], SUBSET SUM [7], SHORTEST PATH [12], etc.

1.1   Reconfiguration Rules and Related Results

The original reconfiguration problem for INDEPENDENT SET was introduced by Hearn and Demaine [5], which employs another reconfiguration rule. Indeed, there are three reconfiguration problems for INDEPENDENT SET (ISRECONF, for short) under different reconfiguration rules, as follows.

- **Token Sliding (TS)** [2,5,6,13]: We can slide a single token only *along an edge* of a graph. In other words, each token can be moved only to its adjacent vertex. This rule corresponds to the original one introduced by Hearn and Demaine [5].

- **Token Jumping (TJ)** [13]: This rule corresponds to TOKEN JUMPING, that is, we can move a single token to any vertex.

- **Token Addition and Removal (TAR)** [8,13,15]: We can either add or remove a single token at a time if it results in an independent set of cardinality at least a given threshold. Therefore, independent sets in the sequence do not have the same cardinality.

We remark that the existence of a desired sequence depends deeply on the reconfiguration rules. For example, Fig. 1 is an yes-instance for TOKEN JUMPING, but it is a no-instance for ISRECONF under the TS rule.

We here explain only the results which are strongly related to TOKEN JUMPING; see the references above for the other results.

Hearn and Demaine [5], [6, Sec. 9.5] proved that ISRECONF under the TS rule is PSPACE-complete for planar graphs of maximum degree three. Then, Bonsma and Cereceda [2] showed that this problem remains PSPACE-complete even for very restricted instances. Indeed, their result implies that TOKEN JUMPING is PSPACE-complete for planar graphs with maximum degree three. (Details will be given in Section 2.3.)

Kamiński *et al.* [13] proved that ISRECONF is PSPACE-complete for perfect graphs under any of the three reconfiguration rules. As the positive results for TOKEN JUMPING, they gave a linear-time algorithm for even-hole-free graphs. Furthermore, their algorithm can find an actual sequence of independent sets with the minimum number of token movements.

1.2 Our Contributions

In this paper, we investigate the parameterized complexity of the TOKEN JUMPING problem.

We first show that the problem is W[1]-hard when parameterized only by the number t of tokens. Therefore, the problem admits no FPT algorithm when parameterized only by t unless FPT = W[1].

We thus consider the problem with two parameters, and give an FPT algorithm for general graphs when parameterized by both the number of tokens and the maximum degree. Recall that the problem remains PSPACE-complete even if the maximum degree is three. (See Section 2.3.) Therefore, it is very unlikely that the problem can be solved in polynomial time even for graphs with bounded maximum degree.

Finally, we show that our FPT algorithm for general graphs can be modified so that it finds an actual sequence of independent sets between I_0 and I_r with the minimum number of token movements. We remark that the sequence of independent sets in Fig. 1 has the minimum length. It is interesting that the token on the vertex u in Fig. 1(a) must be moved twice even though $u \in I_0 \cap I_r$.

2   Preliminaries

In this section, we first introduce some basic terms and notations which will be used throughout the paper. We then formally show the PSPACE-completeness of TOKEN JUMPING in Section 2.3.

2.1   Graph Notations

In TOKEN JUMPING, we may assume without loss of generality that graphs are simple. For a graph G, we sometimes denote by $V(G)$ and $E(G)$ the vertex set and the edge set of G, respectively. Let $n(G) = |V(G)|$ and $m(G) = |E(G)|$. We denote by $\Delta(G)$ the maximum degree of G.

For a vertex v of a graph G, we denote by $N(G; v)$ the set of all neighbors of v in G (which does not include v itself), that is, $N(G; v) = \{w \in V(G) \mid (v, w) \in E(G)\}$. Let $N[G; v] = N(G; v) \cup \{v\}$, and let $N[G; V'] = \bigcup_{v \in V'} N[G; v]$ for a vertex-subset $V' \subseteq V(G)$.

2.2   Definitions for TOKEN JUMPING

Let I_i and I_j be two independent sets of the same cardinality in a graph $G = (V, E)$. We say that I_i and I_j are *adjacent* if there exists exactly one pair of vertices u and v such that $I_i \setminus I_j = \{u\}$ and $I_j \setminus I_i = \{v\}$, that is I_j can be obtained from I_i by *moving* the token on a vertex $u \in I_i$ to another vertex $v \in V \setminus I_i$. We remark that the tokens are unlabeled, while the vertices in a graph are labeled.

A *reconfiguration sequence* between two independent sets I and I' of G is a sequence $\langle I_1, I_2, \ldots, I_\ell \rangle$ of independent sets of G such that $I_1 = I$, $I_\ell = I'$, and I_{i-1} and I_i are adjacent for $i = 2, 3, \ldots, \ell$. We say that two independent sets I and I' are *reconfigurable* each other if there exists a reconfiguration sequence between I and I'. Clearly, any two adjacent independent sets are reconfigurable each other. The *length* of a reconfiguration sequence S is defined as the number of independent sets contained in S. For example, the length of the reconfiguration sequence in Fig. 1 is 6.

The TOKEN JUMPING problem is to determine whether two given independent sets I_0 and I_r of a graph G are reconfigurable each other. We may assume without loss of generality that $|I_0| = |I_r|$; otherwise the answer is clearly "no." Note that TOKEN JUMPING is a decision problem asking the existence of a reconfiguration sequence between I_0 and I_r, and hence it does not ask an actual reconfiguration sequence. We always denote by I_0 and I_r the *initial* and *target* independent sets of G, respectively, as an instance of TOKEN JUMPING.

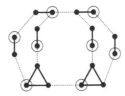

Fig. 2. Graph consisting of token triangles and token edges, where link edges are depicted by thin dotted lines and the vertices in a standard independent set (namely, with tokens) are surrounded by circles

2.3 PSPACE-Completeness

As we have mentioned in Introduction, Bonsma and Cereceda [2] showed that ISRECONF under the TS rule is PSPACE-complete, and their result indeed implies the PSPACE-completeness of TOKEN JUMPING. We here formally explain this fact, as in the following theorem.

Theorem 1. *The* TOKEN JUMPING *problem is* PSPACE-*complete for planar graphs with maximum degree three.*

Proof. The problem is clearly in PSPACE, and hence we show that TOKEN JUMPING is PSPACE-hard for planar graphs with maximum degree three.

Bonsma and Cereceda [2] showed that ISRECONF under the TS rule is PSPACE-complete even for very restricted instances, defined as follows. Every vertex of a graph G is a part of exactly one of *token triangles* (*i.e.*, copies of K_3) and *token edges* (*i.e.*, copies of K_2), as illustrated in Fig. 2. Token triangles and token edges are all mutually disjoint, and joined together by edges called *link edges*. Moreover, $\Delta(G) = 3$ and G has a planar embedding such that every token triangle forms a face. We say that an independent set I of G is *standard* if each of token triangles and token edges contains exactly one token (vertex) in I. The ISRECONF problem under the TS rule remains PSPACE-complete even if G is such a restricted graph and both I_0 and I_r are standard independent sets [2].

Note that a standard independent set of G is a maximal independent set. Then, even under the TJ rule (*i.e.*, in TOKEN JUMPING), each token can jump only to its adjacent vertex. Therefore, I_0 and I_r are reconfigurable each other under the TS rule if and only if they are reconfigurable each other under the TJ rule. Thus, the result follows. □

3   W[1]-Hardness

In this section, we give the hardness result as in the following theorem.

Theorem 2. *The* TOKEN JUMPING *problem is* W[1]-*hard when parameterized by the number of tokens.*

(a) I_0 (b) I_r

Fig. 3. Image of our reduction, where the vertices in independent sets are depicted by large black circles (tokens)

Proof. We give an FPT-reduction from INDEPENDENT SET parameterized by the solution size to TOKEN JUMPING parameterized by the number of tokens. Given a graph G' and a parameter t', the INDEPENDENT SET *problem parameterized by the solution size* is to determine whether there is an independent set I of G' such that $|I| \geq t'$. This problem is known to be W[1]-hard [16, p. 213].

We now construct the corresponding instance of TOKEN JUMPING. (See also Fig. 3.) Let G be the graph which consists of G' and a complete bipartite graph $K_{t'+1,t'+1}$. Therefore, G consists of two connected components. Let $\{U, W\}$ be the bipartition of $V(K_{t'+1,t'+1})$. Let $I_0 = U$ and $I_r = W$, then $|I_0| = |I_r| = |U| = |W| = t' + 1$. Therefore, the parameter (*i.e.*, the number of tokens) for TOKEN JUMPING is $t = t' + 1$. Clearly, the corresponding instance can be constructed in time $O\big(n(G') + t'^2\big)$.

To complete the FPT-reduction, we now show that G' has an independent set I with $|I| \geq t'$ if and only if I_0 and I_r are reconfigurable each other.

Suppose that G' has an independent set I with $|I| \geq t'$. Then, there is a reconfiguration sequence between I_0 and I_r, as follows: first move t' ($= t - 1$) tokens from U to the vertices in I one by one; then move the last token on the vertex in U to any vertex in W; and move t' tokens from I to W one by one. Therefore, I_0 and I_r are reconfigurable each other.

Conversely, suppose that I_0 and I_r are reconfigurable each other, and hence there is a reconfiguration sequence S between I_0 and I_r. Since $K_{t'+1,t'+1}$ is a complete bipartite graph, G has no independent set I' such that both $I' \cap U \neq \emptyset$ and $I' \cap W \neq \emptyset$ hold. Therefore, since we can move only one token at a time, S must contain an independent set I_q of G such that both $I_q \cap U = \{u\}$ and $I_q \cap W = \emptyset$ hold. Then, all vertices in $I_q \setminus \{u\}$ are contained in the component G' of G, and they must form an independent set of G'. Since $|I_q \setminus \{u\}| = t - 1 = t'$, there exists an independent set $I = I_q \setminus \{u\}$ of G' such that $|I| = t'$. □

4   FPT Algorithms

Theorem 2 implies that TOKEN JUMPING admits no FPT algorithm when parameterized only by the number of tokens unless FPT = W[1]. Therefore, in this section, we give an FPT algorithm for general graphs when parameterized by both the number of tokens and the maximum degree. Recall that TOKEN JUMPING remains PSPACE-complete even for planar graphs with bounded maximum degree.

In Section 4.1, we first give an FPT algorithm which simply solves TOKEN JUMPING for general graphs. We then show in Section 4.2 that our FPT algorithm can be modified so that it finds an actual reconfiguration sequence with the minimum length.

4.1 TOKEN JUMPING

The main result of this subsection is the following theorem.

Theorem 3. *Let G be a graph whose maximum degree is bounded by a fixed constant d. Let I_0 and I_r be two independent sets of G such that $|I_0| = |I_r| \le t$ for a fixed constant t. Then, one can determine whether I_0 and I_r are reconfigurable each other in time $O\big((3td)^{2t}\big)$.*

In this subsection, we give such an algorithm as a proof of Theorem 3. We first show in Lemma 1 that, if a graph G has at least $3t(d+1)$ vertices, then I_0 and I_r are always reconfigurable each other. Therefore, one can know that the answer is always "yes" if $n(G) \ge 3t(d+1)$, and hence it suffices to deal with a graph having less than $3t(d+1)$ vertices. For such a graph, we then show in Lemma 2 that there is an $O\big((3td)^{2t}\big)$-time algorithm that determines whether I_0 and I_r are reconfigurable each other.

We first show that any two independent sets are reconfigurable each other if the graph has a sufficiently large number of vertices, as in the following lemma.

Lemma 1. *Let G be a graph with $\Delta(G) \le d$, and let I_i and I_j be an arbitrary pair of independent sets of G such that $|I_i| = |I_j| \le t$. Then, I_i and I_j are reconfigurable each other if $n(G) \ge 3t(d+1)$.*

Proof. Suppose that $n(G) \ge 3t(d+1)$. To prove the lemma, we show that there exists a reconfiguration sequence between I_i and I_j.

Let G^- be the graph obtained from G by deleting all vertices in $N[G; I_i] \cup N[G; I_j]$. Since all neighbors of the vertices in $I_i \cup I_j$ have been deleted from G, no vertex in G^- is adjacent with any vertex in $I_i \cup I_j$. Therefore, if G^- has an independent set I_k with $|I_k| \ge t$, then there is a reconfiguration sequence between I_i and I_j, as follows: move all tokens on the vertices in I_i to the vertices in I_k one by one; and move all tokens on the vertices in I_k to the vertices in I_j one by one.

To complete the proof, we thus show that G^- has an independent set I_k with $|I_k| \ge t$ if $n(G) \ge 3t(d+1)$. Since $\Delta(G) \le d$, we clearly have $\big|N[G; v]\big| \le d+1$ for every vertex v in G. Since $|I_i| \le t$, we thus have

$$\big|N[G; I_i]\big| \le \sum_{v \in I_i} \big|N[G; v]\big| \le t(d+1).$$

Similarly, we have $\big|N[G; I_j]\big| \le t(d+1)$. Therefore,

$$n(G^-) \ge n(G) - \big|N[G; I_i]\big| - \big|N[G; I_j]\big| \ge t(d+1). \tag{1}$$

We now suppose for a contradiction that $|I_{\max}| < t$ holds for a maximum independent set $I_{\max}$ of G^-. Then, we have

$$\left|N[G^-; I_{\max}]\right| \le \sum_{v \in I_{\max}} |N[G; v]| < t(d+1),$$

and hence by Eq. (1)

$$n(G^-) - \left|N[G^-; I_{\max}]\right| \ge 1.$$

Therefore, the graph obtained from G^- by deleting all vertices in $N[G^-; I_{\max}]$ is non-empty, and hence we can add at least one vertex to $I_{\max}$. This contradicts the assumption that $I_{\max}$ is a maximum independent set of G^-. Therefore, $|I_{\max}| \ge t$, and hence G^- has an independent set I_k with $|I_k| \ge t$. □

We then give an FPT algorithm for the case where a given graph G has only a constant number of vertices, as in the following lemma.

Lemma 2. *Suppose that $n(G) < 3t(d+1)$. Then, there is an $O\big((3td)^{2t}\big)$-time algorithm which determines whether I_0 and I_r are reconfigurable each other.*

Proof. We give such an algorithm. For a graph G and a constant $t' = |I_0| = |I_r|$ $(\le t)$, we construct a *configuration graph* $\mathcal{C} = (\mathcal{V}, \mathcal{E})$, as follows:
  (i) each node in $\mathcal{C}$ corresponds to an independent set of G with cardinality exactly t'; and
  (ii) two nodes in $\mathcal{C}$ are joined by an edge if and only if the corresponding two independent sets are adjacent.
For an independent set I of G with $|I| = t'$, we always denote by w_I the node of $\mathcal{C}$ corresponding to I. Clearly, two independent sets I_0 and I_r are reconfigurable each other if and only if there is a path in $\mathcal{C}$ between w_{I_0} and w_{I_r}.

Notice that G has at most the number $\binom{n(G)}{t'}$ of distinct independent sets with cardinality exactly t'. Since $t' \le t$, we thus have

$$|\mathcal{V}| \le \binom{n(G)}{t'} < \binom{3t(d+1)}{t'} = O\big((3td)^t\big).$$

The configuration graph $\mathcal{C}$ above can be constructed in time $O(|\mathcal{V}|^2)$. Furthermore, by the breadth-first search on $\mathcal{C}$ starting from the node w_{I_0}, one can determine whether $\mathcal{C}$ has a path from w_{I_0} to w_{I_r} in time $O(|\mathcal{V}| + |\mathcal{E}|) = O(|\mathcal{V}|^2)$. In this way, our algorithm runs in time $O(|\mathcal{V}|^2) = O\big((3td)^{2t}\big)$ in total. □

Lemmas 1 and 2 complete the proof of Theorem 3. □

4.2   Shortest Reconfiguration Sequence

We now give an FPT algorithm which finds an actual reconfiguration sequence with the minimum length.

Theorem 4. *Let G be a graph whose maximum degree is bounded by a fixed constant d. Let I_0 and I_r be two independent sets of G such that $|I_0| = |I_r| \leq t$ for a fixed constant t. Then, one can find a shortest reconfiguration sequence between I_0 and I_r in time $O\big((4td)^{2t} + n(G) + m(G)\big)$ if there exists.*

We give such an algorithm as a proof of Theorem 4. Let $t' = |I_0| = |I_r| \leq t$. Although our algorithm is based on the proofs in Section 4.1, the number of vertices for the graph classification is slightly changed from $3t(d+1)$ to $4t(d+1)$; this yields that the base of the running time becomes 4 in Theorem 4.

We first consider the case where $n(G) < 4t(d + 1)$.

Lemma 3. *Suppose that $n(G) < 4t(d+1)$. Then, one can find a shortest reconfiguration sequence between I_0 and I_r in time $O\big((4td)^{2t}\big)$ if there exists.*

Proof. As in the proof of Lemma 2, we construct the configuration graph $\mathcal{C} = (\mathcal{V}, \mathcal{E})$ for G and t' in time

$$O(|\mathcal{V}|^2) = O\left(\left(\binom{4t(d+1)}{t'}\right)^2\right) = O\big((4td)^{2t}\big).$$

Recall that the node set of $\mathcal{C}$ corresponds to *all* independent sets in G of cardinality exactly t'. Therefore, a shortest reconfiguration sequence between two independent sets I_0 and I_r corresponds to a shortest path in $\mathcal{C}$ between the two nodes w_{I_0} and w_{I_r}. By the breadth-first search on $\mathcal{C}$ starting from w_{I_0}, one can find a shortest path in $\mathcal{C}$ in time $O(|\mathcal{V}| + |\mathcal{E}|) = O(|\mathcal{V}|^2)$ if there exists. Therefore, if $n(G) < 4t(d+1)$, one can find a shortest reconfiguration sequence in time $O(|\mathcal{V}|^2) = O\big((4td)^{2t}\big)$. $\qquad\square$

We then consider the case where $n(G) \geq 4t(d + 1)$. Notice that, since $n(G)$ is not bounded by a fixed constant, we cannot directly construct the configuration graph $\mathcal{C}$ for G and t' in this case. However, we will prove that only a subgraph of $\mathcal{C}$ having a constant number of nodes is sufficient to find a shortest reconfiguration sequence.

Lemma 1 ensures that there always exists a reconfiguration sequence between I_0 and I_r in this case. Furthermore, in the proof of Lemma 1, we proposed a reconfiguration sequence $\mathcal{S}'$ between I_0 and I_r such that every token is moved exactly twice. Although this is not always a shortest reconfiguration sequence, the minimum length of a reconfiguration sequence between I_0 and I_r can be bounded by the length of $\mathcal{S}'$, that is, $2t'$.

Let G^- be the graph obtained from G by deleting all vertices in $N[G; I_0] \cup N[G; I_r]$. Then, by the counterpart of Eq. (1) we have $n(G^-) \geq 2t(d + 1)$, and hence G^- has an independent set I'_k such that $|I'_k| = 2t'$ ($\leq 2t$). We now give the following lemma.

Lemma 4. *There exists a shortest reconfiguration sequence $\mathcal{S}$ between I_0 and I_r such that $I \subseteq I_0 \cup I'_k \cup I_r$ for all independent sets I in $\mathcal{S}$.*

Proof. Let $\mathcal{S}^* = \langle I_0^*, I_1^*, \ldots, I_\ell^* \rangle$ be an arbitrary shortest reconfiguration sequence between $I_0 = I_0^*$ and $I_r = I_\ell^*$. Then, the proof of Lemma 1 implies that $\ell \leq 2t'$, as we have mentioned above. Note that some independent sets in $\mathcal{S}^*$ may contain vertices in $V(G) \setminus (I_0 \cup I_k' \cup I_r)$. Let

$$V(I_0, I_r; \mathcal{S}^*) = \bigcup_{1 \leq i \leq \ell-1} \left(I_i^* \setminus (I_0 \cup I_r) \right),$$

that is, $V(I_0, I_r; \mathcal{S}^*)$ is the set of all vertices that are not in $I_0 \cup I_r$ but appear in the reconfiguration sequence $\mathcal{S}^*$. Since $\ell \leq 2t'$ and $|I_{i+1}^* \setminus I_i^*| = 1$ for all i, $0 \leq i \leq \ell - 1$, we have $|V(I_0, I_r; \mathcal{S}^*)| < \ell \leq 2t'$.

Therefore, since $|I_k'| = 2t'$, one can replace all vertices in $V(I_0, I_r; \mathcal{S}^*)$ with distinct vertices in I_k'; let $\mathcal{S}$ be the resulting sequence. Recall that I_k' is an independent set of G^-, and hence no vertex in I_k' is adjacent with any vertex in $I_0 \cup I_r$. Therefore, $\mathcal{S}$ is a reconfiguration sequence between I_0 and I_r. Note that any independent set I in $\mathcal{S}$ satisfies $I \subseteq I_0 \cup I_k' \cup I_r$. Furthermore, the length of $\mathcal{S}$ is equal to that of $\mathcal{S}^*$, and hence $\mathcal{S}$ is a shortest reconfiguration sequence. □

We now give the following lemma, which completes the proof of Theorem 4.

Lemma 5. *Suppose that $n(G) \geq 4t(d+1)$. Then, one can find a shortest reconfiguration sequence between I_0 and I_r in time $O\big((4t)^{2t} + n(G) + m(G)\big)$.*

Proof. We first remark that an independent set I_k' of G^- with $|I_k'| = 2t' (\leq 2t)$ can be found in time $O\big(n(G) + m(G)\big)$ by the following simple greedy algorithm: initially, let $I_k' = \emptyset$; choose an arbitrary vertex v in G^-, and add v to I_k'; delete all vertices in $N[G^-; v]$ from G^-, and repeat. Recall that $n(G^-) \geq 2t(d+1)$ and $|N[G^-; v]| \leq d+1$ for every vertex v in G^-. Therefore, this greedy algorithm always finds an independent set I_k' with $|I_k'| = 2t'$.

Let G_{0kr} be the subgraph of G induced by the vertex-subset $I_0 \cup I_k' \cup I_r$. Notice that $n(G_{0kr}) = |I_0 \cup I_k' \cup I_r| \leq 4t'$. Let $\mathcal{C}_{0kr}$ be the configuration graph for G_{0kr} and the constant t'. Since G_{0kr} is an induced subgraph of G, any independent set I of G_{0kr} is an independent set of G. Then, Lemma 4 ensures that there exists a shortest reconfiguration sequence $\mathcal{S}$ between I_0 and I_r such that every independent set I in $\mathcal{S}$ is an independent set of G_{0kr}. Therefore, such a shortest reconfiguration sequence $\mathcal{S}$ between I_0 and I_r can be found as a shortest path in $\mathcal{C}_{0kr}$ between the two nodes w_{I_0} and w_{I_r}. This can be done in time $O\big((4t)^{2t}\big)$, because the number of nodes in $\mathcal{C}_{0kr}$ can be bounded by $\binom{n(G_{0kr})}{t'} = O\big((4t)^t\big)$.

In this way, if $n(G) \geq 4t(d + 1)$, one can find a shortest reconfiguration sequence between I_0 and I_r in time $O\big((4t)^{2t} + n(G) + m(G)\big)$ in total. □

5   Concluding Remarks

In this paper, we mainly gave three results for the parameterized complexity of TOKEN JUMPING. We remark that the running time of each of our FPT algorithms is just a single exponential with respect to the number of tokens; furthermore, the parameter d of maximum degree does not appear in the exponent.

We also remark that the problem parameterized only by the number of tokens is in the class XP, that is, the problem can be solved in polynomial time if the number t of tokens is a fixed constant. To see this, consider the following algorithm: construct the configuration graph $\mathcal{C}$ for a given graph G and the fixed constant t; and find a (shortest) path in $\mathcal{C}$. Since the number of nodes in $\mathcal{C}$ can be bounded by $\binom{n}{t}$, the problem can be solved in time $O(n^{2t})$, where $n = n(G)$. Therefore, the problem for a fixed number of tokens can be solved in polynomial time, while the problem remains PSPACE-complete for a fixed maximum degree.

References

1. Bonamy, M., Johnson, M., Lignos, I., Patel, V., Paulusma, D.: On the diameter of reconfiguration graphs for vertex colourings. Electronic Notes in Discrete Mathematics 38, 161–166 (2011)
2. Bonsma, P., Cereceda, L.: Finding paths between graph colourings: PSPACE-completeness and superpolynomial distances. Theoretical Computer Science 410, 5215–5226 (2009)
3. Cereceda, L., van den Heuvel, J., Johnson, M.: Finding paths between 3-colourings. J. Graph Theory 67, 69–82 (2011)
4. Gopalan, P., Kolaitis, P.G., Maneva, E.N., Papadimitriou, C.H.: The connectivity of Boolean satisfiability: computational and structural dichotomies. SIAM J. Computing 38, 2330–2355 (2009)
5. Hearn, R.A., Demaine, E.D.: PSPACE-completeness of sliding-block puzzles and other problems through the nondeterministic constraint logic model of computation. Theoretical Computer Science 343, 72–96 (2005)
6. Hearn, R.A., Demaine, E.D.: Games, Puzzles, and Computation. A K Peters (2009)
7. Ito, T., Demaine, E.D.: Approximability of the subset sum reconfiguration problem. To appear in J. Combinatorial Optimization, doi:10.1007/s10878-012-9562-z
8. Ito, T., Demaine, E.D., Harvey, N.J.A., Papadimitriou, C.H., Sideri, M., Uehara, R., Uno, Y.: On the complexity of reconfiguration problems. Theoretical Computer Science 412, 1054–1065 (2011)
9. Ito, T., Kamiński, M., Demaine, E.D.: Reconfiguration of list edge-colorings in a graph. Discrete Applied Mathematics 160, 2199–2207 (2012)
10. Ito, T., Kawamura, K., Ono, H., Zhou, X.: Reconfiguration of list $L(2,1)$-labelings in a graph. In: Chao, K.-M., Hsu, T.-S., Lee, D.-T. (eds.) ISAAC 2012. LNCS, vol. 7676, pp. 34–43. Springer, Heidelberg (2012)
11. Ito, T., Kawamura, K., Zhou, X.: An improved sufficient condition for reconfiguration of list edge-colorings in a tree. IEICE Trans. on Information and Systems E95-D, 737–745 (2012)
12. Kamiński, M., Medvedev, P., Milanič, M.: Shortest paths between shortest paths. Theoretical Computer Science 412, 5205–5210 (2011)
13. Kamiński, M., Medvedev, P., Milanič, M.: Complexity of independent set reconfigurability problems. Theoretical Computer Science 439, 9–15 (2012)
14. Makino, K., Tamaki, S., Yamamoto, M.: An exact algorithm for the Boolean connectivity problem for k-CNF. Theoretical Computer Science 412, 4613–4618 (2011)
15. Mouawad, A.E., Nishimura, N., Raman, V., Simjour, N., Suzuki, A.: On the parameterized complexity of reconfiguration problems. In: Gutin, G., Szeider, S. (eds.) IPEC 2013. LNCS, vol. 8246, pp. 281–294. Springer, Heidelberg (2013)
16. Niedermeier, R.: Invitation to Fixed-Parameter Algorithms. Oxford University Press (2006)

Universality of Spiking Neural P Systems with Anti-spikes

Venkata Padmavati Metta and Alica Kelemenová

Institute of Computer Science and Research Institute of the IT4Innovations
Centre of Excellence, Silesian University in Opava, Czech Republic

Abstract. Spiking neural P systems with anti-spikes (in short, SN PA systems) are membrane systems that communicate using two types of objects called spikes and anti-spikes, inspired by neurons communicating through excitatory and inhibitory impulses. This paper shows that computational completeness in an SN PA systems can be achieved with neurons having only two pure spiking rules of the form $a \to a$ and $a \to \bar{a}$ without any forgetting rules. We also construct a small universal SN PA system with 91 simple neurons i.e., neurons having only one rule of the form $a \to \bar{a}$ or $a \to a$.

1 Introduction

Spiking neural P system [5] is a neural-inspired computational model based on the concept of spiking neurons. It consists of a set of neurons placed in the nodes of a directed graph (arcs representing synapses) and neurons communicate with each other using only one kind of objects called spikes, identical electrical impulses. The objects evolve by means of standard spiking rules, which are of the form $E/a^c \to a$, where E is a regular expression over $\{a\}$ and $c \geq 1$. The meaning is that a neuron containing k spikes such that $a^k \in L(E)$, $k \geq c$, can consume c spikes and produce one spike. This spike is sent to all neurons connected by an outgoing synapse from the neuron where the rule was applied. There are also forgetting rules, of the form $a^s \to \lambda$ with the meaning that $s \geq 1$ spikes are removed, provided that the neuron contains exactly s spikes. One neuron is distinguished as the output neuron and its spikes also exit into the environment, thus producing a binary sequence called spike train (moments of time when a spike is emitted by the output neuron are marked with 1, the other moments are marked with 0). The distance between consecutive spikes is the main way to encode information.

Spiking neural P system with anti-spikes [4] works in the same way as standard SN P system but deals with two types of objects called spikes (a) and anti-spikes ($\bar{a}$). The spiking rules are of the form $E/b^c \to b'$, where $b, b' \in \{a, \bar{a}\}$. If $L(E) = \{b^c\}$ then the rules are written as $b^c \to b'$ and are called pure. The system has four categories of spiking rules identified by (a, a), $(a, \bar{a})$ (anti-spikes are produced from usual spikes by means of usual spiking rules), $(\bar{a}, a)$ and $(\bar{a}, \bar{a})$ (rules consuming anti-spikes can produce spikes or anti-spikes). The latter two rules are generally avoided as they are quite unnatural. Each neuron in the system has an implicit

T V Gopal et al. (Eds.): TAMC 2014, LNCS 8402, pp. 352–365, 2014.

annihilation rule of the form $a\bar{a} \rightarrow \lambda$ (if an anti-spike meets a spike in a given neuron, then they annihilate each other (the disappearance of one a and one $\bar{a}$ takes no time), and this happens instantaneously in a maximal way.

The problem which "ingredients" are needed to achieve computational completeness or universality has been a challenging question for these kind of systems also. Several answers have been given, for instance in [4], it was proved that SN PA systems with pure spiking rules of categories (a, a), $(a, \bar{a})$, and $(\bar{a}, a)$ with forgetting rules are universal as number generators. Recently, Song et al. [7] proved that pure spiking rules of categories (a, a) and $(a, \bar{a})$ without forgetting rules , or spiking rules of categories $(\bar{a}, a)$ and $(a, \bar{a})$ without forgetting rules (the neurons change spikes to anti-spikes or change anti-spikes to spikes) are sufficient for universality as number generators. Zeng et al. [9] proved that homogeneous SN PA systems, i.e., SN PA systems where the rules in every neuron are identical, are universal.

All these systems consider spikes to represent the number and the number of spikes present in the neuron corresponding to a register as a function of the number stored in the register. In this paper, we make use of anti-spikes to represent the number stored in the register and number of anti-spikes in the neuron is equal to the number stored in the corresponding register. This avoids the use of rules of the form $\bar{a} \rightarrow a$ in the neuron to check its contents for zero. Since all neurons corresponding to registers are having the same rule $a \rightarrow a$, and no initial spikes/anti-spikes are present in any neurons corresponding to the registers, the output registers can be the subject of SUB instructions also.

This paper proves that only two rules of the form $a \rightarrow a$ and $a \rightarrow \bar{a}$ without any forgetting rules are sufficient for the universality of SN PA systems. It is also a natural and well investigated topic in computer science to look for small universal computing devices of various types. This topic was also considered for SN PA systems. In [8], a universal SN PA system with 75 neurons is constructed as a device of computing functions in which 125 rules, 6 types of neurons and 8 types of rules are used. In this work, the problem of constructing universal SN PA systems with a small number of rules is also investigated. Specifically, a universal SN P system with 91 simple neurons ("simple" in the sense that each neuron has only one rule, so a total of 91 rules) having the rules of the form $a \rightarrow a$ or $a \rightarrow \bar{a}$ is constructed for computing functions.

This paper is organized as follows. We start with Section 2 by giving a brief introduction about the SN P system with anti-spikes. In Section 3, we prove the computational completeness of SN PA systems with neurons having only two rules of the form $a \rightarrow a$ and $a \rightarrow \bar{a}$. Universal SN PA system is constructed in Section 4.

2 Prerequisites

We assume the reader to be familiar with formal languages and automata theory and spiking neural P systems. The reader can find details about them in [2], [1] etc.

For an alphabet V, V^* is the free monoid generated by V with respect to the concatenation operation and the identity λ (the empty string); the set of all non-empty strings over V, that is, $V^* - \{\lambda\}$, is denoted by V^+. The family of Turing computable sets of natural numbers is denoted by NRE (it is the family of length sets of recursively enumerable languages) and the family of Turing computable sets of vectors of natural numbers is denoted by $PsRE$.

We directly introduce the type of SN PA systems we investigate in this paper. (*SN P system with anti-spikes*) A spiking neural P system with anti-spikes, of degree $m \geq 1$, is a construct

$$\Pi = (O, \sigma_1, \sigma_2, \sigma_3, \ldots, \sigma_m, syn, out), \text{ where}$$

1. $O = \{a, \bar{a}\}$ is a binary alphabet. a is called *spike* and $\bar{a}$ is called an *anti-spike*.
2. $\sigma_1, \sigma_2, \sigma_3, \ldots, \sigma_m$ are neurons, of the form

$$\sigma_i = (n_i, R_i), \ 1 \leq i \leq m, \text{ where}$$

   (a) $n_i \in \{0, 1, 2, \ldots\}$ is the initial number of spikes in the neuron σ_i;
   (b) R_i is a finite set of *rules* of the following two forms:
      (i) $E/b^r \to b'$ where $b, b' \in \{a, \bar{a}\}$, $r \geq 1$ and E is either a regular expression over a or $\bar{a}$;
      (ii) $b^s \to \lambda$ for some $s \geq 1$, with the restriction that $b^s \notin L(E)$ for any rule $E/b^r \to b'$ of type (i) from R_i;

   There are four categories of spiking rules identified by $(b, b') \in \{(a, a), (a, \bar{a}), (\bar{a}, a), (\bar{a}, \bar{a})\}$. Here, we allow rules of category $(b, b') \in \{(a, a), (a, \bar{a})\}$ but not the other two types.
3. $syn \subseteq \{1, 2, 3, \ldots, m\} \times \{1, 2, 3, \ldots, m\}$ with $(i, i) \notin syn$ for $1 \leq i \leq m$ (*synapses* among cells);
4. $out \in \{1, 2, 3, \ldots, m\}$ indicates the output neuron.

A rule $E/b^r \to b'$ is applied as follows. If the neuron σ_i contains c spikes/anti-spikes, and $b^c \in L(E)$, $c \geq r$, then the rule can *fire*, and upon application, r spikes/anti-spikes are consumed (thus only $c - r$ remain in σ_i) and a spike/anti-spike is released, which will immediately exit the neuron. The spike/anti-spike emitted by neuron σ_i will pass immediately to all neurons σ_j such that $(i, j) \in syn$. That means transmission of spike/anti-spike takes no waiting time (since the rules do not specify a time delay), the spike/anti-spike will be available in neuron σ_j in the next step. There is an additional restriction that a and $\bar{a}$ cannot stay together, they annihilate each other. If a neuron has either objects a or objects $\bar{a}$, and further objects of either type (maybe both) arrive from other neurons, such that we end with a^q and $\bar{a}^s$ inside, then immediately an annihilation rule $a\bar{a} \to \lambda$ (which is implicit in each neuron), is applied in a maximal manner, so that either a^{q-s} or $(\bar{a})^{s-q}$ remain for the next step, provided that $q \geq s$ or $s \geq q$, respectively. This mutual annihilation of spikes and anti-spikes takes no waiting time and the annihilation rule has priority over spiking and forgetting rules, so each neuron always contains either only spikes or anti-spikes. If we have a rule $E/b^r \to b'$ with $L(E) = \{b^r\}$, then we write it in the simplified form as $b^r \to b'$

and called pure. The rules of the form $b^s \rightarrow \lambda$, are forgetting rules. If neuron contains exactly s spikes/anti-spikes, then forgetting rule $b^s \rightarrow \lambda$ can be applied removing s spikes/anti-spikes from the neuron immediately.

The *configuration* of the system is described by $\mathcal{C} = \langle \beta_1, \beta_2, \ldots, \beta_m \rangle$, where β_i is the number of spikes/anti-spikes present in neuron σ_i. At any moment, if $\beta_i > 0$, it means that there are β_i spikes in neuron σ_i; if $\beta_i < 0$, it indicates that neuron σ_i contains β_i anti-spikes. The initial configuration is $\mathcal{C}_0 = \langle n_1, n_2, \ldots, n_m \rangle$.

A global clock is assumed and in each time unit, each neuron which can use a rule should do it (the system is synchronized), but the work of the system is sequential locally: only (at most) one rule is used in each neuron. For example, if a neuron σ_i has two firing rules, $E_1/b^r \rightarrow b'$ and $E_2/b^c \rightarrow b'$ with $L(E_1) \cap L(E_2) \neq \emptyset$, then it is possible that each of the two rules can be applied, and in that case only one of them is chosen non-deterministically. Thus, the rules are used in the sequential manner in each neuron, but neurons function in parallel with each other. In each step, all neurons which can use a rule of any type, spiking or forgetting, have to evolve, using a rule.

Using the rules in this way, we pass from one configuration of the system to another configuration; such a step is called a transition. For two configurations $\mathcal{C}$ and $\mathcal{C}'$ of Π we denote by $\mathcal{C} \Longrightarrow \mathcal{C}'$, if there is a direct transition from $\mathcal{C}$ to $\mathcal{C}'$ in Π.

A computation of Π is a finite or infinite sequence of transitions starting from the initial configuration, and every configuration appearing in such a sequence is called reachable. A computation halts if it reaches a configuration where no rule can be used. SN PA systems can be used as computing devices in various ways. Here we will use them as generators of numbers. When using an SN PA system in the generative mode, we start from the initial configuration and we define the result of a computation as the number of steps between the first two spikes sent out by the output neuron. The output generated is 0 if no spikes exit the output neuron and the computation halts. The computations and the result of computations are defined in the same way as for usual SN P systems - but we consider the restriction that the output neuron produces only spikes, not also anti-spikes. We denote by $N_2(\Pi)$ the set of numbers computed by Π in this way.

We generalize the SN PA by allowing it to produce k outputs. A k-output SN PA Π has k output neurons, $O_1, \ldots, O_k$. We say that Π generates a k-tuple $(l_1, \ldots, l_k) \in N^k$ if, starting from the initial configuration, there is a sequence of steps such that each output neuron O_i generates exactly two spikes $a\, a$ (the times the pair $a\, a$ are generated may be different for different output neurons) and the time interval between the first a and the second a is l_i. Moreover, after all the output neurons have generated their pair of spikes, the system eventually halts, in the following sense: Π halts if it reaches a configuration where no neurons are fireable. The set of all k-tuples generated is denoted by $Ps_2(\Pi)$. We denote by $N_2S_aNP(cate_y, prule_k, cons_q)$ $[Ps_2S_aNP(cate_y, prule_k, cons_q)]$, the families of all sets $N_2(\Pi)$ $[Ps_2(\Pi), resp.]$ generated by SN PA systems with at most y categories of spiking rules, at most $k \geq 1$ pure rules (only spiking) in each neuron, with all spiking rules $b^r \rightarrow b'$ having $r \leq q$.

In order to compute a function $f : N^k \to N$, k natural numbers $n_1, \ldots, n_k$ are introduced into the system by "reading" from the environment a binary sequence $z = 10^{n_1-1}10^{n_2-1} \ldots 10^{n_k-1}1$. This means that the input neuron of Π receives a spike at each step corresponding to a digit 1 from string z and no spike otherwise. Note that $k+1$ spikes are exactly inputted; that is, it is assumed that no further spike is coming to the input neuron after the last spike. The result of the computation is encoded in the time distance between the first two spikes emitted by the system with the restriction that the system outputs exactly two spikes and halts (immediately after the second spike), hence it produces a spike train of the form $0^b10^{r-1}1$, for some $b \geq 0$ and with $r = f(n_1, \ldots, n_k)$.

3   Computational Completeness of SN PA Systems

We pass now to prove that SN PA systems with neurons having two rules of the form $a \to a$ and $a \to \bar{a}$ are universal as number generators.

In the following proof we use the characterization of NRE by means of register machines [6]. Such a device - in the non-deterministic version - is a construct $M = (m, H, l_0, l_h, I)$, where m is the number of registers, H is the set of instruction labels, l_0 is the start label (labeling an ADD instruction), l_h is the halt label (assigned to instruction $HALT$), and I is the set of instructions; each label from H labels only one instruction from I, thus precisely identifying it. When it is useful, a label can be seen as a state of the machine, l_0 being the initial state, l_h the final/accepting state.

The labeled instructions are of the following forms:

1. $l_i : (ADD(r), l_j, l_k)$ (add 1 to register r and then go to one of the instructions with labels l_j, l_k non-deterministically chosen),
2. $l_i : (SUB(r), l_j, l_k)$ (if register r is non-empty, then subtract 1 from it and go to the instruction with label l_j, otherwise go to the instruction with label l_k),
3. $l_h : HALT$ (the halt instruction).

A register machine M generates a set $N(M)$ of numbers in the following way: we start with all registers empty (i.e., storing the number zero), we apply the instruction with label l_0 and we continue to apply instructions as indicated by the labels (and made possible by the contents of registers). If we reach the halt instruction, then the number n present in register 0 (we assume that the registers are always numbered from 0 to $m-1$) at that time is said to be generated by M. It is known (see, [6]) that register machines generate all sets of numbers which are Turing computable.

It is also possible to consider register machines producing sets of vectors of natural numbers. In this case a distinguished set of k-registers (for some $k \geq 1$) is designated as the output registers. A k-tuple $(l_1, l_2, \ldots, l_k) \in N^k$ is generated if M eventually halts and the contents of the output registers are $l_1, l_2, \ldots, l_k$ respectively. Without loss of generality we may assume that in the halting configuration all the registers, except the output ones, are empty.

We will refer to a register machine with k-output registers (the other registers are auxiliary registers) as a k-output register machine. It is well known that a set S of k-tuples of numbers is generated by a k-output register machine if and only if S is recursively enumerable. Therefore they characterize $PsRE$.

Theorem 1. *Generating spiking neural P systems with anti-spikes with only two types of rules of the form $a \to a$ and $a \to \bar{a}$ are computationally complete, i.e., $N_2 S_a NP(cate_2, prule_2, cons_1) = NRE$.*

Proof. Let $M = (m, H, l_0, l_h, I)$ be a register machine, having the properties specified above; the result of a computation is the number from register 0 and this register can be decremented during the computation.

What we want to do is to have SN PA Π constructed in such a way (1) to simulate the register machine M, and (2) to have its output neuron spiking only twice, at an interval of time which corresponds to a number computed by M.

Instead of specifying all technical details of the construction, we present the three main types of modules of the system Π, with the neurons, their rules, and their synapses represented graphically. In turn, simulating M means to simulate the ADD instructions and the SUB instructions. Thus, we will have a type of modules associated with ADD instructions, one associated with SUB instructions, and one dealing with the spiking of the output neuron (a FIN module). The modules of the three types are given in Figs. 1, 2 and 3 respectively.

For each register r of M, we consider a neuron σ_r in Π whose contents correspond to the contents of the register. Specifically, if the register r holds the number $n > 0$, then the neuron σ_r will contain n anti-spikes.

With each label l_i of an instruction in M, we also associate a neuron σ_{l_i} and some auxiliary neurons $\sigma_{l_{i_q}}$, $q = 1, 2, 3, \ldots$, thus precisely identified by label l_i. Initially, all these neurons are empty, with the exception of the neuron σ_{l_0} associated with the start label of M, which contains a single spike. This means that this neuron is activated. During the computation, the neuron σ_l which receives a spike will become active. Thus, simulating an instruction $l_i : (OP(r), l_j, l_k)$ of M means starting with neuron σ_{l_i} activated, operating the register r as requested by OP, then introducing a spike in one of the neurons σ_{l_j}, σ_{l_k} which becomes in this way active. When activating the neuron σ_{l_h}, associated with the halting label of M, the computation in M is completely simulated in Π; we will then send to the environment two spikes with time gap between them equal to the number stored in the first register of M.

Simulating $l_i : (ADD(r), l_j, l_k)$ (module ADD in Fig. 1).

The initial instruction, that labeled with l_0, is an ADD instruction. Assume that we are in a step t when we have to simulate an instruction $l_i : (ADD(r), l_j, l_k)$, with a spike present in neuron σ_{l_i} (like σ_{l_0} in the initial configuration) and even if some spikes are present in the auxiliary neurons and labels of the previous instruction executed, they will be cleared in the first step when σ_{l_i} fires, so simulating the ADD instruction correctly. Having a spike inside, neuron σ_{l_i} fires producing a spike. This spike will simultaneously go to neurons $\sigma_{l_{i_1}}$, $\sigma_{l_{i_2}}$, $\sigma_{l_{i_3}}$ and $\sigma_{l_{i_4}}$. These four neurons fire at the step $t + 1$ with neuron $\sigma_{l_{i_3}}$ non-deterministically choosing any of its rules $a \to a$ or $a \to \bar{a}$. These rules

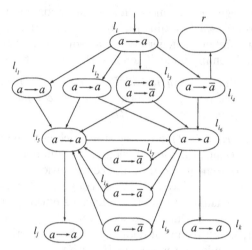

Fig. 1. ADD module: simulation of $l_i : (ADD(r), l_j, l_k)$

determine the non-deterministic choice of the neurons σ_{l_j} or σ_{l_k} to activate. If $a \to a$ is used in $\sigma_{l_{i_3}}$, then $\sigma_{l_{i_5}}$ receives three spikes, $\sigma_{l_{i_6}}$ receives a spike and $\sigma_{l_{i_4}}$ sends an anti-spike to σ_r (thus simulating the increase of the value of register r with 1), $\sigma_{l_{i_6}}$ uses its rule $a \to a$ and sends a spike to $\sigma_{l_{i_7}}, \sigma_{l_{i_8}}, \sigma_{l_{i_9}}$ and σ_{l_k}. At the step $t + 3$, neurons $\sigma_{l_{i_7}}, \sigma_{l_{i_8}}$ and $\sigma_{l_{i_9}}$ fire using their rules $a \to \bar{a}$ and send three anti-spikes to $\sigma_{l_{i_5}}$ (here three spikes and three anti-spikes get annihilated). At the same step σ_{l_k} also becomes active, starting the simulation of the instruction l_k.

If $\sigma_{l_{i_3}}$ uses the rule $a \to \bar{a}$ at $t + 1$, then the anti-spike from $\sigma_{l_{i_3}}$ and spike from $\sigma_{l_{i_2}}$ gets annihilated in both $\sigma_{l_{i_5}}$ and $\sigma_{l_{i_6}}$. Thus at step $t + 2$, $\sigma_{l_{i_5}}$ has one spike and $\sigma_{l_{i_6}}$ has one anti-spike. Neuron $\sigma_{l_{i_5}}$ fires using its rule $a \to a$ sending a spike to $\sigma_{l_{i_6}}$ and σ_{l_j}. In $\sigma_{l_{i_6}}$, the spike gets annihilated with the anti-spike. At time $t + 3$, neuron σ_{l_j} becomes active, thus starting the simulation of the instruction l_j.

Therefore, from the firing of neuron σ_{l_i}, the system adds one anti-spike to neuron σ_r and non-deterministically fires one of neurons σ_{l_j} and σ_{l_k}. Consequently, the simulation of the ADD instruction is possible in Π.

Simulating $l_i : (SUB(r), l_j, l_k)$ (module SUB in Fig. 2).
Assume that we are in a step t when we have to simulate an instruction $l_i : (SUB(r), l_j, l_k)$, with a spike present in neuron σ_{l_i}. Even though some spikes are present in the auxiliary neurons and labels of the previous instruction executed, they will be cleared in the first step when σ_{l_i} fires, so simulating the SUB instruction correctly. Let us examine now Fig. 2, starting from the situation of having a spike in neuron l_i and neuron σ_r, which holds a number of anti-spikes (this number is the value of the corresponding register r). The spike of neuron l_i goes immediately to two neurons, $\sigma_{l_{i_1}}$ and σ_r. If σ_r contains any anti-spikes (this corresponds to the case when register r is non-empty), then the spike gets annihilated with one anti-spike in σ_r, which means the contents of register r is

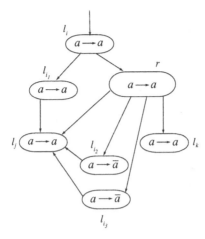

Fig. 2. SUB module: simulation of $l_i : (SUB(r), l_j, l_k)$

decremented by one. In step $t+1$ no spike will come out of σ_r while $\sigma_{l_{i_1}}$ fires and sends a spike to σ_{l_j} and thus activates the neuron σ_{l_j}. In step $t+2$, neuron σ_{l_j} fires, as requested by simulating the SUB instruction.

If in neuron σ_r there is no anti-spike (this corresponds to the case when register r is empty), then the rule $a \rightarrow a$ is used in σ_r at step $t+1$, hence the neuron σ_{l_j} receives two spikes and at the same time neurons $\sigma_{l_{i_2}}$ and $\sigma_{l_{i_3}}$ receive a spike. In the step $t+2$, neurons $\sigma_{l_{i_2}}$ and $\sigma_{l_{i_3}}$ fire and send two anti-spikes to σ_{l_j} and they get annihilated with the two spikes already present in the neuron σ_{l_j}. In the same step σ_{l_k} fires, This means that the simulation of the SUB instruction is correct, we started from l_i and we ended in l_j if the register was non-empty and decreased by one, and in l_k if the register was empty.

Simulating $l_h : (HALT)$ (module FIN in Fig. 3).
Assume now that the computation in M halts, which means that the halting instruction is reached. For Π this means that the neuron l_h gets a spike and fires. Let t be the moment when neuron l_h fires. At that moment, neuron σ_0 contains n anti-spikes, for n being the contents of register 0 of M. The spike of neuron l_h reaches immediately to neurons $\sigma_0, \sigma_{l_{h_1}}, \sigma_{l_{h_2}}$ and $\sigma_{l_{h_3}}$. It is important to remember that this neuron can be involved in a SUB instruction because we have the same rule $a \rightarrow a$ in each neuron that corresponds to any register in M.

If σ_0 has no anti-spikes (when the value in register 0 is 0), at moment $t+1$, four neurons $\sigma_{l_{h_1}}, \sigma_{l_{h_2}}, \sigma_{l_{h_3}}$ and σ_0 fire and all of them spike immediately. Neuron $\sigma_{l_{h_1}}$ sends a spike to $\sigma_{l_{h_6}}$, neuron $\sigma_{l_{h_2}}$ sends a spike to σ_0, neuron σ_0 sends its spike to $\sigma_{l_{h_4}}$ and $\sigma_{l_{h_5}}$, while $\sigma_{l_{h_2}}$ and $\sigma_{l_{h_3}}$ exchange their spikes. At the step $t+2$, neurons $\sigma_{l_{h_4}}, \sigma_{l_{h_5}}, \sigma_{l_{h_6}}$ and σ_0 fire whereas neurons $\sigma_{l_{h_2}}$ and $\sigma_{l_{h_3}}$ will not fire since they have two spikes in each. The spike from neuron σ_0 and the anti-spike from neuron $\sigma_{l_{h_4}}$ are annihilated in $\sigma_{l_{h_5}}$. Neurons σ_{out} and $\sigma_{l_{h_4}}$ receive two spikes each, so cannot fire in the next step and the system halts without sending any spikes to the environment, denoting that the number 0 is generated by the system.

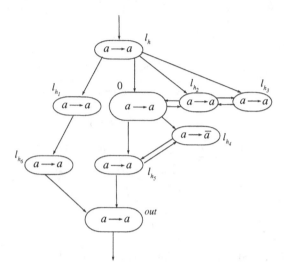

Fig. 3. FIN module: simulation of $l_i : HALT$

If σ_0 has $n > 0$ anti-spikes (when the value of register 0 is $n > 0$), we can observe from the Fig. 3 that at the time $t + 1$, only three neurons $\sigma_{l_{h_1}}, \sigma_{l_{h_2}}$, $\sigma_{l_{h_3}}$ (neuron σ_0 will not fire as the incoming spike is annihilated with one of its anti-spikes). Neuron $\sigma_{l_{h_1}}$ sends a spike to $\sigma_{l_{h_6}}$, neuron $\sigma_{l_{h_2}}$ sends a spike to σ_0, while $\sigma_{l_{h_2}}$ and $\sigma_{l_{h_3}}$ exchange their spikes. At the step $t + 2$, neurons $\sigma_{l_{h_2}}$, $\sigma_{l_{h_3}}$, $\sigma_{l_{h_5}}$ and $\sigma_{l_{h_6}}$ fire. The spike from neuron $\sigma_{l_{h_6}}$ is sent to neuron σ_{out}. So the neuron σ_{out} first fires in step $t + 3$ and sends its spike to the environment. The number of steps from this spike to the next one is the number computed by the system. In each step from $t + 1$ onwards neurons $\sigma_{l_{h_2}}$ and $\sigma_{l_{h_3}}$ exchange their spikes and $\sigma_{l_{h_2}}$ sends one spike to σ_0. The neuron σ_0 does not fire until it has any anti-spikes. This means that the process of removing anti-spikes from neuron σ_0 continues, iteratively having neuron $\sigma_{l_{h_2}}$ sending spikes until σ_0 has no anti-spikes. Thus the neuron σ_0 fires at the step $t + n + 1$ for the first time (for n being the initial number of anti-spikes of neuron σ_0 at time t). Neuron σ_0 sends a spike to $\sigma_{l_{h_2}}, \sigma_{l_{h_4}}$ and $\sigma_{l_{h_5}}$. The remaining steps work in the same way as in the previous case. At the step $t + n + 3$ neurons σ_{out} spikes and the system halts.

The interval between the two spikes of neuron σ_{out} is $(t + n + 3) - (t + 3) = n$, exactly the value of register 0 of M in the moment when its computation halts. Consequently, $N_2(\Pi) = N(M)$ and this completes the proof.   □

This result can have a nice interpretation: it is sufficient for a "brain" (in the form of an SN P system with anti-spikes) to have neurons sending either excitatory or inhibitory impulses which behaves non-deterministically in order to achieve "complete (Turing) creativity".

$l_0 : (SUB(1), l_1, l_2),$ $l_1 : (ADD(7), l_0),$ $l_2 : (ADD(6), l_3),$

$l_3 : (SUB(5), l_2, l_4),$ $l_4 : (SUB(6), l_5, l_3),$ $l_5 : (ADD(5), l_6),$

$l_6 : (SUB(7), l_7, l_8),$ $l_7 : (ADD(1), l_4),$ $l_8 : (SUB(6), l_9, l_0),$

$l_9 : (ADD(6), l_{10}),$ $l_{10} : (SUB(4), l_0, l_{11}),$ $l_{11} : (SUB(5), l_{12}, l_{13}),$

$l_{12} : (SUB(5), l_{14}, l_{15}),$ $l_{13} : (SUB(2), l_{18}, l_{19}),$ $l_{14} : (SUB(5), l_{16}, l_{17}),$

$l_{15} : (SUB(3), l_{18}, l_{20}),$ $l_{16} : (ADD(4), l_{11}),$ $l_{17} : (ADD(2), l_{21}),$

$l_{18} : (SUB(4), l_0, l_h),$ $l_{19} : (SUB(0), l_0, l_{18}),$ $l_{20} : (ADD(0), l_0),$

$l_{21} : (ADD(4), l_{18}),$ $l_h : HALT$

Fig. 4. A universal register machine M_u from Korec [3]

Theorem 1 can be easily extended by allowing more output neurons and then simulating a k-output register machine, producing in this way sets of vectors of natural numbers.

Theorem 2. $Ps_2 S_a NP(cate_2, prule_2, cons_1) = PsRE.$

4 A Small Universal SN P System with Anti-spikes

A register machine specified above can also compute any Turing computable function: we introduce the arguments $n_1, n_2, \ldots, n_k$ in specified registers r_1, $r_2, \ldots, r_k$ (without loss of the generality, we may assume that we use the first k registers), we start with the instruction with label l_0, and if we stop (with the instruction with label l_h), then the value of the function is placed in another specified register, r_t, with all registers different from r_t being empty. The partial function computed in this way is denoted by $M(n_1, n_2, \ldots, n_k)$. In the computing form, the register machines can be considered deterministic, without losing the Turing completeness: all ADD instructions $l_i : (ADD(r), l_j, l_k)$ have $l_j = l_k$ (and the instruction is written in the form $l_i : (ADD(r), l_j)$).

In [3], the register machines are used for computing functions, with the universality defined as follows. Let $(\phi_0, \phi_1, \ldots)$ be a fixed admissible enumeration of the unary partial recursive functions. A register machine M_u is said to be universal if there is a recursive function g such that for all natural numbers x, y we have $\phi_x(y) = M_u(g(x), y)$. In [3], several universal register machines are constructed, with the input (the couple of numbers $g(x)$ and y) introduced in registers 1 and 2, and the result obtained in register 0.

We give now in the notation introduced above the specific universal register machine from [3], which will be used in this section: we have $M_u = (8, H, l_0, l_h, I)$, with the instructions (their labels constitute the set H) presented in Fig. 4 (The

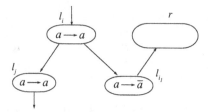

Fig. 5. Deterministic ADD module $l_i : (ADD(r), l_j)$

machine from [3] contains a separate check for zero of register 6, of the form l_8 : if register(6) = 0, then go to l_0, else go to l_{10}; this instruction was replaced in our set up by $l_8 : (SUB(6), l_9, l_0)$, $l_9 : (ADD(6), l_{10})$). Therefore, there are 8 registers (numbered from 0 to 7) and 23 instructions (hence 23 labels), the last instruction being the halting one. The input numbers (the "code" of the partial recursive function to simulate and the argument for this function) are introduced in registers 1 and 2, and the result is obtained in register 0.

In the systems constructed in this work, the neurons are quite "simple" in the sense that each neuron has only one rule.

We proceed now to constructing the universal SN PA system Π_u using pure rules of category (a, a) and $(a, \overline{a})$ without forgetting rules, for computing functions. To this aim, we follow a similar way used in previous section but to simulate a deterministic register machine by an SN PA system. Neurons are associated with each register and with each label of an instruction of the machine. If a register contains a number n, then the associated neuron will contain n anti-spikes. Modules as in Fig. 5 and Fig. 2 are associated with the ADD and the SUB instructions (each of these modules contains auxiliary neurons which do not correspond to registers or to labels of instructions).

The work of the system is triggered by introducing a spike in the neuron σ_{l_0} (associated with the starting instruction of the register machine). In general, the simulation of an ADD or SUB instruction starts by introducing a spike in the neuron with the instruction label (we say that this neuron is activated).

Starting with neurons σ_1 and σ_2 already loaded with $g(x)$ and y spikes, respectively, and introducing a spike in neuron σ_{l_0}, we can compute in our system Π_u in the same way as the universal register machine M_u from Fig. 4; if the computation halts, then neuron σ_0 will contain the $\phi_x(y)$ number of anti-spikes.

There are two additional tasks to solve: to introduce the mentioned anti-spikes in the neurons σ_1, σ_2, and to output the computed number. The first task is covered by module $INPUT$ presented in Fig. 6. The neuron σ_{c_5} converts the spikes it receives from the input neuron into anti-spikes. The neuron σ_{c_8} fires only after receiving the third anti-spike from σ_{c_5}, and then it sends a spike to neuron σ_{l_0}, thus starting the simulation of M_u. At that moment, neurons σ_1 and σ_2 are already loaded: neuron σ_{c_3} sends to neuron σ_1 as many anti-spikes as the number of steps between the first two input spikes, and after that it gets "over flooded" by the second input spike and is blocked (neurons σ_{c_1} and σ_{c_2} supply spikes to σ_{c_3} till they receive second spike through σ_{in}); in turn, neuron σ_{c_5}

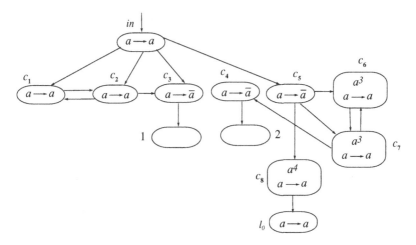

Fig. 6. $INPUT$ module

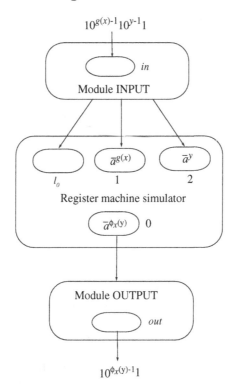

Fig. 7. The general design of the universal SN PA system

sends anti-spikes to neurons σ_{c_6}, σ_{c_7} and they start working only after collecting two anti-spikes. Neurons σ_{c_6} and σ_{c_7} supply one spike in each step to neuron σ_{c_4}, which loads σ_2 with as many anti-spikes as the number of steps between

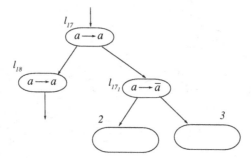

Fig. 8. A module simulating two consecutive ADD instructions

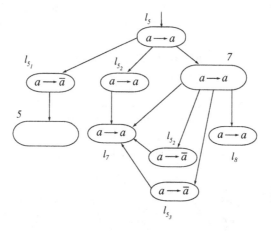

Fig. 9. A module simulating $ADD - SUB$ instructions

the last two input spikes and all three neurons stop working after receiving the third anti-spike from σ_{c_5}.

In this construction, we do not need to modify the universal register machine as in [8] for not allowing subtraction operations on the neuron where we place the result. So the result will be in the neuron σ_0 which corresponds to the register 0 of M_u.

Having the result of the computation in register 0, we can output the result by means of the module $OUTPUT$ which is same as FIN module in Fig. 3 (the working of this module is explained in the previous section). The overall design of the system is given in Fig. 7.

We can check that each neuron in the system Π_u has only one rule; that is, the system Π_u is simple. The system Π_u has 8 neurons for the 8 registers, 23 neurons for the 23 labels, 9 neurons for the 9 ADD instructions, 39 neurons for 13 SUB instructions, 9 neurons in the $INPUT$ module and 6 neurons in the $OUTPUT$ module which comes to a total of 94 neurons. This number can be slightly decreased, by some "code optimization", exploiting some particularities of the register machine M_u.

First, let us observe that the sequence of two consecutive ADD instructions: $l_{17} : (ADD(2), l_{21}), l_{21} : (ADD(3), l_{18})$, without any other instruction addressing the label l_{21}, can be simulated by the module from Fig. 8, and in this way we save a neuron associated with l_{21}.

The module from Fig. 9 can simulate the consecutive $ADD - SUB$ instructions $l_5 : (ADD(5), l_6), l_6 : (SUB(7), l_7, l_8)$. A similar module can be constructed to simulate the consecutive ADD-SUB instructions $l_9 : (ADD(6), l_{10})$, $l_{10} : (SUB(4), l_0, l_{11})$. So two neurons (associated with the labels l_6 and l_{10}) are saved. We save a total of 3 neurons and get the improvement from 94 to 91 neurons. We state this result in the form of a theorem in order to stress its importance:

Theorem 3. *There exists a universal simple SN PA system with 91 neurons for computing functions.*

5   Conclusion

By using the characterization of types of rules, we were able to show that for obtaining computational completeness of spiking neural P system with anti-spikes only two rules of the form $a \to a$, $a \to \bar{a}$ are needed. In this work, the problem of constructing universal SN PA systems with a small number of rules is also investigated. The systems constructed in this work has 91 rules of the form $a \to a$ or $a \to \bar{a}$. It is possible to use less neurons to construct universal SN PA systems provided that neurons have more types of spiking rules.

References

1. Păun, G., Rozenberg, G., Salomaa, A. (eds.): Handbook of Membrane Computing. Oxford University Press (2010)
2. Rozenberg, G., Salomaa, A. (eds.): Handbook of Formal Languages, 3 volumes. Springer, Berlin (1998)
3. Korec, I.: Small universal Turing machines. Theoretical Computer Science 168, 267–301 (1996)
4. Pan, L., Păun, G.: Spiking Neural P Systems with Anti-Spikes. International Journal of Computers, Communications and Control 4(3), 273–282 (2009)
5. Ionescu, M., Păun, G., Yokomori, T.: Spiking Neural P Systems. Fundamenta Informaticae 71, 279–308 (2006)
6. Minsky, M.: Computation finite and infinite machines. Prentice Hall, Englewood Cliffs (1967)
7. Song, T., Pan, L., Wang, J., Venkat, I., Subramanian, K.G., Abdullah, R.: Normal Forms of Spiking Neural P systems with anti-spikes. IEEE Transactions on Nanobioscience 11(4), 352–359 (2012)
8. Song, T., Jiang, Y., Shi, X., Zeng, X.: Small Universal Spiking Neural P Systems with Anti-Spikes. Journal of Computational and Theoretical Nanoscience 10(4), 999–1006 (2013)
9. Zeng, X., Zhang, X., Pan, L.: Homogeneous spiking neural P systems. Fundamenta Informaticae 97(12), 1–20 (2009)

Self-stabilizing Minimal Global Offensive Alliance Algorithm with Safe Convergence in an Arbitrary Graph

Yihua Ding, James Z. Wang, and Pradip K. Srimani

School of Computing
Clemson University
Clemson, SC 29634, USA
{yihuad,jzwang,srimani}@clemson.edu

Abstract. In a graph or a network $G = (V, E)$, a set $S \subseteq V$ is a global offensive alliance if each node $i \in \{V - S\}$ has $|N[i] \cap S| \geq |N[i] - S|$. A global offensive alliance S is called minimal when there does not exist a node $i \in S$ such that the set $S - \{i\}$ is a global offensive alliance. In this paper, we propose a new self-stabilizing algorithm for minimal global offensive alliance. It has safe convergence property under synchronous daemon in the sense that starting from an arbitrary state, it quickly converges to a global offensive alliance (a safe state) in two rounds, and then stabilizes in a minimal global offensive alliance (the legitimate state) in $O(n)$ rounds without breaking safety during the convergence interval, where n is the number of nodes. Space requirement at each node is $O(\log n)$ bits.

Keywords: Self-stabilization, Minimal Global Offensive Alliance, Safe Convergence, Synchronous Daemon.

1 Introduction

Self-stabilization is an optimistic paradigm to provide autonomous adaptability against an unlimited number of transient faults (transient fault is an event of corrupting the data but not the program code) in the distributed systems. An algorithm is self-stabilizing iff it reaches some legitimate state starting from an arbitrary state [1]. In a self-stabilizing algorithm, each node maintains a set of local variables, which is defined as *local state*. The product of the local states of all nodes in the system is defined as *global state*. A self-stabilizing algorithm is usually written as a set of rules at each node. Each rule at a node consists of a *condition* and an *action*. A condition is a boolean predicate involving the states of the node and its neighbors. If one or more conditions on a node are satisfied, i.e., some boolean predicates are true, we say the node is *privileged*. If a privileged node is selected by the daemon (run time scheduler), it makes a *move* by changing its local variables. A detailed exposition of self-stabilizing algorithm can be found in [2].

T V Gopal et al. (Eds.): TAMC 2014, LNCS 8402, pp. 366–377, 2014.
© Springer International Publishing Switzerland 2014

Recently, a new concept of *safe convergence* has been introduced in [3]. In a traditional self-stabilizing algorithm, the desired global property (hence, the relevant service in the system) is not guaranteed during the convergence interval; the concept of safe convergence was introduced to limit this inconvenience to a minimum possible. A self-stabilizing algorithm has the safe convergence property iff it first converges to a safe state quickly ($O(1)$ time is expected), and then converges to a legitimate state without breaking safety in the process. Safe convergence property is especially attractive since it provides a measure of safety during the convergence interval of the self-stabilizing algorithm. Various self-stabilizing algorithms with safe convergence have been proposed in the literature, such as minimal independent dominating set, connected dominating set and so on [3–6]. Note that all self-stabilizing algorithms with safe convergence have assumed synchronous daemon; it is not possible to reach a safe state in constant time using either central or distributed daemon.

In this paper, we are interested in the global offensive alliance of a network graph. Intuitively, a node is said to be a *defender* of an adjacent node, if both the nodes are in the alliance, or both the nodes are not in the alliance. A node is considered a defender of itself. A node is said to be a *attacker* of an adjacent node, if one of them is in the alliance but the other one is not. A node is called *defended* if the number of its defenders is not less than the number of its attackers; and *attacked* if the number of its attackers is not less than the number of its defenders. An alliance is called *global offensive alliance* if all nodes outside the alliance are attacked [7, 8].

To the best of our knowledge, only [8] presents a self-stabilizing algorithm for constructing minimal global offensive alliance in an arbitrary network graph; the algorithm assumes a central daemon (only one privileged node is selected to move at each step), and uses 1 bit of space at each node. The stabilization time $O(n^2)$, where n is the number of nodes in the network. It does not enjoy the safe convergence property.

In this paper, we assume a synchronous daemon and nodes with unique IDs and propose the first self-stabilizing algorithm with safe convergence to compute the minimal global offensive alliance of an arbitrary network graph; starting from an arbitrary state, the proposed algorithm first converges to a global offensive alliance (a *safe* state, not necessarily the legitimate state) in two rounds, and then stabilizes to a minimal one (the *legitimate* state) in $O(n)$ rounds without breaking safety rule during the convergence interval, where n is the number of nodes in the network.

2 Model and Terminology

A network or a distributed system is modeled by an undirected graph $G = (V, E)$, where V is the set of nodes, and E is the set of edges. For a node i, $N(i)$, its *open neighborhood*, denotes the set of nodes adjacent to node i; $N[i] = N(i) \cup i$ denotes its *closed neighborhood*. For a node i, $N^2(i) = \cup_{j \in N[i]} N(j) - \{i\}$, its *2-hop open neighborhood*, denotes the set of nodes that are at most distance of

2 from node i. Each node $j \in N(i)$ is called a *neighbor* of node i and each node $j \in N^2(i)$ is called a *2-neighbor* of node i.

Consider any graph $G = (V, E)$, where $|V| = n$ and $|E| = m$. A set $\mathcal{S} \subseteq V$ is a *global offensive alliance* if each node $i \in \{V - \mathcal{S}\}$ is attacked, i.e., $|N[i] \cap \mathcal{S}| \geq |N[i] - \mathcal{S}|$. A global offensive alliance $\mathcal{S}$ is called *minimal* iff there does not exist a node $i \in \mathcal{S}$ such that $\mathcal{S} - \{i\}$ is a global offensive alliance [9]. We assume that each node has a unique identifier 1 through n. For example, consider the graph shown in Figure 1, where the values inside the nodes give the identifiers. $n = 6$, $m = 8$. $N(1) = \{2, 4\}$, $N[1] = \{1, 2, 4\}$. In this graph, $\{2, 4, 6\}$, $\{1, 2, 3, 5, 6\}$, and $\{1, 2, 4, 5, 6\}$ all are global offensive alliances, but only $\{2, 4, 6\}$ is minimal among these three global offensive alliances.

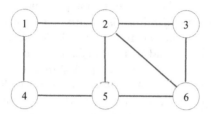

Fig. 1. A Graph with 6 Nodes

Execution Model: Execution of the protocol at each node is managed by a synchronous scheduler (daemon), that selects all privileged nodes in a system state to move synchronously and atomically in each step, called a *round*. A node is privileged in a given system state iff it is enabled to move by at least one rule of the protocol. If a privileged node is selected by the daemon, it executes the entire code. The protocol terminates in a system state when no node is privileged. The protocol assumes a *shared-memory model* and each node knows only its own state and the local states of its immediate neighbors as is customary in self-stabilizing algorithms. The proposed algorithm does not need to know the size of the network graph.

3   Minimal Global Offensive Alliance with Safe Convergence

In our proposed self-stabilizing minimal global offensive alliance algorithm with safe convergence (we call it algorithm MGOASC), each node $i, 1 \leq i \leq n$, maintains the following variables:

- A boolean flag s_i; at any time (system state) $\mathcal{S}$ is the current set of nodes with $s_i = 1$.
- An integer variable d_i denotes the difference between the number of nodes in the closed neighborhood of node i in $\mathcal{S}$ and those outside of $\mathcal{S}$, i.e., $d_i = |N[i] \cap \overline{\mathcal{S}}| - |N[i] - \mathcal{S}|$, at any given system state.

- A pointer p_i (which may be null) that points to a node $j \in N[i]$, indicated by $p_i = j$. If $p_i = i$ for a node i, we say node i has a **self-pointer**.
- A boolean flag $\textbf{\textit{mutex}}_i$; node i sets this flag to obtain mutually exclusive (with respect to its neighbors) right for some activity.

Definition 1. *A global system state is the union of the local states of all nodes. A system state is* **safe** *if* $\mathcal{S} = \{i \in V : s_i = 1\}$ *denotes a global offensive alliance, i.e., each node* $j \in \{V - \mathcal{S}\}$ *is attacked* $(d_j \geq 0)$. *A system state is* **legitimate** *if* $\mathcal{S}$ *denotes a minimal global offensive alliance. We denote a global system state by* Σ_i, $i = 0, 1, 2, \cdots$, *where* Σ_0 *denotes the initial arbitrary state and* Σ_r *denotes the system state after the r-th round of the protocol,* $r = 1, 2, \cdots$; *r-th round operates on* Σ_{r-1} *to generate* Σ_r.

Definition 2. *In any system state, $minN_i$ of a node i, $1 \leq i \leq n$, is defined as the smallest ID node in the* <u>closed neighborhood</u> *of node i that points to itself, i.e.,*

$$minN_i = \min\{j | j \in N[i] \wedge p_j = j\}, where \ \min\{\} = null$$

The value of the logical variable $minN_i$ is locally computable at node i in any system state.

Definition 3. *For a node i, a Boolean predicate* $\texttt{enter}_i$ *is defined as:*

$$\texttt{enter}_i \stackrel{\text{def}}{\equiv} (s_i = 0) \wedge (d_i < 0)$$

Definition 4. *For a node i, a Boolean predicate* $\texttt{leave}_i$ *is defined as:*

$$\texttt{leave}_i \stackrel{\text{def}}{\equiv} (s_i = 1) \wedge (d_i \geq 2) \wedge (\forall j \in N(i) - \mathcal{S} : d_j \geq 2)$$

Definition 5. *A node i is called* **favored** *node iff the Boolean predicate* $\texttt{favored}_i = 1$, *where*

$$\texttt{favored}_i \stackrel{\text{def}}{\equiv} \forall j \in N[i] : p_j = i$$

Definition 6. *A node i is called* **ready** *iff the Boolean predicate* $\texttt{ready}_i = 1$, *where*

$$\texttt{ready}_i \stackrel{\text{def}}{\equiv} \texttt{leave}_i \wedge \texttt{favored}_i \wedge (\nexists j \in N(i) : mutex_j = 1)$$

Remark 1. In any system state,

1. For a node i, $\texttt{enter}_i = 1$, iff $i \notin \mathcal{S}$ and i has less attackers than defenders (i.e., node i is not attacked).
2. For a node i, $\texttt{leave}_i = 1$, iff $i \in \mathcal{S}$ has at least two more defenders than attackers, and each neighbor $j \in N(i) - \mathcal{S}$ has at least two more attackers than defenders (it is possible that d_j's are erroneous in some system states).
3. A node i is favored ($\texttt{favored}_i$ is true), iff all nodes in the closed neighborhood of node i point to node i. Thus, if a node i is favored in a system state, no node in $N^2(i)$ is favored in the same state.

4. A node i is ready (ready_i is true), iff 3 conditions are simultaneously true: (i) i is a favored node, (ii) leave_i is true, and (iii) there does not exist a neighbor $j \in N(i)$ with $mutex_j = true$. Thus, if a node i is ready in a system state, no node in $N^2(i)$ is ready in the same state.

The objective of algorithm MGOASC is to quickly converge to a safe state and thereafter to transition through safe states to reach the legitimate state to obtain the minimal global offensive alliance. We assume a synchronous daemon where at any round all privileged nodes make their moves simultaneously. The underlying approach is as follows. If a node i is out of S and not attacked, it enters S. A node $i \in S$ leaves S iff it can ensure each neighbor $j \in N(i) - S$ at the beginning of the current round is still attacked after the current round. The complete pseudo code of algorithm MGOASC is shown in Figure 2. We highlight a few simple characteristics of the algorithm in the following remark.

RA: **if** $d_i \neq |N[i] \cap S| - |N[i] - S|$
       then $d_i \leftarrow |N[i] \cap S| - |N[i] - S|$;   **Update_Difference**

RB: **if** $(p_i \neq minN_i) \wedge (mutex_i = 1)$
       then $\{p_i \leftarrow minN_i; mutex_i \leftarrow 0;\}$ **Update_Pointer**

RC: **if** $\text{leave}_i \wedge (\forall j \in N[i] : p_j = null)$
       then $p_i \leftarrow i$;   **Get_Self-pointer**

RD: **if** $(mutex_i = 0) \wedge \big((p_i \neq minN_i) \vee (\text{leave}_i \wedge \text{favored}_i) \big)$
       then $mutex_i \leftarrow 1$;   **Get_mutex**

RE: **if** $(mutex_i = 1) \wedge \text{ready}_i$
       then $\{s_i \leftarrow 0; p_i \leftarrow null; mutex_i \leftarrow 0;\}$ **Leave** S

RF: **if** enter_i
       then $s_i \leftarrow 1$;   **Enter** S

RG: **if** $\neg\text{leave}_i \wedge (p_i = i)$
       then $p_i \leftarrow null$;   **Release_Self-pointer**

RH: **if** $(mutex_i = 1) \wedge (p_i = minN_i) \wedge \neg\text{ready}_i$
       then $mutex_i \leftarrow 0$;   **Release_mutex**

Fig. 2. The Algorithm MGOASC at Node i, $1 \leq i \leq n$

Definition 7. *In a given round a node i is called **privileged** if it is enabled by any of the rules of the algorithm. Execution of the algorithm **terminates** when in a system state no node is privileged.*

Remark 2. In a given round r, $r \geq 1$, of execution,

1. Each node i updates its variable d_i; in system state Σ_r, each d_i denotes correct difference as of Σ_{r-1}.
2. A node i enters S by executing rule RF iff enter_i is true.
3. If leave_i is true, node i cannot immediately leave S. A node i can leave S (by executing rule RE) iff it is ready (Definition 6) and $mutex_i = 1$.

4. A node cannot acquire a self-pointer by executing rule $\mathbb{RB}$ (Update_Pointer). A node i can acquire a self-pointer ($p_i = i$) only by executing rule $\mathbb{RC}$; a node i can execute rule $\mathbb{RC}$ only when all its neighbors have *null* pointers.

5. A node i must have $mutex_i = 1$ in order to execute either rule $\mathbb{RB}$ (Update_Pointer) or rule $\mathbb{RE}$ (Leave $\mathcal{S}$).

6. If $mutex_i = 0$ and node i is eligible either to update its pointer and/or to leave $\mathcal{S}$, node i will execute rule $\mathbb{RD}$ (Get_mutex).

7. If $mutex_i = 1$, then node i will either leave $\mathcal{S}$ or update its pointer, iff it is eligible to do so; in any case it will release the mutex (rule $\mathbb{RB}$, or $\mathbb{RE}$ or $\mathbb{RH}$).

8. If a node i executes rule $\mathbb{RE}$ (Leave $\mathcal{S}$) in a round, no node $j \in N^2(i)$ can concurrently execute rule $\mathbb{RE}$ (Leave $\mathcal{S}$) in the same round since both cannot be simultaneously ready (Remark 1.4).

9. Consider two adjacent node i and j: if node i leaves $\mathcal{S}$ by executing rule $\mathbb{RE}$, node j cannot update its pointer by executing rule $\mathbb{RB}$ in the same round (since $ready_i$ and $mutex_j$ cannot be simultaneously be true by Remark 1.4) and vice versa. Note that two adjacent nodes can update their pointers in the same round.

4 Correctness

In this section, we first prove that $\mathcal{S}$ is a minimal global offensive alliance when algorithm MGOASC terminates, and then we show the algorithm has safe convergence property in the sense that starting from an arbitrary state, it first converges to a safe state (in which a global offensive alliance is computed) in two rounds, and then stabilizes to a legitimate state (in which a minimal global offensive alliance is computed) in $O(n)$ rounds without breaking safety, where n is the number of nodes.

Lemma 1. *If algorithm* MGOASC *terminates, then for each node* $i \in V$:

(a) d_i *is correct, i.e.,* $d_i = |N[i] \cap \mathcal{S}| - |N[i] - \mathcal{S}|$.
(b) $mutex_i = 0$.
(c) $p_i = null$.
(d) $\mathtt{enter}_i = 0$ *and* $\mathtt{leave}_i = 0$.

Proof. (a) This is obvious since no node is privileged by the rule $\mathbb{RA}$.

(b) Assume, by contradiction, there exists at least one node j such that $mutex_j = 1$. We must have $p_j = minN_j$ (node j is not privileged by rule $\mathbb{RB}$) and $ready_j = 0$ (node j is not privileged by rule $\mathbb{RE}$); thus, node j must be privileged by rule $\mathbb{RH}$, a contradiction.

(c) Assume, by contradiction, there exist some nodes, each pointing to itself and consider, among those nodes, the node j with minimum ID. For each $k \in N(j)$, we must have $p_k = j$; if not, for such a node k, $p_k \neq minN_k = j$ (since j is the minimum ID node pointing to itself) and node k would be privileged by rule $\mathbb{RD}$. Thus, node j is a favored node; also since it is not privileged by rule

$\mathbb{RG}$, we must have $\texttt{leave}_j$ true. Thus, we get node j is privileged by rule $\mathbb{RD}$ since $mutex_j = 0$ (by part (b)), a contradiction.

(d) No node i is privileged by rule $\mathbb{RC}$ or rule $\mathbb{RF}$; the claim follows from part (a) and (c).

Theorem 1. *Starting from an arbitrary system state, if algorithm* MGOASC *terminates using synchronous daemon, then* $\mathcal{S}$ *is a minimal global offensive alliance.*

Proof. First, we show $\mathcal{S}$ is a global offensive alliance. Assume, by contradiction, $\mathcal{S}$ is not a global offensive alliance, i.e., there is at least one node $i \in \{V - \mathcal{S}\}$ $(s_i = 0)$ such that $d_i < 0$. Then, $\texttt{enter}_i = 1$ and node i is privileged by the rule $\mathbb{RF}$ (Enter $\mathcal{S}$), a contradiction. Thus $\mathcal{S}$ is a global offensive alliance.

Next, we claim $\mathcal{S}$ is minimal. Assume otherwise, i.e., there exists a node $i \in \mathcal{S}$ such that $\mathcal{S} - \{i\}$ is a global offensive alliance. Since $i \in \mathcal{S}$ (i.e., $s_i = 1$), and $\texttt{leave}_i = 0$ (Lemma 1(d)), either (1) node i has $d_i < 2$, or (2) there exists at least one neighbor $j \in N(i) - \mathcal{S}$ with $d_j < 2$. Thus, the removal of i from $\mathcal{S}$ (i.e., s_i becomes 0) would make either $d_i < 0$ (thus $\texttt{enter}_i = 1$) or $d_j < 0$ for some neighbor(s) $j \in N(i) - \mathcal{S}$ (thus $\texttt{enter}_j = 1$). Thus either node i or j is privileged by the rule $\mathbb{RF}$ (Enter $\mathcal{S}$), a contradiction.

Lemma 2. *In any system state, if either (a) $p_i = j$ for some $j \in N(i)$ and node j is not enabled to execute rule $\mathbb{RE}$ (Leave $\mathcal{S}$), or (b) node i is enabled by rule $\mathbb{RB}$ to update its pointer p_i, then $d_i \leq |N[i] \cap \mathcal{S}| - |N[i] - \mathcal{S}|$ after (at the end of) the current round.*

Proof. (a) Since $p_i = j$, no node $k \in N(i) - \{j\}$ is favored ($\texttt{favored}_k$ is false) in the current state and can leave $\mathcal{S}$; coupled with the fact that node j is not enabled to leave $\mathcal{S}$, it follows that $d_i \leq |N[i] \cap \mathcal{S}| - |N[i] - \mathcal{S}|$ at the end of the current round. (b) the argument is similar along with Remark 2.9. **Note:** When node i updates its pointer, it is possible that some neighbor of node i enters $\mathcal{S}$ (hence the inequality).

Lemma 3. *Beginning with round 2 of execution, whenever a node i leaves $\mathcal{S}$ (by executing rule $\mathbb{RE}$) in any round, each neighbor $j \in N(i) - \mathcal{S}$ remains attacked $(|N[j] \cap \mathcal{S}| - |N[j] - \mathcal{S}| \geq 0)$ at the end of the round.*

Proof. When a node i leaves $\mathcal{S}$ in round 1, it is possible that $d_j \leq |N[j] \cap \mathcal{S}| - |N[j] - \mathcal{S}|$ is false for some neighbor(s) j of node i since initial values of d-variables are arbitrary in the initial arbitrary illegitimate state.

Assume node i is privileged to execute $\mathbb{RE}$ to leave $\mathcal{S}$ in round $r \geq 2$; $\texttt{favored}_i$ is true, i.e., $p_j = i$ for all $j \in N[i]$. For any such j, consider the round when $p_j = i$ was set for the last time: node i has not executed rule $\mathbb{RE}$ from that point until now; thus, $d_j \leq |N[j] \cap \mathcal{S}| - |N[j] - \mathcal{S}|$ (Lemma 2) from that point until now, since no neighbor of node j can make a Leave move when $p_j = i$. Thus, $d_j \leq |N[j] \cap \mathcal{S}| - |N[j] - \mathcal{S}|$ for all neighbors $j \in N(i)$. When node i leaves $\mathcal{S}$, all its neighbors $j \in N(i) - \mathcal{S}$ still has $N[j] \cap \mathcal{S}| - |N[j] - \mathcal{S}| \geq 0$ since no other neighbors of node j can leave $\mathcal{S}$ in the same round (Remark 2.8).

Theorem 2. *Starting from any initial illegitimate state, algorithm MGOASC converges to a safe state ($\mathcal{S}$ denotes a global offensive alliance, i.e., each node $i \in \{V - \mathcal{S}\}$ is attacked) after 2 rounds.*

Proof. We show that each node $i \in \{V - \mathcal{S}\}$ is attacked (i.e., $N[i] \cap \mathcal{S}| - |N[i] - \mathcal{S}| \geq 0$) after 2 rounds of execution; we consider two cases:

(a) Consider a node i with $s_i = 0$ before and after round 2. Since s_i remains 0 after the second round, $\mathtt{enter}_i$ is false (i.e., node i is attacked) at the beginning of the round. If any $\mathcal{S}$ neighbor of node i leaves $\mathcal{S}$ in the second round, by Lemma 3 node i remains attacked at the end of round 2.

(b) Consider a node i that changes s_i from 1 to 0 by executing rule $\mathbb{RE}$ in round 2. At the beginning of round 2, $\mathtt{leave}_i$ is true and thus has $d_i \geq 2$; since node i leaves $\mathcal{S}$ in the current round, no neighbor of i can leave $\mathcal{S}$ in the same round (Remark 2.8); thus, node i will have $N[i] \cap \mathcal{S}| - |N[i] - \mathcal{S}| \geq 0$ (i.e., node i is attacked) after round 2.

Theorem 3. *After round 2, algorithm MGOASC maintains safety in all subsequent rounds before converging to a legitimate state.*

Proof. It suffices to show that starting from a safe state, the system transitions to another safe state in each round until reaching the legitimate state. The proof is similar to the one for Theorem 2, we here omit the details.

Lemma 4. *Starting from a safe state, no node will ever execute rule $\mathbb{RF}$ (Enter $\mathcal{S}$) in subsequent rounds.*

Proof. In a safe state, each node in $V - \mathcal{S}$ is attacked, i.e., $\mathtt{enter}_i$ (Definition 3) is false for each node i; thus, no node i can execute rule $\mathbb{RF}$ (Enter $\mathcal{S}$) in the current round. The algorithm MGOASC always in the safe state after the second round (Theorem 3), thus each node in $V - \mathcal{S}$ always remains attacked after the second round. The lemma holds.

Lemma 5. *In any system state, two adjacent nodes i and j have self-pointers (i.e., $p_i = i$ and $p_j = j$), then the larger ID node will lose the self-pointer (i.e., if say $i < j$, $p_j \neq j$) in at most 2 rounds.*

Proof. In the worst case, neither node i nor j has the mutex lock; in the next round node j will acquire mutex lock by executing rule $\mathbb{RD}$, since $p_j \neq minN_j$ and in next round node j gets $p_j \neq j$ by executing rule $\mathbb{RB}$.

Lemma 6. *In a safe system state (i.e., Σ_r, $r \geq 2$),*

(a) *if $p_i = null$, then $d_i \leq |N[i] \cap \mathcal{S}| - |N[i] - \mathcal{S}|$.*
(b) *If two adjacent nodes i and j have self-pointers (i.e. $p_i = i$ and $p_j = j$), then nodes i and j must have concurrently acquired the self-pointers in a previous round r', $r' \leq r$.*

Proof. (a) If $p_i = null$ in Σ_r, then either (i) $p_i = null$ in Σ_r and node i did not change p_i in round r; or (ii) node i executed either rule $\mathbb{RB}$, $\mathbb{RE}$, or $\mathbb{RG}$ in round

r. In either case, no node $j \in N(i)$ can leave $\mathcal{S}$ in round r (Remark 2.9). The claim follows (Remark 2.1). [Note: It is possible that some neighbor of node i enters $\mathcal{S}$ in round r (possible only in rounds 1 and 2), hence the inequality.]

(b) If $p_i = i$ but $p_j \neq j$, then node j cannot execute rule $\mathbb{RC}$ to get the self-pointer in the next state (Remark 2.4).

Lemma 7. *In a safe state Σ_r, $r \geq 2$, for two adjacent nodes i and j, if $\mathtt{leave}_j = 1 \wedge p_j = j$ and $p_i = i$, then $\mathtt{leave}_i = 1$.*

Proof. Assume otherwise, i.e., $\mathtt{leave}_i = 0$. It follows from Lemmas 5 and 6(b) that nodes i and j had concurrently executed rule $\mathbb{RC}$ to get self-pointers either in round $r - 1$ or r. There are 2 possible scenarios:

Case 1 [Nodes i and j executed rule $\mathbb{RC}$ in the round $r - 1$]: In Σ_{r-2}, $\mathtt{leave}_i = 1$, $\mathtt{leave}_j = 1$, $p_i = p_j = null$, and for each $k \in N(i) \cup N(j)$, $p_k = null$ (nodes i and j are enabled for rule $\mathbb{RC}$; Definition 4). We argue that:

(a) Node i can not leave $\mathcal{S}$ (execute rule $\mathbb{RE}$) in either round $r - 1$ or round r since $\mathtt{favored}_i = 0$ in Σ_{r-2} ($p_i = null$) and in Σ_{r-1} ($p_j = j$).
(b) Any node $k \in N(i)$ cannot leave $\mathcal{S}$ (execute rule $\mathbb{RE}$) in either round $r - 1$ or round r since $\mathtt{favored}_k = 0$ in Σ_{r-2} ($p_k = null$) and in Σ_{r-1} ($p_i = i$).
(c) Any neighbor k' of $k \in N(i)$ cannot leave $\mathcal{S}$ (execute rule $\mathbb{RE}$) in either round $r - 1$ or round r since $\mathtt{favored}_{k'} = 0$ in Σ_{r-2} ($p_k = null$) and $\mathtt{ready}_{k'} \wedge \mathtt{mutex}_{k'} = 0$ in Σ_{r-1} [there are two probabilities: (i) $\mathtt{favored}_{k'} = 0$ in Σ_{r-1}: $\mathtt{ready}_{k'} = 0$ in Σ_{r-1} by Remark 1.4; (ii) $\mathtt{favored}_{k'} = 1$ in Σ_{r-1}: $p_{k'} = minN_{k'} = k'$ must be true in Σ_{r-2} and the neighbors of k' executed rule $\mathbb{RB}$ to change their pointers to k' in round $r - 1$ such that $\mathtt{favored}_{k'}$ becomes 1 in Σ_{r-1}. Coupled with the fact that $\mathtt{favored}_{k'} = 0$ in Σ_{r-2}, $\mathtt{mutex}_{k'}$ lock cannot be acquired in round $r - 1$; if node k' had $\mathtt{mutex}_{k'}$ lock in Σ_{r-2}, it must have reset it in round $r - 1$ by executing rule $\mathbb{RH}$].

Thus, $\mathtt{leave}_i$ remains 1 in Σ_{r-1} and Σ_r (by similar reasoning, $\mathtt{leave}_j$ remains 1 in Σ_{r-1} and Σ_r); we arrive at a contradiction.

Case (2) (Nodes i and j executed rule $\mathbb{RC}$ in the round r): In Σ_{r-1}, $\mathtt{leave}_i = 1$, $\mathtt{leave}_j = 1$, $p_i = p_j = null$, and for each $k \in N(i) \cup N(j)$, $p_k = null$ (nodes i and j are enabled for rule $\mathbb{RC}$; Definition 4). Again, it can be shown that both $\mathtt{leave}_i$ and $\mathtt{leave}_j$ remains 1 in Σ_r after round r (by similar reasoning as in round $r - 1$ of the previous case 1), a contradiction.

Definition 8. *In any system state,*

(a) We define an island $\mathcal{I}$ to be a maximal set of nodes $\{i \in V | \mathtt{leave}_i = 1 \wedge p_i = i\}$ such that the subgraph of G induced by the set $\mathcal{I}$ is connected.
(b) We use α to denote the number of islands and β to denote the number of nodes i with $\mathtt{leave}_i = 1$.

Remark 3. In any system state:

1. An island may consist of a single or multiple nodes; a node i with $\mathtt{leave}_i = 1$ and $p_i \neq i$ is not a member of any island.

2. For a node i in an island of size ≥ 2, $\mathtt{favored}_i = 0$ since it has a neighbor j with $p_j = j \neq i$ (Definition 5).
3. $\alpha \leq \beta$; $\alpha < n$; $\beta \leq n$;
4. After round 2, β is non increasing in subsequent rounds (Definition 4 and Lemma 4); β cannot decrease in a round unless at least one node executes rule $\mathbb{RE}$ in the round.
5. When algorithm $\mathtt{MGOASC}$ terminates, $\alpha = \beta = 0$.

Lemma 8. *If a node i leaves $\mathcal{S}$ (by executing rule $\mathbb{RE}$) in a round, node i constitutes a single node island at the beginning of the round.*

Proof. Node i leaves $\mathcal{S}$; thus $\mathtt{leave}_i = 1$ and $\mathtt{ready}_i = 1$ (rule $\mathbb{RE}$) and hence $\mathtt{favored}_i = 1$. Since $p_i = i$, node i belongs to an island (Definition 8); node i does not belong to an island of size ≥ 2 (Remark 3.2).

Lemma 9. *In any round r, $r \geq 3$ (starting from a safe state Σ_{r-1}), (a) α can not decrease if β remains constant; (b) α decreases at least by 1 and at most by ℓ, if β decreases by ℓ $(1 \leq \ell \leq \beta)$.*

Proof. (a) If β remains constant, no node i changes $\mathtt{leave}_i$ from 1 to 0. (1) Any island $\mathcal{I}$ cannot disappear since the smallest ID node in $\mathcal{I}$, say node i, cannot change its pointer in round r ($p_i = i = minN_i$ in Σ_{r-1} [no neighbor j of i with $\mathtt{leave}_j = 0$ has a self-pointer by Lemma 7 and node i does not have any island node neighbor with a smaller ID]). (2) Two islands cannot merge into one: consider any 2 islands $\mathcal{I}_1$ and $\mathcal{I}_2$; since $\mathcal{I}_1 \cup \mathcal{I}_2 = \emptyset$, for the two islands to merge, there must be a node $j \in N(\mathcal{I}_1 \cup \mathcal{I}_2)$ such that $\mathtt{leave}_j = 1$ in Σ_{r-1} and node j acquires self-pointer in Σ_r (j becomes an island node in Σ_r) by executing rule $\mathbb{RC}$ in round r; this is impossible since j has neighbor(s) with non null pointers in Σ_{r-1} (see rule $\mathbb{RC}$).

(b) Starting in a safe state Σ_{r-1}, if β decreases by ℓ in Σ_r, ℓ nodes have changed their $\mathtt{leave}$ bits from 1 to 0. Consider any node i whose $\mathtt{leave}_i$ is changed from 1 to 0. At least one of the three must occur in round r: (1) node i leaves $\mathcal{S}$ by executing rule $\mathbb{RE}$; (2) some neighbor $j \in N(i)$ leaves $\mathcal{S}$ by executing rule $\mathbb{RE}$ such that $d_i < 2$; or (3) at least one neighbor of node $j \in N(i) - \mathcal{S}$ leaves $\mathcal{S}$ by executing rule $\mathbb{RE}$ such that $d_j < 2$ in (Definition 4). If all ℓ nodes change their $\mathtt{leave}$ from 1 to 0 because of (1), then α is decreased by ℓ by Lemma 8; If some node(s) i changes $\mathtt{leave}_i$ from 1 to 0 because of (2) or (3), then it is possible that node i does not belong to any island. although the change of $\mathtt{leave}_i$ on node i causes β to decrease in Σ_r, it does not cause α to decrease (if node i is not an island node in Σ_{r-1}); but, for the possibilities (2) or (3), at least some other node must leave $\mathcal{S}$ (change s bit to 0) by executing rule $\mathbb{RE}$ in the round, thereby causing α to decrease (Lemma 8). Thus, α decreases by at most ℓ and at least by 1 in Σ_r.

Lemma 10. *Starting from a safe state with $\beta \neq 0$,*

(a) either α increases in at most 4 next rounds, if β remains constant;
(b) or β decreases in at most 5 next rounds.

Proof. Starting from a safe state Σ_r, $r \geq 2$, any node i with $\texttt{leave}_i = 0$ and $p_i = i$ must execute rule $\mathbb{RG}$ in round $r + 1$ to make $p_i = null$ in Σ_{r+1}. Then, Σ_{r+1} does not have a node j with $\texttt{leave}_j = 0 \wedge p_j = j$ (otherwise, node j had $\texttt{leave}_j = 1$ in Σ_r and hence β has decreased by at least 1 in one round). There are two possibilities:

(1) **There is no island node:** In Σ_{r+1}, each node i has $minN_i = null$ (no node with self-pointer and Definition 2). Also, since $\beta \neq 0$, there must be a node k with $\texttt{leave}_k = 1$; node k, in the worst case, must execute rules $\mathbb{RD}$ (Get_mutex), $\mathbb{RB}$ (Update_Pointer) and $\mathbb{RC}$ (Get_Self-pointer) in that sequence to get $p_k = k$. Thus, there is a new island $\{k\}$, i.e., α has increased in at most 4 rounds starting in Σ_r.

(2) **There is at least one island node:** If there are multiple such island nodes, let i be the node with minimum ID among those. In the worst case, each node $j \in N(i)$ executes rule $\mathbb{RD}$ in round $r + 2$ to get mutex in Σ_{r+2} and rule $\mathbb{RB}$ in round $r + 3$ to update their pointers to i; node i becomes favored, i.e., $\texttt{favored}_i = 1$ [Definition 5], in Σ_{r+3}. Now, there are two possibilities:

(i) At least one $j \in N(i)$ has $minN_j = k$ in Σ_{r+3} where $k \in N(j)$ (Definition 2), $p_k = k$, and $k < i$. Node k must have acquired its self-pointer by executing rule $\mathbb{RC}$ in round $r + 3$ and so, $\texttt{leave}_k = 1 \wedge (\forall k' \in N(k) : p_{k'} = null)$ in Σ_{r+2}, i.e., node k is not connected to any island nodes. Thus, $\{k\}$ is a newly formed single node island in Σ_{r+3}, i.e., α increases in at most 3 rounds.
(ii) Each $j \in N(i)$ has $minN_j = i$ in Σ_{r+3}; in round $r + 4$, node i executes rule $\mathbb{RD}$ to get mutex (no neighbor of node i can get mutex); thus, node i is ready in Σ_{r+4}. Node i executes rule $\mathbb{RE}$ to leave $\mathcal{S}$ in round $r + 5$; so $\texttt{leave}_i = 0$ in Σ_{r+5}, i.e., β decreases in at most 5 rounds.

Lemma 11. *After round 2, algorithm MGOASC reaches a safe state with $\alpha = \beta = 0$ in at most $O(n)$ rounds under the synchronous daemon.*

Proof. In a safe state where $\alpha = \beta$, if β decreases by 1, α must decrease by 1 (Lemma 9); $\beta \leq n$ and β is non-increasing (Remarks 3.3 and 3.4). Recall that β decreases by at least 1 in at most 5 rounds (Lemma 10(b)). Thus, from any safe system state with $\alpha = \beta$, the system will be in a safe state with $\alpha = \beta = 0$ in at most $5n$ rounds. Also, if β remains constant, α must increase by 1 in at most 4 rounds (Lemma 10(a)); in at most $4n$ rounds, α will be equal to β. Thus, the system will be in a safe state with $\alpha = \beta = 0$ in at most $5n + 4n = 9n$ rounds.

Theorem 4. *Starting in an arbitrary state, the algorithm MGOASC terminates in $O(n)$ rounds under the synchronous daemon, where n is the number of nodes in the graph.*

Proof. The system reaches a safe state with $\alpha = \beta = 0$ in $9n + 2$ rounds in the worst case (Theorem 2 and Lemma 11). In the next at most 2 rounds all node pointers will be *null* and all *mutex* variables will be 0.

Theorem 5. *The algorithm MGOASC is a self-stabilizing algorithm with safe convergence.*

Proof. This readily follows from Theorems 2, 3, and 4.

5   Conclusion

We have proposed the first self-stabilizing minimal global offensive alliance algorithm with safe convergence. It is assumed to face a synchronous daemon. Starting in an arbitrary initial state, the algorithm quickly converges to a safe state (a state which denotes a global offensive alliance, not necessarily a minimal one) in two rounds, and then stabilizes to a legitimate state (denoting a minimal global offensive alliance) in $O(n)$ rounds (n is the number of nodes in the network) without breaking safety during the intermediate state transitions. The behavior of the nodes is managed by a synchronous scheduler (daemon) where all nodes enabled by the protocol execute actions simultaneously in a round.

Acknowledgement. The research was partly supported by NSF awards # CCF-0832582, # DBI-0960586, and # DBI-0960443.

References

1. Dijkstra, E.W.: Self-stabilizing systems in spite of distributed control. Communications of the ACM 17(11), 643–644 (1974)
2. Dolev, S.: Self stabilization. MIT Press (2000)
3. Kakugawa, H., Masuzawa, T.: A self-stabilizing minimal dominating set algorithm with safe convergence. In: 20th IEEE International Parallel and Distributed Processing Symposium, pp. 25–29 (2006)
4. Cobb, J., Gouda, M.: Stabilization of general loop-free routing. Journal Parallel Distributed Computing 62(5), 922–944 (2002)
5. Kamei, S., Kakugawa, H.: A self-stabilizing approximation for the minimum connected dominating set with safe convergence. In: Baker, T.P., Bui, A., Tixeuil, S. (eds.) OPODIS 2008. LNCS, vol. 5401, pp. 496–511. Springer, Heidelberg (2008)
6. Kamei, S., Kakugawa, H.: A self-stabilizing 6-approximation for the minimum connected dominating set with safe convergence in unit disk graphs. Theoretical Computer Science 428, 80–90 (2012)
7. Hedetniemi, S.M., Hedetniemi, S.T., Kristiansen, P.: Alliance in graphs. Journal of Combinatorial Mathematics and Combinatorial Computing 48, 157–177 (2005)
8. Srimani, P., Xu, Z.: Distributed protocols for defensive and offensive alliances in network graphs using self-stabilization. In: International Conference on Computing: Theory and Applications, Kolkata, pp. 27–31 (March 2007)
9. Rodríguez, J., Sigarreta, J.: Global offensive alliances in graphs. Electronic Notes in Discrete Mathematics 25, 157–164 (2006)

A Local-Global Approach
to Solving Ideal Lattice Problems⋆

Yuan Tian, Rongxin Sun, and Xueyong Zhu

Software School, Dalian University of Technology, P.R. China
{tianyuan_ca,zhuxueyong}@sina.com,
sunrongxin7666@163.com

Abstract. We construct an innovative SVP(CVP) solver for ideal lattices in case of any relative extension of number fields L/K of degree n where L is totally real(i.e., all L's conjugations are contained in R). The solver reduces solving SVP(CVP) of the input ideal A in field L to solving a set of (at most n) SVP(CVP) of the ideals A_i in field L_i with relative degree $1 \le n_i < n$ and $\sum_i n_i = n$. Both the solver's space-complexity and its time-complexity's explicit dependence on the dimension (relative extension degree n) are polynomial. Precisely, our solver's time-complexity is $poly(n,|S|,N_{PG},N_{PT},N_d,N_l)$ where $|S|$ is bit-size of the input data and N_{PG}, N_{PT}, N_d, N_l are the time-complexities to implement some oracles for significantly simpler problems. If such oracles can be implemented by efficient algorithms, which is indeed possible in some situations, our solver will operate in this case only with polynomial time-complexity.

1   Introduction

Lattice problems take important roles in combinatorial optimization, public-key cryptography and many other fields in computer science[1–5]. In the shortest lattice vector problem (SVP), a non-zero lattice vector x in $\mathbf{B}\mathbf{Z}^n$ is to be found to minimize $|x|$ on input the lattice basis matrix $\mathbf{B}$ with respect to some specific norm $||$ in R^n. In the closest lattice vector problem (CVP), a lattice vector x is to be found to minimize $|u - x|$ on input the basis matrix $\mathbf{B}$ and a target vector u in R^n. In recent years, lots of innovative cryptographic schemes and protocols have been devised with proofs of security under the assumption that there is not (probabilistic and sometimes quantum) polynomial-time algorithm to solve arbitrary instances of variants of SVP and CVP.

From a computational hardness perspective, SVP, CVP and other related variants are NP-hard under deterministic (e. g.,CVP) or randomized (e. g.,SVP) reductions[4]. Even some approximation variants of these problems are proven to be NP-hard if the approximation factor is within some specific range. Despite of these facts, finding new algorithms to solve lattice problems exactly are still interesting and meaningful both because many applications (e. g., in mathematics and communication theory) involve lattices in relatively small dimensions, and

⋆ This work is supported by China NSF(61370144).

T V Gopal et al. (Eds.): TAMC 2014, LNCS 8402, pp. 378–400, 2014.
© Springer International Publishing Switzerland 2014

because approximation algorithms for high dimensional lattices for which the exact solution is infeasible typically involve the exact solution of low dimensional sub-problems.

Recently a sub-category of lattices, the ideal lattice, is discovered to have indispensable values in innovative cryptography applications, e. g., the wonderful fully homomorphic encryption scheme for secure cloud computing[2], stimulating lots of works in cryptography theory and practices. On the one hand, such schemes are based-on some computational hardness hypothesis on some problems in ideal lattices, e. g., SVP or CVP's hardness, on the other hand, few deep knowledge is known on these points. Since the ideal lattice has rich intrinsic algebraic properties the general lattice doesn't have, it's reasonable to ask whether its related problems, e. g., SVP and CVP, are really as hard as those of general lattices, or "how easy" are they in comparison with their counterparts in general lattices? No matter what the answer (positive or negative) to this question would be, it will have fundamental significance to ideal lattice theory and applications.

In this paper we work on this question in case of SVP and CVP problems in an algorithmic approach.

2   Related Works

To find the exact solution to lattice problems, so far three main families of SVP and CVP solvers exist [3, 6–12]. With our knowledge, there're no generic algorithms for ideal lattice problems, except some ones modified from the solvers for non-ideal lattices which doesn't essentially exploit the ideal lattice's algebraic properties.

Among the solvers in [3, 6–12], MV and Kannan algorithms are deterministic while AKS algorithms are randomized. All algorithms work in ℓ^2-norm (AKS algorithm can work in other norms, e. g., ℓ_∞). The core of MV algorithm[6] is to compute the Voronoi cell of the lattice[1], whose knowledge facilitates the tasks to solve SVP and CVP. Kannan algorithm[8] relies on a deterministic procedure to enumerate all lattice vectors below a prescribed norm or within a prescribed distance to the target vector. This procedure uses the Grahm-Schmidt orthogonalization of the input lattice basis to recursively bound the integer coordinates of the candidate solutions.The AKS algorithm[9] is the first single-exponential time (random) algorithm for SVP.Recently this algorithm has been significantly improved and the currently best time complexity is $2^{2.465n+o(n)}$[3]. However, the AKS variant solver for CVP only finds the $(1+\varepsilon)$-approximate solution for arbitrary $\varepsilon > 0$ in time complexity bounded by $(2 + 1/\varepsilon)^{O(n)}$[7, 11].

It's already known that when the lattice dimension n is fixed, there are polynomial time-complicated solvers for lattice problems, e. g.,SVP/CVP. i. e., lattice problem's computational hardness only depends on dimension n([3, 6–12]).

Some related works show that there are important differences in computational complexity between the lattice problems of general and ideal lattices. For example, some decisional problems of the ideal lattice family with constant root discriminant is in P while the counterparts of general lattice are NP-hard[13].

However, (with our knowledge) there is not search or optimization SVP/CVP (see the concepts in Sect.3.1) solver exploiting the ideal lattice algebraic features and performing significantly better than the best known solvers for general lattices.

Overview on Innovations of our Approach: Construction and Performance

Our algorithms constructed in this paper are to find the exact solutions to SVP and CVP in ideal lattices. In this paper we only deal with the case of totally real number field, i. e., all conjugations of the number field to which the input ideal belongs are contained in R.

Our solver works on the input $(L/K, A)$ where L/K is a totally-real and finite-degree extension of number field with degree n, A is an (fractional) ideal in L, K is fixed (and contained in R) and (L, A) is arbitrarily given. In other words, our solver can work for any finite-degree relative extension, not only the special case of L/Q (where Q is the rational number field). Furthermore, with some improvements (in full-version paper) this solver can also work on any Abelian extension L/K of arbitrary real ground field K (i.e., the Galois group $G_{L/K}$ is commutative and K is contained in R) while L may not be totally real.

In construction aspects, our solver, by exploiting the relationships between the so-called local and global number fields, reduces solving SVP(CVP) of the input ideal A in field L to solving a set of (at most n) SVP(CVP) of the ideals A_i in field L_i with relative degree $1 \leq n_i < n$ and $\sum_i n_i = n$. Roughly speaking, by tensor-producting L with a local field K_P where P is an appropriately selected (not unique) prime ideal in the ground field K, the tensor product (as a n-dimensional vector space over the local field K_P) can be always decomposed into a set of sub-spaces of dimension $n_i < n$ which are orthogonal each other and $\sum_i n_i = n$. Furthermore, this orthogonal decomposition is metric-preserving and by constructing appropriate injective homomorphisms all operations in intermediate local fields can be replaced by those in some intermediate global fields(i. e., ordinary number fields), so that the solution to the original problem can be efficiently reconstructed from the solutions to the sub-problems. This procedure can proceed recursively down to a set of (at most n) sub-problems of ideal lattices with dimensions as low as possible. In particular, in case of Galois extension L/K, each recursion can decrease the problem's dimension by at least half.

In performance aspects, our SVP(CVP) solver's space-complexity is polynomial. Its time-complexity's explicit dependence on the dimension (relative extension degree n of the number fields) is also polynomial. More precisely, our solver's time-complexity is

$$poly(n, |S|, N_{PG}, N_{PT}, N_d, N_l)$$

where $|S|$ is bit-size of the input data and N_{PG}, N_{PT}, N_d, N_l are the time complexities to implement oracles for some relatively simpler problems (some of them are decisional, e. g., ideal's primality testing in the extended field L). This feature implies that if such oracles can be implemented by efficient algorithms (with time-complexity polynomial in n and $|S|$), which is really possible in some situations, our solver will perform in this case with polynomial time-complexity. Even if in general there are no efficient implementations for these oracles, this solver's time-complexity may still be significantly lower than those for general lattices ([3, 6–12]), because the oracles implementations may be only sub-exponential in time-complexity or even not hard against the quantum computer (more details in Appendix B).

3   Preliminaries

In this section we present all basic notions and facts fundamental to our work in this paper. For more details we refer readers to [4](for general theory on lattices), [14–17](for algebraic number theory) and [18] (for abstract algebra, e. g., the general notions and facts on (DedeKind) rings, ideals, unique factorization domains, fields and Galois theory).

3.1   Lattices, SVP and CVP

The set of rational integers is denoted by Z and rational numbers by Q. A lattice is a finitely generated discrete subset in Euclidean space. More explicitly, in the Euclidean space R^n with a positive non-singular bilinear form $<.,.>$, a n-dimensional *rational lattice*, denoted $\Lambda(\mathbf{B})$ where $\mathbf{B}$ is a matrix of rank n with column vectors $(\boldsymbol{b}_1,\ldots,\boldsymbol{b}_n)$, is the set of vectors $\{x_1\boldsymbol{b}_1+\ldots+x_n\mathbf{b}_n : x_1,\ldots,x_n \in Z\}$ where the values $< \boldsymbol{b}_i, \boldsymbol{b}_j >$ are all rational numbers. The lattice with basis $\boldsymbol{b}_1,\ldots,\boldsymbol{b}_n$ is denoted $Z\boldsymbol{b}_1 + \ldots + Z\boldsymbol{b}_n$. A lattice is called *integral* if $< \boldsymbol{b}_i, \boldsymbol{b}_j >$ are all integers.

For any vector $\boldsymbol{u} = (u_1,\ldots,u_n)$ in R^n, its norm $< \boldsymbol{u}, \boldsymbol{u} >^{1/2}$ is denoted $|\boldsymbol{u}|$. It's easy to verify that the squared norm of any lattice vector in an integral lattice is always an integer.

Lattice Problems. Given a lattice $\Lambda(\mathbf{B}) = Z\boldsymbol{b}_1 + \ldots + Z\boldsymbol{b}_n$, let

$$\lambda_1(\Lambda) \equiv \min\{|\boldsymbol{x}| : \boldsymbol{x} \text{ in } \Lambda \text{ and non-zero}\} \tag{1}$$

be the minimal value of the norms of non-zero lattice vectors in $\Lambda(\mathbf{B})$. The optimization shortest vector problem with respect to the norm $||$ is to find $\lambda_1(\Lambda)$. The search shortest vector problem is to find a lattice vector $\boldsymbol{x}$ in Λ such that $|\boldsymbol{x}| = \lambda_1(\Lambda)$. Given a lattice $\Lambda(\mathbf{B})$ and a rational target vector $\boldsymbol{u}$ in Q^n, let

$$dist(\Lambda; \boldsymbol{u}) \equiv min\{|\boldsymbol{x} - \boldsymbol{u}| : \boldsymbol{x} \text{ in } \Lambda\} \tag{2}$$

be the minimum distance between $\boldsymbol{u}$ and all lattice vectors in Λ. The *optimization closest vector problem* with respect to the norm $||$ is to find $dist(\Lambda; \boldsymbol{u})$. The *search closest vector problem* is to find a lattice vector $\boldsymbol{x}$ in Λ such that $|\boldsymbol{x} - \boldsymbol{u}| = dist(\Lambda; \boldsymbol{u})$.

There are many other lattice-related problems[19, 20]. For example, the *covering radius of a lattice*, $\mu(\Lambda)$, is defined as the maximal distance between any vector and the lattice. The covering radius problem is to find

$$\mu(\Lambda) \equiv \max\{dist(\Lambda; \boldsymbol{u}) : \boldsymbol{u} \text{ in } Q^n\} \tag{3}$$

In this paper we focus on the algorithms to solve SVP and CVP. It has been known that these problems are computationally hard[4, 20].We focus on constructing the algorithms for SVP and CVP (both in optimization and search version) for ideal lattices, a sub-category of the general lattices with rich algebraic structures originating from number theory.

3.2   Number Field: Relative Extension and Prime Ideal Decomposition

Let K be a number field with its integral ring O_K(or more generally, a fractional field of a *Dedekind domain* O_K[17, 18]), $L = K(\alpha)$ is an extension of K by adding a root α of a polynomial $f(x) \in O_K[x]$. $f(x)$ is called α's *minimal polynomial* if it has the minimal degree among the polynomials in $O_K[x]$ with α as a root. Such a polynomial is unique up to a constant factor in O_K and is prime (irreducible) in $O_K[x]$. The minimal polynomial $f(x)$'s degree is called the degree of field extension $L/K$and denoted by $[L : K]$.

L/K is called *relative extension* from the ground field K. An equivalent (isomorphic) picture about the arithmetic in L is to regard it as the quotient set of $O_K[x]/(f(x))$ with the operations as polynomial addition, subtraction and multiplication modulo $f(x)$.

Regarding L as a vector space on K with dimension $n = [L : K]$, we can introduce the *relative trace* and norm for any element z in L and denote these as $Tr_{L/K}(z)$ and $N_{L/K}(z)$ respectively[15–17].

Given z's minimal polynomial $g(t) = (-1)^m g_0 + g_1 z + \ldots + g_{m-1} t^{m-1} + t^m$ in $K[t]$(hence $m|n$), z's trace and norm can be computed by

$$Tr_{L/K}(z) = -(n/m)g_{m-1}, \quad N_{L/K}(z) = g_0^{n/m} \qquad (4)$$

For relative extension L/K, there is an important subset, called O_K's integral closure in L, defined as:

$$O_L \equiv \{z \text{ in } L : \text{ there exist } a_0, \ldots, a_{n-1} \text{ in } O_K$$

$$\text{such that } a_0 + a_1 z + \ldots + a_{n-1}z^{n-1} + z^n = 0\} \quad (5)$$

O_L is a ring with the following important properties[13, 15–17]:

(1)L is O_L's fractional field.

(2)For any relative extension L/K of degree n, there are exactly n field (relative) embeddings (injective homomorphisms) mapping L into C which are fixed in K element-wise, among which $\rho_1, \ldots, \rho_{r_1}$ embed L into R and the other $2r_2$ ones $\tau_1, \ldots, \tau_{2r_2}$ (where each τ_j is complex conjugate to τ_{j+r_2}) embed L into C.

With these n K-embeddings $\sigma_1, \ldots, \sigma_n$, the trace and norm of an element can be computed by

$$Tr_{L/K}(z) = \sigma_1(z) + \ldots + \sigma_n(z), \quad N_{L/K}(z) = \sigma_1(z) \ldots \sigma_n(z) \qquad (6)$$

As long as K is a number field, L is also a number field with degree $[L : Q] = [L : K][K : Q]$ and O_L defined in (5) is exactly the set of $\{z$ in L: there exist

$a_0, \ldots, a_{m-1}$ in Z such that $a_0 + a_1 z + \ldots + a_{m-1} z^{m-1} + z^m = 0\}$ where $m = [L : Q]$. Therefore any ideal A in O_L can be regarded, by the number field L's embeddings into C, as a lattice of dimension $[L : Q]$ in $R^{[L:Q]}$ with the positive-definite and non-degenerate bilinear form

$$< x, y > \equiv \sigma_1(x)\bar{\sigma}_1(y) \ldots + \sigma_n(x)\bar{\sigma}_n(y) \tag{7}$$

where $\bar{z}$ denotes z's complex conjugation. When L is totally real, i.e., $r_2 = 0$ then $< x, y > = Tr_{L/K}(xy)$.

Because L is a number field, as a result, every prime ideal M in O_L is maximal and O_L/M is a (finite) field. The important property of the unique factorization on prime ideals is true for any ideal in O_L.

(3)Let P be a prime ideal in O_L , generally the ideal PO_L may be no longer prime in O_L . As an ideal in O_L , there is the following law about PO_L's decomposition:

For any prime ideal P in O_K , there exist a finite set of prime ideals $M_1, \ldots, M_r$ in O_L such that $M_1 \cap O_K = \ldots = M_r \cap O_K = P$ and PO_L decomposes into prime ideals multiplication on and only on these $M_1, \ldots, M_r$:

$$PO_L = M_1^{e_1} \ldots M_r^{e_r} \tag{8}$$

Furthermore, $e_1 f_1 + \ldots + e_r f_r = [L : K]$ where $f_i = [O_L/M_i : O_K/P] =$the degree of the extension from the finite field O_K/P to O_L/M_i. Integers $e_1, \ldots, e_r$ are called *ramification indices* for P on $M_1, \ldots, M_r$ (or $M_1, \ldots, M_r$ on P).

Remarks on Galois Extension: When L/K is a Galois extension, the decomposition law (8) can be further refined. In this case we always have $e_1 = \ldots = e_r \equiv e$ and $f_1 = \ldots = f_r \equiv f$. Furthermore, Galois group $G_{L/K}$ is transitive on $M_1, \ldots, M_r$, i.e., for any M_i, M_j there exists g in $G_{L/K}$ such that $M_i = g(M_j)$.

3.3 Valuations, p-adic Completions and Local-Global Relations

Section 3.2 presented number theory on the so-called *global field*. Now we turn to number theory on the so-called *local field*.

General Notions and Facts. Let K be a field, R^+ be the set of all non-negative real numbers, a (multiplicative) valuation on K is a mapping $|.|: K \to R^+$ with the following properties:

$$|xy| = |x||y|; |x| = 0 \; if f \; x = 0; \quad |x + y| \leq |x| + |y| \; for \; any \; x \; and \; y \; in \; K$$

When $|n| \leq 1$ for all $n = 0, \pm 1, \pm 2, \pm 3, \ldots$, $|.|$ is called *non-Archimedean* valuation, otherwise called *Archimedean*. For non-Archimedean valuation, the third property in the above is equivalent to the inequality

$$|x + y| \leq \max(|x|, |y|) \; for \; any \; x \; and \; y \; in \; K, \quad Or \; equivalently$$
$$|x + y| = \max(|x|, |y|) \; for \; any \; x \; and \; y \; in \; K \; and \; |x| \neq |y| \tag{9}$$

An equivalent non-Archimedean valuation model is the index valuation, i. e., a mapping $w : |.|:K \rightarrow R$ satisfying $w(xy) = w(x) + w(y); w(x) = +\infty$ iff $x = 0; w(x + y) \geq \min(w(x), w(y))$ for any x and y in K. Obviously, for any $a > 1$ $w(x) = -\log_a |x|$ gives the correspondance between these two models.

Note: Hereafter we freely interchange the use of these two valuation models at convenience.

Two (multiplicative) valuations $|.|_1$ and $|.|_2$ on filed K is called *equivalent* if there exists a positive real number $a > 0$ such that $|x|_1 = |x|_2^a$ for all x in K. For two non-Archimedean valuations $|.|_1$ and $|.|_2$, this definition equals the statement that $|x|_1 \leq 1$ iff $|x|_2 \leq 1$ for any x in K.

A valuation $|.|$ is called *discrete* if the image of $|.|$ is discrete in R.

Given a non-Archimedean valuation $|.|$(or its equivalent index valuation w) on field K, the subset

$$J_K \equiv \{x \ in \ K : |x| \leq 1\} = \{x \ in \ K : w(x) \geq 0\} \tag{10a}$$

is a ring with the unique maximal ideal[14, 17]:

$$M_K \equiv \{x \ in \ K : |x| < 1\} = \{x \ in \ K : w(x) > 0\} \tag{10b}$$

The field J_K/M_K is called the valuation's *residue class field*.

For number field K/Q with degree $n = [K : Q]$ we have the following important general facts about valuations on it[14]:

(1)Each (real or complex) Q-embedding σ_j: $K \rightarrow C$ derives an Archimedean (multiplicative) valuation on K by $|x|_j \equiv |\sigma_j(x)|$ where the latter $|.|$ is the ordinary complex valuation $|z| = ((Rez)^2+(Imz)^2)^{1/2}$. Furthermore, two derived Archimedean valuations $|.|_j$ and $|.|_i$ are equivalent iff $\sigma_j(.)$ and $\sigma_i(.)$ are complex conjugate each other.

(2)Each prime ideal P in O_K derives a discrete non-Archimedean (index) valuation on K by

$$w_P(x) \equiv e \ where \ P^e|(x) \ and \ P^{e+1} \nmid (x)$$

This is called the P-adic valuation. Furthermore, different prime ideals P_i, P_j derive distinct (inequivalent) P-adic valuations w_{P_i}, w_{P_j}.

(3)The valuations presented in (1) and (2) enumerates all valuations on the number field K. As a result, there are finite (exactly r_1+r_2) number of Archimedean valuations and infinite distinct non-Archimedean valuations, each corresponding to a prime ideal.

Completeness and Local Field. Let K be a field with a (Archimedean or non-Archimedean) valuation $|.|$. Since $|.|$ derives a metric on K by $d(x,y) \equiv |x - y|$, the standard metric-completion procedure derives a $|.|$-completion on K, denoted $K_{||}$, which is also a field with K as a dense subfield in it.

Let K be a number field. The completion by anyone of its Archimedean valuations is R or C, depending on whether K is a subfield in R or not. Let P be a prime ideal in O_K, the P-adic completion of K, denoted K_P and called K's *localization* on P (local field), has the following properties[14, 17]:

(1)K_P is a complete and discrete valued field. Further more, K_P/Q_p is a finite-degree extension where p is a prime number such that $(p) \equiv pZ = P \cap Z$ and Q_p is the p-adic completion of the field of rational numbers Q.

(2)For K_P's valuation ring (r.f., (10a)) we have

$$J_{K,P} \equiv \{x \ in \ K_P : |x|_P \leq 1\} \equiv \{x \ in \ K_P : w_P(x) \geq 0\}$$
$$= \{x \ in \ K_P : \exists \ a_0, \ldots, a_{n-1} \ in \ Q_p \ such \ that \ w_p(a_i) \geq 0 \qquad (11a)$$
$$for \ all \ and \ a_0 + a_1 x + \ldots + a_{n-1}x^{n-1} + x^n = 0\}$$

Furthermore $J_{K,P}$ is a principal ideal domain with the unique maximal ideal:

$$M_{K,P} = \{x \ in \ K_P : |x|_P < 1\} = \{x \ in \ K_P : w_P(x) > 0\} \qquad (11b)$$

Hence there exists a element π, called K_P's *prime element*, such that $M_{K,P} = (\pi)$. Actually π can be any element in $M_{K,P}$ with the greatest $|.|_P$-value.

(3)Given an ideal B in $J_{K,P}$, there exists a unique integer $m \geq 0$ such that $B = M_{K,P}^m$. In consequence, all integral ideals in $J_{K,P}$ constitute a chain $\ldots \subset M_{K,P}^4 \subset M_{K,P}^3 \subset M_{K,P}^2 \subset M_{K,P}$.

(4)The residue class field of K_P, i.e., $J_{K,P}/M_{K,P}$, is a finite field with characteristic p (the p specified in (1)) and isomorphic to O_K/P.

(5)There is a homomorphism Ω mapping the ideals in O_K to ideals in $J_{K,P}$, defined as:

$$\Omega(A) = M_{K,P}^e, \ if \ P^e|A \ but \ P^{e+1} \nmid A; \quad \Omega(A) = J_{K,P} \ if \ P \ and \ A \ are \ co\text{-}prime \qquad (12)$$

It's easy to verify that $\Omega(AB) = \Omega(A)\Omega(B)$ and Ω can be easily prolonged onto the multiplicative group of fractional ideals on K. Ω *"localizes"* an ideal A in global field K to a (principal) ideal $\Omega(A)$ in K_P and this localization is non-trivial iff P is a prime factor of A.

Local-Global Relations. Now back to Sect.3.2(3), let both L and K be number fields and L/K a field extension of degree $n = [L : K]$, P a prime ideal in O_K. There exist a finite set of prime ideals $M_1, \ldots, M_r$ in O_L and integers $e_1, \ldots, e_r \geq 1$ such that :

$$M_1 \cap O_K = \ldots = M_r \cap O_K = P$$
$$PO_L = M_1^{e_1} \ldots M_r^{e_r} \qquad (13)$$
$$e_1 f_1 + \ldots + e_r f_r = n$$

where $f_i = [O_L/M_i : O_K/P]$=the degree of the extension from the finite field O_K/P to O_L/M_i.

Let L_{M_j} be the M_j-adic completion of L with its valuation ring denoted by J_{M_j}, prime element η_j (i.e., $J_{M_j} = (\eta_j)$), $j = 1, \ldots, r$, K_P be the P-adic completion of K with its valuation ring denoted by $J_{K,P}$ and prime element π, now we can state more important and deep details about this decomposition law[14]:

(1)Each L_{M_j} is an extension of K_P and $[L_{M_j} : K_P] = e_j f_j$, $j = 1, \ldots, r$. In particular, each L_{M_j} is a vector space on local field K_P in dimension $e_j f_j$.

(2)For each j, the residue class field of L_{M_j} is an extension of the residue class field of K_P with degree f_j, i.e., $[J_{M_j}/(\eta_j) : J_{K,P}/(\pi)] = f_j$.

(3)For each j, the ground field prime element π is decomposed in the extended local field with ramification index e_j, i.e., $(\pi) = (\eta_j)^{e_j}$ in J_{M_j}.

(4)For each j, there is a prolongation from the P-adic valuation on K_P to L_{M_j} specified by

$$|y|_{M_j} = |N_{L_{M_j}/K_P}(y)|_P^{1/e_j f_j} \ \text{for any } y \text{ in } L_{M_j}. \tag{14}$$

where $|.|_P$ denotes the P-adic multiplicative valuation on K_P. It's easy to see that $|y|_{M_j} = |y|_P$ when y is in K_P. Furthermore, $|y|_{M_j}$ in (14) is the only prolongation of $|.|_P$ onto L_{M_j}.

(5)For each j, L_{M_j}'s valuation ring J_{M_j} is exactly the integral closure of K_P's valuation ring $J_{K,P}$, i.e.,

$$\begin{aligned} J_{M_j} &\equiv \{y \text{ in } L_{M_j} : |y|_{M_j} \leq 1\} \equiv \{y \text{ in } L_{M_j} : w_{M_j}(y) \geq 0\} \\ &= \{y \text{ in } L_{M_j} : \exists\ a_0, \ldots, a_{m-1} \text{ such that } w_P(a_i) \geq 0 \\ &\quad \text{for all } i \text{ and } a_0 + a_1 x + + a_{m-1} y^{m-1} + y^m = 0\} \end{aligned} \tag{15}$$

(6)Let $L = K\omega_1 + \ldots + K\omega_n$ and w.l.o.g., all ω_i's are in O_K. Denote the vector space $K_P\omega_1 + \ldots + K_P\omega_n$ on field K_P by $K_P \otimes_K L$ (tensor product on K) and denote the direct summation between vector spaces by $\oplus$, there is a K_P-linear isomorphism ψ between $K_P \otimes L$ and $L_{M_1} \oplus \ldots \oplus L_{M_r}$ where each L_{M_i} is a (distinct) vector space on K_P in dimension $e_j f_j$:

$$\psi : K_P \otimes_K L \cong L_{M_1} \oplus \ldots \oplus L_{M_r} \tag{16}$$

Furthermore, denote the element corresponding in (16) as $y \cong (y_1, \ldots, y_r)$ then for any y in L we have

$$Tr_{L/K}(y) = Tr_{L_{M_1}/K_P}(y_1) + \ldots + Tr_{L_{M_r}/K_P}(y_r) \tag{17a}$$

$$N_{L/K}(y) = N_{L_{M_1}/K_P}(y_1) \ldots N_{L_{M_r}/K_P}(y_r) \tag{17b}$$

Let $y^{(1)} \cong (y_1^{(1)}, \ldots, y_r^{(1)})$ and $y^{(2)} \cong (y_1^{(2)}, \ldots, y_r^{(2)})$, at element level the isomorphism has:

$$y^{(1)} \pm y^{(2)} \cong (y_1^{(1)} \pm y_1^{(2)}, \ldots, y_r^{(1)} \pm y_r^{(2)}) \tag{18a}$$

$$y^{(1)} y^{(2)} \cong (y_1^{(1)} y_1^{(2)}, \ldots, y_r^{(1)} y_r^{(2)}) \tag{18b}$$

Combined with (17a) and (18a) we have

$$Tr_{L/K}(xy) = Tr_{L_{M_1}/K_P}(x_1 y_1) + \ldots + Tr_{L_{M_r}/K_P}(x_r y_r) \tag{19}$$

for L's any element $x \cong (x_1, \ldots, x_r)$ and $y \cong (y_1, \ldots, y_r)$. In other words, (16) presents an orthogonal decomposition of the K_P-vector space $K_P \otimes_K L$.

(7)Let A be any (integral or fractional) ideal in L, then A is a finitely generated module on the Dedekind domain. There exist L's K-basis $\omega_1, \ldots, \omega_n$ and a set of K's ideals $I_1, \ldots, I_n$ such that[18, 21]

$$A = I_1\omega_1 + \ldots + I_n\omega_n \tag{20}$$

Such $\omega_1, \ldots, \omega_n$ are called A's pseudo-basis and in general they are not in A. Different pseudo-basis share the same cardinality n and it is known how to transform from one pseudo-basis to another[21].

Let A has a pseudo-basis representation in (20), define $J_{K,P} \otimes_K A \equiv I_P^{(1)}\omega_1 + \ldots + I_P^{(n)}\omega_n$ where $I_P^{(i)} = I_i$'s image under the localization mapping Ω in (12) in K. Let Ω_j be the localization mapping in (12) in L_{M_j}, i.e., mapping the ideals in L to ideals in L_{M_j}, then we have the following fact.

Theorem 1 *[22] If A's pseudo-basis $\omega_1, \ldots, \omega_n$ are in O_L, then the K_P-linear isomorphism ψ in (16) deduces:*

$$\psi : J_{K,P} \otimes_K A \cong \Omega_1(A) \oplus \ldots \oplus \Omega_r(A) \tag{21}$$

4   Local-Global Algorithm to Solve SVP and CVP in Ideal Lattices: High Level Descriptions

In this section we construct our algorithms to solve SVP and CVP in ideal lattices. Only the search version is considered because the optimization version can be solved in exactly the same way. Furthermore, we only focus on SVP because the same approach can be easily applied to CVP.

4.1   Problem

The search shortest vector problem in ideal lattice is presented in the following. Instead of only dealing with the case K/Q, our algorithm works for any finite-degree relative extension L/K where K is fixed and L is arbitrary, both are number fields.

Problem SVP(A,L/K) ─────────────────────

Parameter: *A* number field K.
Input: K's extended field $L = K(\alpha)$ with the generator α's minimal polynomial $f(t) = t^n + a_1 t^{n-1} + \ldots + a_{n-1}t + a_0$ in $O_K[t]$, and an ideal A in L.
Note: In this paper we only deal with the case of totally real number field, i.e., K is contained in R and all the roots of $f(t)$ are real.

For the ideal A on input, we always assume a given pseudo-basis representation, i.e., a set of L's K-basis $\omega_1, \ldots, \omega_n$ in O_L and a set of K's ideals $I_1, \ldots, I_n$ such that $A = I_1\omega_1 + \ldots + I_n\omega_n$.
Output: An element y^* in A such that

$$Tr_{L/K}(y^{*2}) = \min\{Tr_{L/K}(y^2) : \text{all non-zero } y\text{'s in A}\}$$

Algorithm for SVP(A,L/K): High-Level(1)

(1)Given L and A on input, find a prime ideal P in O_K such that:

$$PO_L \text{ is not prime in } O_L; \tag{22a}$$

$$P \text{ is unramified in } O_L, \text{ i.e.,} \tag{22b}$$

all its ramification indices $e_1 = \ldots = e_r = 1$;

$$P \nmid \sharp(O_L/O_K[\alpha]). \tag{22c}$$

(2)Given L and P obtained from last step, find the local fields $L_{M_1}, \ldots, L_{M_r}$ associated with P's all decomposition prime ideals $M_1, \ldots, M_r$ in O_L, integers $f_1, \ldots, f_r \geq 1$ s.t. :

$$PO_L = M_1 \ldots M_r$$

$$f_i = [O_L/M_i : O_K/P] = [L_{M_i}\text{'s residue class field} : K_P\text{'s residue class field}]$$

Secondly, find K_P-linear isomorphism ψ and its component mappings $\psi_1, \ldots, \psi_r$ in (16)-(18) where each $\psi_i : L \to L_{M_i}$, i.e., $y \cong (y_1, \ldots, y_r)$ means $\psi(y) = (\psi_1(y), \ldots, \psi_r(y))$. (3)Given L, P and $L_{M_1}, \ldots, L_{M_r}$, integers $f_1, \ldots, f_r \geq 1$ obtained from last step, find K's extended fields $L_1, \ldots, L_r$ and field embeddings $\varphi_1, \ldots, \varphi_r$ with $\varphi_i : \psi_i(L) \to L_i$, s.t.: Each L_i is a global field with extension degree

$$[L_i : K] = f_i; \tag{23a}$$

For each i and y in L:

$$Tr_{L_i/K}(\varphi_i\psi_i(y)) = Tr_{L_{M_i}/K_P}(\psi_i(y)); \tag{23b}$$

(4)Given all the results obtained, for each i set $\lambda_i \equiv \varphi_i\psi_i : L \to L_i$ and $A_i \equiv \lambda_i(A)$ which is an ideal in L_i. Do: For each $i = 1, \ldots, r$ find a non-zero x_i^* in A_i s.t.

$$Tr_{L_i/K}(x_i^{*2}) = \min\{Tr_{L_i/K}(x^2) : \text{all non-zero } x \text{' s in } A_i\} \tag{24}$$

ie, solve the SVP for ideal lattice A_i in field L_i in a strictly lower dimension $f_i(< n)$;

Find a x_m^* among $x_1^*, \ldots, x_r^*$ s.t.

$$Tr_{L_m/K}(x_m^{*2}) = \min\{Tr_{L_i/K}(x_i^{*2}) : i = 1, \ldots, r\};$$

Find y^* in A such that

$$\lambda_m(y^*) = x_m^* \text{ and } \lambda_i(y^*) = 0 \text{ for all } i \neq m; \tag{25}$$

Output(y^*).

4.2   High Level Algorithm

Before going to the technically involved solver construction, we briefly present the motivation. The idea comes from a simple fact that, although lattice problems (e.g., SVP, CVP, CRP, etc) are computationally hard in general cases, a subset of them, in particular the problems of the orthogonal lattice family, can be always solved with polynomial-complexity algorithms. Of course for a general lattice in

R^n neither it is always orthogonal nor it can be even decomposed to a set of sub-lattices orthogonal each other, however, for ideal lattices originating from number field, (16)-(21) shows that there exists some "orthogonal decomposition" structure exploitable to develop a solver more efficient than those of general lattice problems. Doing such exploitations as far as possible is exactly what will proceed in this paper.

Now we present the whole algorithm's logic at a high level, then working out all technical details in sequel. In the following, all notations are inherited from Sect.3 and "s.t." means "such that". By "global field" we mean any number field and "local field" means the completion of a number field under some of its prime ideal induced valuation.

Such obtained y^* is indeed the solution because (note that all the intermediate global fields L_i's are totally real, r.f., remark in B.2):

$$
\begin{aligned}
&\min\{Tr_{L/K}(y^2) : \text{ all non-zero } y\text{'s in } A\} \\
=&\min\{Tr_{L_{M_1}/K_P}(y_1^2) + \ldots + Tr_{L_{M_r}/K_P}(y_r^2) : \text{any } y_i \text{ in } \psi_i(A) \\
&\quad \text{and } y_i = 0 \text{ doesn't hold simultaneously}\} \ by(17), (18) and (21) \\
=&\min\{Tr_{L_1/K}(x_1^2) + \ldots + Tr_{L_r/K}(x_r^2) : \text{any } x_i \text{ in } A_i \\
&\quad \text{and } x_i = 0 \text{ doesn't hold simultaneously}\} \ by(23) \\
=&\min_{1\leq i\leq r} \min\{Tr_{L_i/K}(x_i^2) : \text{ any } x_i \text{ in } A_i \text{ and non-zero}\} \\
=&\min_{1\leq i\leq r} Tr_{L_i/K}(x_i^{*2})
\end{aligned}
$$

Note that in step♯2 we don't need P's decomposition prime ideals per se, but just some information about their local fields $L_{M_1}, \ldots, L_{M_r}$ where each $L_{M_i} = K_P[t]/(f_i(t))$ with some irreducible polynomial $f_i(t)$ in $K_P[t]$. In the low-level constructions it can be seen that even $f_i(t)$ is not needed but just the polynomial $h_i(t) = f_i(t) \bmod P$ in $K[t]$ instead

In conclusion, this algorithm reduces an ideal lattice SVP instance of dimension n to a set of $r(\leq n)$ ideal lattice SVP instances of strictly lower dimensions. It's already known that lattice problem's computational hardness is only dominated by its dimension n (in other words, there are known algorithms in polynomial time and space complexity to solve lattice problems like SVP and CVP for any fixed dimension[3–5, 7–9, 23]), this feature of our algorithm is significantly helpful to raise the solver's efficiency in solving ideal lattice SVP.

For all those derived sub-instances the ground field are all K, the same as that in the original SVP instance, so (24) in step♯4 can be recursively solved by this algorithm down to some appropriately lower dimensions, calling some existed solver at these levels or continue the recursion down to 1-dimensional SVP sub-instances. More details are discussed in Appendices.

4.3 Complexity

To construct the complete algorithm, we introduce the following oracles:

Oracle-PG_K where K is a number field: Generates a prime ideal at random in O_K. The input is void and each output is probabilistically independent of any others.

Oracle-$PT_L(M)$ where L is a number field: on input any ideal M in O_L, tests whether M is prime or not.

Oracle-$d_K(L)$: On input any L where L/K is a number field extension of finite degree, outputs the relative discriminant $d_{L/K}$, an integral ideal in O_K which is the greatest common divisor of $det(Tr_{L/K}(a_i a_j))$ of all K-linear independent integers $a_1, \ldots, a_n$ in O_L.

Oracle-$l_K(L, \alpha)$: On input any $L = K(\alpha)$ where L/K is a number field extension with finite degree n, outputs $\sharp(O_L/O_K[\alpha])$, the cardinality of the finite quotient set $O_L/O_K[\alpha]$.

How to use these oracles is presented in the Appendix B.

Let $|S|$ denote the input size of the ideal lattice Problem $SVP(A, L/K)$, N_{PG}, N_{PT}, N_d and N_l denote the time-complexities to implement Oracle-PG_K, Oracle-PT_L, Oracle-d_K and Oracle-l_K in the algorithm. From the constructions in Sect.4 and Sect.B, it's easy to see that all the subroutines and operations in the algorithm are only those with time and space complexity polynomial in the input size, except the above four oracles which intrinsic computational complexity may be non-polynomial. Furthermore the recursion depth of the (high level) algorithm, hence the number of calls to all those oracles, is only $O(n)$ where $n = [L : K]$=the dimension of the input ideal lattice A. In summary, we can have the following conclusions (details in the appendix B).

Theorem 2 *(1)Given any number field K contained in R, there exists the algorithm to solve (exactly) SVP on input any totally real extended field L and ideal A in O_L with time complexity*

$$poly(n, |S|, N_{PG}, N_{PT}, N_d, N_l) \tag{26}$$

and space complexity $poly(n, |S|)$ where $n = [L : K]$. (2)For CVP of ideal lattices, we have exactly the same conclusion. □

For the case of L/Q where Q is the rational number field, some of the oracles are known to have efficient implementations or can be efficiently reduced to others (r.f. B.4-B.5), so we have:

Corollary 3 *(1)There exists the algorithm to solve (exactly) SVP on input any totally real number field L and ideal A in O_L with time complexity*

$$poly(n, |S|, N_{PT}, N_d)$$

and space complexity $poly(n, |S|)$ where $n = [L : Q]$. (2)For CVP of ideal lattices, we have exactly the same conclusion. □

Notice that in this case only the oracle to do ideal's primality testing in the extended field and the oracle to compute the extended field's discriminant are needed, the former's hardness is reasonably expected significantly lower than SVP/CVP and the latter's hardness is known at most as hard as rational integer factorization.

5 Conclusion and Future Works

We construct an innovative SVP(CVP) solver for ideal lattices in case of any relative extension of number fields L/K of degree n where $L = K(\alpha)$ is totally real. By this construction, solving SVP/CVP of ideal lattices is efficiently reduced to solving SVP/CVP of strictly lower dimensional ideal lattices and the problems of generating prime ideals in the ground field K, testing the ideal's primality in the extended field L, calculating the relative discriminant $d_{L/K}$ and the cardinality of $O_L/O_K[\alpha]$. The solver's space-complexity is polynomial and its time-complexity's explicit dependence on the dimension n is also polynomial.

As a result, the first open problems are to construct the algorithms to implement the above oracles, which also have independent values in theory and applications. The second and more interesting open problem is that, for some of the oracles computationally hard to implement, whether its hardness can be still preserved against the quantum computing model. An answer to this problem will imply whether the ideal lattice problems' hardness is solid for post-quantum cryptography, although the answer is believed to be Yes for the general lattice problems' hardness.

References

[1] Sloane, N.J., Conway, J., et al.: Sphere packings, lattices and groups, 3rd edn. Springer (1998)
[2] Gentry, C.: Fully homomorphic encryption using ideal lattices. In: Proc. 41st ACM STOC, pp. 169–178 (2009)
[3] Hanrot, G., Pujol, X., Stehlé, D.: Algorithms for the shortest and closest lattice vector problems. In: Chee, Y.M., Guo, Z., Ling, S., Shao, F., Tang, Y., Wang, H., Xing, C. (eds.) IWCC 2011. LNCS, vol. 6639, pp. 159–190. Springer, Heidelberg (2011)
[4] Micciancio, D., Goldwasser, S.: Complexity of Lattice Problems: a cryptographic perspective. Kluwer Academic Publishers, Boston (2002)
[5] Nguyen, P.Q., Valle, B.: The LLL algorithm: survey and applications. Springer (2009)
[6] Micciancio, D., Voulgaris, P.: A deterministic single exponential time algorithm for most lattice problems based on voronoii cell computations. SIAM J. Comput. (2012) (Special Issue on STOC 2010)
[7] Ajtai, M., Kumar, R., Sivakumar, D.: Sampling short lattice vectors and the closest lattice vector problem. In: IEEE Conference on Computational Complexity, pp. 53–57 (2002)
[8] Kannan, R.: Minkowski's convex body theorem and integer programming. Mathematics of Operations Research 12(3), 415–440 (1987)

[9] Ajtai, M., Kumar, R., Sivakumar, D.: A sieve algorithm for the shortest lattice vector problem. In: Proceedings of the Thirty-third Annual ACM Symposium on Theory of Computing, pp. 601–610. ACM (2001)

[10] Agrawal, M., Kayal, N., Saxena, N.: Primes is in P. Annals of Mathematics, 781–793 (2004)

[11] Nguyen, P.Q., Vidick, T.: Sieve algorithms for the shortest vector problem are practical. Journal of Mathematical Cryptology 2(2), 181–207 (2008)

[12] Kannan, R.: Lattice translates of a polytope and the frobenius problem. Combinatorica 12(2), 161–177 (1992)

[13] Peikert, C., Rosen, A.: Lattices that admit logarithmic worst-case to average-case connection factors. In: Proc. STOC, pp. 478–487 (2007)

[14] Hasse, H.: Number theory, 3rd edn. Springer (1969)

[15] Hecke, E.: Lectures on the theory of algebraic numbers. Springer (1981)

[16] Ireland, K., Rosen, M.I.: A classical introduction to modern number theory. Springer (1990)

[17] Lang, S.: Algebraic number theory, 2nd edn. Springer (1994)

[18] Rotman, J.J.: Advanced modern algebra. Prentice-Hall Inc. (2002)

[19] Lyubashevsky, V., Peikert, C., Regev, O.: On ideal lattices and learning with errors over rings. Springer (2010)

[20] Micciancio, D.: Efficient reductions among lattice problems. In: Proc. SODA 2008, pp. 84–93 (2008)

[21] Cohen, H.: Advanced topics in computational number theory. Springer (2000)

[22] Li, W.C.W.: Number theory with applications. World Scientific, Singapore (1996)

[23] Haviv, I., Regev, O.: Hardness of the covering radius problem on lattices. In: IEEE CCC 2006, pp. 145–158 (2006)

[24] Pohst, M., Zassenhaus, H.: Algorithmic algebraic number theory. Cambridge University Press (1989)

[25] Roblot, X.F.: Polynomial factorization algorithms over number fields. Journal of Symbolic Computation 2002(11), 1–14 (2002)

[26] Kedlaya, K.S., Umans, C.: Fast polynomial factorization and modular composition. SIAM Journal on Computing 40(6), 1767–1802 (2011)

[27] Cohen, H.: A course in computational algebraic number theory. Springer (1993)

[28] Schoof, R.: Four primality testing algorithms. In: Algorithmic Number Theory: Lattices, Number Fields, Curves and Cryptography, pp. 101–126. Cambridge University Press (2008)

Appendix

A   Remarks on the High-Level Algorithm

(1)For CVP of ideal lattices, i. e., on input the totally real extended field $L = K(\alpha)$, ideal A in L and an element z in L, to find y^* in A s.t.

$$Tr_{L/K}((z - y^*)^2) = \min\{Tr_{L/K}((z - y)^2) : \text{ all } y's \text{ in } A\}$$

Because:

$$\min\{Tr_{L/K}((z-y)^2) : \text{ all non-zero } y\text{'s in } A\}$$
$$= \min\{Tr_{L_{M_1}/K_P}((z-y_1)^2) + \ldots + Tr_{L_{M_r}/K_P}((z-y_r)^2) :$$
$$\textit{any } y_i \textit{ in } \psi_i(A)\} \textit{ by}(17), (18), (21)$$
$$= \min\{Tr_{L_1/K}((\lambda_1(z)-x_1)^2) + \ldots + Tr_{L_r/K}((\lambda_r(z)-x_r)^2) :$$
$$\textit{any } x_i \textit{ in } A_i\} \textit{ by}(23)$$
$$= \min_{1 \le i \le r} \min\{Tr_{L_i/K}((\lambda_i(z)-x_i)^2) : \textit{any } x_i \textit{ in } A_i\}$$
$$= \min_{1 \le i \le r} Tr_{L_i/K}((\lambda_i(z)-x_i^*)^2)$$

solving the CVP of ideal lattices can be done by a similar algorithm following the logics of that for SVP. For this reason, we will only focus on solving SVP hereafter.

(2)In the first step, if such a prime ideal P is found that completely splits in the extended field L, i.e., $PO_L = M_1 \ldots M_n$, then solving the SVP instance in this case is reduced to solving n 1-dimensional SVP instances of some ideals in K.

(3)In case of Galois extension L/K, the decomposition law (22) will have $e_1 = \ldots = e_r \equiv e = 1$ and $f_1 = \ldots = f_r \equiv f$. Since $ref = n$ and $r \ge 2$, we always have $f = n/r \le n/2$, i.e., each reduction can decrease the instance's dimension by at least half and at most $O(\log n)$ recursions are needed.

For example, supposing that each recursion reduces the dimensions (the intermediate fields' extension degrees on K) by half, then after m recursions the original n-dimensional SVP instance will be decomposed to 2^m number of $n/2^m$-dimensional SVP instances. As a result, the time complexity would be at most $2^m 2^{O(n/2^m)}$ by calling some single-exponential time-complexity generic solvers on these $n/2^m$-dimensional SVP instances(e.g., the elegant solver in [6]), substantially more efficient than the time-complexity of $2^{O(n)}$ if the n-dimensional original instance is directly solved.

(4)In the case of Galois extension L/K, the fact that Galois group $G_{L/K}$ is transitive on $M_1, \ldots, M_r$ in (22), i.e., for any M_i, M_j there exists g in $G_{L/K}$ such that $M_i = g(M_j)$, can significantly simplify lots of details in our algorithm's construction.

(5)It can be proven that (i)If the input number field L is a Galois extension of the ground field K then all the intermediate global fields L_i's in the algorithm can be constructed as K's Galois extension. (ii)For such input complex number field L that L/K is Galois and $\sigma(\bar{y}) = \bar{\sigma}(y)$ for all the σ's in the Galois group $G_{L/K}$ and all y's in L, all the intermediate global fields L_i's in the algorithm can be constructed with the same above properties as L. In the above situation, the Euclidean metric on the ideal lattice A is $Tr_{L/K}(y\bar{y})$ and as a result our solver can apply to such input $(L/K, A)$, particularly when L/K is the Abelian extension and K is real (contained in R).

All the above details are further elaborated in the full-version paper.

B   Low Level Details in the Algorithm

Now we turn from the high-level descriptions to low-level technical details, each step discussed in a subsection. We begin with the relatively easy step♯2, ♯3 and ♯4 and finally end with step♯1.

B.1   Solving Subproblems in Step♯2

In this step we solve such problems: Given $L = K(\alpha)$ with α's minimal monic polynomial $f(t) \in K[t]$ and unramified prime ideal P in O_K where $P \nmid \sharp(O_L/O_K[\alpha])$, firstly, find irreducible polynomials $h_1(t), \ldots, h_r(t) \in K[t]$ s.t. $h_i(t) = f_i(t) \mod P$ where $L_{M_i} = K_P[t]/(f_i(t))$'s are local fields associated with P's all decomposition prime ideals $M_1, \ldots, M_r \subset O_L$. Note that in this situation naturally (i. e.,due to P's unramification) each $deg h_i(t) = deg f_i(t) = [O_L/M_i : O_K/P]=[L_{M_i}$'s residue class field : K_P's residue class field]. Secondly, find K_P-linear isomorphism ψ and its component mappings $\psi_1, \ldots, \psi_r$ in (16)-(18) where each $\psi_i : L \to L_{M_i}$, i.e., $y \cong (y_1, \ldots, y_r)$ means $\psi(y) = (\psi_1(y), \ldots, \psi_r(y))$.

> **Solution to the 1st sub-problem:**
>
> Decompose $f(t) \mod P$ by calling any appropriate polynomial factorization algorithm modulo the prime ideal(e. g., those in [21, 24, 25]), i. e., to compute distinct monic irreducible polynomials $h_1(t), \ldots, h_r(t) \in (O_K/P)[t]$(hence irreducible in $K[t]$) s.t.
>
> $$f(t) = h_1(t), \ldots, h_r(t) \mod P \tag{27}$$

Proof of the solution's correctness: Suppose in $K_P[t]$ there is the factorization of $f(t)$:

$$f(t) = f_1(t) \ldots f_s(t) \tag{28}$$

where $f_1(t), \ldots, f_s(t) \in K_P[t]$ are distinct monic irreducible polynomials. By the famous Hensel's lemma [14](and O_K/P is the residue class field of K_P) it follows that $r = s$ and $h_i(t) = f_i(t) \mod P$ for $i = 1, \ldots, r$ and since $h_i(t)$'s and $f_i(t)$'s are all monic, we have $deg f_i(t) = deg h_i(t)$.

In addition, under the condition $P \nmid \sharp(O_L/O_K[\alpha])$ we have [24, 25]:

$$PO_L = M_1 \ldots M_r$$

where each prime ideal factor $M_i = (P, h_i(\alpha))$ in O_L. In particular, each M_i-adic local field $L_{M_i} = K_P[t]/(f_i(t))$ and $[L_{M_i} : K_P] = deg f_i(t) = deg h_i(t)$.   □

Remark on the Polynomial Factorization Algorithm for (27)
(27) can be solved via lots of algorithms, for example, the algorithm in [25] is a good solver which time-complexity is polynomial in the degree n and the number of basic arithmetic operations in the (finite) field O_K/P. It's also an elegant random algorithm which success probability is at least 4/9. The most recently available factorization algorithm is [26], e.g.,in case of $K=Q$ (here P is a prime integer) its time-complexity is only $O(n^{1.5}log P + n(log P)^2)$.

Solution to the 2nd sub-problem:

For any $y(t) \in K_P[t]/(f(t)) = K_P \otimes_K L$, set $\psi(y) = (\psi_1(y), \ldots, \psi_r(y))$ where

$$\psi_i(y(t)) \equiv y(t) \mod f_i(t) \tag{29}$$

Proof of the solution's correctness: By $K_P \otimes_K L = K_P \otimes_K K[t]/(f(t)) = K_P[t]/(f(t))$ and (28), it follows from the Chinese Remainder Theorem that there is isomorphism

$$K_P \otimes_K L \cong K_P[t]/(f_1(t)) \oplus \ldots \oplus K_P[t]/(f_r(t)) = L_{M_1} \oplus \ldots \oplus L_{M_r}$$

where the K_P-linear isomorphism ψ's components $\psi_i(y(t)) \equiv y(t) \mod f_i(t)$, $i = 1, \ldots, r$ and obviously $\psi_i(y \pm z) \equiv \psi_i(y) \pm \psi_i(z)$, $\psi_i(yz) \equiv \psi_i(y)\psi_i(z)$ in L_{M_i} for any $y = y(t)$, $z = z(t)$ in L_{M_i} and i. Furthermore, $T(y)z \equiv yz = y_1z_1 \oplus \ldots \oplus y_r z_r = T(y_1)z_1 \oplus \ldots \oplus T(y_r)z_r$, i.e., there is always the diagonalization $T(y) = T(y_1) \oplus \ldots \oplus T(y_r)$ so (17) holds (but (17b) is not needed). In particular, each ψ_i's restriction on L can be computed by:

$$\psi_i(y(t)) \equiv y(t) \mod h_i(t), \; for \; any \; y(t) \in K[t]/(f(t)) = L \tag{30}$$

and $\psi_i(L) = L_i$. $\qquad\square$

B.2 Solving Subproblems in Step♯3

Given L, unramified prime ideal P in O_K and local fields associated with PO_L's all prime factors $M_1, \ldots, M_r$ in O_L, i.e., $L_{M_1}, \ldots, L_{M_r}$ each with an irreducible polynomial $h_i(t)$ in $K[t]$ s.t. $h_i(t) = f_i(t) \mod P$ and $L_{M_i} = K_P[t]/(f_i(t))$, integers $f_1, \ldots, f_r \geq 1$, we need to find K's extended fields $L_1, \ldots, L_r$ and field embeddings $\varphi_1, \ldots, \varphi_r$ satisfying (23). We solve this for each $i = 1, \ldots, r(r \leq n)$ so the sub-problem is re-specified as:

Given L, unramified prime ideal P in O_K and a local field associated with one of PO_L's prime factor M in O_L, i.e., L_M with an irreducible polynomial $h_M(t)$ in $K[t]$ s.t. $h_M(t) = f_M(t) \mod P$ and $L_M = K_P[t]/(f_M(t))$ of extension degree $f \geq 1$, find K's extended (global) field L^* of degree $[L^* : K] = [L_M : K_P]$ and a field embedding $\varphi_M : \psi_M(L) \to L_M$ where ψ_M denotes the ψ's component-mapping on L_M s.t. $Tr_{L^*/K}(\varphi_M\psi_M(y)) = Tr_{L_M/K_P}(\psi_M(y))$ for any y in L^*.

Solution:

$$Set \; L^* \equiv K[t]/(h_M(t)) \; and \; \varphi_M = id. \tag{31}$$

Proof of the solution's correctness: Obviously L^* is global because $h_M(t) \in K[t]$. Now prove L^* is dense in L_M and $[L^* : K] = [L_M : K_P]$. Since $L_M = K_P[t]/(f_M(t))$ is a unramified (local field) extension with extension degree $f = deg f_M(t)$ and $f_M(t)$ is irreducible in $K_P[t]$ with leading coefficient 1,

$h_M(t) = f_M(t) \mod P$ so by Hensel lemma $h_M(t)$ is irreducible in $(O_K/P)[t]$ with the same degree f and leading coefficient 1. In consequence[13, Chapter 14; 16, Chapter2], this unramified extension L_M/K_P induces a finite field extension $O_L/M = (O_K/P)[t]/(h_M(t))$ of the same degree f and vice versa, a one-to-one correspondence up to isomorphism.

As a result, we have $K_P[t]/(f_M(t)) = K_P[t]/(h_M(t))$ and in particular $h_M(t)$ is irreducible in $K[t]$ so $[L^* : K] = f = [L_M : K_P]$. By definition $L^* = K[t]/(h_M(t))$ we have that L^* is densely contained in the field $K_P[t]/(h_M(t)) = K_P[t]/(f_M(t)) = L_M$. Furthermore, $K_P \otimes_K L^* = K_P[t]/(h_M(t))$ $= K_P[t]/(f_M(t)) = L_M$ so $Tr_{L^*/K}(y) = Tr_{L_M/K_P}(y)$ for any y in L^* by (16).
Finally, $\psi_M(L) = L^*$ so $\varphi_M = id$. □

Remark: If L is totally real, so is L^*. In fact, every image of L under the conjugate embedding σ is real so for L's any prime ideal M, $\sqrt{-1}$ is not in the $\sigma(M)$-adic completeness of $\sigma(L)$, i.e., $\sqrt{-1}$ is not in $\sigma(L_M)$. As a result, $\sqrt{-1}$ is not in $\sigma(L^*)$ which is dense in $\sigma(L_M)$, i.e., L^* is real.

B.3   Solving Subproblems in Step♯4

In step♯4 we need to solve two sub-problems. Firstly, given an ideal A (with its pseudo-basis) in L and a surjective homomorphism $\psi_m : L \to L_m$, compute the ideal $\psi_m(A)$ in L_m. Secondly the sub-problem (25), i.e., given $x^*(t)$ *in* $L_m = K[t]/(h_m(t))$ find $y^*(t)$ in L s.t.

$$y^*(t) = x^*(t) \mod h_m(t), \ y^*(t) = 0 \mod h_j(t) \ for \ all \ j \neq m \qquad (32)$$

Solution to the 1st sub-problem

(Ideal's homomorphism image).On input an ideal A with the pseudo-basis representation,i.e., a set of L' s K-basis $\omega_1, \ldots, \omega_n$ in O_L and a set of K's ideals $I_1, \ldots, I_n$ such that $A = I_1\omega_1 + \ldots + I_n\omega_n$, do:

Compute $b_i = \psi_m(\omega_i)$, $i = 1, \ldots, n$.

Find the maximal subset of K-linear independent members, w.l.o.g., denoted $b_1, \ldots, b_{d_m}$ where $d_m = [L_m : K]$, and the integers β, λ_{ij} in O_K s.t.

$$\beta b_i = \sum_{1 \leq j \leq d_m} \lambda_{ij} b_j \quad i = d_m + 1, \ldots, n$$

(e.g., this step can be accomplished by Gauss elimination algorithm regarding the b_i's as vectors in the d_m-dimensional K-linear space L_m);

Compute the ideal $J_j = \beta I_j + \sum_{1+d_m \leq i \leq n} \lambda_{ij} I_i$ for each $j = 1, \ldots, d_m$;

Compute and output the ideal

$$\psi_m(A) = \sum_{1 \leq j \leq d_m} J_j b_j / \beta.$$

Proof of the solution's correctness: Since $A = I_1\omega_1 + \ldots + I_n\omega_n$ and ψ_m is a K-homomorphism, we have $\psi_m(A) = I_1b_1 + \ldots + I_nb_n$ so

$$\beta\psi_m(A) = \beta I_1b_1 + \ldots + \beta I_{d_m}b_{d_m} + I_{d_m+1}\beta b_{d_m+1} + \ldots + I_n\beta b_n$$
$$= \beta I_1b_1 + \ldots + \beta I_{d_m}b_{d_m}$$
$$+ I_{d_m+1}\sum_{1\leq j\leq d_m}\lambda_{d_m+1,j}b_j + \ldots + I_n\sum_{1\leq j\leq d_m}\lambda_{n,j}b_j$$
$$= \sum_{1\leq j\leq d_m}(\beta I_j + \sum_{1+d_m\leq i\leq n}\lambda_{ij}I_i) = \sum_{1\leq j\leq d_m}J_jb_j.$$

and note that b_i's are all in O_{L_m} since ω_i's are all in O_L. $\square$

Solution to the 2nd sub-problem:

Find $g^*(t)$ in $K[t]$ s.t.

$$g^*(t) = x^*(t) \mod h_m(t), g^*(t) = 0 \mod h_j(t) \text{ for all } j \neq m$$

by the standard algorithm derived from Chinese Remainder Theorem. Then set $y^*(t) \equiv g^*(t) \mod f(t)$

The solution's correctness can be verified by direct calculations.

B.4   Solving Subproblems in Step♯1

Now we turn to this problem: given L and A on input, find a prime ideal P in O_K such that:

$$PO_L \text{ is not prime in } O_L; \tag{33a}$$

$$P \text{ is unramified in } O_L, i.e., \text{ all its} \tag{33b}$$
$$\text{ramification indices } e_1 = \ldots = e_r = 1;$$

$$P \nmid \sharp(O_L/O_K[\alpha]). \tag{33c}$$

Before constructing the solver, we specify the following oracles at first.

Oracle-PG_K where K is a number field: Generates a prime ideal at random in O_K. The input is void and each output is probabilistically independent of any others.

Orcale-$PT_L(M)$ where L is a number field: on input any ideal M in O_L, tests whether M is prime or not.

Oracle-$d_K(L)$: On input any L where L/K is a number field extension of finite degree, outputs the relative discriminant $d_{L/K}$, an integral ideal in O_K which is the greatest common divisor of $det(Tr_{L/K}(a_ia_j))$ of all K-linear independent integers $a_1, \ldots, a_n$ in O_L.

Oracle-$l_K(L, \alpha)$: On input any $L = K(\alpha)$ where L/K is a number field extension with finite degree n, outputs $\sharp(O_L/O_K[\alpha])$, the cardinality of the finite quotient set $O_L/O_K[\alpha]$.

Solution

(I) Compute the relative discriminant $d_{L/K}=$ **Oracle-$d_K(L)$**;
Compute $l=$**Oracle-$l_K(L,\alpha)$**:

(II) Do{
$P=$**Oracle-PG_K**; /*generate prime ideal P in O_K*/
}while ($P|d_{L/K}$ or $P|l$); /*equivalently, $d_{L/K}$ is a subset of P or $l \in P$.*/

(III) If **Orcale-$PT_L(PO_L)$** is true /*i. e., P is prime in O_L */
Then goto II;
output(P);

Proof of the solution's correctness By general algebraic number theory, a prime ideal P in O_K is ramified in the integral closure O_L of the field extension L/K iff it divides the relative discriminant $d_{L/K}$[14, 15, 17]. As a result, the output prime ideal P is unramified in O_L and obviously satisfies all other requirements in (33). □

Remarks on Implementation of the Oracles: In general, how to implement all the above oracles is not completely clear with our best knowledge. However, in the important case that $L = Q(\alpha) \cong Q[t]/(f(t))$ where the polynomial $f(t)$ is monic and irreducible in $Z[t]$, $K = Q$ (hence $O_K = Z$), we can have further arguments about their implementations.

(1)**Oracle-$l_Q(L,\alpha)$** can be completely implemented by **Oracle-$d_Q(L)$**. If fact, in this case there is the formula

$$\sharp(O_L/Z[\alpha]) = |N_{L/Q}(f'(\alpha))/d_{L/Q}|^{1/2} \tag{34}$$

where $|.|$ is the ordinary absolute value.

Proof. When $L = Q(\alpha) \cong Q[t]/(f(t))$, (due to the fact that $O_K = Z$ is a principal ideal domain) there exist a set of integral basis $\xi_1,\ldots,\xi_n$ s.t. $O_L = Z\xi_1 + \ldots + Z\xi_n$ and the determinant $d_{L/Q} = det(Tr_{L/Q}(\xi_i\xi_j))$, i. e., $|d_{L/Q}|$ is the squared volume of the lattice O_L's fundamental domain. Note that $Z[\alpha] = Z + Z\alpha + Z\alpha^2 + \ldots + Z\alpha^{n-1}$ ($\alpha \in O_L$) is a sub-lattice in O_L so its squared fundamental domain's volume

$$|det(Tr_{L/Q}(\alpha^{i-1}\alpha^{j-1}))| = \sharp(O_L/Z[\alpha])^2|d_{L/Q}|$$

On the other hand, $|det(Tr_{L/Q}(\alpha^{i-1}\alpha^{j-1}))| = |det(\alpha^{(i)j-1})|^2 =$ the square of the Vandermond determinant of α's conjugates $\alpha^{(1)}, \alpha^{(2)},\ldots,\alpha^{(n-1)} = |\prod_{1\leq i<j\leq n}(\alpha^{(i)} - \alpha^{(j)})|^2 = |f'(\alpha^{(1)})\ldots f'(\alpha^{(n)})| = |N_{L/Q}(f'(\alpha))|$, which proves (34).

(2) **Oracle-$d_Q(L)$**: In this case there exist the algorithms to compute O_L's integral basis $\xi_1,\ldots,\xi_n$ and the discriminant $d_{L/Q}$, e. g., the algorithm 6.1.8 in [27]. It's worthwhile to note that the performance-dominating step in this algorithm is to factorize the rational integer[27] which bit-size in our algorithm's

context is $O(nlogn)$, as a result, this oracle's intrinsic complexity may be only as hard as integer factorization which has implementation of sub-exponential time complexity and even not hard against the quantum computer.

For relative extension L/K where $K \neq Q$, it's worthwhile to mention the special case $d_{L/K} = O_K$ (which can never happen if K is Q) and hence $d_{L_j/K} = O_K$ for all the intermediate fields L_j during the algorithm's recursion, e.g., K's Hilbert class field $L = K(\mu^{1/q})$ where q divides K's class number $h(K)$. In this situation the oracle-$d_K(.)$ is trivial and the decision $P|d_{L/K}$ (always false) can be simply omitted from the algorithm. As a result, the algorithm's complexity can be significantly reduced (r.f., Appendix.B.5).

(3) **Oracle-$PT_L(M)$**: On input any ideal M in O_L, decide whether M is prime or not. For this oracle's counterpart in rational number field Q, i.e., rational integer's primality testing, there are not only practically efficient but also deterministic polynomial time-complexity algorithms [10, 28]. Although so far it's unknown how to efficiently implement Orcale-$PT_L(.)$ in arbitrary number field L, it's reasonable to expect that it's complexity would be lower than SVP/CVP.

(4)**Oracle-PG_Q**: Generates a prime number at random in Z, a problem with known efficient solvers.

B.5 Computational Complexity

Let $|S|$ denote the input size of the ideal lattice Problem $SVP(A, L/K)$, N_{PG}, N_{PT}, N_d and N_l denote the time-complexities to implement Oracle-PG_K, Oracle-PT_L, Oracle-d_K and Oracle-l_K in the algorithm. From the constructions in Sect.4 and Sect.B, it's easy to see that all the subroutines and operations in the algorithm are only those with time and space complexity polynomial in the input size, except the above four oracles which intrinsic computational complexity may be non-polynomial. Furthermore the recursion depth of the (high level) algorithm, hence the number of callings to all those oracles, is only $O(n)$ where $n = [L : K]$=the dimension of the input ideal lattice A. In summary, we can have the following conclusions(details in the full-version paper).

Theorem 4 *(1)Given any number field K contained in R, there exists the algorithm to solve (exactly) SVP on input any totally real extended field L and ideal A in O_L with time complexity*

$$poly(n, |S|, N_{PG}, N_{PT}, N_d, N_l) \tag{35}$$

and space complexity $poly(n, |S|)$ where $n = [L : K]$. (2)For CVP of ideal lattices, we have exactly the same conclusion. □

Corollary 5 *(1)Given any number field K contained in R, there exists the algorithm to solve (exactly) SVP on input any totally real extended field L_μ and ideal A_μ in O_{L_μ} from the family $\{(L_\mu, A_\mu) : d_{L_\mu/K} = O_K\}$ with time complexity $poly(n_\mu, |S|, N_{PG}, N_{PT}, N_l)$ and space complexity $poly(n_\mu, |S|)$ where $n_\mu = [L_\mu : K]$. (2)For CVP of ideal lattices, we have exactly the same conclusion.*

Remark: It is known that there exists the infinite family (e. g.the Hilbert class field extension tower) $\{(L_\mu, A_\mu) : d_{L_\mu/K} = O_K\}$ which extension degree n_μ is upper-boundless. For such input family, the algorithm constructed in this paper would be efficient (polynomial in time) as long as the Oracle-PG_K, Orcale-PT_L and Oracle-l_K can be implemented efficiently, which possibility seems positive.

Now back to the case of L/Q, because there exist efficient algorithms to implement Oracle-PG_Q, i. e., to efficiently generate prime integers, we have:

Corollary 6 *(1)There exists the algorithm to solve (exactly) SVP on input any totally real number field L and ideal A in O_L with time complexity*

$$poly(n, |S|, N_{PT}, N_d)$$

and space complexity $poly(n, |S|)$ where $n = [L : Q]$. (2)For CVP of ideal lattices, we have exactly the same conclusion. □

Modular Form Approach
to Solving Lattice Problems*

Yuan Tian, Xueyong Zhu, and Rongxin Sun

Software School, Dalian University of Technology, P.R. China
{tianyuan_ca,zhuxueyong}@sina.com,
sunrongxin7666@163.com

Abstract. We construct new randomized algorithms to find the exact solutions to the shortest and closest vector problems (SVP and CVP) in ℓ^2-norm for integral lattices. Not only the minimal norm of non-zero lattice vectors in SVP and the minimal distance in CVP, but also how many lattice vectors reach those minimums can be simultaneously computed. Our approach is based on special properties of the generating function of lattice vectors' ℓ^2-norms, the lattice-associated theta function. In computational complexity perspective and take our SVP solver as an example, for the integral lattice family $\{\Lambda_n\}$ of dimension $dim\Lambda_n = n$ and level $h_n = l(\Lambda_n)$ (the minimal positive integer such that the dual lattice Λ_n^* scaled by $h_n^{1/2}$ is integral), this algorithm can find the minimal ℓ^2-norm of non-zero lattice vectors and the number of such shortest vectors in Λ_n with success probability 1-ε in the space-complexity of polynomial in n and time-complexity of $(\log \log n^2 h_n)^{O(n)} \log(1/\varepsilon)$.

1 Introduction

Lattice problems take important roles in public-key cryptography, combinatorial optimization and many other fields in computer science[1–8]. In the shortest lattice vector problem (SVP), a non-zero lattice vector $\boldsymbol{x}$ in $\mathbf{B}Z^n$ is to be found to minimize $|\boldsymbol{x}|$ on input the lattice basis matrix $\mathbf{B}$ with respect to some specific norm $||$ in R^n. In the closest lattice vector problem (CVP), a lattice vector $\boldsymbol{x}$ is to be found to minimize $|\boldsymbol{u} - \boldsymbol{x}|$ on input the basis matrix $\mathbf{B}$ and a target vector $\boldsymbol{u}$ in R^n. In recent years, lots of cryptographic schemes and protocols have been devised with proofs of security under the assumption that there is no (probabilistic and sometimes quantum) polynomial-time algorithm to solve arbitrary instances of variants of SVP and CVP.

From a computational hardness perspective, SVP, CVP and other related variants are NP-hard under deterministic (e.g., CVP) or randomized (e.g., SVP) reductions[4, 7, 9, 10]. Even some approximation variants of these problems are proven to be NP-hard if the approximation factor is within some specific range. Despite of these facts, finding new algorithms to solve lattice problems exactly are still interesting and meaningful both because many applications (e.g.,

---

* This work is supported by China NSF(61370144).

T V Gopal et al. (Eds.): TAMC 2014, LNCS 8402, pp. 401–421, 2014.
© Springer International Publishing Switzerland 2014

in mathematics and communication theory) involve lattices in relatively small dimensions, and because approximation algorithms for high dimensional lattices for which the exact solution is infeasible typically involve the exact solution of low dimensional sub-problems. In this paper we develop randomized algorithms to find the exact solutions to SVP and CVP.

1.1   Basic Results

We develop new randomized algorithms to find the exact solutions to SVP and CVP in Euclidean norm (ℓ^2) for any integral lattice. Not only the minimal ℓ^2-norm of non-zero lattice vectors in SVP and the ℓ^2-minimal distance in CVP, but also how many lattice vectors reach those minimums(e.g., the kissing number in SVP) can be simultaneously computed by the algorithms. More concretely and take SVP as an example, for the integral lattice family $\{\Lambda_n\}$ with dimension $dim\Lambda_n = n$ and level $h_n = l(\Lambda_n)$ (the minimal positive integer such that the dual lattice $\Lambda_n{}^*$ scaled by $h_n^{1/2}$ is integral), this algorithm can find the minimal ℓ^2-norm of non-zero lattice vectors and the number of such shortest vectors in Λ_n with success probability $1 - \varepsilon$ in the asymptotic space-complexity of polynomial in n and asymptotic time-complexity of $(\log \log n^2 h_n)^{O(n)} \log(1/\varepsilon)$. Interestingly, the only contribution to the algorithm's exponential time complexity $(\log \log n^2 h_n)^{O(n)} \log(1/\varepsilon)$ comes from independently repeating a randomized lattice vector sampler $(\log \log n^2 h_n)^{O(n)} \log(1/\varepsilon)$ times. All the rest of operations contribute to the time-complexity with only an additive polynomial in n. Similar situations occur when solving the exact CVP by our algorithm. As a result, our solvers can be (very easily) parallelized to be polynomial in time-complexity. Due to the same feature, a variant of our CVP solver can solve the closest lattice vector problem with preprocessing (CVPP) in polynomial time and $(\log \log n^2 h_n)^{O(n)} \log(1/\varepsilon)$ space complexity.

1.2   A Sketch on Our Approach

Our approach is based on some special properties of the generating function of lattice vectors' ℓ^2-norms. This function is a measure used in previous works mainly for hardness analysis on lattice and related problems[9, 11, 12] but rarely for computational purposes. For SVP, such function is defined as:

$$\vartheta(\tau; \Lambda) \equiv \sum_{x \in \Lambda} \exp(2\pi i \tau |x|^2)$$

where $|x|$ denotes the vector x's ℓ^2-norm and $\tau = \sigma + it$ is a complex variable on the upper-half complex plane(i.e., $t > 0$). If Λ is integral, i.e., all $|x|^2$'s are integers for any x in Λ (an assumption without any loss in generality when we only deal with rational lattices), this function can be equivalently represented as a Fourier expansion (with complex variable τ)

$$\vartheta(\tau; \Lambda) = \sum_{m \geq 0} a(m) \exp(2\pi i \tau m)$$

where $a(0) = 1$ and $a(m)$ is the number of lattice vectors in Λ which squared ℓ^2-norms equal m. From this viewpoint, solving SVP on Λ reduces to finding its theta function's first non-zero Fourier coefficient $a(m)$ among its non-constant items.

The technical support to the above idea comes from the fact that, as a function of complex variable τ $(Im\tau > 0)$, $\vartheta(\tau; \Lambda)$ is a so-called modular form of weight $n/2$ (details in section 2.2) and therefore has a series of special properties. As a result, the theta function can be expanded on a polynomial (in the lattice's level h and dimension n) number of base functions and then its Fourier coefficients $a(m)$ can be efficiently computed from the linear combination of a set of the basis' Fourier coefficients.

For CVP, when restricting the target vector u to be the integral vector (without any loss in generality when we only work in the rational number field), the same idea applies to the non-homogenous theta function

$$\vartheta(\tau; \Lambda, u) \equiv \sum_{x \in \Lambda} exp(2\pi i\tau|x - u|^2) = \sum_{m \geq 1} b(m)exp(2\pi i\tau m)$$

which is also a modular form, where $b(0) = 0$ (except for the trivial case that $u \in \Lambda$) and $b(m)$ is the number of lattice vectors in Λ which squared ℓ^2-distance to u is m. From this viewpoint, solving CVP on input Λ and u reduces to finding the non-homogenous theta function's first non-zero Fourier coefficient $b(m)$.

1.3 Related Works

To find the exact solutions to lattice problems, so far three main families of SVP and CVP solvers exist which are listed in Table1 together with our algorithms developed in this paper in comparison.

Among these solvers, MV and Kannan algorithms are deterministic while AKS (and our) algorithms are randomized. All algorithms work in ℓ^2-norm (only AKS algorithm can work in other norms, e.g., ℓ_∞). The core of MV algorithm[13] is to compute the Voronoi cell of the lattice[1], whose knowledge facilitates the tasks to solve SVP and CVP. Kannan algorithm[5, 6] relies on a deterministic procedure to enumerate all lattice vectors below a prescribed norm, or within a prescribed distance to the target vector. This procedure uses the Grahm-Schmidt orthogonalization of the input lattice basis to recursively bound the integer coordinates of the candidate solutions.

The AKS algorithm[14] was the first single-exponential time algorithm for SVP which can be described as follows: Let $\gamma < 1$ be a constant and S be a set of N lattice vectors sampled in the ℓ^2-ball of radius $R = 2^{O(n)}\lambda_1(\Lambda)$ where $\lambda_1(\Lambda)$ is the minimal norm of non-zero lattice vectors in Λ. For sufficiently large N, there exists a pair of lattice vectors u, v such that $|u - v| < \gamma R$, so $u - v$ is shorter in Λ. The core of the algorithm is to chose a subset C in S such that $|C|$ is not too large and for any u in $S \backslash C$ there exists v in C such that $|u - v| < \gamma R$. This is used to produce a set of lattice vectors S_1 in the ball $\gamma R B_2^n$ with $|S_1| = |S| - |C|$. This procedure can be applied a polynomial number of times to obtain lattice vectors of norms less than $a\lambda_1(\Lambda)$ for some constant a. Recently this algorithm has been significantly improved and the currently best time complexity is $2^{2.465n+o(n)}$[3]. However, the AKS variant solver for CVP only finds the $(1+\varepsilon)$-approximate solution for arbitrary $\varepsilon > 0$ in time complexity bounded by $(2+1/\varepsilon)^{O(n)}$[15, 16].

As a randomized algorithm, our solver outperforms the sieve algorithms in the aspects that it has space complexity only polynomial in n and can solve both SVP and CVP precisely. In time-complexity, our solver is only slightly inferior to AKS and MV solvers for a wide range of SVP/CVP instances. For example, up to $h_n = 2^{O(n)}$ our solver can operate with time-complexity of $(\log n)^{O(n)}$ where the O-constant is $1+\delta(\delta > 0)$, only suffering a slight loss in comparison to, e.g., AKS solver's time-complexity of $2^{O(n)}$. Another characteristic of our algorithm is its ability to be parallelized to be polynomial in time complexity. As noticed in Appendix A, when sampling the lattice by calling N independent Gaussian samplers in concurrency rather than in sequence, the whole algorithm to solve SVP or CVP becomes polynomial in time complexity (but exponential in parallelism). Another variant of our CVP solver can solve CVPP in polynomial time and $n^{O(n)}log(1/\varepsilon)$ space complexity with success probability 1-ε. Such characteristics will be valuable in practices, e.g., in solving lattice problems of moderately high dimensions. So far with our understanding no other solvers can be parallelized to be polynomial in time complexity. For example, the critical component in the elegant MV algorithm[13] is an iterative subroutine to operate at most 2^n times, which is hard to be parallelized due to its iterative nature. The core of AKS algorithm and its variants[14–16], the sieve subroutine which dominates the algorithm's time complexity, is also hard to be parallelized to be polynomial. Similar situations occur for Kannan algorithm. The last (but not the least) important feature of our approach is its potential to apply to SVP and CVP for the ideal lattices in algebraic number fields where the associated theta functions have more special properties to exploit.

1.4   Roadmap

In section 2 we give necessary backgrounds in lattice geometry and modular forms. In section 3, we give a sketch on our approach which technical details are elaborated in Appendices. The complete algorithms to solve SVP and CVP are presented in Appendix A-C and the complexity analysis is given in Appendix D.

2   Preliminaries

2.1   Lattices

General: The set of integers is denoted by Z and rational numbers by Q. In the Euclidean space R^n, a n-dimensional rational lattice, denoted $\Lambda(\mathbf{B})$ where $\mathbf{B}$ is a matrix with column vectors $(\boldsymbol{b}_1,\dots,\boldsymbol{b}_n)$, is the set of vectors $\{x_1\boldsymbol{b}_1+\dots+x_n\boldsymbol{b}_n : x_1,\dots,x_n \in Z\}$ where the scalar products $<\boldsymbol{b}_i,\boldsymbol{b}_j>$ are all rational numbers.

Table 1. Comparing the existed families of SVP and CVP solvers and our algorithms

Solvers	Time complexity upper bound	Space complexity upper bound	Remarks
Kannan [3, 5, 6]	$n^{O(n)}$	$poly(n)$	deterministic; the O-constant is improved as small as $1/2e$
MV[3, 13]	$2^{2n+o(n)}$	$2^{O(n)}$	deterministic
AKS [3, 14–16]	SVP: $2^{2.465n+o(n)}$ CVP: $(2+1/\varepsilon)^{O(n)}$	SVP: $2^{1.325n+o(n)}$ CVP: $(1+1/\varepsilon)^{O(n)}$	randomized; solves $(1+\varepsilon)$-CVP only
Our algorithm	$(\log\log n^2 h)^{O(n)}$	$poly(n)$	h is the lattice's level. The O-constant is $1+\delta$, $\delta > 0$. Easy to be parallelized to be polynomial-time to solve SVP, CVP and CVPP.

The lattice with basis $b_1, \ldots, b_n$ is also denoted $Zb_1 + \ldots + Zb_n$. Without loss of generality in computer science, in this work we only consider the integral lattice in which $< b_i, b_j >$ are all integers.

For any lattice $\Lambda = Zb_1 + \ldots + Zb_n$, the lattice $\Lambda^* \equiv Zb_1^* + \ldots + Zb_n^*$ where $< b_i^*, b_j > = \delta_{ij}$ for all $i, j = 1, \ldots, n$ is called Λ's dual lattice. Equivalently, Λ^* is a discrete set of vectors $\mathbf{y}$ such that $< x, y > \in Z$ for all x's in Λ. The dual Λ^* of a rational lattice Λ is always rational, but Λ^* may not be integral even when Λ is integral. When Λ has a base matrix $\mathbf{B} = (b_1, \ldots, b_n)$, its dual lattice Λ^* will have a base matrix $\Lambda^* = (b_1^*, \ldots, b_n^*) = \mathbf{B}^{-T}$ so both Λ and Λ^* are integral iff $det(\mathbf{B}) = det(\mathbf{B}^*) = \pm 1$. Another important property is that $\Lambda^{**} = \Lambda$.

For any vector $u = (u_1, \ldots, u_n)$ in R^n, its ℓ^2-norm $< u, u >^{1/2} = (u_1^2 + \ldots + u_n^2)^{1/2}$ is denoted $|u|$. The squared ℓ^2-norm of any lattice vector in an integral lattice is always an integer.

Lattice Problems: Given a lattice $\Lambda(B) = Zb_1 + \ldots + Zb_n$, let

$$\lambda_1(\Lambda) \equiv \min\{|x| : x \ in \ \Lambda \ \text{and non-zero}\} \tag{1}$$

be the minimal value of ℓ^2-norms of non-zero lattice vectors in $\Lambda(\mathbf{B})$. The *optimization* $(\ell^2$-$)$ *shortest vector problem*, SVP(Λ) in brief, is to find $\lambda_1(\Lambda)$. The *search* $(\ell^2$-$)$ shortest vector problem, s-SVP(Λ) in brief, is to find a lattice vector x in Λ such that $|x| = \lambda_1(\Lambda)$.

Given a lattice $\Lambda(B)$ and a rational target vector u in Q^n, let

$$dist(\Lambda; u) \equiv \min\{|x - u| : x \ in \ \Lambda\} \tag{2}$$

be the minimum ℓ^2-distance between u and all lattice vectors in Λ. The *optimization* $(\ell^2$-$)$*closest vector problem*, CVP(Λ, u) in brief, is to find $dist(\Lambda; u)$. The *search* $(\ell^2$-$)$closest vector problem, s-CVP(Λ, u) in brief, is to find a lattice vector x in Λ such that $|x - u| = dist(\Lambda; u)$.

The covering radius of a lattice, $\mu(\Lambda)$, is defined as the maximal distance between any vector and the lattice. The covering radius problem, CRP(Λ) in brief, is to find

$$\mu(\Lambda) \equiv \max\{dist(\Lambda; \boldsymbol{u}) : \boldsymbol{u} \ in \ Q^n\} \tag{3}$$

In this paper we focus on the algorithm to solve SVP and CVP problems. It has been known that these problems are computationally hard[4, 7, 9, 10, 17]. However, there is:

Theorem 1 *[7, 10, 17] (1)s-SVP can be solved in polynomial time given the oracle to solve s-CVP. (2)s-CVP can be solved in polynomial time given the oracle to solve (optimization) CVP.* □

In consequence, the algorithm for optimization CVP can be used as the cornerstone to solve both search problems. In this paper we focus on constructing the randomized algorithms for optimization SVP and CVP with similar ideas and techniques.

General Bounds: For any n-dimensional lattice Λ, one of the most important general fact is the Minkowski's inequality[1, 10]:

$$Vol(B_2^n)\lambda_1(\Lambda)^n \leq 2^n|det(\Lambda)|$$

where $Vol(B_2^n)$ is the n-dimensional volume of the unit Euclidean ball B_2^n , e.g., $\pi^{n/2}/(n/2)!$, and $|det(\Lambda)|$ is the determinant of the lattice's base matrix $\mathbf{B}$, numerically equal to the lattice's elementary parallelotope's volume. It follows that

$$\lambda_1(\Lambda) \leq cn^{1/2}|det(\Lambda)|^{1/n} \tag{4}$$

where $c(\leq 1)$ is some absolute constant.

Another important general property is the transference theorem[10, 12]

$$\lambda_1(\Lambda^*)\mu(\Lambda) \leq dn \tag{5}$$

where $d(\leq 1/2)$ is some absolute constant. In particular, let h be some positive integer such that the lattice $h^{1/2}\Lambda^*$ is integral, then due to $\lambda_1(h^{1/2}\Lambda^*) \geq 1$ we have

$$\mu(\Lambda) \leq dn/\lambda_1(\Lambda^*) \leq dnh^{1/2}/\lambda_1(h^{1/2}\Lambda^*) \leq dnh^{1/2} \tag{6}$$

Lattice Level: Let Λ be an integral lattice. In this case the dual lattice Λ^* is rational so there exists a positive integer h such that $h^{1/2}\Lambda^*$ is integral.

Definition 2 *Given an integral lattice Λ, the level of this lattice, denoted $l(\Lambda)$, is defined as the minimal positive integer h such that $h^{1/2}\Lambda^*$ is integral.* □

It's easy to see that $l(\Lambda)$ is an invariant of Λ, i.e., independent of Λ's basis choice.

Let $\mathbf{B}$ and $\mathbf{B}^*$ be Λ's and Λ^*'s base matrix respectively (so $\mathbf{B}^* = \mathbf{B}^{-T}$), so the dual lattice Λ^*'s Grahm matrix $\mathbf{A}^* = \mathbf{B}^{*T}\mathbf{B}^* = \mathbf{B}^{-1}\mathbf{B}^{-T} = \mathbf{A}^{-1}$, the inverse of the lattice Λ's Grahm matrix. Notice that $h^{1/2}\Lambda^*$ is integral means that $h\mathbf{A}^*$ is an integral matrix, and since $(det\mathbf{A})\mathbf{A}^* = \mathbf{A}^{adj}$ is always integral (because the adjoint matrix $\mathbf{A}^{adj}$'s entries are all integers), it follows that $h|det(\mathbf{A})$. On the other hand, the fact that $\mathbf{M} = h\mathbf{A}^*$ is an integral matrix deduces that $h\mathbf{I} = \mathbf{MA}$ and then we have $det\mathbf{A}|h^n$. In summary, the level h satisfies $h|(det\Lambda)^2|h^n$.

Moreover, the level h can be computed by $h = det(\mathbf{A})/g$ where $g = \gcd(\mathbf{A}^{adj})=$ the greatest common divisor of all the entries in $\mathbf{A}$'s adjoint matrix $\mathbf{A}^{adj}$.

2.2   Modular Forms

In this section we present a very brief description about the concepts and facts of one-variable modular forms needed in our work.

General: Let

$$SL_2(Z) \equiv \{\gamma = \begin{bmatrix} a & b \\ c & d \end{bmatrix} : a, b, c, d \in Z \text{ and } ad - bc = 1\}$$

be the group of 2×2 integer matrices with determinant 1. For any γ in $SL_2(Z)$ there is an related action on the upper-half complex plane $H \equiv \{\sigma + it : t > 0\}$ defined as:

$$\gamma(\tau) \equiv (a\tau + b)/(c\tau + d) : H \rightarrow H$$

Notice that $\pm\gamma$ induces the same action $\gamma(\tau)$. $SL_2(Z)$ is a finitely generated group with two generators[18]

$$\gamma_1 = \begin{bmatrix} 1 & 1 \\ 0 & 1 \end{bmatrix} , \quad \gamma_2 = \begin{bmatrix} 0 & -1 \\ 1 & 0 \end{bmatrix}$$

i.e., any action $\gamma(z)$ can be composed by the actions $\gamma_1(\tau) = \tau + 1$ and $\gamma_2(\tau) = -1/\tau$.

Instead of $SL_2(Z)$, in our work we consider its congruence subgroup of a given positive integer N:

$$\Gamma(N) \equiv \{\gamma \in SL_2(Z) : \gamma = \begin{bmatrix} 1 & 0 \\ 0 & 1 \end{bmatrix} \text{ mod } N\}, \tag{7}$$

$$\Gamma_1(N) \equiv \{\gamma \in SL_2(Z) : \gamma = \begin{bmatrix} 1 & b \\ 0 & 1 \end{bmatrix} \text{ mod } N\} \tag{8}$$

Both $\Gamma(N)$ and $\Gamma_1(N)$ are finite-index subgroups in $SL_2(Z)$ [18, 19].

Definition 3 *[18, 20] Let k be some positive integer or half-integer, Γ be a subgroup in $SL_2(Z)$ and $\Gamma(N) \subseteq \Gamma$, $f(\tau) : H \rightarrow C$ be a complex function holomorphic on the upper-half plane H. Let $f[\gamma]_k \equiv (c\tau + d)^{-k} f(\gamma(\tau))$ for γ in $SL_2(Z)$, f is defined as a modular form of weight k with respect to Γ, if both the following properties hold:*

(1) $f[\gamma]_k$ is bounded at the infinity point, i.e., $\lim_{t\to\infty} |f[\gamma]_k(\sigma + it)|$ exists for any γ in Γ and real number σ;

(2) $f[\gamma]_k = f$, i.e., $f(\gamma(\tau)) = (c\tau + d)^k f(\tau)$ for any γ in Γ and τ in H $\square$

The set of such functions is a linear space and is denoted $M_k(\Gamma)$.

Example: Consider the case $\Gamma = SL_2(Z)$, then $f(\tau)$ is in $M_k(SL_2(Z))$ iff it satisfies the above conditions (1) and (2) for all γ's in $SL_2(Z)$. Because $SL_2(Z)$ is generated by two generators $\gamma_1(\tau) = \tau + 1$ and $\gamma_2(\tau) = -1/\tau$, the modularity condition (2) is equivalent to the transformation law $f(\tau + 1) = f(\tau)$ and $f(-1/\tau) = (1/\tau)^k f(\tau)$.

Finiteness of Modular Form Space's Dimension: The transformation law under the congruence group's action imposed on the modular forms is a very strong restriction, so strong as to imply lots of special properties of the modular forms. One of the most important consequences followed is that the function space $M_k(\Gamma)$ is finite dimensional.

Theorem 4 *[18, 20, 21] For any positive integer N, positive integer or half-integer k, $M_k(\Gamma)$ is a finite-dimensional linear space on the complex field with*

$$dim_C M_k(\Gamma) \leq dim_C M_k(\Gamma(N)) = O(kN^3)$$

2.3   Lattice-Associated Theta Function and Its Modularity

One of the relations between (integral) lattices and modular forms is through the theta function, defined as

$$\vartheta(\tau; \Lambda) \equiv \sum_{x \in \Lambda} \exp(2\pi i \tau |x|^2) \tag{9}$$

where $\|$ denotes the ℓ^2 norm and $\tau = \sigma + it$ is a complex variable on the upper-half complex plane. Since Λ is integral, its Fourier expansion is

$$\vartheta(\tau; \Lambda) = \sum_{m \geq 0} a(m) q^m, \text{ where } q = \exp(2\pi i \tau)$$

where $a(0) = 1$ and $a(m)$ is the number of lattice vectors in Λ which squared ℓ^2-norms equal m. From this viewpoint, solving SVP on Λ reduces to finding its theta function's first non-zero Fourier coefficient $a(m)$ among non-constant items.

It's easy to prove that such theta function absolutely and uniformly converges in any compact subset of the upper half-plane H and is bounded at $+i\infty$, as a result, holomorphic on H. Another obvious property is

$$\vartheta(\tau + 1; \Lambda) = \vartheta(\tau; \Lambda) \text{ due to } \Lambda's \text{ integrality} \tag{10}$$

Let $n = dim\Lambda$ and Λ^* be the dual lattice of Λ. By Poisson formula, we have

$$\vartheta(\tau; \Lambda) = (i/2\tau)^{n/2} det\Lambda^* \vartheta(-1/4\tau; \Lambda^*) \tag{11a}$$

or equivalently

$$\vartheta(\tau; \Lambda^*) = (i/2\tau)^{n/2} det\Lambda \vartheta(-1/4\tau; \Lambda) \tag{11b}$$

Let h be any positive integer such that $h^{1/2}\Lambda^*$ is also an integral lattice. Since $h|y|^2$ is an integer for any y in Λ^*, for any η in H we have

$$\vartheta(\eta + h; \Lambda^*) = \sum_{y \in \Lambda^*} \exp(2\pi i (\eta + h)|y|^2) = \sum_{y \in \Lambda^*} \exp(2\pi i \eta |y|^2) = \vartheta(\eta; \Lambda^*)$$

let $\xi \equiv -(h + 1/4\tau)$,then by (11a) and the above h-periodicity

$$\vartheta(\tau/(4h\tau + 1); \Lambda) = \vartheta(-1/4\xi; \Lambda)$$
$$= (2\xi/i)^{n/2} det\Lambda^* \vartheta(\xi; \Lambda^*) = (2\xi/i)^{n/2} det\Lambda^* \vartheta(-1/4\tau; \Lambda^*)$$
$$= (2\xi/i)^{n/2} det\Lambda^* (2\tau/i)^{n/2} det\Lambda \vartheta(\tau; \Lambda) = (-4\xi\tau)^{n/2} \vartheta(\tau; \Lambda)$$
$$= (4h\tau + 1)^{n/2} \vartheta(\tau; \Lambda)$$

A more general relation between (integral) lattices and modular forms is through the following parameterized theta function, defined as Let $\vartheta(\tau; \Lambda, \boldsymbol{u}) \equiv \sum_{x \in \Lambda} \exp(2\pi i \tau |x - \boldsymbol{u}|^2)$,the same calculation derives that

$$\vartheta(\tau/(4h\tau + 1); \Lambda, \boldsymbol{u}) = (4h\tau + 1)^{n/2} \vartheta(\tau; \Lambda, -\boldsymbol{u}) = (4h\tau + 1)^{n/2} \vartheta(\tau; \Lambda, \boldsymbol{u}) \quad (12)$$

In summary, we have proven:

Lemma 5 *For any n-dimensional integral lattice Λ, the integer h such that that $h^{1/2}\Lambda^*$ is integral and an integral vector $\boldsymbol{u}$ in Z^n, $\vartheta(\tau; \Lambda, \boldsymbol{u})$ is a modular form of weight $n/2$ with respect to the congruence subgroup generated by*

$$\begin{bmatrix} 1 & 1 \\ 0 & 1 \end{bmatrix} \quad and \quad \begin{bmatrix} 1 & 0 \\ 4h & 1 \end{bmatrix}$$

Remarks: Let such generated congruence subgroup be denoted $J(h)$. The lemma states that

$$\vartheta(\tau; \Lambda, \boldsymbol{u}) \in M_{n/2}(J(h))$$

Since $\Gamma(4h) \subset J(h) \subset \Gamma_1(4h)$, it follows that $M_{n/2}(\Gamma_1(4h)) \subset M_{n/2}(J(h)) \subset M_{n/2}(\Gamma(4h))$ and by the dimension formulas (theorem 4) when n is even we have

$$dim_C M_{n/2}(J(h)) \leq dim_C M_{n/2}(\Gamma(4h)) \leq O(nh^3)$$

When n is odd we have the same upper-bound by the dimension formulas of the space of modular forms with weight half-integer $n/2$[20].

In practice, h can be selected as the lattice level $l(\Lambda)$, a lattice invariant efficiently computable (section 2.1).

3   The Modular Form Approach to Solving SVP and CVP

In this section we present our approach in a heuristic way, leaving technical details in the Appendices. To make the idea clear and easy to understand, we present this approach at first to solve two basic problems in section 3.1 and section 3.2 then apply the subroutines to solve the optimization SVP and CVP in section 3.

3.1 Basic Problems

Definition 6 *Given an integral lattice $\Lambda(\boldsymbol{B}) = \boldsymbol{Zb}_1 + \ldots + \boldsymbol{Zb}_n$ in Q^n and a positive integer m, the ℓ^2-vector counting problem, $VCP(\Lambda, m)$ in brief, is to find the number of lattice vectors in $\Lambda(\boldsymbol{B})$ which squared ℓ^2-norms equal m, i.e., to find $a(m) = |\{\boldsymbol{x} \text{ in } \Lambda : |\boldsymbol{x}|^2 = m\}|$.* □

Definition 7 *Given an integral lattice $\Lambda(\boldsymbol{B}) = \boldsymbol{Zb}_1 + \ldots + \boldsymbol{Zb}_n$ in Q^n, a vector $\mathbf{u}$ in Z^n such that $2\boldsymbol{Bu}$ also in Z^n, a positive integer m, the ℓ^2- non-homogenous vector counting problem, $n\text{-}VCP(\Lambda, m, \boldsymbol{u})$ in brief, is to find the number of lattice vectors in $\Lambda(\boldsymbol{B})$ which squared ℓ^2-distances to $\mathbf{u}$ equal m, i.e., to find $b(m) = |\{\boldsymbol{x} \text{ in } \Lambda : |\boldsymbol{x} - \mathbf{u}|^2 = m\}|$.* □

Remark: As long as both the lattice matrix $\mathbf{B}$ and the target vector $\boldsymbol{u}$ have only rational entries, it's easy to satisfy all the above requirements by scaling the original lattice $\Lambda(\mathbf{B})$ and $\boldsymbol{u}$ simultaneously with some appropriately large integer. In this case,i.e., for an integral lattice $\Lambda(\mathbf{B}) = \boldsymbol{Zb}_1 + \ldots + \boldsymbol{Zb}_n$ in Q^n and a vector $\boldsymbol{u}$ in Z^n such that $2\mathbf{B}\boldsymbol{u}$ also in Z^n, the squared distance $|\boldsymbol{x} - \boldsymbol{u}|^2 = |\mathbf{B}\boldsymbol{z} - \boldsymbol{u}|^2 = \boldsymbol{z}^T\mathbf{B}^T\mathbf{B}\boldsymbol{z} - 2\boldsymbol{z}^T\mathbf{B}\boldsymbol{u} + \boldsymbol{u}^T\boldsymbol{u}$ ($\boldsymbol{z}$ in Z^n) is always an integer.

3.2 Solving the Basic Problems

Given an integral lattice $\Lambda(\mathbf{B}) = \boldsymbol{Zb}_1 + \ldots + \boldsymbol{Zb}_n$ of dimension n and a positive integer m, consider how to solve $VCP(\Lambda, m)$ at first. Assume the level of Λ is h. By lemma 5 the lattice-associated theta function $\vartheta(\tau; \Lambda)$ (9) is in $M_{n/2}(J(h))$, it follows that

$$\vartheta(\tau; \Lambda) = \sum_{\alpha=1}^{M} h_\alpha(\Lambda)\varphi_\alpha(\tau) \tag{13}$$

where $M = dim M_{n/2}(J(h))$ and $\varphi_\alpha(\tau)$'s are basis of the space $M_{n/2}(J(h))$. Let

$$\varphi_\alpha(\tau) = \sum_{m=0}^{\infty} a_\alpha(m)q^m, \ where \ q = \exp(2\pi i\tau) \tag{14}$$

The basis $\{\varphi_\alpha(\tau) : \alpha = 1, \ldots, M\}$ and therefore their Fourier coefficients $a_\alpha(m)$ only depend on the congruence subgroup $J(h)$, which can be determined even in preprocessing when h is fixed. Then the m-th Fourier coefficient $a(m)$, i.e., the solution to the problem $VCP(\Lambda, m)$, can be computed by the formula

$$a(m) = \sum_{\alpha=1}^{M} h_\alpha(\Lambda)a_\alpha(m) \tag{15}$$

In this viewpoint, as long as the linear combination coefficients $h_\alpha(\Lambda)$ are known, the solution $a(m)$ is obtained.

Then arises the second question: how to compute $\{h_\alpha(\Lambda)\}_{\alpha=1,\ldots,M}$? Suppose we know $M(= dim M_{n/2}(J(h)))$ points $\tau_1, \ldots, \tau_M$ on the upper-half plane H and the values of the theta function at these points, $\vartheta(\tau_1; \Lambda), \ldots, \vartheta(\tau_M; \Lambda)$. As long as $det(\varphi_\alpha(\tau_\beta)) \neq 0$, by solving the linear system of equations

$$\vartheta(\tau_\alpha; \Lambda) = \sum_{\beta=1}^{M} h_\beta(\Lambda)\varphi_\beta(\tau_\alpha) \quad \alpha = 1, \ldots, M \tag{16}$$

all of $h_\alpha(\Lambda)$, $\alpha = 1, \ldots, M$ can be efficiently obtained.

Now the third question: for a given lattice Λ and any given point τ on the upper-half plane H, how to determine the value of $\vartheta(\tau; \Lambda)$? By definition $\vartheta(\tau; \Lambda)$ depends on the norms of all lattice vectors in Λ including those to be found in question, how to determine such an object prior to determining some of its unknown constituents? It is to solve this (and only this) sub-problem that the randomness in our algorithm is introduced.

The idea is to estimate $\vartheta(\tau; \Lambda)$ by appropriate random sampling over the lattice Λ. Note that when $t > 0$:

$$1/\vartheta(it; \Lambda) = 1/\sum_{x \in \Lambda} \exp(-2\pi|x|^2 t) = \underset{x \leftarrow D_{\Lambda, 1/t}}{E[\delta(x)]}$$

where $\delta(x)$ is the delta-function on Λ, vanishing at all non-zero lattice vectors and having the value 1 at $x = 0$:

$$\delta(x) = 1 \; if \; x = 0; \; \delta(x) = 0 \; if \; x \neq 0 \tag{17}$$

and $D_{\Lambda, 1/t}(x)$ is the discrete Gaussian probabilistic distribution over lattice Λ:

$$D_{\Lambda, 1/t}(x) \equiv \frac{\exp(-2\pi|x|^2 t)}{\sum_{x' \in \Lambda} \exp(-2\pi|x'|^2 t)} \quad for \; x \; in \; \Lambda$$

As a result, $1/\vartheta(it; \Lambda)$ might be estimated by statistical averaging over a set of $\delta(x_j)$'s where each x_j is a lattice vector independently sampled from Λ with distribution $D_{\Lambda, 1/t}$. However, the existed Gaussian samplers[22, 23] requires that t in this case be sufficiently small, potentially incompatible with some other requirements and practical considerations in our algorithm construction. Instead, we consider another way to estimate $\vartheta(it; \Lambda)$. The starting point is Poisson formula (11a):

$$\vartheta(it; \Lambda) = (1/2t)^{n/2} det\Lambda^* \vartheta(i/4t; \Lambda^*)$$

$$1/\vartheta(i/4t; \Lambda^*) = 1/\sum_{y \in \Lambda^*} \exp(-2\pi|y|^2/4t) = \underset{y \leftarrow D_{\Lambda^*, 4t}}{E[\delta(y)]} \tag{18}$$

where $\delta(y)$ is the delta-function on the dual lattice Λ^* and $D_{\Lambda^*, 4t}(y)$ is the discrete Gaussian distribution over the dual lattice Λ^*:

$$D_{\Lambda^*, 4t}(x) \equiv \frac{\exp(-2\pi|y|^2/4t)}{\sum_{y' \in \Lambda^*} \exp(-2\pi|y'|^2/4t)} \quad for \; y \; in \; \Lambda^* \tag{19}$$

Borrowing the techniques developed in [22, 23], $1/\vartheta(i/4t; \Lambda^*)$ hence $1/\vartheta(it; \Lambda)$ can be estimated by efficient random sampling algorithms as long as t is appropriately large. In this case $\vartheta(it; \Lambda) = O(1)$, i.e. $1/\vartheta(it; \Lambda)$ is not too small hence $\vartheta(i/4t; \Lambda^*)$ can be estimated by $1/\vartheta(i/4t; \Lambda^*)^{-1}$ with sufficiently small errors. Once $\vartheta(i/4t; \Lambda^*)$ can be estimated with given $t > 0$, by the following equation (derived in Appendix A)

$$\theta(\sigma + it; \Lambda) = (\frac{i}{2(\sigma + it)})^{\frac{n}{2}} det(\Lambda^*)\theta(\frac{it}{4(\sigma^2 + t^2)}; \Lambda^*) \cdot \underset{y \leftarrow D_{\Lambda^*, \frac{4(\sigma^2 + t^2)}{t}}}{E} [\exp(\frac{-2\pi i\sigma|y|^2}{4(\sigma^2 + t^2)})]$$

(20)

$\vartheta(\tau; \Lambda)$ can be estimated at $\tau = \sigma + it(t > 0)$ where in the expectation (replaced by statistical average when doing estimation) lattice vectors are distributed with the probability $D_{\Lambda^*, 4(\sigma^2+t^2)/t}(y)$ over Λ^*(with appropriately large t). Up to this point, the basic problem $VCP(\Lambda, m)$ is completely solved.

Similar steps are taken to solve the non-homogenous vector counting problem $n - VCP(\Lambda; m, \boldsymbol{u})$ by using Fourier expansions in space $M_{n/2}(J(h))$ and estimating the theta function $\vartheta(\tau; \Lambda, \boldsymbol{u}) = \sum_{\boldsymbol{x} \in \Lambda} \exp(2\pi i\tau|\boldsymbol{x} - \boldsymbol{u}|^2)$ in a similar randomized method.

Remark: As long as $\vartheta(\tau; \Lambda)$ can be estimated at sufficiently many points $\tau_j = \sigma_j + it_j$, it seems that its Fourier coefficient $a(m)$ can be computed directly by approximating the integral.

$$a(m) = \exp(2\pi mt) \int_0^1 d\sigma\hat{\vartheta}(\sigma + it; \Lambda) \exp(-2\pi im\sigma)$$

other than by (15), where $\hat{\vartheta}(\sigma+it; \Lambda)$ is the estimation for $\vartheta(\sigma+it; \Lambda)$ and $t > 0$. However, a complete analysis (details see Appendix D) concludes that this direct method has the time complexity at least $exp(2\pi n^2)$, inferior to the approach we take in (17)-(20) which is at most $n^{O(n)} = exp(nlogn)$ in time complexity.

3.3   Solving SVP and CVP

For a given (integral) lattice $\Lambda(\boldsymbol{B})$ in Q^n, let $m^* = \lambda_1(\Lambda)^2$ and by $\vartheta(\tau; \Lambda)$'s Fourier expansion

$$\vartheta(\tau; \Lambda) = \sum_{m=0}^{\infty} a(m)exp(2\pi im\tau) = 1 + a(m^*)exp(2\pi im^*\tau) + \dots$$

solving $SVP(\Lambda)$ reduces to computing the first non-zero $a(m)$ which can be achieved by repeatedly calling the subroutine $VCP(\Lambda, m)$ described in last section from $m=1,2,\dots$ up to some appropriate upper-bound, e.g., the upper-bound $cn|det(\Lambda)|^{2/n} = O(nl(\Lambda))$ derived from Mincowski's theorem (section 2.1).

Similarly, let $d^* = dist(\Lambda; \boldsymbol{u})^2$, $\boldsymbol{u}$ be an integral vector such that $2\boldsymbol{Bu}$ in Z^n and $\boldsymbol{u} \notin \Lambda$, by $\vartheta(\tau; \Lambda, \boldsymbol{u})$'s Fourier expansion

$$\vartheta(\tau; \Lambda, \boldsymbol{u}) = \sum_{m=1}^{\infty} b(m)exp(2\pi im\tau) = b(d^*)exp(2\pi id^*\tau) + \dots$$

solving $CVP(\Lambda; \boldsymbol{u})$ reduces to computing the first non-zero $b(m)$ which can be achieved by repeatedly calling the subroutine $n\text{-}VCP(\Lambda, m, u)$ described in last section from $m=1,2,\ldots$ up to some appropriate upper-bound, e.g., the upper-bound $O(n^2 l(\Lambda))$ derived from the transference theorem ((5)-(6)).

In summary, our algorithms to solve $SVP(\Lambda)$ and $CVP(\Lambda, \boldsymbol{u})$ in n dimension can be sketched in the following steps.

(1)Call the Gaussian sampler N times independently to get dual lattice vectors $y_1, \ldots, y_N$ in Λ^*. N needs to be large enough to make the error sufficiently small (Appendix D).

(2)Estimate the lattice-associated theta function $\vartheta(\tau; \Lambda)$ (in case of solving SVP) or $\vartheta(\tau; \Lambda, \boldsymbol{u})$ (CVP and CVPP) by $y_1, \ldots, y_N$ and $\boldsymbol{u}$ at sufficiently many points $\tau_j = \sigma_j + it_j$ with all $t_j > 0$. This step is only $poly(n)$ in time and space complexity.

(3)Compute the linear combination coefficients of the theta function on appropriately selected basis in the modular form space. This step is also $poly(n)$ in time and space complexity.

(4) Search the first non-zero Fourier coefficient in the theta function's Fourier expansions.

We note that step (1) can be completely parallelized, i.e., all N Gaussian samplers can work completely in concurrency (each sampler only performs in polynomial time and space complexity[22, 23]). For solving $CVPP(\Lambda, \boldsymbol{u})$, this step can be even performed totally in preprocessing. Since each Fourier coefficient can be computed independently, step (4) can also operate in concurrency of $O(nl(\Lambda))$(for SVP) and $O(n^2 l(\Lambda))$(for CVP and CVPP) where $l(\Lambda)$ is the level of lattice Λ. As a result, the whole algorithm can be easily parallelized to be polynomial in time complexity.

4   Conclusions

So far the framework to solve (integral) lattice optimization problems SVP and CVP has been established. The complete algorithms and computational complexity analysis are elaborated in the appendices. The main results are presented at below with detailed proofs in Appendix D.

Theorem 8 *There exists a randomized algorithm to solve the optimization SVP with correctness probability at least 1-ε for any integral lattice instance $\Lambda(\boldsymbol{b}_1, \ldots, \boldsymbol{b}_n)$ of dimension n and level $l(\Lambda)$, in time complexity of*

$$n^{16}(\log \log n^2 l(\Lambda))^n \cdot poly(n, S) \log(1/\varepsilon) + poly(n, S, l(\Lambda))$$

and space complexity of $poly(n, S)$ where $S = \max_{1 \leq i \leq n} bit\text{-}size$ of each entry in $\boldsymbol{b}_i$.

For solving the optimization CVP, the same result holds. $\square$

References

[1] Conway, J.H., Sloane, N.J.A.: Sphere packings, lattices and groups, 3rd edn. Springer (1998)

[2] Gentry, C.: Fully homomorphic encryption using ideal lattices. In: Proc. 41st ACM STOC, pp. 169–178 (2009)

[3] Hanrot, G., Pujol, X., Stehlé, D.: Algorithms for the shortest and closest lattice vector problems. In: Chee, Y.M., Guo, Z., Ling, S., Shao, F., Tang, Y., Wang, H., Xing, C. (eds.) IWCC 2011. LNCS, vol. 6639, pp. 159–190. Springer, Heidelberg (2011)

[4] Haviv, I., Regev, O.: Hardness of the covering radius problem on lattices. In: IEEE CCC 2006, pp. 145–158 (2006)

[5] Kannan, R.: Mincowski 's convex body theorem and integer programming. Math. Oper. Res. 12(3), 415–440 (1987)

[6] Kannan, R.: Improved algorithms for integer programming and related lattice problems. In: Proc. STOC 1983, pp. 193–206 (1983)

[7] Micianccio, D., Goldwasser, S.: Complexity of Lattice Problems: a cryptographic perspective. Kluwer Academic Publishers, Boston (2002)

[8] Nguyen, P.Q., Vallee, B. (eds.): The LLL Algorithm: Survey and Applications. Springer (2009)

[9] Aharonov, D., Regev, O.: Lattice problems in NP∩coNP. J. ACM 52(5), 749–765 (2005)

[10] Regev, O.: Lecture notes of lattices in computer science (2004), http://www.cs.tau.il/~odedr

[11] Micianccio, D., Regev, O.: Worst-case to average-case reductions based-on gaussian measures. SIAM J. Comput. 37(1), 267–302 (2007)

[12] Banaszczk, W.: New bounds in some transference theorems in the geometry of numbers. Mathematische Annalen 296(4), 625m–635m (1993)

[13] Micciancio, D., Voulgaris, P.: A deterministic single exponential time algorithm for most lattice problems based on voronoii cell computations. SIAM J. Comput. (2012) (Special Issue on STOC 2010)

[14] Ajati, M., Kumar, R., Sivakumar, D.: A sieve algorithm for the shortest lattice vector problem. In: STOC 2001, pp. 601–610 (2001)

[15] Ajati, M., Kumar, R., Sivakumar, D.: Sampling short lattice vectors and the closest lattice vector problem. In: IEEE Conference on Computational Complexity, pp. 53–57 (2002)

[16] Nguyen, P.Q., Vidick, T.: Sieve algorithms for the shortest vector problem are practical. Journal of Mathematical Cryptology 2, 181–207 (2008)

[17] Maccianccio, D.: Efficient reductions amomg lattice problems. In: Proc. SODA 2008, pp. 84–93 (2008)

[18] Diamond, F., Shurman, J.: A first course in modular forms. Springer, Berlin (2005)

[19] Koblitz, N.: Introduction to elliptic curves and modular forms. Springer (1993)

[20] Wang, X., Pei, D.: Modular forms with integral and half-integral weights (in English). Science Press, Beijing (2011)

[21] Li, W.Q.: Number theory with applications. World Science Publication (1996)

[22] Gentry, C., Peikert, C., Vaikuntananthan, V.: How to use a short basis: trapdoors for hard lattices and new cryptographic constructions. In: Proc. 40th ACM STOC, pp. 197–206 (2008)

[23] Peikert, C.: An efficient and parallel gaussian sampler for lattices. In: Rabin, T. (ed.) CRYPTO 2010. LNCS, vol. 6223, pp. 80–97. Springer, Heidelberg (2010)

Appendix

A   Randomized Algorithms for Estimating the Values of Lattice-Associated Theta Functions $\vartheta(\tau; \Lambda)$ and $\vartheta(\tau; \Lambda, u)$

Recall the algorithm framework developed in section 3.2, the goal of estimating theta function's values is to compute the linear coefficients $h_\alpha(\Lambda), \alpha = 1, \ldots, M$ via solving the linear system of equations (16). Therefore, it's adequate to do the estimation at a finite number of points $\tau_\alpha, \alpha = 1, \ldots, M$ where $M = \dim M_{n/2}(J(h))$. In particular, these points in H can be selected according to computational efficiency considerations. Our approach to do the estimation is based upon the techniques developed in [22, 23] which basic result is presented in theorem 9. For all technical details, see section 4 in [22] and [23].

Theorem 9 *[22]:There is a probabilistic polynomial-time algorithm that, given the basis $B = (\mathbf{b}_1, \ldots, \mathbf{b}_n)$ of n -dimensional lattice Λ, a parameter $s > \omega(\log(n))$ $\max_j |\tilde{b}_j|^2$ where $\tilde{b}_1, \ldots, \tilde{b}_n$ is the Gram-Schmidt orthogonalization of $\mathbf{b}_1, \ldots, \mathbf{b}_n$, and a vector $\mathbf{u}$ in R^n, outputs a sample from the distribution which is statistically close to the discrete Gaussian distribution*

$$D_{\Lambda,s,u}(\mathbf{x}) \equiv \frac{\exp(-2\pi|x - u|^2/s)}{\sum_{x' \in \Lambda} \exp(-2\pi|x' - u|^2/s)} \quad \textit{for } \mathbf{x} \textit{ in } \Lambda \qquad \square$$

When $u = 0$, $D_{\Lambda,s,u}(x)$ is simply denoted $D_{\Lambda,s}(x)$. Hereafter the sampler in theorem 9 is denoted $SampD(\Lambda(\mathbf{B}), s, u)$. The original sampler[22]'s efficiency is significantly improved in [23] at a mild price of a larger t. In this paper we neglect such efficiency differences and call $SampD$ as a black-box. In the following we almost always apply $SampD$ to sample on the dual lattice $\Lambda^*(b_1^*, \ldots, b_n^*)$ of the input lattice $\Lambda(b_1, \ldots, b_n)$, in this case the original condition $s > \omega(\log n) \max_j |\tilde{b}_j|^2$ becomes $s > \omega(\log n) \max_j |\tilde{b}_j|^{-2}$ because of the relationship between the Gram-Schmidt orthogonalization $\tilde{b}_1, \ldots, \tilde{b}_n$ of the basis $b_1, \ldots, b_n$ and $d_n, \ldots, d_1$, that of the dual basis $(b_1^*, \ldots, b_n^*):d_j = \tilde{b}_j/|\tilde{b}_j|^2$. In particular $|d_j| = |\tilde{b}_j|^{-1}$ for $j = 1, \ldots, n$.

Our algorithms to do the estimation are presented in the full version paper.

B   Computing Linear Combination Coefficients $h_\alpha(\Lambda)$ and $h_\alpha(\Lambda; u)$

In solving SVP, let $\{\varphi_\alpha(\tau) : \alpha = 1, \ldots, M\}$ be the basis of space $M_{n/2}(J(h))$, $M = dim M_{n/2}(J(h))$, $\hat{\vartheta}(\tau_i; \Lambda)$be theta-function value estimations at points τ_1, $\ldots \tau_M$ on the upper-half plane H. As long as $det(\varphi_\alpha(\tau_\beta))_{1 \le \alpha, \beta \le M} \neq 0$, by solving the system of linear equations

$$\hat{\vartheta}(\tau_\alpha; \Lambda) = \sum_{\beta=1}^{M} \hat{h}_\beta(\Lambda)\varphi_\beta(\tau_\alpha) \qquad \alpha = 1, \ldots, M \qquad (21)$$

all $h_\alpha(\Lambda)$'s estimations $\hat{h}_\alpha(\Lambda)$ can be obtained.

In solving CVP the situation is similar with the only difference that we need to solve the linear system of equations

$$\hat{\vartheta}(\tau_\alpha; \Lambda, u) = \sum_{\beta=1}^{M} \hat{h}_\beta(\Lambda, u)\varphi_\beta(\tau_\alpha) \qquad \alpha = 1, \ldots, M. \tag{22}$$

For simplicity hereafter we only use the notation h_α instead of $h_\alpha(\Lambda)$ and $h_\alpha(\Lambda; u)$.

In essence, what is really needed in our algorithms is not any specific basis of space $M_{n/2}(J(h))$, but just a set of points $\tau_1 \ldots \tau_M$ such that $det(\varphi_\alpha(\tau_\beta))_{1\le\alpha,\beta\le M} \ne 0$, a set of function values $\{\varphi_\alpha(\tau_\beta))\}_{1\le\alpha,\beta\le M}$ and a set of Fourier coefficients of these basis (see Appendix C). Moreover, notice the fact that $M_k(J(h))$ is a subspace in $M_k(\Gamma(4h))$, in practice we can even use the basis of the much better understood space $M_k(\Gamma(4h))$ with only moderate prices in time and space complexity. Given any positive integer N, the space $M_k(\Gamma(N))$ has an orthogonal decomposition (with respect to the Petersson inner product) [18–20] where $S_k(\Gamma(N))$ is the so called cusp form subspace:

$$M_k(\Gamma(N)) = S_k(\Gamma(N)) \oplus E_k(\Gamma(N))$$

For instance, the basis of subspace $E_k(\Gamma(N))$ can be selected to be the Eisenstein functions

$$G_k^{(u,v)}(\tau) = \sum_{(c,d):(c,d)=(u,v)mod\ N} 1/(c\tau + d)^k \ = \sum_{m\ge 0} g_k^{(u,v)}(m)exp(2\pi im\tau/N)$$

for all integer-pairs (u, v) of order N in $Z_N \times Z_N$. It's well known that these basis have Fourier coefficients[18, 20]

$$g_k^{(u,v)}(m) = ((-2\pi i)^k/(k-1)!) \sum_{j=-m,\ldots,+m, j|m,\ m/j=u\ mod\ N,\ j\ne 0} sgn(j)j^{k-1}exp(2\pi ivj/N) \qquad m \ge 1$$

where $sgn(j)=1$ when $j > 0$, -1 when $j < 0$(we neglect $g_k^{(u,v)}(0)$ which is not needed in our algorithm). It's clear from the formulas that the m-th Fourier coefficient can be computed in at most $poly(m, logk)$ time complexity. In the proceeding applications to solve lattice problems, both m and k are O(n) where n is the lattice's dimension. Similar situation holds for $S_k(\Gamma(N))$.

Since this paper is only concentrated on the algorithm's logic and complexity analysis, we defer to discuss all numerical computation related details in a separate paper, only pointing out that for integer or half-integer k there exit efficient algorithms (polynomial in $logk$ and m) to output the m-th Fourier coefficient of the basis in space $M_k(\Gamma(N))$.

To complete the computation, we need to confirm that the condition

$$det(\varphi_\alpha(\tau_\beta))_{1\le\alpha,\beta\le M} \ne 0$$

can be really satisfied. The following lemma guarantees the existence of such points $\tau_1, \ldots, \tau_M$ on the upper-half plane H.

Lemma 10 *Let m be a positive integer, D be a domain in the upper-half plane H, $\varphi_1(\tau), \ldots, \varphi_m(\tau)$ be complex-valued functions holomorphic in D. If $\varphi_1(\tau), \ldots,$ $\varphi_m(\tau)$ are linearly independent over the complex field, then there exist m points $\tau_1, \ldots, \tau_m$ in D such that $det(\varphi_\alpha(\tau_\beta))_{1\leq\alpha,\beta\leq m} \neq 0$.*

Proof. (by induction on m) For $m = 1$ the result is trivial. Now suppose the lemma is true for m. For $m + 1$ complex linearly independent functions $\varphi_0(\tau)$, $\varphi_1(\tau), \ldots, \varphi_m(\tau)$ holomorphic in D, by induction there exist points $\tau_1, \ldots, \tau_m$ in D such that $det(\varphi_\alpha(\tau_\beta))_{1\leq\alpha,\beta\leq m} \neq 0$. Because $\varphi_0(\tau)$ is holomorphic and not identically zero in D, we can always assume (by slightly changing some τ_β's if needed) that at least one of the $\varphi_0(\tau_\beta)$'s is non-zero. As a result, there exist (obtained by solving the following linear system of equations) complex values $a_1, \ldots, a_m$ such that

$$\varphi_0(\tau_\beta) = a_1\varphi_1(\tau_\beta) + \ldots + a_m\varphi_m(\tau_\beta) \qquad for \ all \ \beta = 1, \ldots, m \qquad (23)$$

and at least one of the a_β's is non-zero. By complex linear independency among the functions $\varphi_0, \varphi_1, \ldots, \varphi_m,$ $\varphi_0 \neq a_1\varphi_1 + \ldots + a_m\varphi_m$ so there exists a point τ_0 in D such that

$$\varphi_0(\tau_0) \neq a_1\varphi_1(\tau_0) + \ldots + a_m\varphi_m(\tau_0) \qquad (24)$$

in consequence, $det(\varphi_\alpha(\tau_\beta))_{0\leq\alpha,\beta\leq m} \neq 0$ (otherwise the following matrix

$$\begin{bmatrix} \varphi_0(\tau_0) & \varphi_1(\tau_0) & \cdots & \varphi_m(\tau_0) \\ \varphi_0\tau_1) & \varphi_1(\tau_1) & \cdots & \varphi_m(\tau_1) \\ \cdots\cdots\cdots\cdots\cdots\cdots\cdots\cdots\cdots \\ \varphi_0(\tau_m) & \varphi_1(\tau_m) & \cdots & \varphi_m(\tau_m) \end{bmatrix}$$

is singular so there exist $a_1, \ldots, a_m$ such that

$$\varphi_0(\tau_\beta) = a_1\varphi_1(\tau_\beta) + \ldots + a_m\varphi_m(\tau_\beta) \qquad for \ all \ \beta = 0, 1, \ldots, m$$

But due to $det(\varphi_\alpha(\tau_\beta))_{1\leq\alpha,\beta\leq m} \neq 0$, these $a_1, \ldots, a_m$'s are exactly those in (23), a contradiction to (24)

Remark: It's easy to derive an efficient algorithm from the lemma's proof to output a sequence of points $\tau_1, \ldots, \tau_m$ in D such that $det(\varphi_\alpha(\tau_\beta))_{1\leq\alpha,\beta\leq m} \neq 0$, given the functions $\varphi_1, \ldots, \varphi_m$ and domain D satisfying the conditions specified in this lemma.

C The Complete Algorithms

Now we integrate all the components to construct the complete algorithms to solve the optimization lattice problems. As indicated before, these algorithms find not only the classical solutions to the optimization SVP and CVP but also the number of lattice vectors which reach the minimums.

To make the algorithm's structure clear, we introduce an oracle to help collect necessary information.

Oracle-M(h, k, m^*, t_0)

Input: A positive integer h, a positive integer or half-integer k and two positive real numbers m^*, t_0.

Output:

(1) A collection of Fourier coefficients $\{a_\alpha(m) : \alpha = 1, \ldots, M, m = 1, \ldots, m^*\}$ where $M = dim_C M_k(J(h))$ with respect to some basis $\{\varphi_\alpha(\tau) : \alpha = 1, \ldots, M\}$ of the space $M_k(J(h))$. $a_\alpha(m)$ denotes the m-th Fourier coefficient of $\varphi_\alpha(\tau)$.

(2) A collection of points $\tau_1, \ldots, \tau_M$ on the upper-half complex plane H such that

$$Im\tau_\alpha > t_0 \ for \ each \ 1 \le \alpha \le M$$
$$det(\varphi_\alpha(\tau_\beta))_{1 \le \alpha, \beta \le M} \ne 0$$

(3) A collection of values $\{\Phi_{\alpha\beta} : 1 \le \alpha, \beta \le M, \ the \ matrix \ (\Phi_{\alpha\beta}) = (\varphi_\alpha(\tau_\beta))^{-1}\}$

Remark: The oracle-M can be implemented based on and only on the knowledge about the congruence subgroup $J(h)$ or, as explained in Appendix A.2, the group $\Gamma(4h)$. As explained in Appendix B, any basis of space $M_k(J(h))$ or even $M_k(\Gamma(4h))$ is sufficient for our algorithmic goals so we can always select the most appropriate and efficient basis in practice. In summary, each Fourier coefficient $a_\alpha(m)$ can be computed with time complexity polynomial in k and m, and the points $\tau_1, \ldots, \tau_M$ can be also determined efficiently.

Now we present our algorithms to solve the optimization lattice problem SVP and CVP.

Algorithm to Solve Optimization SVP

Input: an integral lattice $\Lambda(\mathbf{B}) = Z\mathbf{b}_1 + \ldots + Z\mathbf{b}_n$ in Q^n.

Parameters: Positive absolute constants $c \le 1$ and $c_0 > (2\pi)^{-1/2}$.

Output: $\lambda_1(\Lambda) \equiv min\{|\mathbf{x}| : \mathbf{x} \ in \ \Lambda \ and \ non\text{-}zero\}$ and $a^*(\Lambda) = |\{\mathbf{x} \ in \ \Lambda : |\mathbf{x}| = \lambda_1(\Lambda)\}|$.

Operations:

(1) Compute $h = l(\Lambda)$, the level of lattice Λ, as stated in the paragraph following definition 2.

(2) Set $m^* = cn|det(\Lambda)|^{2/n}$ and $t_0 > max(c_0^2 n, \omega(logn)max_j|\tilde{b}_j|^{-2})$. Call *Oracle-M*$(h, \ n/2, \ m^*, \ t_0)$ to obtain:

A collection of Fourier coefficients $\{a_\alpha(m) : \alpha = 1, \ldots, M, \ m = 1, \ldots, m^*\}$ where $M = dim_C M_{n/2}(J(h))$ with respect to some basis $\{\varphi_\alpha(\tau) : \alpha = 1, \ldots, M\}$ of space $M_{n/2}(J(h))$ and $a_\alpha(m)$ denotes the m-th Fourier coefficient of $\varphi_\alpha(\tau)$;

A collection of points $\{\tau_\beta = \sigma_\beta + it_\beta : \beta = 1, \ldots, M\}$ such that $t_\beta > t_0$ for each $1 \le \beta \le M$ and $det(\varphi_\alpha(\tau_\beta))_{1 \le \alpha, \beta \le M} \ne 0$;

A collection of values $\{\Phi_{\alpha\beta} : 1 \le \alpha, \beta \le M, \ the \ matrix \ (\Phi_{\alpha\beta}) = (\varphi_\alpha(\tau_\beta))^{-1}\}$.

(3) For each $\beta = 1, \ldots, M$ call *EstimTheta* version#2 with input $(\tau_\beta, \Lambda(\mathbf{B}))$ and parameter N to obtain $\hat{\vartheta}(\tau_\beta; \Lambda)$($N$ depends on $n = dim\Lambda(\mathbf{B})$ and its value will be determined according to complexity analysis in Appendix D).

(4) compute $\hat{h}_\beta = \sum_{\alpha=1}^{M} \Phi_{\alpha\beta}\hat{\vartheta}(\tau_\alpha; \Lambda)$ for each $\beta = 1, \ldots, M$.

(5)For each $m = 1, 2, \ldots, m^*$ do: Compute $\hat{a}(m) = \sum_{\beta=1}^{M} \hat{h}_\beta a_\beta(m)$; if $\hat{a}(m) > 1/2$

then break;

(6)Output $(m^{1/2}, [\hat{a}(m)])$ where $[x]$ denotes the integer nearest to x.

Algorithm to Solve Optimization CVP

Input: an integral lattice $\Lambda(\mathbf{B}) = Z\boldsymbol{b}_1 + \ldots + Z\boldsymbol{b}_n$ in Q^n, a vector $\boldsymbol{u}$ in $Z^n \backslash \Lambda(\mathbf{B})$ such that $2\mathbf{B}\boldsymbol{u}$ in Z^n.

Parameters: Positive absolute constants $d \leq 1/2$ *and* $c_0 > (2\pi)^{-1/2}$.

Output: $dist(\Lambda; u) \equiv \min\{|x - u| : x \in \Lambda\}$ and $b^*(\Lambda) = |\{x \in \Lambda : |x - u| = dist(\Lambda; u)\}|$

Operations:

(1) and (2): The same as steps (1) and (2) in the algorithm to solve the optimization SVP, except that $m^* = dn^2 |det(\Lambda)|^{2/n}$. All notations are inherited.

(3)For each $\beta = 1, \ldots, M$ call *EstimTheta* version#4 with input $(\tau_\beta, \Lambda(\mathbf{B}), \boldsymbol{u})$ and parameter N to obtain $\hat{\vartheta}(\tau_\beta; \Lambda, u)$ (N depends on $n = dim\Lambda(\mathbf{B})$ and its value will be determined according to complexity analysis).

(4)Compute $\hat{h}_\beta = \sum_{\alpha=1}^{M} \Phi_{\alpha\beta} \hat{\vartheta}(\tau_\alpha; \Lambda, u)$ *for each* $\beta = 1, \ldots, M$.

(5)For each $m = 1, 2, \ldots, m^*$ do :Compute $\hat{b}(m) = \sum_{\beta=1}^{M} \hat{h}_\beta a_\beta(m)$; if $\hat{b}(m) > 1/2$

then break;

(6)Output $(m^{1/2}, [\hat{b}(m)])$ where $[x]$ denotes the integer nearest to x.

D   Complexity Analysis

Before delve into the algorithm's complexity, we need a fact about the modular form's Fourier coefficient's asymptotic increasing degree.

Lemma 11 *[20] Let $\Gamma(r)$ be the congruence subgroup in $SL_2(Z)$ and $\varphi(\tau)$ be the Eisenstein basis for the space $M_k(\Gamma(r))$ with Fourier coefficients $a(m)$, $m \geq 1$, then*

$$|a(m)| \leq A(\log|\log(m^2/r)|)^k \text{ for any } m \geq 1.$$

where A *is a constant irrelevant with k. For the basis $\varphi(\tau)$ of the space $S_k(\Gamma(r))$ with Fourier coefficients $a(m)$, $m \geq 1$, the inequality is*

$$|a(m)| \leq A(\log|\log(m^2/r)|)^{k/2} \text{ for any } m \geq 1.$$

$\square$

Let $h = l(\Lambda)$, the level of lattice Λ. Now consider the algorithm for SVP. According to step (5), for any $1 \leq m \leq m^*$ we have

$$|the\ error\ of\ \hat{a}(m)| \leq M \max_{1\leq\beta\leq M} |\hat{h}_\beta| \max_{1\leq\beta\leq M} |a_\beta(m)|$$

$$\leq M^2 \max_{1\leq\alpha\leq M} \hat{e}_\alpha \max_{1\leq\alpha,\beta\leq M} |\Phi_{\alpha\beta}| A(\log\log(m^2/4h))^{n/2}$$

$$\leq C \cdot M^2 (\log\log(m^2/4h))^{n/2} \max_{1\leq\alpha\leq M} \hat{e}_\alpha$$

$$\leq C \cdot n^2 h^6 (\log\log(n^2 h))^{n/2} \max_{1\leq\alpha\leq M} \hat{e}_\alpha$$

where C is a constant and $\hat{e}_\alpha = |the\ estimation\ error\ of\ \hat{\vartheta}(\tau_\alpha; \Lambda)|$. The second inequality is derived by step(4) and the upper-bound for $|a_\beta(m)|$. The fourth inequality is from $m \leq m^* = O(n|det(\Lambda)|^{2/n})$ and $det(\Lambda)^2|h^n$.

Notice that the exact value of each $a(m)$, the Fourier coefficient of the theta function $\vartheta(\tau; \Lambda)$, is a non-negative integer so it is sufficient to get the correct solution as long as $|the\ error\ of\ \hat{a}(m)| < 1/2$. As a result, we need $|$the estimation error of $\hat{\vartheta}(\tau_\alpha; \Lambda)| = O(n^{-2}h^{-6}(\log\log(n^2 h))^{-n/2})$ for all τ_α's in step(3).

Direct calculation shows (details in the full version paper) that this requires the number of (dual) lattice vector samples N, i.e., the times for the Gaussian sampler to be independently called, should be $N = O(n^{16}(\log\log(n^2 h))^n \log(1/\varepsilon_2))$ to make $P[|\hat{\vartheta}_N(\sigma + it; \Lambda) - \vartheta(\sigma + it; \Lambda)| < \varepsilon_1] > 1 - \varepsilon_2$, equivalently, to make $P[|\hat{a}(m) - a(m)| < 1/2] > 1 - \varepsilon_2$. In summary, we have proven:

Theorem 12 *For the algorithm in Appendix C to solve the optimization SVP for integral lattice Λ, $n = dim(\Lambda)$, $h = l(\Lambda)$ and $1 > \varepsilon > 0$, it holds that the probability of the algorithm terminating with the correct solution $(\lambda_1(\Lambda), a^*(\Lambda))$ is at least $1-\varepsilon$, if the number of lattice vector samples $N = O(n^{16}(\log\log(n^2 h))^n \log\frac{1}{\varepsilon})$. For the algorithm to solve the optimization CVP, the result is the same.* □

Now we can estimate the algorithm's time and space complexity. Let $T(i)$ and $S(i)$ denote the time and space complexity in step i respectively, $n = dim(\Lambda)$, $h = l(\Lambda)$, $S = \max_{1\leq i\leq n}$the bit size of each entry in $\boldsymbol{b}_i$, poly denote some (multivariate) polynomial. It's easy to verify that:both $T(1)$ and $S(1)$ are $poly(n, S)$ according to the analysis after definition 2.

$T(2) = \sum_{1\leq m\leq m^*} poly(m, S) = poly(n, S, h)$ according to $m^* = O(n|det(\Lambda)|^{\frac{2}{n}})$ and $det(\Lambda)^2|h^n$, the analysis in Appendix B and remarks on oracle-M. $S(2)$=the space to store the outputs from the $oracle\text{-}M = O(m^*M + M^2)poly(S) = O(n^2h^6)poly(S) = poly(n, S, h)$ according to remarks on lemma 5, i.e., $M = O(nh^3)$.

$T(3) = Npoly(n, S)$ where $N = O(n^{16}(\log\log(n^2 h))^n log(1/\varepsilon))$ as stated in theorem 13 and $S(3) = poly(n, S)$.

$T(4) = Mpoly(S) = poly(n, S, h)$ and $S(4) = poly(n, S)$.

$T(5) = Mm^*poly(S) = poly(n, S, h)$ and $S(5) = poly(nS)$.

In summary we obtain the central result in this paper:

Theorem 13 *There exists a randomized algorithm to solve the optimization SVP with correctness probability at least $1 - \varepsilon$ for integral lattice instance $\Lambda(\boldsymbol{b}_1, \ldots, \boldsymbol{b}_n)$ of dimension n and level $l(\Lambda)$, in time complexity of*

$$n^{16}(\log\log(n^2 l(\Lambda)))^n \cdot poly(n, S)log(1/\varepsilon) + poly(n, S, l(\Lambda))$$

and space complexity of $poly(n, S, l(\Lambda))$ where $S = \max_{1 \leq i \leq n} bit\text{-}size$ of each entry in $\boldsymbol{b}_i$. For solving the optimization CVP, the same result holds.

Remarks: It is not really necessary for the algorithms in Appendix C to store all the outputs from the oracle in a batch. Instead they can get these outputs in sequence when needed. As a result, the space complexity for both algorithms can be actually reduced to only $poly(n, S)$, independent of the level $l(\Lambda)$, while the time complexity's asymptotic bounds are unchanged.

Before ending the section we make a brief analysis on why the Fourier coefficient $a(m)$ is not computed directly by approximating the following integral

$$a(m) = \exp(2\pi mt) \int_0^1 d\sigma \hat{\vartheta}(\sigma + it; \Lambda)exp(-2\pi im\sigma) \ m = 1, 2, \ldots, m^*$$

where $\hat{\vartheta}(\sigma + it; \Lambda)$ is the estimation for $\vartheta(\sigma + it; \Lambda)$ and $t > 0$. The reason is that, for the error of all such computed $a(m)$'s to be within $1/2$, the estimation error of $\hat{\vartheta}(\sigma + it; \Lambda)$ needs to be within $O(\exp(-2\pi m^* t) = O(\exp(-2\pi l(\Lambda)n^2))$ implying that the number of lattice vector samples in step (3), N, needs to be $N = O(\exp(4\pi l(\Lambda)n^2)log(1/\varepsilon))$ to make the correctness probability at least $1-\varepsilon$, significantly inferior to the performance concluded in theorem 13.

Author Index